Characters in Low-Dimensional Topology

CONTEMPORARY MATHEMATICS

760

Centre de Recherches Mathématiques Proceedings

C R M

Characters in Low-Dimensional Topology

A Conference Celebrating the Work
of
Steven Boyer
June 2–6, 2018
Université du Québec à Montréal, Montréal,
Québec, Canada

Olivier Collin
Stefan Friedl
Cameron Gordon
Stephan Tillmann
Liam Watson
Editors

AMS AMERICAN MATHEMATICAL SOCIETY

2010 *Mathematics Subject Classification.* Primary 57M25, 57M27, 57M50, 57N10, 53C25, 11F06, 20F06, 20E08, 20F65, 20F67.

Library of Congress Cataloging-in-Publication Data

Names: Collin, Olivier, 1971– editor.

Title: Characters in low-dimensional topology : a conference celebrating the work of Steven Boyer, June 2–6, 2018, Université du Québec à Montréal, Montréal, Québec, Canada / Olivier Collin, Stefan Friedl, Cameron Gordon, Stephan Tillmann, Liam Watson, editors.

Description: Providence, Rhode Island : American Mathematical Society ; Montréal, Quebec, Canada : Centre de Recherches Mathématiques, [2020] | Series: Contemporary mathematics, 0271-4132 ; volume 760 | Centre de Recherches Mathématiques Proceedings. | Includes bibliographical references. |

Identifiers: LCCN 2020016572 | ISBN 9781470452094 (paperback) | ISBN 9781470461355 (ebook)

Subjects: LCSH: Low-dimensional topology–Congresses. | AMS: Manifolds and cell complexes – Low-dimensional topology – Knots and links in S^3. | Manifolds and cell complexes – Low-dimensional topology – Invariants of knots and 3-manifolds. | Manifolds and cell complexes – Low-dimensional topology – Geometric structures on low-dimensional manifolds. | Manifolds and cell complexes – Topological manifolds – Topology of general 3-manifolds. | Differential geometry – Global differential geometry. | Number theory – Discontinuous groups and automorphic forms – Structure of modular groups and generalizations; arithmetic groups. | Group theory and generalizations – Special aspects of infinite or finite groups – Cancellation theory; application of van Kampen diagrams. | Group theory and generalizations – Structure and classification of infinite or finite groups – Groups acting on trees. | Group theory and generalizations – Special aspects of infinite or finite groups – Geometric group theory. | Group theory and generalizations – Special aspects of infinite or finite groups – Hyperbolic groups and nonpositively curved groups.

Classification: LCC QA612.14 .C45 2020 | DDC 514/.22–dc23

LC record available at https://lccn.loc.gov/2020016572

Contemporary Mathematics ISSN: 0271-4132 (print); ISSN: 1098-3627 (online)

DOI: https://doi.org/10.1090/conm/760

♾ The paper used in this book is acid-free and falls within the guidelines
established to ensure permanence and durability.
Visit the AMS home page at `https://www.ams.org/`

10 9 8 7 6 5 4 3 2 1 25 24 23 22 21 20

Contents

Preface

In June 2018, the 'Boyerfest' was held at the Université du Québec à Montréal. It was organised by the editors of this volume to celebrate Steven Boyer's major contributions in the area of low-dimensional topology and his important role in the Canadian mathematical community on the occasion of his 65th birthday. The high standard of the 18 talks at the conference contributed greatly to its success. They covered a broad range of topics related to the topology and geometry of 3–manifolds, properties of their fundamental groups and associated representation varieties. The conference was attended by over a hundred mathematicians, and about a quarter of these were graduate students or postdoctoral fellows. It is the hope of the organisers that the conference inspired these young scientists to engage in exploring the many exciting research problems in this area of mathematics.

These proceedings mirror the spirit of the event. The papers in this volume were written by speakers and participants of the conference, as well as a number of contributors who could not attend. Included are papers by collaborators, former students of Steven Boyer and former postdoctoral fellows who have enjoyed his mentorship. We hope the proceedings capture the mathematical endeavours and recognition of Steve.

The editors would like to thank the authors for the work they have put into their contributions, and the referees for their commitment invested in the verification and improvement of these papers. The editors thank Christine Thivierge for her assistance and unrelenting patience in preparing this volume.

The organisers would like to acknowledge the generous support of the 'Boyerfest' by the Centre interuniversitaire de recherches en géométrie et topologie (CIRGET), the Centre de recherches mathématiques (CRM), the Fields Institute for Research in Mathematical Sciences, the National Science Foundation, the Unité Mixte Internationale CRM-CNRS, and the Université du Québec à Montréal. CIRGET and the Département de mathématiques at the Université du Québec à Montréal provided a wonderful conference environment, and the organisers are indepted to Alexandra Haedrich for precient and cheerful administrative support.

The Editors
March 2020

Talks at the Boyerfest

Michel Boileau (Université d'Aix-Marseille)
Grothendieck rigidity of 3-manifold groups
Jeff Danciger (University of Texas at Austin)
Exotic real projective Dehn surgery space
Nathan Dunfield (University of Illinois at Urbana-Champaign)
An $\mathrm{SL}(2,\mathbb{R})$ *Casson-Lin invariant and applications*
Paolo Ghiggini (Université de Nantes)
Knot Floer homology from a dynamical viewpoint
Bill Goldman (University of Maryland)
Topology, Geometry and Dynamics of Character Varieties
Cameron Gordon (University of Texas at Austin)
Cyclic branched covers of L-space knots
Josh Greene (Boston College)
On curves intersecting at most once
Eli Grigsby (Boston College)
Braids, surfaces, and homological invariants
Marc Lackenby (University of Oxford)
Every knot has characterising slopes
Ana Lecuona (University of Glasgow)
Torus knots and rational homology balls
Joseph Maher (CUNY Staten Island)
The compression body graph has infinite diameter
Alan Reid (Rice University)
Profinite rigidity from representation rigidity
Rachel Roberts (Washington University)
Nontrivial connected sums of fibered knots are persistently foliar
Saul Schleimer (Warwick University)
From veering triangulations to pseudo-Anosov flows
Genevieve Walsh (Tufts University)
Relatively hyperbolic groups with planar boundaries
Liam Watson (Université de Sherbrooke)
Heegaard Floer homology via immersed curves
Dani Wise (McGill University)
Orderability, Coherence, and Counting Cycles in Graphs
Xingru Zhang (SUNY Buffalo)
On real characters

Biographical sketch of Steven Boyer

First steps. Steven Patrick Boyer was born in 1953 in South Charleston, West Virginia and moved to Montréal at the age of six. From a family of three boys, three girls, and many pets, he spent a good deal of his early years playing with friends and participating in organized sports. Access to a large and diverse home library coupled with lively family conversations nurtured his developing interests in a variety of subjects, among them history, natural sciences, and mathematics.

It was during his undergraduate studies at the University of New Brunswick in Fredericton that his passion for mathematics blossomed, and after completing a BA and MA in mathematics he joined the PhD program at Cornell University. Those formative years saw him shift his interests from geometric complex analysis to 4-dimensional geometric topology under the supervision of Peter Jack Kahn. His thesis, written in the post-Freedman era of 4-dimensional topology, investigated the representation of 2-dimensional homology classes by locally flat 2-spheres.

Steve met his wife, Maryse Camille Desrochers, at Cornell, where she was an algebra graduate student working under Stephen Chase. After obtaining their PhD's in 1983, they spent two years at the University of Cambridge as postdoctoral fellows. Steve's interest in 3-manifold topology was sparked by a Part III course given by Raymond Lickorish and solidified through daily discussions with Daniel Lines, who was also a postdoctoral fellow at Cambridge during this period. Interactions with visitors such as Tim Cochran, Mark Feighn, Rob Kirby, Bill Menasco, Marty Scharlemann, Hamish Short, and Abigail Thompson also contributed to his shift in research direction.

A mathematical correspondence with Kunio Murasugi led to an instructorship at the University of Toronto, which he took up on leaving Cambridge. The following two years were productive on various fronts. On the personal side, Steve and Maryse's daughter Stephanie was born soon after they arrived in Toronto. Mathematically, he produced a topological classification of compact, simply-connected 4-manifolds with connected boundaries, as well as a study of proper powers in free products of groups motivated by applications to problems in 3-dimensional surgery theory. He also collaborated with Andy Nicas, another young researcher at the University of Toronto, on generalisations of Casson's invariant to rational homology spheres.

The early years at UQAM. A chance meeting with François Lalonde led to a return to Montréal in 1987 to take up a tenure track position at Université du Québec à Montréal, one of a group of young academics recruited from mathematical centres around the world who aspired to create an international research pole in that city. Amongst other things, this led to the creation of the geometry-topology research centre CIRGET (Centre Interuniversitaire de Recherche en Géométrie et Topologie).

Mathematically, Steve had started to focus on applications of representation-theoretic methods to 3-manifold topology and in the paper *Surgery formulae for Casson's invariant and extensions to homology lens spaces*, he and Daniel Lines showed how Dedekind sums could be used to produce an extension of Casson's invariant which was particularly well-suited to study problems in surgery theory. On the personal side, Steve's son Anthony was born soon after he returned to Montréal.

Between 1992 and 1994, Steve had the great fortune of supervising a Chinese postdoctoral fellow, Xingru Zhang, who has had a significant and lasting impact on him. They were heavily influenced by the work of Peter Shalen, especially his seminal contributions with Marc Culler to $SL(2,\mathbb{C})$-character variety theory and the application of the theory to Dehn surgery problems with Culler, Cameron McA. Gordon, and John Luecke. Steve's collaboration with Xingru led to a sequence of articles published between 1996 and 2001 exploited, with increasing sophistication and strength, the $SL(2,\mathbb{C})$-character variety theory of 3-manifold groups, culminating in the paper *A proof of the finite surgery conjecture.* Towards the end of the 1990's, Steve began a long-term collaboration with Cameron Gordon, whose camaraderie and complementary approach on mathematical problems has proved an unending source of inspiration.

CIRGET director years at UQAM. By 2001, Steve had established himself among the most influential topologists in Canada, having built an international network of collaborators and bringing much activity in the field to Montréal. When the founding director of CIRGET, François Lalonde, left UQAM for the Université de Montréal, Steve took on the challenge of pursuing CIRGET's development and increasing its international impact. The successes the centre has enjoyed over the years are due in no small part to Steve's dedication, patience, and pure tenacity. Few other colleagues were prepared to navigate the murky waters of university administration in the quest for recognition and resources. Much to the amusement of his colleagues, he did so with a determination and gusto undimmed by the inevitable disappointments. His efforts paid off though, as he secured long-term infrastructure support for the centre, the creation a research chair in geometry, and a suite of purpose-designed research rooms from the university. Under Steve's directorship, CIRGET's recruitment of postdoctoral fellows and research visitors has been transformed from the ad hoc to a stable program. Many of the young mathematicians who spent formative years in Montréal greatly benefited from Steve's non-prescriptive but responsible mentoring.

The last two decades have witnessed a constant expansion of methods and topics covered, and collaborators in Steve's research. In work with Culler, Shalen, and Zhang, and later with Gordon and Zhang, a powerful machine for studying exceptional surgery problems based on the Jaco-Shalen-Johannson theory of characteristic submanifolds was developed. With Zhang and Ian Agol he published a paper on the virtual fibring conjecture in 2008. A research collaboration with Michel Boileau was begun in the early 2000's which to date has resulted in a number of publications on a diverse list of topics including non-zero degree maps between 3-manifolds, commensurability between hyperbolic knot complements, the existence of foliations on graph manifold integer homology spheres, L–space branched covers of links in the 3-sphere, and the Tits alternative for PD(3) groups. Much of this work was in collaboration with others including his doctoral student Radu Cebanu and postdoctoral fellow Genevieve Walsh, and colleagues Alan Reid, Dale Rolfsen, and Shicheng Wang.

Steve's interest in the orderability of the fundamental groups of 3-manifolds grew from discussions with fellow Canadian topologist Dale Rolfsen, and their 2005 article *Orderable 3-manifold groups*, written with Bert Wiest, was a great example of a work finely blending research aims with survey and reference purposes. Later discussions with his doctoral student Liam Watson and Cameron Gordon led to what in the early 2010's seemed like a very bold conjecture: an irreducible rational homology sphere is not a Heegaard Floer homology L–space if and only if its fundamental group is left-orderable. Inspired by a question of Ozsváth and Szabó, asking about a converse to their result that L–spaces do not admit taut foliations, András Juhász further contended that these conditions were equivalent to the existence of a co-oriented taut foliation on the 3-manifold, thus leading to what is now known as the *L–space conjecture*, one of the most important problems in contemporary 3-manifold topology

Work on the L–space conjecture gave Steve the opportunity to turn his attention to foliations on 3-manifolds. With Boileau he had constructed co-oriented taut foliations on graph manifold integer homology spheres with infinite fundamental groups, thus verifying the L–space conjecture for this class of manifolds. The 2017 paper *Foliations, orders, representations, L–spaces and graph manifolds*, with his postdoctoral fellow Adam Clay, introduced a relative form of the L–space conjecture and used it to establish the equivalence of the existence of a co-oriented taut foliation on a general graph manifold W with the left-orderability of its fundamental group. This relative approach was the model used by Jonathan Hanselmann, Jake Rasmussen, Sarah Rasmussen, and Liam Watson to formulate their Heegaard Floer gluing theorems, which allowed them to complete the proof of the L–space conjecture for graph manifolds. A recent collaboration with his postdoctoral fellow Ying Hu verified the L-space conjecture for families of branched covers by producing foliations with zero Euler class.

Approach to mathematics. Steve's approach to mathematics has been consistent since his early years. He is motivated primarily by the aesthetic pleasure it affords, the depths and unexpected connections it reveals, and the comradeship it engenders. He likes to say that intuition should lead technique, but not by too much, and finds that working on problems that allow him to assimilate new techniques is a great way to expand his mathematical horizons, though the choice of topics to explore is often a case of one thing leading to another. In the end, he sees that the research mathematician's lot is "to chase the shadows of that which interests, the challenge being in the choosing of the light."

Contemporary Mathematics
Volume **760**, 2020
https://doi.org/10.1090/conm/760/15283

Hyperbolic four-manifolds with vanishing Seiberg-Witten invariants

Ian Agol and Francesco Lin

ABSTRACT. We show the existence of hyperbolic 4-manifolds with vanishing Seiberg-Witten invariants, addressing a conjecture of Claude LeBrun. This is achieved by showing, using results in geometric and arithmetic group theory, that certain hyperbolic 4-manifolds contain L-spaces as hypersurfaces.

Introduction

In [**9**, Conjecture 1.1], Claude LeBrun asked whether the Seiberg-Witten invariants of hyperbolic 4-manifolds vanish. This question stems from his result that for a hyperbolic 4-manifold, Seiberg-Witten basic classes satisfy much stronger constraints than one would expect; furthermore, it turns out to be related to several problems in low-dimensional topology [**18**, §4]. Here, we show that there exist certain hyperbolic 4-manifolds with vanishing Seiberg-Witten invariants.

THEOREM 0.1. *There exist closed arithmetic hyperbolic 4-manifolds with vanishing Seiberg-Witten invariants.*

In the statement, we consider all possible Seiberg-Witten invariants coming from evaluating elements of the cohomology ring $\Lambda^* H^1(X;\mathbb{Z}) \otimes \mathbb{Z}[U]$ of the space of configurations. Theorem 0.1 is proved by exhibiting hyperbolic 4-manifolds admitting separating L-spaces, using the main result of [**6**]; under mild additional conditions, this implies that such manifolds admit finite covers with vanishing Seiberg-Witten invariants. Our construction will show in fact that there are infinitely many commensurability classes of arithmetic hyperbolic 4-manifolds containing representatives with vanishing Seiberg-Witten invariants. Furthermore, by interbreeding as in [**4**], one can also obtain non-arithmetic examples.

1. A vanishing criterion for the Seiberg-Witten invariants

We discuss a vanishing result for the Seiberg-Witten invariants of four-manifolds containing a separating hypersurface. This is well-known to experts, but the exact form we will need is only implicitly stated in [**7**], so we will point it out for the reader's convenience. Most of our discussion is based on formal properties of the invariants, and we will follow closely follow the exposition of [**7**, Chapter 3].

2010 *Mathematics Subject Classification.* Primary 57K41, 57N16, 22E40.
Key words and phrases. Hyperbolic geometry, Seiberg-Witten invariants, arithmetic groups.

Consider a spin^c structure $\mathfrak{s}_X$ on a closed oriented 4-manifold X. For a cohomology class $u \in \Lambda^* H^1(Y;\mathbb{Z}) \otimes \mathbb{Z}[U]$, we define the Seiberg-Witten invariant $\mathfrak{m}(u|X, \mathfrak{s}_X)$ to be the evaluation of u on the moduli space of solutions to the Seiberg-Witten equations. This is a topological invariant provided that $b_2^+ \geq 2$. The latter is not a particularly restrictive assumption in our case; hyperbolic 4-manifolds have signature zero by [**2**, Theorem 3] and the Hirzebruch signature formula. Hence

$$\chi(X) = 2(1 - b_1(X) + b_2^+(X)).$$

If $b_2^+(X) \leq 1$, we would have $\chi(X) \leq 4$; on the other hand, in all known examples of closed orientable hyperbolic 4-manifolds $\chi \geq 16$ [**11, 14**] (recall that by Chern-Gauss-Bonnet, volume and Euler characteristic are proportional).

We discuss a vanishing criterion for $\mathfrak{m}(u|X, \mathfrak{s}_X)$. Let Y be a closed, oriented three-manifold. To this, in [**7**, Section 3.1] it is defined for each spin^c structure $\mathfrak{s}$ on Y the monopole Floer homology groups fitting in the exact triangle of graded $\mathbb{Z}[U]$-modules

$$\cdots \longrightarrow \overline{HM}_*(Y, \mathfrak{s}) \xrightarrow{i_*} \widecheck{HM}_*(Y, \mathfrak{s}) \xrightarrow{j_*} \widehat{HM}_*(Y, \mathfrak{s}) \xrightarrow{p_*} \overline{HM}_*(Y, \mathfrak{s}) \longrightarrow \cdots \tag{1.1}$$

where U has degree -2 (notice that this convention differs from the one in the four-dimensional literature; this is because we identify U with the corresponding capping operation in homology). The reduced Floer group $HM_*(Y, \mathfrak{s})$ is defined to be the image of j_* in $\widehat{HM}_*(Y, \mathfrak{s})$ [**7**, Definition 3.6.3]. We will be particularly interested in the case in which Y is a rational homology sphere. In this case we have an identification of $\mathbb{Z}[U]$-modules (up to grading shift) with Laurent series [**7**, Proposition 35.3.1]

$$\overline{HM}_*(Y, \mathfrak{s}) \cong \mathbb{Z}[U^{-1}, U].$$

DEFINITION 1.1 ([**8**]). We say that a rational homology sphere Y is an *L-space* if, up to grading shift, $\widehat{HM}_*(Y, \mathfrak{s}) \cong \mathbb{Z}[U]$ as $\mathbb{Z}[U]$-modules for all spin^c structures $\mathfrak{s}$.

As the map p_* in equation (1.1) is an isomorphism in degrees low enough [**7**, Section 22.2], for an L-space $HM_*(Y, \mathfrak{s}) = 0$ for all spin^c structures $\mathfrak{s}$

PROPOSITION 1.2. *Let X be a four-manifold given as $X = X_1 \cup_Y X_2$. Suppose that the separating hypersurface Y is an L-space (so that in particular $b_1(Y) = 0$), and that $b_2^+(X_i) \geq 1$. Then all the Seiberg-Witten invariants of X vanish.*

REMARK 1.3. A simpler vanishing criterion is the following: if $b_1(X) = 0$ and $b_2^+(X)$ is even, then all Seiberg-Witten invariants are zero. In fact, under this assumption all Seiberg-Witten moduli spaces are odd dimensional [**7**, Theorem 1.4.4], while all classes in our cohomology ring are even dimensional. On the other hand, we are not aware of examples of hyperbolic 4-manifolds satisfying these conditions.

PROOF OF PROPOSITION 1.2. All we need to do is to discuss the results of [**7**, Chapter 3] while keeping track of the specific spin^c structures. First of all, notice that as $b_1(Y) = 0$, a spin^c structure $\mathfrak{s}_X$ on X is determined by the restrictions $\mathfrak{s}_i = \mathfrak{s}_X|_{X_i}$. This follows from the injectivity of the map $H^2(X;\mathbb{Z}) \to H^2(X_1;\mathbb{Z}) \oplus H^2(X_2;\mathbb{Z})$ in the Mayer-Vietoris sequence, and the fact the these groups classify spin^c structures. Let $\mathfrak{s} = \mathfrak{s}_X|_Y$. It is sufficient to show that $\mathfrak{m}(u|X, \mathfrak{s}_X) = 0$ for classes $u = u_1 u_2$ where u_i is a cohomology class in the configuration space of X_i. Recall from [**7**, Section 3.4] that a cobordism W from Y_0 to Y_1 induces a map in homology fitting with the exact triangle; furthermore, if $b_2^+(W) \geq 1$, we have that $\overline{HM}_*(u|W, \mathfrak{s}) = 0$ [**7**, Proposition 3.5.2]. Given data as above, we can define the relative invariant $\psi_{(u_1|X_1,\mathfrak{s}_1)} \in \widehat{HM}_*(Y, \mathfrak{s})$ obtained as follows: let W_1 be the cobordism obtained from X_1 by removing a ball, and consider the induced map

$$\widehat{HM}_*(u_1|W_1, \mathfrak{s}_1) : \widehat{HM}_*(S^3) \equiv \mathbb{Z}[U] \to \widehat{HM}_*(Y, \mathfrak{s}).$$

Then $\psi_{(u_1|X_1,\mathfrak{s}_1)} = \widehat{HM}_*(u_1|W_1, \mathfrak{s}_1)(1)$. On the other hand, we have the commutative diagram

$$\begin{array}{ccc} \widehat{HM}_*(S^3) & \xrightarrow{p_*} & \overline{HM}_*(S^3) \\ {\scriptstyle \widehat{HM}_*(u_1|W_1,\mathfrak{s}_1)}\Big\downarrow & & \Big\downarrow{\scriptstyle \overline{HM}_*(u_1|W_1,\mathfrak{s}_1)} \\ \widehat{HM}_*(Y, \mathfrak{s}) & \xrightarrow{p_*} & \overline{HM}_*(Y, \mathfrak{s}) \end{array}$$

and as $b_2^+(W_1) \geq 1$, the vertical map on the right vanishes; in turn, this implies that $\psi_{(u_1|X_1,\mathfrak{s}_1)} \in \ker(p_*) \cong HM_*(Y, \mathfrak{s})$. Similarly, using the map induced in cohomology by W_2, we obtain an element $\psi_{(u_2|X_2,\mathfrak{s}_2)} \in HM_*(-Y, \mathfrak{s})$; this last group is by Poincaré duality identified with $HM^*(Y, \mathfrak{s})$. The general gluing theorem in [**7**, Equation 3.22], when keeping track of the spin^c structures, is then

$$\mathfrak{m}(u|X, \mathfrak{s}_X) = \langle \psi_{(u_1|X_1,\mathfrak{s}_1)}, \psi_{(u_2|X_2,\mathfrak{s}_2)} \rangle,$$

where the angular brackets denote the natural pairing

$$HM_*(Y, \mathfrak{s}) \times HM^*(Y, \mathfrak{s}) \to \mathbb{Z}.$$

Under our assumptions, the group $HM_*(Y, \mathfrak{s})$ vanishes, so this pairing is zero, and the result follows. □

REMARK 1.4. In fact, for our purposes of understanding the gluing formula for Seiberg-Witten invariants, it suffices to consider the reduced invariants with rational coefficients, $HM_*(Y, \mathfrak{s}; \mathbb{Q})$. In particular, the previous discussion only relies on the vanishing of this group. Furthermore, via the universal coefficients theorem, this is implied by the vanishing of $HM_*(Y, \mathfrak{s}; \mathbb{Z}/2\mathbb{Z})$, so that our main result actually applies for the reduced Floer homology group with $\mathbb{Z}/2\mathbb{Z}$-coefficients.

Our examples will be based on the following.

COROLLARY 1.5. *Suppose X is a 4-manifold with $b_2^+ \geq 1$ which admits an embedded non-separating L-space Y. Then X admits infinitely many covers which have all vanishing Seiberg-Witten invariants.*

PROOF. Consider the double cover $\tilde{X}$ of X formed by gluing together two copies W_1 and W_2 of the cobordism from Y to Y obtained by cutting X along Y, see Figure 1. Consider a properly embedded path $\gamma \subset W_1$ between the two copies of Y, and denote by T its tubular neighborhood. We then have the decomposition $X = (W_1 \setminus T) \cup (W_2 \cup T)$, where the two manifolds are glued along a copy of $Y \# \overline{Y}$; here $\overline{Y}$ denotes Y with the opposite orientation. The latter is an L-space [**10**, Section 4], and both $W_1 \setminus T$ and $W_2 \setminus T$ have $b_2^+ \geq 1$, so the conclusion follows. Finally, to obtain infinitely many examples, for any $N \geq 3$ we cyclically glue together copies W_i for $i = 1, \ldots, N$ of the cobordism from Y to Y obtained by

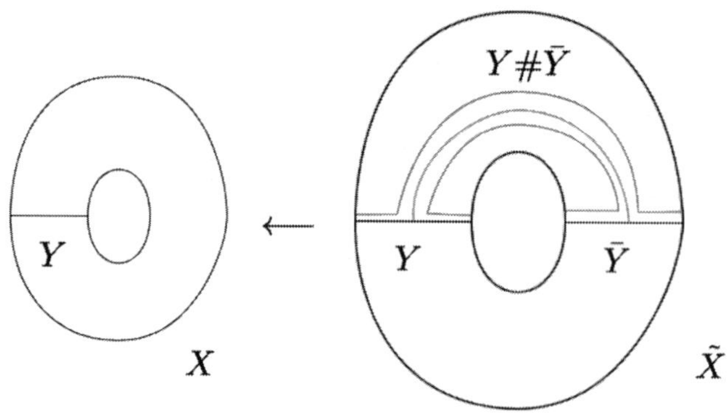

FIGURE 1. A double cover of X contains a separating L-space.

cutting X along Y; this again contains a separating hypersurface diffeomorphic to $Y\#\overline{Y}$ with the required properties. □

2. Geodesic hypersurfaces in arithmetic hyperbolic 4-manifolds

In this section, we will discuss various properties of arithmetic hyperbolic lattices. For the general case of arithmetic lattices, see [**20**], and for the 3-dimensional case, consult [**13**]. We first review the definitions and construction of arithmetic manifolds of simplest type.

DEFINITION 2.1. Let G be a group, $H_1, H_2 \leq G$ be subgroups. We say that H_1 is *commensurable in* G with H_2 if $[H_1 : H_1 \cap H_2] < \infty, [H_2 : H_1 \cap H_2] < \infty$.

DEFINITION 2.2. Consider a non-degenerate quadratic form $q : k^{n+1} \to k$ for a totally real number field $k \subset \mathbb{R}$ with ring of integers $\mathcal{O}_k$. Assume that q is Lorentzian, i.e. has signature $(n,1)$ over $\mathbb{R}$. Moreover, for each non-trivial embedding $\sigma : k \to \mathbb{R}$, assume that $\sigma \circ q$ is positive definite. Let $O(q;k)$ denote the group of matrices preserving q, i.e. linear transforms $A : k^{n+1} \to k^{n+1}$ such that $q \circ A = q$. Then the subgroup $O(q;\mathcal{O}_k) \subset O(q;k) \subset O(q;\mathbb{R})$ is a lattice, and acts discretely on the hyperboloid of two sheets $\mathcal{H} = \{x \in \mathbb{R}^{n+1} | q(x) = -1\}$. Up to isometry, the group $O(q;\mathbb{R}) \cong O(n,1;\mathbb{R})$, the orthogonal group associated to the quadratic form $-x_0^2 + x_1^2 + \cdots + x_n^2$. Projectivizing, $PO(q;\mathcal{O}_k)$ acts discretely on hyperbolic space $\mathbb{H}^n$, which is the quotient of the hyperboloid $\mathcal{H}$ by the antipodal map. A hyperbolic orbifold $\mathbb{H}^n/\Gamma$ is said to be of *simplest type* if Γ is commensurable (up to conjugacy) with $PO(q;\mathcal{O}_k)$ for some such q.

Example: Let $q_n : k^{n+1} \to k$ be defined by $q_n(x_0, x_1, \ldots, x_n) = -\sqrt{2}x_0^2 + x_1^2 + \cdots + x_n^2$ over the field $k = \mathbb{Q}(\sqrt{2})$. Let $\sigma : k \to k$ be the Galois automorphism induced by $\sigma(\sqrt{2}) = -\sqrt{2}$. Then $\sigma \circ q_n(x_0, \ldots, x_n) = \sqrt{2}x_0^2 + x_1^2 + \cdots + x_n^2$ is positive definite. Hence $PO(q_n;\mathbb{Z}[\sqrt{2}])$ is a discrete arithmetic lattice acting on $\mathbb{H}^n$. See [**20**, §6.4].

DEFINITION 2.3. Let G be a group. Then $G^{(2)} = \langle g^2 | g \in G \rangle$.

If G is finitely generated, then $G^{(2)}$ is finite-index in G, and $G/G^{(2)}$ is an elementary abelian 2-group.

THEOREM 2.4. *Let M^3 be an orientable hyperbolic arithmetic 3-manifold of simplest type with $H_1(M;\mathbb{Z}/2) = 0$ and not defined over $\mathbb{Q}$. Then M embeds as a totally geodesic non-separating submanifold in a compact arithmetic hyperbolic 4-manifold.*

PROOF. Let $\Gamma = \pi_1(M) \leq Isom^+(\mathbb{H}^3)$. Since M is a $\mathbb{Z}/2\mathbb{Z}$-homology sphere, $\Gamma^{(2)} = \Gamma$. By [**6**, Theorem 1.1 (2)], $\mathbb{H}^n/\Gamma^{(2)} \cong M$ embeds as a totally geodesic submanifold of a closed orientable hyperbolic 4-manifold W (the fact that M is not defined over $\mathbb{Q}$ implies that W is compact). Briefly, this is proved by showing that $\Gamma^{(2)} \leq PO(q;k)$ so that it is commensurable with $PO(q;\mathcal{O}_k)$ for some Lorentzian quadratic form $q : k^4 \to k$. Taking the quadratic form $Q_d = dy^2 + q, d \in \mathbb{N}$, we get an embedding of $PO(q;\mathcal{O}_k) < PO(Q_d;\mathcal{O}_k) < PO(Q_d;\mathbb{R}) \cong PO(4,1;\mathbb{R})$. Then a subgroup separability result allows one to embed Γ in a torsion-free lattice $\Lambda < PO(Q_d;k)$ so that $W = \mathbb{H}^4/\Lambda$. By [**1**, Theorem 2], there exists a further finite-sheeted cover $\tilde{W} \to W$, and a lift $M \to \tilde{W}$ such that the lift of M is non-separating in $\tilde{W}$. This is achieved again by a subgroup separability result. □

3. Examples

The *Fibonacci manifold* M_n is the cyclic branched n-fold cover over the figure-eight knot. For $n = 2$ we obtain a lens space, for $n = 3$ the Hantzche-Wendt manifold, while for $n \geq 4$ it is hyperbolic.

For every n the Fibonacci manifold M_n is an L-space. To see this, recall from [**19**] that M_n is the branched double cover over the closure of the 3-braid $(\sigma_1\sigma_2^{-1})^n$ (see Figure 2), which is alternating. Using the surgery exact triangle [**8**], these can be shown to be L-spaces as in the context of Heegaard Floer homology [**15**], with the caveat that in our setting the computation only holds with coefficients in $\mathbb{Z}/2\mathbb{Z}$; on the other hand this is enough for our purposes, see Remark 1.4. Notice also that for $n \neq 0$ modulo 3, the closure is a knot, so that M_n is a $\mathbb{Z}/2\mathbb{Z}$-homology sphere.

By [**5**], M_n is arithmetic when $n = 4, 5, 6, 8, 12$. Of these examples, $n = 4, 5, 8$ are $\mathbb{Z}/2\mathbb{Z}$ homology spheres. The only one of these three which is simplest type and not defined over $\mathbb{Q}$ is M_5. This is example [**13**, 13.7.4(a)(iii)], which has invariant trace field a quartic field. As they point out, this is commensurable with a tetrahedral group [**13**, 13.7.4(a)(i)] which is simplest type and not defined over $\mathbb{Q}$ by [**12**, Theorem 1]. It is defined over a quadratic form over the field $\mathbb{Q}(\sqrt{5})$.

Thus, by Theorem 2.4, M_5 has a non-separating embedding into a closed orientable hyperbolic 4-manifold W. We may assume that $\chi(W) > 2$ (by passing to a 2-fold cover if needed), and hence $b_2^+(W) > 1$. Thus by Corollary 1.5, these embed into a hyperbolic 4-manifold with vanishing Seiberg-Witten invariants. This completes the proof of Theorem 0.1.

REMARK 3.1. One may also get other examples by cutting and doubling or using the interbreeding technique of Gromov-Piatetskii-Shapiro to get non-arithmetic examples. One can isometrically embed this L-space M_5 in infinitely many incommensurable hyperbolic 4-manifolds via the method of [**6**] by taking the forms Q_1 and Q_d in the proof of Theorem 2.4 so that d is square-free in $k = \mathbb{Q}(\sqrt{5})$, and then cut and cross-glue to give a closed non-arithmetic manifold containing M_n as a non-separating hypersurface [**4**, §2.9].

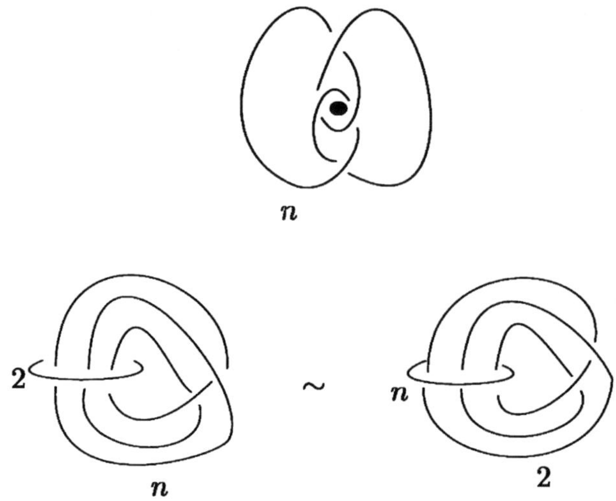

FIGURE 2. In this picture, the numbers indicate the branching. The top picture has an obvious order 2 rotational symmetry along the axis depicted by the big dot. The quotient is the link in S^3 depicted on the bottom left. This is isotopic to the link on the right (which is topologically the same, but with different branchings). Now, the curve with branching 2 is the 3-braid $\sigma_1\sigma_2^{-1}$, so that taking the n-fold branched cover along the other component we see that M_n is the branched double cover over $(\sigma_1\sigma_2^{-1})^n$.

4. Conclusion

We conclude by pointing out some natural questions related to our method.

(1) Can one find an explicit hyperbolic example (such as the Davis manifold or the manifolds described in [**11**]) that satisfies the properties of Proposition 1.2? Recall that the Davis manifold has $b_1 = 24$ and $b_2^+ = 36$ [**16**], so that all moduli spaces have odd dimension.

(2) Can one embed any orientable hyperbolic 3-manifold of simple type as a geodesic hypersurface in an orientable hyperbolic 4-manifold? More generally, can one show that orientable hyperbolic 3-manifolds have quasiconvex embeddings into orientable hyperbolic 4-manifolds?

(3) Can one use bordered Floer theory to compute the Seiberg-Witten invariants of Haken hyperbolic 4-manifolds (in the sense of [**3**])?

(4) Which commensurability classes of compact hyperbolic 3-manifold of the simplest type contain L-spaces? Note that it is not even known if there are infinitely many commensurability classes of arithmetic rational homology 3-spheres.

Acknowledgments

The first author was funded by a Simons Investigator award. The second author was partially funded by NSF grant DMS-1807242. We thank Alan Reid for comments on an earlier draft.

References

[1] N. Bergeron, *Premier nombre de Betti et spectre du laplacien de certaines variétés hyperboliques* (French, with English summary), Enseign. Math. (2) **46** (2000), no. 1-2, 109–137. MR1769939

[2] S.-s. Chern, *On curvature and characteristic classes of a Riemann manifold*, Abh. Math. Sem. Univ. Hamburg **20** (1955), 117–126, DOI 10.1007/BF02960745. MR75647

[3] B. Foozwell and H. Rubinstein, *Introduction to the theory of Haken n-manifolds*, Topology and geometry in dimension three, Contemp. Math., vol. 560, Amer. Math. Soc., Providence, RI, 2011, pp. 71–84, DOI 10.1090/conm/560/11092. MR2866924

[4] M. Gromov and I. Piatetski-Shapiro, *Nonarithmetic groups in Lobachevsky spaces*, Inst. Hautes Études Sci. Publ. Math. **66** (1988), 93–103. MR932135

[5] H. M. Hilden, M. T. Lozano, and J. M. Montesinos-Amilibia, *The arithmeticity of the figure eight knot orbifolds*, Topology '90 (Columbus, OH, 1990), Ohio State Univ. Math. Res. Inst. Publ., vol. 1, de Gruyter, Berlin, 1992, pp. 169–183. MR1184409

[6] A. Kolpakov, A. W. Reid, and L. Slavich, *Embedding arithmetic hyperbolic manifolds*, Math. Res. Lett. **25** (2018), no. 4, 1305–1328, DOI 10.4310/MRL.2018.v25.n4.a12. MR3882165

[7] P. Kronheimer and T. Mrowka, *Monopoles and three-manifolds*, New Mathematical Monographs, vol. 10, Cambridge University Press, Cambridge, 2007. MR2388043

[8] P. Kronheimer, T. Mrowka, P. Ozsváth, and Z. Szabó, *Monopoles and lens space surgeries*, Ann. of Math. (2) **165** (2007), no. 2, 457–546, DOI 10.4007/annals.2007.165.457. MR2299739

[9] C. LeBrun, *Hyperbolic manifolds, harmonic forms, and Seiberg-Witten invariants*, Proceedings of the Euroconference on Partial Differential Equations and their Applications to Geometry and Physics (Castelvecchio Pascoli, 2000), Geom. Dedicata **91** (2002), 137–154, DOI 10.1023/A:1016222709901. MR1919897

[10] F. Lin, Pin(2)*-monopole Floer homology, higher compositions and connected sums*, J. Topol. **10** (2017), no. 4, 921–969, DOI 10.1112/topo.12027. MR3705144

[11] C. Long, *Small volume closed hyperbolic 4-manifolds*, Bull. Lond. Math. Soc. **40** (2008), no. 5, 913–916, DOI 10.1112/blms/bdn077. MR2439657

[12] C. Maclachlan and A. W. Reid, *The arithmetic structure of tetrahedral groups of hyperbolic isometries*, Mathematika **36** (1989), no. 2, 221–240 (1990), DOI 10.1112/S0025579300013097. MR1045784

[13] C. Maclachlan and A. W. Reid, *The arithmetic of hyperbolic 3-manifolds*, Graduate Texts in Mathematics, vol. 219, Springer-Verlag, New York, 2003. MR1937957

[14] M. Conder and C. Maclachlan, *Compact hyperbolic 4-manifolds of small volume*, Proc. Amer. Math. Soc. **133** (2005), no. 8, 2469–2476, DOI 10.1090/S0002-9939-05-07634-3. MR2138890

[15] P. Ozsváth and Z. Szabó, *On the Heegaard Floer homology of branched double-covers*, Adv. Math. **194** (2005), no. 1, 1–33, DOI 10.1016/j.aim.2004.05.008. MR2141852

[16] J. G. Ratcliffe and S. T. Tschantz, *On the Davis hyperbolic 4-manifold*, Topology Appl. **111** (2001), no. 3, 327–342, DOI 10.1016/S0166-8641(99)00221-7. MR1814232

[17] Alan W. Reid, *Arithmetic kleinian graups and their fuchsian subgroups*, Ph.D. thesis, Aberdeen, 1987.

[18] A. W. Reid, *Surface subgroups of mapping class groups*, Problems on mapping class groups and related topics, Proc. Sympos. Pure Math., vol. 74, Amer. Math. Soc., Providence, RI, 2006, pp. 257–268, DOI 10.1090/pspum/074/2264545. MR2264545

[19] A. Yu. Vesnin and A. D. Mednykh, *Fibonacci manifolds as two-sheeted coverings over a three-dimensional sphere, and the Meyerhoff-Neumann conjecture* (Russian, with Russian summary), Sibirsk. Mat. Zh. **37** (1996), no. 3, 534–542, ii, DOI 10.1007/BF02104848; English transl., Siberian Math. J. **37** (1996), no. 3, 461–467. MR1434698

[20] D. W. Morris, *Introduction to arithmetic groups*, Deductive Press, [place of publication not identified], 2015. MR3307755

University of California, Berkeley, 970 Evans Hall #3840, Berkeley, California 94720-3840

Email address: `ianagol@math.berkeley.edu`

Columbia University, Department of Mathematics, 2990 Broadway, New York, New York 10027

Email address: `flin@math.columbia.edu`

Contemporary Mathematics
Volume **760**, 2020
https://doi.org/10.1090/conm/760/15284

Farey recursion and the character varieties for 2-bridge knots

Eric Chesebro

ABSTRACT. We describe the (P)$SL_2\mathbb{C}$ character varieties of all 2-bridge knots and the diagonal character varieties for all 2-bridge links in terms of a set of polynomials defined using *Farey recursion*.

1. Introduction

When M is a hyperbolic 3-manifold with a single cusp, the (P)$SL_2\mathbb{C}$ character variety for M has proven to be a valuable tool in study the topology and geometry of M. There are too many examples to list here, but some classics are [**7**], [**8**], and [**32**]. Although the theory of character varieties is well-developed and very useful, it is notoriously difficult to compute character varieties in specific examples, particularly as the complexity of the manifolds increases.

A large class of link complements in which some computational progress has been made is the class of 2-bridge knots. Independent of the study of character varieties, these attractive links have a long history of mathematical interest. A highlight in this history, is the connection between 2-bridge links, continued fractions, and the Farey graph. In 1956, Schubert showed that 2-bridge links are classified by rational numbers and showed that continued fraction expansions of rational numbers give a natural way to diagram their corresponding link as the closure of a 4-strand braid. Famously in [**14**], Hatcher and Thurston used this classification and its relationship to the Farey graph to classify the essential surfaces in 2-bridge link complements. Sakuma and Weeks [**30**] also used the relationship between these links and the Farey graph to construct canonical ideal triangulations of the complements of 2-bridge links. Independently, [**2**] and [**12**] showed later that these triangulations are geometric and canonical in the sense of [**9**].

In [**27**] and [**28**], Riley began a thorough study of the (P)$SL_2\mathbb{C}$ representations of 2-bridge knot groups. Among other things, he introduced the Riley polynomial for 2-bridge knots, whose solutions correspond to representations which take peripheral elements to parabolic matrices. Riley's formulas have been used in [**16**], [**17**], [**22**], and [**25**] to collectively give recursive formulas for the (P)$SL_2\mathbb{C}$ character varieties of 2-bridge links obtained by doing $1/n$ Dehn filling on unknotted components of the links 6_1^3, 6_2^3, 6_3^2, 8_2^4, and 8_9^3. This is a large class of links with arbitrarily large crossing numbers. However, being Dehn fillings on a hand full of links, it is

---

2010 *Mathematics Subject Classification.* Primary 57M27.
Key words and phrases. Character variety, two bridge link.

quick to check that the volumes for these examples are bounded by 7.4, while the volumes of 2-bridge links can be arbitrarily large. Nonetheless, including Riley's original results, these formulas have enjoyed a variety of applications including [**5**], [**6**], [**11**], [**17**], [**18**], [**19**], and [**26**].

Here, we study the (P)$\mathrm{SL}_2\mathbb{C}$ character varieties of all 2-bridge knots together with the *diagonal* character varieties of the 2-bridge links. This makes it possible to see an attractive relationship between the defining polynomials for all of these algebraic sets. We refer to this relationship as *Farey recursion.*

By identifying rational numbers p/q with the point $p/q+i/q$ in the upper half $\mathbb{H}^2$ of $\mathbb{C}$, we can connect the rationals in a collection of interlocking triangles to form the Stern-Brocot diagram, see Figures 2 and 3. Each of these triangles contains exactly one rational number $p/q+i/q$ in its interior, which we call its *center.* The vertices that occur along the edges of this triangle form a bi-infinite sequence of rational numbers, which we refer to as the *boundary sequence* for p/q. In Section 5, we argue that there is a unique function

$$\mathcal{T}\colon\ \mathbb{Q}\cup\{\infty\}\to\mathbb{Z}[x,z]$$

with the following properties.

(1) $\mathcal{T}(0)=x$, $\mathcal{T}(\infty)=0$, and $\mathcal{T}(1)=z$;
(2) Given $p/q\in\mathbb{Q}$, and three successive terms $\gamma_1,\gamma_2,\gamma_3$ in the boundary sequence for p/q,

$$\mathcal{T}(\gamma_3)=-\mathcal{T}(\gamma_1)+\mathcal{T}(p/q)\,\mathcal{T}(\gamma_2).$$

Property (2) is called the *Farey recursion* condition. We can imagine $\mathcal{T}(\mathbb{Q})$ as a set of polynomials that are "linearly" recursive on the Stern-Brocot diagram, rather than along a line. Property (2) also makes it possible to compute the values of $\mathcal{T}$ relatively efficiently with a computer. Lastly, property (2) often allows us to make elementary arguments regarding $\mathcal{T}$. For instance, if we define $f\colon\ \mathbb{Q}\cup\{\infty\}\to\mathbb{Z}[x,z]$ by

$$f(p/q)=\begin{cases} x & \text{if } p \text{ is even and } q \text{ is odd}\\ z & \text{if } p \text{ and } q \text{ are odd}\\ xz & \text{if } p \text{ is odd and } q \text{ is even}\end{cases}$$

and set $X=x^2$, and $Z=z^2$, an elementary inductive argument (Lemma 5.4) shows that

$$\mathcal{T}_0(\alpha)=\frac{\mathcal{T}(\alpha)}{f(\alpha)}\in\mathbb{Z}[X,Z]$$

for every $\alpha\in\mathbb{Q}\cup\{\infty\}$.

The main theorem of this paper shows that the $\mathrm{PSL}_2\mathbb{C}$ character variety for the 2-bridge link $L(\alpha)$ associated to α is naturally identified with the complex affine algebraic set determined by $\mathcal{T}_0(\alpha)$.

THEOREM 7.4. *A point $\chi=(X_0,Z_0)\in\mathbb{C}^2$ is an irreducible character in $\overline{\mathsf{D}}_\alpha$ if and only if $X_0Z_0\neq 0$ and χ satisfies the polynomial $\mathcal{T}_0(\alpha)$.*

Here $\overline{\mathsf{D}}_\alpha$ is the *diagonal* $\mathrm{PSL}_2\mathbb{C}$ character variety for the link $L(\alpha)$. In the case that $L(\alpha)$ is a knot, $\overline{\mathsf{D}}_\alpha$ is the full character variety of irreducible characters.

It is relatively efficient to compute the character variety polynomials $\mathcal{T}_0(\alpha)$. With a simple program implemented in CoCalc [**29**], we are able to compute all

6079 values for $\mathcal{T}_0$ on the set

$$\left\{ \frac{p}{q} \in \mathbb{Q} \cap [0, 1/2] \,\middle|\, q < 200 \right\}$$

in roughly 8 minutes. Our program builds a list of polynomials up to denominator q, starting with $\mathcal{T}(0)$, $\mathcal{T}(\infty)$, and $\mathcal{T}(1)$. It then uses condition (2) above to compute the polynomials with denominator $q+1$ in terms of polynomials from the existing list. (It is also straightforward to write a program which computes $\mathcal{T}(\alpha)$ by following only the recursive steps needed for this particular α.)

The main theorem of [**24**] gives conditions on α' which guarantee existence of an epimorphism from the link group of $L(\alpha)$ to that of $L(\alpha')$. In Subsection 7.1, we employ a natural action of the modular group on the set of polynomials $\mathcal{T}(\mathbb{Q}\cup\{\infty\})$ to understand when the factors of $\mathcal{T}_0(\alpha)$ must divide $\mathcal{T}_0(\alpha')$, see Corollary 7.7. Ultimately, this provides an elementary argument which reproduces the result in [**24**], see Corollary 7.8.

In Subsection 7.2, we describe the relationship between our formulas and the formulas in [**22**].

In [**27**], Riley defines a polynomial in $\mathbb{Z}[X]$ for each rational number with odd denominator. These polynomials have since been refered to as *Riley polynomials.* Among other things, Riley proved that his polynomials are monic and their roots give rise to normalized *p-reps* for the corresponding 2-bridge knot groups. We discuss this in Section 8, where we also show that by setting $Z = -X$ in $\mathcal{T}_0(\alpha)$ we recover the Riley polynomial and thus extend the definition for Riley polynomials to rational numbers with even denominators. Our approach provides an efficient way to compute Riley polynomials as well as access to straightforward inductive arguments concerning their properties.

Again using CoCalc [**29**], we are able to compute Λ_α for each of the 13,662 numbers in

$$\left\{ \frac{p}{q} \in \mathbb{Q} \cap [0, 1/2] \,\middle|\, q < 300 \right\}$$

in roughly 12 minutes.

To demonstrate the inductive facility of our setup, we give a quick proof that Λ_α is monic for every α.

2. Surfaces

The material in this section is standard. Roughly, we follow the approach in [**1**]. Let $\mathbb{X}$ be the punctured plane $\mathbb{C} - \mathbb{Z}[i]$ equipped with its usual Euclidean geometry. Define subgroups $\Gamma_O, \Gamma_S, \Gamma_T \subset \mathrm{Isom}(\mathbb{X})$ by

$$\Gamma_O = \left\langle \text{ order-2 rotations centered at the points in } \frac{1}{2}\mathbb{Z}[i] \right\rangle$$

$$\Gamma_S = \left\langle \text{ order-2 rotations centered at the points in } \mathbb{Z}[i] \right\rangle$$

$$\Gamma_T = \langle \text{ horizontal and vertical unit translations } \rangle.$$

Then $\mathbb{O} = \mathbb{X}/\Gamma_O$ is a $(2,2,2,\infty)$-pillowcase orbifold, $\mathbb{S} = \mathbb{X}/\Gamma_S$ is a 4-punctured sphere, and $\mathbb{T} = \mathbb{X}/\Gamma_T$ is a once punctured torus. We also have a commutative diagram of covers.

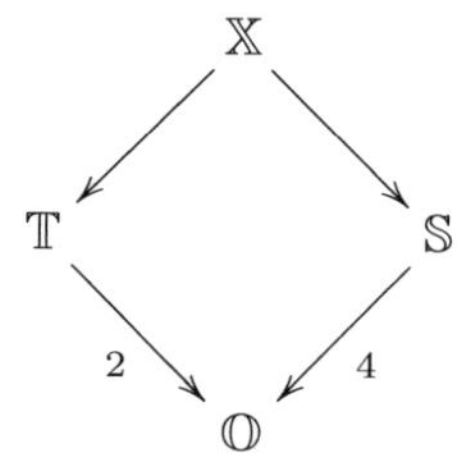

Take $\tilde{x} = (1/4, 1/4)$ as a basepoint in $\mathbb{X}$. Consider the paths

$$\frac{e^{\frac{3\pi i}{4}}}{2\sqrt{2}} e^{t\pi i} + \frac{1}{2} \qquad \frac{e^{\frac{5\pi i}{4}}}{2\sqrt{2}} e^{t\pi i} + \frac{1}{2}(1+i) \qquad \frac{e^{\frac{7\pi i}{4}}}{2\sqrt{2}} e^{t\pi i} + \frac{i}{2}$$

in $\mathbb{X}$ based at $\tilde{x}$ and let $\tilde{p}$, $\tilde{q}$, and $\tilde{r}$ be their respective homotopy classes (rel ∂). Then, if p, q, and r are the corresponding elements of $\pi_1(\mathbb{O})$,

$$\pi_1(\mathbb{O}) = \langle p, q, r \mid p^2, q^2, r^2 \rangle .$$

Let $k_0 = rpq$ and observe that k_0 is primitive and peripheral.

Define

$$a = rq \qquad \text{and} \qquad b = qp.$$

We also let $A = a^{-1}$ and $B = b^{-1}$. As a subgroup of $\pi_1(\mathbb{O})$, $\pi_1(\mathbb{T})$ is the rank-2 free group generated by a and b. Moreover, the corresponding deck transformations are translations by -1 and i respectively. The commutator $[a, b] = k_0^2$ is primitive and peripheral in $\pi_1\mathbb{T}$.

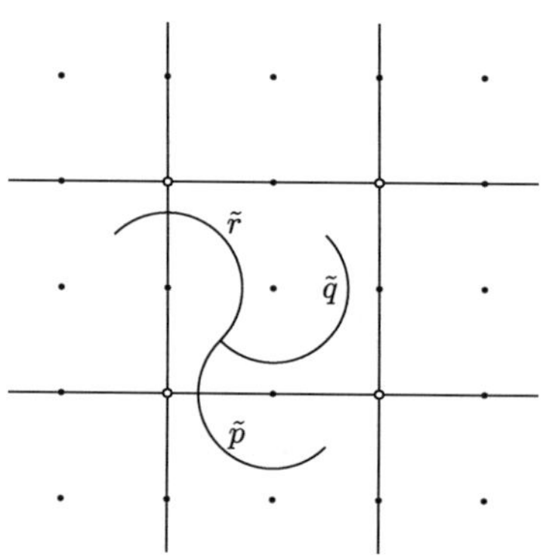

FIGURE 1. Paths in $\mathbb{X}$.

Define distinguished conjugates

$$k_1 = pk_0p \qquad k_2 = qk_0q \qquad k_3 = rk_0r$$

of k_0. As a subgroup of $\pi_1(\mathbb{O})$, $\pi_1(\mathbb{S})$ is the rank-3 free group generated by the peripheral elements k_0, k_1, and k_2. The product $k_0k_2k_1k_3$ is trivial and

$$a^2 = k_3^{-1}k_2^{-1} \qquad B^2 = k_3k_0 \qquad (aB)^2 = k_3^{-1}k_2k_3k_0.$$

3. $\mathbf{PSL_2\mathbb{C}}$ representations

Given $\mathbf{z} = (x, z) \in \mathbb{C}^2$ we can find $t \in \mathbb{C} - \{0\}$ so that

$$x^2 + z^2 + \left(t - t^{-1}\right)^2 = 0. \tag{3.1}$$

We refer to the pair $\mathbf{z}, t$ as a **parameter pair**. Every parameter pair $\mathbf{z}, t$ determines a representation $\rho_{\mathbf{z},t} \colon \pi_1\mathbb{O} \to \mathrm{PSL}_2\mathbb{C}$ as follows. If $x = 0$, define

$$\rho_{\mathbf{z},t}(p) = \begin{bmatrix} 0 & 1 \\ -1 & 0 \end{bmatrix} \qquad \rho_{\mathbf{z},t}(q) = \begin{bmatrix} i & 0 \\ 0 & -i \end{bmatrix} \qquad \rho_{\mathbf{z},t}(r) = \begin{bmatrix} 0 & it \\ \frac{i}{t} & 0 \end{bmatrix}.$$

If $x \neq 0$ and $z = 0$, define

$$\rho_{\mathbf{z},t}(p) = \begin{bmatrix} i & 0 \\ 0 & -i \end{bmatrix} \qquad \rho_{\mathbf{z},t}(q) = \begin{bmatrix} 0 & 1 \\ -1 & 0 \end{bmatrix} \qquad \rho_{\mathbf{z},t}(r) = \begin{bmatrix} 0 & it \\ \frac{i}{t} & 0 \end{bmatrix}.$$

If $xz \neq 0$, define

$$\begin{aligned} \rho_{\mathbf{z},t}(p) &= \frac{1}{x} \begin{bmatrix} t^{-1} - t & -1 \\ -z^2 & t - t^{-1} \end{bmatrix} \\ \rho_{\mathbf{z},t}(q) &= \frac{1}{z} \begin{bmatrix} 0 & 1 \\ -z^2 & 0 \end{bmatrix} \\ \rho_{\mathbf{z},t}(r) &= \frac{1}{x} \begin{bmatrix} -z\,t^{-1} & z + t^2\,z^{-1} - z^{-1} \\ z\,(1 - t^{-2}) & z\,t^{-1} \end{bmatrix}. \end{aligned}$$

Straightforward calculations verify that for all x, y, and z

$$\mathrm{Tr}\big(\rho_{\mathbf{z},t}(a)\big) = \pm x \qquad \mathrm{Tr}\big(\rho_{\mathbf{z},t}(b)\big) = 0 \qquad \mathrm{Tr}\big(\rho_{\mathbf{z},t}(aB)\big) = \pm z.$$

Also, if $x = 0$, then

$$\rho_{\mathbf{z},t}(k_0) = \rho_{\mathbf{z},t}(k_1^{-1}) = \begin{bmatrix} t & 0 \\ 0 & t^{-1} \end{bmatrix},$$

if $x \neq 0$ and $z = 0$, then

$$\rho_{\mathbf{z},t}(k_0) = \rho_{\mathbf{z},t}(k_1) = \begin{bmatrix} t & 0 \\ 0 & t^{-1} \end{bmatrix},$$

and, if $xz \neq 0$, then

$$\rho_{\mathbf{z},t}(k_0) = \begin{bmatrix} t & 1 \\ 0 & t^{-1} \end{bmatrix} \qquad \rho_{\mathbf{z},t}(k_1) = \begin{bmatrix} t & 0 \\ z^2 & t^{-1} \end{bmatrix}.$$

Note that $\rho_{\mathbf{z},t}$ is irreducible if and only if $xz \neq 0$.

4. Slopes and essential simple closed curves

Lines in $\mathbb{X}$ with rational slopes descend to essential simple closed curves on $\mathbb{O}$, $\mathbb{T}$, and $\mathbb{S}$. In fact, there are bijections from $\widehat{\mathbb{Q}} = \mathbb{Q} \cup \{1/0\}$ to the set of conjugacy classes in $\pi_1(\mathbb{O})$, $\pi_1(\mathbb{T})$, and $\pi_1(\mathbb{S})$ whose corresponding free homotopy classes are represented by essential simple closed curves. For example, in $\pi_1(\mathbb{T})$ the slope of the class containing a is zero and the slope of the class containing b is infinite.

Given $\alpha \in \mathbb{Q}_0 = \mathbb{Q} \cap [0,1]$ and an $\mathrm{SL}_2\mathbb{C}$ representation of $\pi_1(\mathbb{O})$, $\pi_1(\mathbb{S})$, or $\pi_1(\mathbb{T})$, it is surprisingly easy to find elements of $\pi_1(\mathbb{S})$ and $\pi_1(\mathbb{T})$ which represent the slope α and to compute their traces under the representation. To show how this is done, it will be helpful to first discuss the Farey sum and the Stern-Brocot diagram.

4.1. Farey sums and the Stern-Brocot diagram. Throughout the remainder of the paper, when we write elements of $\mathbb{Q}$ as integer quotients, we will always write them in lowest terms with positive denominators. If p/q and r/s are elements of $\widehat{\mathbb{Q}}$ such that $ps - rq = \pm 1$, then $\{p/q, r/s\}$ is called a **Farey pair**. The **Farey sum** of p/q and r/s is defined as

$$\frac{p}{q} \oplus \frac{r}{s} = \frac{p+r}{q+s}.$$

It is remarkable that if $\{p/q, r/s\}$ is a Farey pair then $p+r$ and $q+s$ are relatively prime. Notice that every Farey pair is a subset of exactly two **Farey triples**. In particular, if $\{\alpha, \gamma\}$ is a Farey pair, then $\{\alpha, \gamma, \alpha \oplus \gamma\}$ is a Farey triple. So, when $\{\alpha, \gamma\}$ is a Farey pair, we can define the notation

$$\left(\oplus_0^0 \alpha\right) \oplus \gamma = \gamma, \text{ and}$$
$$\left(\oplus_0^j \alpha\right) \oplus \gamma = \alpha \oplus \left(\left(\oplus_0^{j-1} \alpha\right) \oplus \gamma\right)$$

inductively.

There is a closely related geometric graph $\mathcal{G}$ embedded in the upper-half space $\mathbb{H}^2 \subset \mathbb{C}$. The vertices of $\mathcal{G}$ correspond to $\mathbb{Q}$ and the edges correspond to Farey pairs. More precisely, for $\frac{p}{q} \in \mathbb{Q}$, define $v_{\frac{p}{q}} = \frac{p}{q} + \frac{i}{q} \in \mathbb{H}^2$. The set

$$\{v_\alpha \,|\, \alpha \in \mathbb{Q}\}$$

constitutes the vertices of $\mathcal{G}$ and the edges of $\mathcal{G}$ are the straight line segments between Farey pairs of vertices. (We will often blur the distinction between a number $\alpha \in \mathbb{Q}$ and its vertex v_α.) See Figure 2 below as well as Figure 1.3 of [**13**]. Here, we refer to $\mathcal{G}$ as the **Stern-Brocot diagram**. Elementary properties of the Farey sum and Farey pairs imply that every intersection between a distinct pair of edges in $\mathcal{G}$ occurs at a vertex of $\mathcal{G}$.

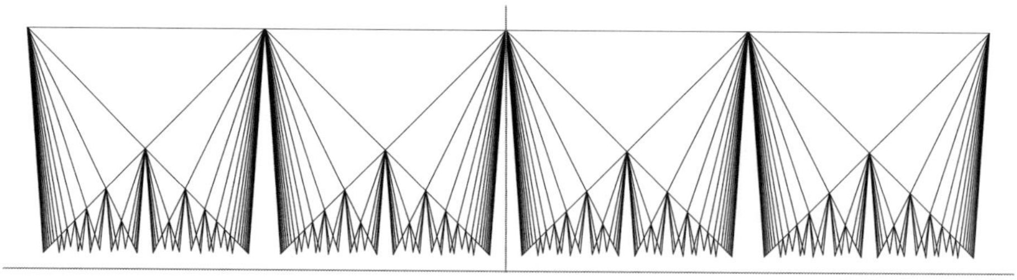

FIGURE 2. The portion of $\mathcal{G}$ spanned by fractions in $[-2, 2]$ with denominator at most 15.

Suppose $\alpha \in \mathbb{Q}$. The graph theoretical neighborhood $\Delta(\alpha)$ of v_α in $\mathcal{G}$ is, by definition, the subgraph spanned by the vertices which are Farey pairs with α, see Figure 3.

If $\alpha \in \mathbb{Q} - \mathbb{Z}$ then $\Delta(\alpha)$ corresponds to a triangle with an ideal vertex at $\alpha \in \mathbb{R}$. This triangle contains exactly one vertex of $\mathcal{G}$ in its interior, namely v_α. For this reason, we refer to $\Delta(\alpha)$ as the **triangle centered at $\mathbf{v}_\alpha$**. Its ideal vertex α is called its **tip**. Emanating from the tip of $\Delta(\alpha)$ are its left and right sides whose slopes are $\pm q$, where q is the denominator of α. The vertices at the tops of these edges are called the left and right **corners** of $\Delta(\alpha)$. The **positive** (resp. negative) **vertex**

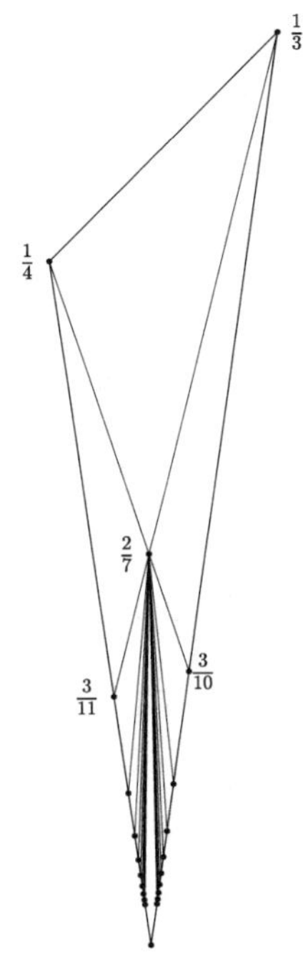

FIGURE 3. $\Delta\left(\frac{2}{7}\right)$.

sequence $\partial_+(\alpha)$ (resp. $\partial_-(\alpha)$) of $\Delta(\alpha)$ is defined to be the ordered sequence of vertices down the side of $\Delta(\alpha)$ whose slope is positive (resp. negative). In fact, if $\kappa_\pm$ is the corner on this side, then

$$\partial_\pm(\alpha) = \left\{\left(\oplus_0^j \alpha\right) \oplus \kappa_\pm\right\}_{j=0}^{\infty}. \tag{4.1}$$

We write $\partial(\alpha)$ to denote the bi-infinite sequence formed by reversing $\partial_-(\alpha)$ and juxaposing the sequence $\partial_+(\alpha)$ and we refer to $\partial(\alpha)$ as the **boundary sequence** for α.

The Stern-Brocot diagram and our notions of triangles, edges, and corners all extend nicely to include an additional vertex for $1/0$. To do this, add a vertex $v_{1/0}$ and an edge from $v_{1/0}$ to each integer vertex of $\mathcal{G}$. The triangles centered at integers and at $1/0$ involve the new vertex $v_{1/0}$. They are exceptional in that they each have two sides but only one corner. We explicitly define their vertex sequences by listing their terms. Let $n \in \mathbb{Z}$ and define

$$\begin{aligned}
\partial_-(1/0) &= \{0, -1, -2, -3, \ldots\} \\
\partial_+(1/0) &= \{0, 1, 2, 3, \ldots\} \\
\partial_-(n) &= \left\{\frac{1}{0}, n-1, \frac{2n-1}{2}, \frac{3n-1}{3}, \ldots\right\} \\
\partial_+(n) &= \left\{\frac{1}{0}, n+1, \frac{2n+1}{2}, \frac{3n+1}{3}, \ldots\right\}.
\end{aligned}$$

In the boundary sequences $\partial(1/0)$ and $\partial(n)$, we do not repeat the duplicate corner, for instance, $\partial(1/0)$ is the ordered sequence of integers. Also, if we allow $\frac{1}{0} = \frac{-1}{0}$ and $\frac{0}{1} = \frac{0}{-1}$ and choose appropriate representatives, then the Farey sum formulas (4.1) still apply in these exceptional cases.

There is a direct correlation between $\mathcal{G}$ and the Farey Graph. In [**15**], Hatcher points out that if the vertices of $\mathcal{G}$ are pushed straight down to $\mathbb{R} \subset \mathbb{H}^2$ and the edges follow their endpoints, ultimately forming geodesics in $\mathbb{H}^2$, the result is the Farey graph.

The next lemma will be useful later in this paper.

LEMMA 4.1. *Given $\alpha \in \mathbb{Q} - \mathbb{Z}$ there is a number $\gamma \in \mathbb{Q}$ where α is at least the third term of a vertex sequence for $\Delta(\gamma)$.*

PROOF. If the denominator of α is two, then α is the third term of a sequence $\partial_+(n)$ for some $n \in \mathbb{Z}$. So, we may assume that the denominator of α is larger than two.

Since the denominator of α is larger than two, $\Delta(\alpha)$ has two corners, $\kappa_\pm \in \mathbb{Q}$, each of which lies above α in $\mathcal{G}$. The Farey pair condition forces the denominators of κ_+ and κ_- to be distinct. Without loss of generality, assume that the denominator of κ_- is smaller than that of κ_+.

We know that α comes after κ_+ in the sequence $\partial_+(\kappa_-)$. Because the denominator of κ_+ is larger than that of κ_-, we also know that κ_+ is not the first term of this sequence. Hence, α is at least the third term of $\partial_+(\kappa_-)$. $\square$

Let G be the extended modular group

$$G = \mathrm{PSL}_2\mathbb{Z} \oplus \left\langle \begin{bmatrix} -1 & 0 \\ 0 & 1 \end{bmatrix} \right\rangle.$$

Every element of G determines an invertible Möbius transformation and these transformations preserve Farey pairs when applied to $\widehat{\mathbb{Q}}$. This gives an action of G on $\mathcal{G}$. Although this action preserves triangles, it does not preserve their corners. For instance, if $\phi = \left[\begin{smallmatrix} -1 & 1 \\ -4 & 3 \end{smallmatrix}\right]$ then ϕ fixes $\frac{1}{2}$ and preserves $\partial\left(\frac{1}{2}\right)$. On the other hand, ϕ acts non-trivially on $\partial\left(\frac{1}{2}\right)$ taking each term to its neighbor on the left.

4.2. Words of slope α. There is a function $\Omega\colon \widehat{\mathbb{Q}}_0 \to \pi_1(\mathbb{O})$ which is uniquely determined by the following two properties

(1) If $\{\alpha, \beta\}$ is a Farey pair in $\widehat{\mathbb{Q}}_0$ and $\alpha < \beta$ then
$$\Omega(\alpha \oplus \beta) = \Omega(\alpha)\,\Omega(\beta).$$
(2) $\Omega(0) = a$ and $\Omega(1/0) = B$.

Here, we use the convention that $0/1 < 1/0$.

If α is a positive rational number which is not in $\mathbb{Q}_0$, we define $\Omega(\alpha)$ to be the word obtained from $\Omega(1/\alpha)$ by swapping a's and B's. If α is a negative rational number, we define $\Omega(\alpha)$ to be the word obtained from $\Omega(-\alpha)$ by replacing B's with b's. This determines an extension $\Omega\colon \widehat{\mathbb{Q}} \to \pi_1(\emptyset)$. The next proposition is established in [**23**] and [**31**].

PROPOSITION 4.2. *$\Omega(\alpha)$ has slope α in $\mathbb{O}$ and $\mathbb{T}$.*

REMARK 4.3. If we consider the cover $\mathbb{S} \to \mathbb{O}$, we see that $\Omega(\alpha)^2$ has slope α in $\mathbb{S}$.

5. Trace functions and Farey recursion.

5.1. Farey recursion. Suppose that R is a commutative ring and $d\colon \widehat{\mathbb{Q}} \to R$ is a function. A function $\mathcal{F}\colon \widehat{\mathbb{Q}} \to R$ is a **Farey recursive function** (FRF) with **determinant** d if, whenever α and γ make a Farey pair, then

$$\mathcal{F}(\alpha \oplus \alpha \oplus \gamma) = -d(\alpha)\,\mathcal{F}(\gamma) + \mathcal{F}(\alpha)\,\mathcal{F}(\alpha \oplus \gamma). \tag{5.1}$$

Notice that if $\{\alpha, \gamma\}$ is a Farey pair then $\{\alpha, \gamma, \alpha \oplus \gamma, \alpha \oplus \alpha \oplus \gamma\}$ are the vertices of a pair of adjacent triangles in the Farey graph.

Suppose that $\mathcal{F}$ is an FRF with determinant d and $\alpha \in \widehat{\mathbb{Q}}$. Choose $\epsilon \in \{\pm 1\}$ and let $a_j \in R$ be the element obtained by applying $\mathcal{F}$ to the jth element of $\partial_\epsilon(\alpha)$. Then the sequence $\{a_j\}_0^\infty$ is linearly recursive and satisfies

$$\begin{pmatrix} 0 & 1 \\ -d(\alpha) & \mathcal{F}(\alpha) \end{pmatrix} \begin{pmatrix} a_{j-1} \\ a_j \end{pmatrix} = \begin{pmatrix} a_j \\ a_{j+1} \end{pmatrix} \tag{5.2}$$

for every $j \geq 1$. We call the 2×2 matrix in Equation 5.2 the **recursion matrix** at α for $\mathcal{F}$.

Henceforth, we will assume that R is a ring with unity 1_R and also that determinant functions are always constant with value 1_R. For example, an FRF $\mathcal{F}\colon \widehat{\mathbb{Q}} \to \mathbb{C}$ with determinant 1 which satisfies the Markov condition

$$\mathcal{F}(0)^2 + \mathcal{F}(1/0)^2 + \mathcal{F}(1)^2 = \mathcal{F}(0)\,\mathcal{F}(1/0)\,\mathcal{F}(1)$$

is called a **Markov map** in [**3**]. As in [**3**], if $\{\alpha, \beta, \gamma\}$ is a Farey triple and $\{a, b, c\} \subset R$ then there is a unique FRF $\mathcal{F}\colon \widehat{\mathbb{Q}} \to R$ with

$$\mathcal{F}(\alpha) = a \qquad \mathcal{F}(\beta) = b \qquad \mathcal{F}(\gamma) = c.$$

The next lemma shows that the sequence $\mathcal{F}(\partial(\alpha))$ is a bi-infinite linear recursive sequence with recursion matrix

$$W_\alpha = \begin{pmatrix} 0 & 1 \\ -1 & \mathcal{F}(\alpha) \end{pmatrix}$$

and the reverse of this sequence is also recursive with the same recursion matrix.

LEMMA 5.1. *Suppose that $\mathcal{F}$ is an FRF and $\alpha \in \widehat{\mathbb{Q}}$. Let $\{a_j\}_{-\infty}^\infty$ be the ordered bi-infinite sequence obtained by applying $\mathcal{F}$ to $\partial(\alpha)$. Then, for every integer j,*

$$a_{j+1} = -a_{j-1} + \mathcal{F}(\alpha)\,a_j \quad \textit{and} \quad a_{j-1} = -a_{j+1} + \mathcal{F}(\alpha)\,a_j.$$

PROOF. First, we assume that the indices for $\{a_j\}$ are shifted so that $\mathcal{F}(\kappa_-) = a_0$ and $\mathcal{F}(\kappa_+) = a_1$, where $\kappa_\pm$ are the corners of $\Delta(\alpha)$.

Notice that the first equation in the statement of the lemma holds if and only if the second does. So, to prove the lemma, it suffices to verify the equalities

$$a_1 = -a_{-1} + \mathcal{F}(\alpha)\,a_0 \quad \text{and} \quad a_2 = -a_0 + \mathcal{F}(\alpha)\,a_1$$

occurring at the corners of $\partial(\alpha)$.

The second term of $\partial_-(\alpha)$ coincides with the term in $\partial_+(\kappa_-)$ immediately following κ_+ and α. Similarly, the second term of $\partial_+(\alpha)$ immediately follows κ_- and α in the sequence $\partial_-(\kappa_+)$. Hence, the Farey recursive condition implies that

$$a_{-1} = -a_1 + \mathcal{F}(\alpha)\,a_0 \quad \text{and} \quad a_2 = -a_0 + \mathcal{F}(\alpha)\,a_1.$$

These are evidently equivalent to the equalities above. □

Define the **generic FRF** to be the FRF

$$\mathcal{U}\colon \widehat{\mathbb{Q}} \to \mathbb{Z}[x,y,z]$$

determined by

$$\mathcal{U}(0) = x \qquad \mathcal{U}(1/0) = y \qquad \mathcal{U}(1) = z.$$

Consider a composition of $\mathcal{U}$ with a ring homomorphism $\mathbb{Z}[x,y,z] \to R$. Since the homomorphism preserves the Farey recursive conditions, the composition will also be an FRF. On the other hand, if $\mathcal{F}\colon \widehat{\mathbb{Q}} \to R$ is an FRF there is a unique ring homomorphism $\psi\colon \mathbb{Z}[x,y,z] \to R$ making the diagram

$$\begin{array}{ccc} \widehat{\mathbb{Q}} & \xrightarrow{\;\mathcal{U}\;} & \mathbb{Z}[x,y,z] \\ & {\scriptstyle\mathcal{F}}\searrow & \downarrow{\scriptstyle\psi} \\ & & R \end{array}$$

commute. Hence, every FRF can be viewed as a specialization of $\mathcal{U}$.

Recall, from Subsection 4.1, the action of the group G on $\widehat{Q}$. Since an element $\phi \in G$ preserves Farey pairs, the map $\mathcal{U} \circ \phi$ is an FRF. Hence, we have a homomorphism $\psi_\phi\colon \mathbb{Z}[x,y,z] \to \mathbb{Z}[x,y,z]$ and a commutative diagram

$$\begin{array}{ccc} \widehat{\mathbb{Q}} & \xrightarrow{\;\mathcal{U}\;} & \mathbb{Z}[x,y,z] \\ {\scriptstyle\phi}\downarrow & & \downarrow{\scriptstyle\psi_\phi} \\ \widehat{\mathbb{Q}} & \xrightarrow{\;\mathcal{U}\;} & \mathbb{Z}[x,y,z] \end{array}$$

The homomorphism ψ_ϕ is given by evaluating x, y, z at the polynomials

$$\mathcal{U}(\phi \cdot 0), \quad \mathcal{U}(\phi \cdot (1/0)), \quad \mathcal{U}(\phi \cdot 1).$$

Because ϕ is a bijection, ψ_ϕ is an isomorphism with inverse $\psi_{\phi^{-1}}$.

In the cases of

$$\sigma = \begin{bmatrix} -1 & 1 \\ 0 & 1 \end{bmatrix} \quad \text{and} \quad \zeta = \begin{bmatrix} 0 & 1 \\ -1 & 1 \end{bmatrix},$$

$\mathcal{U}(\gamma)$ and $\mathcal{U}(\sigma\cdot\gamma)$ differ by interchanging x and z, while $\mathcal{U}(\gamma)$ and $\mathcal{U}(\zeta\cdot\gamma)$ are related by the cyclic permutation $(x\ z\ y)$ of the three variables. The $\langle\sigma,\zeta\rangle$-orbit of any $\alpha \in \widehat{Q}$ intersects $\mathbb{Q}_0$, which is part of the reason for our focus on this portion of $\mathcal{G}$.

As the denominator of α grows, the polynomial $\mathcal{U}(\alpha)$ gets complicated. The next result is a simple application of the relationship between $\mathcal{U}$ and the action $G \times \widehat{\mathbb{Q}} \to \widehat{\mathbb{Q}}$ which shows that the isomorphism class of the affine variety determined by $\mathcal{U}(\alpha)$ is constant.

PROPOSITION 5.2. *If $p \in \mathcal{U}(\widehat{\mathbb{Q}})$ then p is irreducible over $\mathbb{C}$. More specifically, the affine algebraic variety determined by p is isomorphic to $\mathbb{C}^2$.*

PROOF. Suppose $p = \mathcal{U}(\alpha)$ for some $\alpha \in \widehat{\mathbb{Q}}$. Then there exists $\phi \in G$ with $\phi \cdot 0 = \alpha$. By definition, the isomorphism $\psi_\phi\colon \mathbb{Z}[x,y,z] \to \mathbb{Z}[x,y,z]$ takes x to $\mathcal{U}(\alpha)$. So we have isomorphic varieties

$$\mathbb{C}^2 \cong \{(x,y,z) \in \mathbb{C}^3 \mid x = 0\} \cong \{(x,y,z) \in \mathbb{C}^3 \mid p(x,y,z) = 0\}.$$

□

5.2. Traces for $\Omega(\alpha)$. Suppose that $\rho\colon \pi_1(\mathbb{T}) \to \mathrm{SL}_2\mathbb{C}$ is a representation with character χ_ρ. Let

$$x_0 = \chi_\rho(a), \quad y_0 = \chi_\rho(B), \quad \text{and} \quad z_0 = \chi_\rho(aB).$$

Define the homomorphism $\psi_\rho\colon \mathbb{Z}[x,y,z] \to \mathbb{C}$ to be evaluation at $(x_0, y_0, z_0) \in \mathbb{C}^3$.

Take $\beta \in \partial(\alpha)$. Using the Cayley-Hamilton theorem, the definition of Ω, and the fact that χ_ρ is constant on conjugacy classes, we find that

$$(\chi_\rho \circ \Omega)(\alpha \oplus \alpha \oplus \beta) = -(\chi_\rho \circ \Omega)(\beta) + (\chi_\rho \circ \Omega)(\alpha)\,(\chi_\rho \circ \Omega)(\alpha \oplus \beta).$$

In other words, the trace function $\chi_\rho \circ \Omega$ is Farey recursive and

$$(\chi_\rho \circ \Omega)(\alpha) = \psi_\rho \circ \mathcal{U}(\alpha).$$

The set of all such FRFs is precisely the set of Markov maps from [**3**].

Any two elements of $\pi_1(\mathbb{T})$ which are represented by loops freely homotopic to a closed curve with slope α are conjugage in $\pi_1(\mathbb{T})$. So, if $\omega \in \pi_1\mathbb{T}$ is such an element, then $\chi_\rho(\omega) = \psi_\rho \circ \mathcal{U}(\alpha)$. If instead, ρ is a representation into $\mathrm{PSL}_2\mathbb{C}$, then x_0, y_0, and z_0 are only defined up to sign, as is $\mathrm{Tr}(\rho(w))$. If we choose these signs arbitrarily, then $\mathrm{Tr}(\rho(\omega)) = \pm(\psi_\rho \circ \mathcal{U}(\alpha))$.

In this paper, we are concerned with representations of 2-bridge link groups. As we will see, this leads us to consider representations with $y_0 = 0$. Let

$$\psi_{\mathcal{T}}\colon \mathbb{Z}[x,y,z] \to \mathbb{Z}[x,z]$$

be the evaluation $y = 0$. Define the FRF

$$\mathcal{T}\colon \widehat{\mathbb{Q}} \to \mathbb{Z}[x,z]$$

by $\mathcal{T} = \psi_{\mathcal{T}} \circ \mathcal{U}$.

The function $\mathcal{T}$ has predictable factors of x and z. If we cancel these factors, the resulting polynomial has exponents which are exclusively even. The next two lemmas make this precise. To begin, define functions $f_j\colon \widehat{\mathbb{Q}} \to \{0,1\}$ by

$$f_1(p/q) \equiv pq + 1 \pmod 2 \qquad \text{and} \qquad f_2(p/q) \equiv p \pmod 2.$$

Both functions preserve Farey sums in the sense that, if $\{\alpha, \beta\}$ are a Farey pair, then

$$f_j(\alpha \oplus \beta) \equiv f_j(\alpha) + f_j(\beta) \pmod 2. \tag{5.3}$$

Define $X = x^2$ and $Z = z^2$. Also, for $\alpha \in \widehat{\mathbb{Q}}$, define $f(\alpha) = x^{f_1(\alpha)} z^{f_2(\alpha)}$.

LEMMA 5.3. *If $\{\alpha, p/q\}$ is a Farey pair in $\widehat{\mathbb{Q}}$, then*

$$f(p/q) = f(\alpha \oplus \alpha \oplus p/q) \quad \text{and} \quad \frac{f(\alpha)\, f(\alpha \oplus p/q)}{f(p/q)} = \begin{cases} 1 & \text{if } q \text{ is even} \\ X & \text{if } pq \text{ is odd} \\ Z & \text{if } p \text{ is even.} \end{cases}$$

PROOF. The first assertion is a consequence of Equation 5.3 above. By considering the different possibilities for the parities of p and q, the second follows. □

Define the **2-bridge character function** as

$$\mathcal{T}_0(\alpha) = \frac{\mathcal{T}(\alpha)}{f(\alpha)}.$$

It is easy to check that $\mathcal{T}_0$ is not Farey recursive.

LEMMA 5.4. *For $p/q \in \widehat{\mathbb{Q}}$, $\mathcal{T}_0(p/q) \in \mathbb{Z}[X, Z]$.*

PROOF. The proof is by induction on q. Since $\mathcal{T}(1/0) = 0$, the lemma holds for $q = 0$. The recursion matrix for $\mathcal{T}$ on $\partial(1/0)$ is $\left(\begin{smallmatrix} 0 & 1 \\ -1 & 0 \end{smallmatrix}\right)$ and it follows that $\{\mathcal{T}(n)\}_{n\in\mathbb{Z}}$ is the repeated concatenation of the sequence $\{x, z, -x, -z\}$ shifted so that $\mathcal{T}(0) = x$. Since $f(n)$ is x if n is even and y otherwise, the lemma holds for $q = 1$. So, we assume that $q \geq 2$ and that the lemma holds for elements of $\widehat{\mathbb{Q}}$ whose denominator is smaller than q.

As in the previous lemma, Lemma 4.1 provides a Farey pair $\{\alpha, \beta\}$ with $\alpha \in \mathbb{Q}$ and $\beta \in \widehat{\mathbb{Q}}$ so that

$$p/q = \alpha \oplus \alpha \oplus \beta$$

and the denominators of α, β, and $\alpha \oplus \beta$ are are all less than q.

By definition of $\mathcal{T}$,

$$\mathcal{T}\,(p/q) = -\mathcal{T}(\beta) + \mathcal{T}(\alpha)\,\mathcal{T}(\alpha \oplus \beta).$$

From Lemma 5.3, $f(p/q) = f(\beta)$. So, dividing through by this, we have

$$\mathcal{T}_0\,(p/q) = -\mathcal{T}_0(\beta) + \frac{f(\alpha)f(\alpha \oplus \beta)}{f(\beta)}\,\mathcal{T}_0(\alpha)\mathcal{T}_0(\alpha \oplus \beta).$$

Lemma 5.3 and the inductive assumption imply that this is an element of $\mathbb{Z}[X, Z]$. □

6. 2-bridge links

Every 2-bridge link is determined by a number $\alpha \in \mathbb{Q}_0$. We denote this link as $L(\alpha)$. Its complement is $M(\alpha) = S^3 - L(\alpha)$. It is well known that the compact manifold formed by removing an open tubular neighborhood of $L(\alpha)$ from S^3 can also be obtained by attaching a pair of 2-handles to a thickened 4-holed sphere along a pair of simple closed curves with slopes $1/0$ and α. If we attach only the slope ∞ handle, the result is a handle body H. Define

$$\Gamma_H = \pi_1(\mathbb{S})/\langle\!\langle b^2 \rangle\!\rangle = \left\langle k_0, k_1, k_2, k_3 \mid k_2 = k_1^{-1},\ k_3 = k_0^{-1} \right\rangle = \langle k_0, k_1 \rangle$$

and

$$\Gamma_\alpha = \left\langle k_0, k_1 \,\middle|\, \Omega(\alpha)^2 \right\rangle.$$

Then the inclusions of the 4-holed sphere into H and $M(\alpha)$ induce natural identifications of Γ_H and Γ_α with the fundamental groups of H and $M(\alpha)$.

REMARK 6.1.

(1) $L(\alpha)$ is a knot when the denominator of α is odd and is a 2-component link otherwise.
(2) If $\alpha \in \mathbb{Q}_0$ then there is a link complement M_α such that the set of knot complements in $\{M(\gamma) \mid \gamma \in \partial(\alpha)\}$ is the same as the set of $(1, n)$-Dehn fillings on a particular unknotted component (crossing circle) of M_α. Also, by adjusting M_α with a half twist through this crossing circle, we get a similar statement for the 2-component link complements in this set. See Figure 4.

The next lemma shows that the representations $\rho_{\mathbf{z},t}$ from Section 3 are relevant to our discussion of Γ_H and Γ_α.

LEMMA 6.2. *For every parameter pair $(\mathbf{z}, t)$, the representation $\rho_{\mathbf{z},t}$ descends to Γ_H. Moreover, if $\mathbf{z}$ satisfies $\mathcal{T}(\alpha)$ then $\rho_{\mathbf{z},t}$ descends further to Γ_α.*

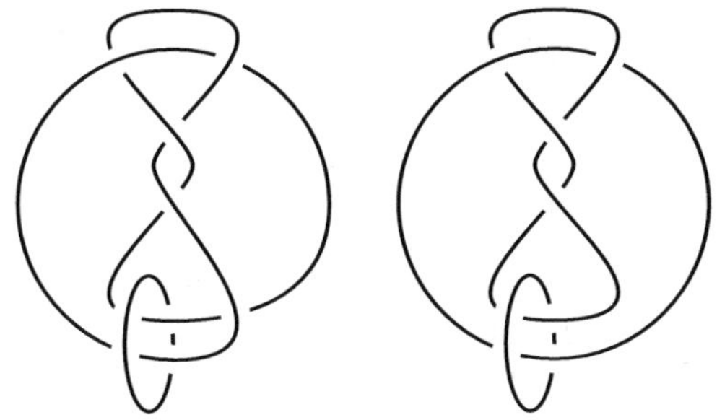

FIGURE 4. The manifolds obtained by doing $(1,n)$-Dehn filling on the unknotted cusps are precisely the 2-bridge links which correspond to the terms of $\partial(2/5) \subset \mathcal{G}$.

PROOF. $\mathrm{Tr}\,(\rho_{\mathbf{z},t}(b)) = 0$, so $\rho_{\mathbf{z},t}(b)$ has order two and $\rho_{\mathbf{z},t}$ descends to Γ_H. If $\mathbf{z}$ satisfies $\mathcal{T}(\alpha)$ then $\mathrm{Tr}\,(\rho_{\mathbf{z},t}\,(\Omega(\alpha))) = 0$ and $\rho_{\mathbf{z},t}$ descends further to Γ_α. □

7. Character varieties

We refer readers to [**7**] for background on $\mathrm{SL}_2\mathbb{C}$ character varieties and to [**4**], [**10**], or [**21**] for background on $\mathrm{PSL}_2\mathbb{C}$ character varieties.

Let X_H and X_α be the affine algebraic sets of $\mathrm{SL}_2\mathbb{C}$ characters of Γ_H and Γ_α whose algebraic components contain irreducible characters. Because Γ_α is a quotient of Γ_H, we may regard X_α as a subset of X_H.

As in [**7**] and [**10**], it is typical to use the functions

$$u = \chi(k_0), \quad w = \chi(k_1), \quad \text{and} \quad v = \chi(k_0k_1)$$

as coordinates for χ in X_H. In this paper, we are interested in the **diagonal subvarieties** $\mathsf{D}_H \subseteq \mathsf{X}_H$ and $\mathsf{D}_\alpha \subseteq \mathsf{X}_\alpha$ cut out by the polynomial $u - w$.

REMARK 7.1.

(1) If $L(\alpha)$ is a knot then k_0 and k_1 are conjugate in Γ_α and $\mathsf{D}_\alpha = \mathsf{X}_\alpha$.
(2) If $L(\alpha)$ is a link, then every component of X_α has dimension two. Since the components of D_α have dimension one, D_α is a proper subset of X_α.
(3) Whenever $L(\alpha)$ is hyperbolic, D_α contains its holonomy characters.

The projection $(u, w, v) \to (u, v)$ from D_H to $\mathbb{C}^2$ is a bijection onto its image. So, by using (u, v) as coordinates, we regard D_H and D_α as affine subsets of $\mathbb{C}^2$.

LEMMA 7.2. $\mathsf{D}_H = \mathbb{C}^2$. *Moreover, a character* $(u_0, v_0) \in \mathsf{D}_H$ *is reducible if and only if* $(2 - v_0)(2 + v_0 - u_0^2) = 0$.

PROOF. Let $(u_0, v_0) \in \mathbb{C}^2$, take $\mathbf{z} = (x, z)$ so that

$$x^2 = 2 - v_0 \qquad\qquad z^2 = 2 + v_0 - u_0^2,$$

and take $t \in \mathbb{C}$ to satisfy Equation 3.1 from Section 3. Lemma 6.2 implies that $\rho_{\mathbf{z},t} \colon \Gamma_H \to \mathrm{PSL}_2\mathbb{C}$ is a representation. Using Equation 3.1, it is possible to choose lifts of $\rho_{\mathbf{z},t}(k_0)$ and $\rho_{\mathbf{z},t}(k_1)$ to obtain a lift of $\rho_{\mathbf{z},t}$ to $\mathrm{SL}_2\mathbb{C}$ whose character χ satisfies $\chi(k_0) = u_0$ and $\chi(k_0k_1) = v_0$. The second part of the lemma holds by definition of $\rho_{\mathbf{z},t}$ (see the last sentence in Section 3). □

Let $\widehat{\mathcal{T}_0}(\alpha) \in \mathbb{Z}[u,v]$ be the polynomial obtained from $\mathcal{T}_0(\alpha)$ by setting $X = 2-v$ and $Z = 2+v-u^2$.

THEOREM 7.3. *A character $\chi = (u_0, v_0) \in \mathbb{C}^2$ is an irreducible character in D_α if and only if χ satisfies $\widehat{\mathcal{T}_0}(\alpha)$ but not $(2-v)(2+v-u^2)$.*

PROOF. By Lemma 7.2, χ is reducible if and only if it satisfies $(2-v)(2+v-u^2)$. Together with Lemma 6.2, the proof of Lemma 7.2 shows that if χ satisfies $\widehat{\mathcal{T}_0}(\alpha)$ then $\chi \in \mathsf{D}_\alpha$.

It remains only to show that every irreducible character in D_α satisfies $\widehat{\mathcal{T}}(\alpha)$. Assume, for a contradiction, that this is not the case. Using Proposition 3.2.1 of [**7**], there is an irreducible plane curve $\mathsf{C} \subset \mathsf{D}_\alpha$ on which $\widehat{\mathcal{T}_0}(\alpha)$ and $(2-v)(2+v-u^2)$ each have at most finitely many zeros. Suppose that $(u_0, v_0) \in \mathbb{C}$ avoids these zeros.

As in the proof of Lemma 7.2, take $\mathbf{z} = (x_0, z_0)$ so that

$$x_0^2 = 2 - v_0 \qquad\qquad z_0^2 = 2 + v_0 - u_0^2,$$

and take $t \in \mathbb{C}$ to satisfy Equation 3.1 with x_0 and z_0. As before, the restriction of $\rho_{\mathbf{z},t}$ to Γ_H lifts to $\mathrm{SL}_2\mathbb{C}$. This lift represents the character (u_0, v_0) and so b^2 and $\Omega(\alpha)^2$ must map trivially under $\rho_{\mathbf{z},t}$. But because, $\widehat{\mathcal{T}_0}(\alpha) \neq 0$, the image of $\Omega(\alpha)$ cannot have order 2. Therefore, $\Omega(\alpha) \in \ker \rho_{\mathbf{z},t}$.

Since b^2 and $\Omega(\alpha)$ map trivially under $\rho_{\mathbf{z},t}$, the image of $\pi_1\mathbb{T}$ must be abelian. The expression $(2-v_0)(2+v_0-u_0^2)$ is not zero and the commutator $[a,b]$ is k_0^2, so the definition of $\rho_{\mathbf{z},t}$ implies that $t = \pm i$. By Equation 3.1, we must have $u_0 = 0$. Hence, C must be the plane curve $\{(0,v) \mid v \in \mathbb{C}\}$.

We've seen that $\Omega(\alpha) \in \ker \rho_{\mathbf{z},t}$, Thus, $\mathcal{T}(x_0, z_0)^2 = 4$. This implies that, after substituting $x^2 = 2-v$ and $z^2 = 2+v$ into $f(\alpha)^2$, the function $f(\alpha)^2\,\widehat{\mathcal{T}_0}(\alpha)^2$ is never zero on C. But $f(\alpha)^2 \in \{2-v, 2+v, 4-v^2\}$, so $f(\alpha)^2$ has at least one zero on C. This is a contradiction. $\square$

If $\rho\colon \Gamma_\alpha \to \mathrm{PSL}_2\mathbb{C}$ is a representation, the traces of $\rho(k_0)$ and $\rho(k_1)$ are only defined up to sign. On the other hand, $\mathrm{Tr}\,(\rho(k_0))^2$ and $\mathrm{Tr}\,(\rho(k_0k_1))$ are well-defined. In fact, the polynomial map (u^2, v) takes characters in D_α to their induced characters in the $\mathrm{PSL}_2\mathbb{C}$ character variety $\overline{\mathsf{X}}_\alpha$ for Γ_α. Let $\overline{\mathsf{D}}_\alpha$ be the image of D_α under this map. From [**10**], we know that every $\mathrm{PSL}_2\mathbb{C}$ representation for Γ_α lifts to $\mathrm{SL}_2\mathbb{C}$. This implies that, as in Remark 7.1, if $L(\alpha)$ is a knot, then $\overline{\mathsf{D}}_\alpha$ is equal to the entire $\mathrm{PSL}_2\mathbb{C}$ character variety for Γ_α.

It is natural to use $X = 2-v$ and $Z = 2+v-u^2$ to change coordinates on $\overline{\mathsf{D}}_\alpha$ and we do so for the remainder of this paper. The next result follows immediately from Theorem 7.3.

THEOREM 7.4. *A point $\chi = (X_0, Z_0) \in \mathbb{C}^2$ is an irreducible character in $\overline{\mathsf{D}}_\alpha$ if and only if $X_0 Z_0 \neq 0$ and χ satisfies the polynomial $\mathcal{T}_0(\alpha)$.*

7.1. Common factors and epimorphisms. Define G_α to be the stabilizer in G of $\alpha \in \widehat{\mathbb{Q}}$ and define $\widehat{G}_\alpha$ to be the subgroup in G generated by G_∞ and G_α. Let $\widehat{G}_\alpha\{\alpha, \infty\}$ be the union of the $\widehat{G}_\alpha$-orbits of α and ∞. The main result of [**24**], states that if $\alpha' \in \widehat{G}_\alpha\{\alpha, \infty\}$ then there is an epimorphism $\Gamma_\alpha \to \Gamma_{\alpha'}$ which takes $k_j \in \Gamma_\alpha$ to $k_j \in \Gamma_{\alpha'}$. As an application of Corollary 7.4 and the action $G \times \widehat{\mathbb{Q}} \to \widehat{\mathbb{Q}}$, we give an elementary argument reproducing this result.

For $\phi \in G$ let $\theta_\phi\colon \mathbb{Z}[x,z] \to \mathbb{Z}[x,z]$ be the ring homomorphism determined by

$$\theta_\phi(x) = \mathcal{T}(\phi \cdot 0) \quad \text{and} \quad \theta_\phi(z) = \mathcal{T}(\phi \cdot 1).$$

For a polynomial $p \in \mathbb{C}[x,z]$, let $\pi_p\colon \mathbb{C}[x,z] \to \mathbb{C}[x,z]/(p)$ be the natural quotient homomorphism. The next pair of results describe the relationship between $\mathcal{T}$ and the action $G \times \widehat{\mathbb{Q}} \to \widehat{\mathbb{Q}}$.

LEMMA 7.5. *If $\phi \in G$, $p \in \mathbb{C}[x,z]$, and $\mathcal{T}(\phi \cdot (1/0)) \in \ker \pi_p$ then*

$$\pi_p \circ \mathcal{T} \circ \phi = \pi_p \circ \theta_\phi \circ \mathcal{T}.$$

PROOF. The map $\pi_p \circ \mathcal{T} \circ \phi$ is Farey recursive, so there is a unique ring homomorphism $\psi\colon \mathbb{Z}[x,y,z] \to \mathbb{C}[x,z]/(p)$ such that $\pi_p \circ \mathcal{T} \circ \phi = \psi \circ \mathcal{U}$. The map ψ is uniquely determined by

$$\psi(x) = \pi_p \circ \mathcal{T}(\phi \cdot 0), \quad \psi(y) = \pi_p \circ \mathcal{T}(\phi \cdot \infty) = 0, \quad \text{and} \quad \psi(z) = \pi_p \circ \mathcal{T}(\phi \cdot \infty).$$

Hence $\psi = \pi_p \circ \theta_\phi \circ \psi_{\mathcal{T}}$. Since $\mathcal{T} = \psi_{\mathcal{T}} \circ \mathcal{U}$, this proves the lemma. □

PROPOSITION 7.6. *If $\alpha \in \widehat{\mathbb{Q}}$ and $p \in \mathbb{C}[x,z]$ such that $\mathcal{T}(\alpha) \in \ker \pi_p$ then*

$$\pi_p \circ \mathcal{T} \circ \phi = \pm \pi_p \circ \mathcal{T}$$

for every $\phi \in \widehat{G}_\alpha$.

PROOF. To prove the theorem, it is enough to show that the conclusion holds separately for $\phi \in G_\infty$ and for $\phi \in G_\alpha$.

Let

$$\phi_1 = \begin{bmatrix} -1 & 0 \\ 0 & 1 \end{bmatrix} \quad \text{and} \quad \phi_2 = \begin{bmatrix} -1 & 2 \\ 0 & 1 \end{bmatrix}$$

and notice that ϕ_1 and ϕ_2 generate G_∞. Also,

$$\begin{aligned} \phi_1 \cdot 0 &= 0 & \qquad \phi_2 \cdot 0 &= 2 \\ \phi_1 \cdot 1 &= -1 & \qquad \phi_2 \cdot 1 &= 1. \end{aligned}$$

By the second sentence in the proof of Lemma 5.4, θ_{ϕ_1} fixes x and reverses the sign of z. Similarly, θ_{ϕ_2} fixes z and reverses the sign of z. For $j = 1, 2$, $\mathcal{T}(\phi_j \cdot \infty) = 0$, so Lemmas 5.4 and 7.5 imply that

$$\pi_q \circ \mathcal{T} \circ \phi_j = \pm \pi_q \circ \mathcal{T} \tag{7.1}$$

for every $q \in \mathbb{C}[x,z]$. Since $G_\infty = \langle \phi_1, \phi_2 \rangle$, the same formula holds if we replace ϕ_j with any $\phi \in G_\infty$.

Now take $\psi \in G$ so that $\psi \cdot \alpha = \infty$ and note that $G_\alpha = \psi^{-1} G_\infty \psi$. Let θ denote both the isomorphism $\theta_{\psi^{-1}}$ as well as the induced map on quotient rings.

$$\begin{array}{ccc} \mathbb{C}[x,z] & \xrightarrow{\quad\theta\quad} & \mathbb{C}[x,z] \\ {\scriptstyle \pi_p}\big\downarrow & & \big\downarrow{\scriptstyle \pi_{\theta(p)}} \\ \mathbb{C}[x,z]/(p) & \xrightarrow{\quad\theta\quad} & \mathbb{C}[x,z]/(\theta(p)) \end{array}$$

Since $\psi^{-1} \cdot \infty = \alpha$, Lemma 7.5 and the diagram above give

$$\begin{aligned} \pi_p \circ \mathcal{T} \circ \psi^{-1} &= \pi_p \circ \theta \circ \mathcal{T} \\ &= \theta \circ \pi_{\theta(p)} \circ \mathcal{T}. \end{aligned} \tag{7.2}$$

In particular,

$$\begin{aligned}\pi_p \circ \mathcal{T} &= \pi_p \circ \mathcal{T} \circ \psi^{-1} \circ \psi \\ &= \theta \circ \pi_{\theta(p)} \circ \mathcal{T} \circ \psi.\end{aligned}$$

Take $\xi \in G_\alpha$. Then $\xi = \psi^{-1}\phi\psi$ for some $\phi \in G_\infty$. Then, using Equations 7.1 and 7.2, we have

$$\begin{aligned}\pi_p \circ \mathcal{T} \circ \xi &= \theta \circ \pi_{\theta(p)} \circ \mathcal{T} \circ \phi \circ \psi \\ &= \pm\theta \circ \pi_{\theta(p)} \circ \mathcal{T} \circ \psi \\ &= \pm\pi_p \circ \mathcal{T}.\end{aligned}$$

□

The next corollary is immediate.

COROLLARY 7.7. *If $\alpha' \in \widehat{G}_\alpha\{\alpha, \infty\}$ then every factor of $\mathcal{T}(\alpha)$ divides $\mathcal{T}(\alpha')$.*

COROLLARY 7.8. *Suppose that $\alpha \in \widehat{\mathbb{Q}}$ and $M(\alpha)$ is hyperbolic. If $\alpha' \in \widehat{G}_\alpha\{\alpha, \infty\}$ then there is an epimorphism $\Gamma_\alpha \to \Gamma_{\alpha'}$ taking $k_j \in \Gamma_\alpha$ to $k_j \in \Gamma_{\alpha'}$.*

PROOF. Since $M(\alpha)$ is hyperbolic, we may take $\chi = (X_0, Z_0) \in \overline{\mathsf{D}}_\alpha$ to be the character of a holonomy representation for Γ_α. By Corollary 7.4, χ satisfies $\mathcal{T}_0(\alpha)$ and $X_0Z_0 \neq 0$. Proposition 7.6 implies that χ also satisfies $\mathcal{T}_0(\alpha')$. Corollary 7.4 implies that χ also corresponds to a character for a representation for $\Gamma_{\alpha'}$.

Since χ is an irreducible character for Γ_H, we may choose a single representation $\rho\colon \Gamma_H \to \mathrm{PSL}_2\mathbb{C}$ which represents χ and factors through both Γ_α and $\Gamma_{\alpha'}$. The corresponding representation $\Gamma_\alpha \to \mathrm{PSL}_2\mathbb{C}$ is faithful, so we obtain an epimorphism as claimed in the statement of the corollary. □

QUESTION. Lee and Sakuma show in [**20**] that the converse of Corollary 7.7 holds. Is there an elementary proof of this using Farey recursion?

7.2. Relationship to [22]. Our work here can be easily reconciled with that in [**22**], wherein the authors discuss the the character varieties of double twist knots. They describe these knots with a pair of integers k and l, where l is even and $k > 0$, and denote them as $J(k, l)$. The following facts can be shown with straightforward induction arguments.

(1) The group presentation used by [**22**] for the knot $J(k, l)$ corresponds exactly to our presentation for $\Gamma_{\gamma_{k,l}}$ where

$$\gamma_{k,l} = \frac{1 + l(k-1)}{1 + lk}.$$

For fixed k, these numbers form the sequence obtained from $\partial\left(\frac{k-1}{k}\right)$ by deleting every other term in such a way that the corner v_1 survives.

(2) The polynomial given in Proposition 3.8 of [**22**] which determines the $\mathrm{PSL}_2\mathbb{C}$ character variety for $J(k, l)$ becomes $\pm\mathcal{T}_0(\gamma_{k,l})$ after letting $y = 2 - X - Z$ and $r = 2 - Z$.

In view of the recursion matrices for $\mathcal{T}$, the substitution $s = \mathcal{T}(\alpha)$ is natural while considering the terms of $\partial(\alpha)$. Lemma 5.4 tells us that the value $S = s^2$ can be expressed as a polynomial in $\mathbb{Z}[X, Z]$. For $J(k, l)$, we have $S = \mathcal{T}\left(\frac{k-1}{k}\right)^2$. Since the numbers $\frac{k-1}{k}$ are in $\partial(1)$, their images under $\mathcal{T}$ satisfy the recursion given by

$\left(\begin{smallmatrix} 0 & 1 \\ -1 & z \end{smallmatrix}\right)$ and induction shows that $S = \mathcal{T}\left(\frac{k-1}{k}\right)^2$ is linear in X. So, as shown in [**22**], the map $(X, Z) \mapsto (S, Z)$ gives a birational change of coordinates for $\overline{\mathsf{D}}_{\gamma_{k,l}}$.

(3) In Section 4 of [**22**], the authors introduce polynomials in $\mathbb{Z}[r, t]$ whose corresponding plane curves have projective closures in $\mathbb{P}^1 \times \mathbb{P}^1$ which realize the smooth models for the double twist knots. If we set $t = 2 - S$ and $r = 2 - Z$ in their polynomials, the result is the same (up to sign) as when we change X to S in $\mathcal{T}_0(\gamma_{k,l})$ as described above.

Item (3) suggests that it might be fruitful to consider the maps $(X, Z) \mapsto (S, Z)$ and the techniques of [**22**] more generally across the 2-bridge link family.

8. Riley polynomials and holonomy representations

Suppose $\alpha \in \mathbb{Q}_0$. A **parabolic representation** for Γ_α, or **p-rep** for short, is an irreducible $\mathrm{PSL}_2\mathbb{C}$ representation for Γ_α whose character satisfies $u^2 = w^2 = 4$ (or equivalently $Z = -X$). In particular, characters of p-reps for Γ_α lie on $\overline{\mathsf{D}}_\alpha$.

Define $\psi_\Lambda \colon \mathbb{Z}[X, Z] \to \mathbb{Z}[X]$ to be the ring homomorphism defined by $X \mapsto X$ and $Z \mapsto -X$. Define

$$\Lambda_\alpha = \psi_\Lambda \circ \mathcal{T}_0(\alpha).$$

The next result is a direct consequence of Corollary 7.4.

COROLLARY 8.1. *A point $(X_0, -X_0) \in \mathbb{C}^2$ is the character of of an irreducible p-rep for Γ_α if and only if $X_0 \neq 0$ and $\Lambda_\alpha(X_0) = 0$. When this is the case, $(X_0, -X_0)$ is a character for the representation determined by*

$$k_0 \mapsto \begin{bmatrix} 1 & 1 \\ 0 & 1 \end{bmatrix} \qquad\qquad k_1 \mapsto \begin{bmatrix} 1 & 0 \\ -X_0 & 1 \end{bmatrix}.$$

The **Riley polynomial** for a number $\alpha \in \mathbb{Q}_0$ with odd denominator is defined in [**27**]. In Theorem 2 of [**27**], Riley shows that the Riley polynomial for α is monic and has degree $\frac{1}{2}(q-1)$, where q is the denominator of α. Also, as part of this this theorem, Riley shows that X_0 satisfies the Riley polynomial for α if and only if the assignment

$$k_0 \mapsto \begin{bmatrix} 1 & 1 \\ 0 & 1 \end{bmatrix} \qquad\qquad k_1 \mapsto \begin{bmatrix} 1 & 0 \\ -X_0 & 1 \end{bmatrix}$$

determines a p-rep for Γ_α.

It is easy to use induction to compute the degree of Λ_α. We do this primarily to show that Λ_α is the Riley polynomial. Here, $\lfloor\cdot\rfloor$ denotes the greatest integer function.

PROPOSITION 8.2.

$$\deg\left(\Lambda_{p/q}\right) = \left\lfloor \frac{q-1}{2} \right\rfloor.$$

PROOF. Let $\nu \colon \mathbb{Z}[x, z] \to \mathbb{Z}[i][x]$ be the ring homomorphism determined by $x \mapsto x$ and $z \mapsto ix$. Define $T = \nu \circ \mathcal{T}$.

First, we induct on q to show that $\deg(T(p/q)) = q$.

Since $T(0) = T(1) = x$ and $T(\infty) = 0$, $\deg(T(p/q)) = q$ when $q \in \{0, 1\}$. Assume now that $q \geq 2$ and the det $(T(r/s)) = s$ whenever $s < q$. By Lemma 4.1, we can find a Farey pair $\{\alpha, \beta\}$ with $\alpha \in \mathbb{Q}_0$ and $\beta \in \widehat{\mathbb{Q}}_0$ such that

$$p/q = \alpha \oplus \alpha \oplus \beta$$

and the denominators of α, β, and $\alpha \oplus \beta$ are all less than q.

By definition of T,

$$T(p/q) = -T(\beta) + T(\alpha)T(\alpha \oplus \beta).$$

By assumption, the degree of $T(\beta)$ is the same as the denominator of β and the degree of $T(\alpha)T(\alpha \oplus \beta)$ is q. Since the first of these quantities is smaller than the second, $\deg(T(p/q)) = q$ as claimed.

By definition,

$$\Lambda_{p/q} = \frac{T(p/q)}{(\nu \circ f)(p/q)}$$

after setting $x^2 = X$. Since,

$$(\nu \circ f)(p/q) = \begin{cases} x & \text{if } p \text{ is even} \\ ix^2 & \text{if } q \text{ is even} \\ ix & \text{otherwise} \end{cases}$$

it follows that $\deg\left(\Lambda_{p/q}\right) = \left\lfloor \frac{q-1}{2} \right\rfloor$. □

COROLLARY 8.3. *If $\alpha \in \mathbb{Q}_0$ has odd denominator then, up to sign, Λ_α is the Riley polynomial.*

Suppose $\alpha = \frac{p}{q} \in \mathbb{Q}_0$. It follows from Thurston's geometrization theorems that $M(\alpha)$ admits a complete hyperbolic structure with finite volume if and only if $p \notin \{1, q-1\}$. The corresponding holonomy representations for Γ_α are always p-reps. In particular, there is a root of $\mathcal{T}_0(\alpha)$ which provides the holonomy as in Corollary 8.1. It is well-known that the traces of the holonomy for a knot complement $M(\alpha)$ are algebraic integers. In fact, whenever the holonomy representation of a hyperbolic 3-manifold has non-integer traces, the manifold contains a closed essential surface. Hatcher and Oertel show in [**13**] that no 2-bridge link contains a closed essential surface and so their holonomy representations are integral. According to Alan Reid, Riley was also aware that the p-rep polynomials for two component 2-bridge links are monic.

In the present setting, it is not difficult to reproduce this result. Here, we show that Λ_α is monic for every $\alpha \in \mathbb{Q}_0$.

THEOREM 8.4. *If $\alpha \in \mathbb{Q}_0$ then Λ_α is monic.*

PROOF. Let d be the denominator of α. Since $\Lambda_0 = \Lambda_1 = \Lambda_{1/2}$ the theorem holds when $d \leq 2$. Suppose then, that $d \geq 3$ and that the theorem holds for elements of $\mathbb{Q}_0$ whose denominators are smaller than d.

As usual, Lemma 4.1 provides a Farey pair $\{\gamma, \beta\}$ with $\gamma \in \mathbb{Q}_0$ and $\beta \in \widehat{\mathbb{Q}}_0$ so that

$$\alpha = \gamma \oplus \gamma \oplus \beta$$

and the denominators of γ, β, and $\gamma \oplus \beta$ are all less than d. Also, because $d \geq 3$, $\beta \neq \frac{1}{0}$. Write $\gamma = p/q$ and $\beta = r/s$. Define

$$Q = \begin{cases} 1 & \text{if } s \text{ is even} \\ X & \text{if } rs \text{ is odd} \\ -X & \text{if } r \text{ is even.} \end{cases}$$

From Lemma 5.3 and the proof of Lemma 5.4,

$$\Lambda_\alpha = -\Lambda_\beta + Q\,\Lambda_\gamma\,\Lambda_{\gamma\oplus\beta}.$$

Using Proposition 8.2 to compute degrees, we have

$$\deg\Lambda_\alpha - \deg\Lambda_\beta = \left\lfloor\frac{2q+s-1}{2}\right\rfloor - \left\lfloor\frac{s-1}{2}\right\rfloor = q$$

which is positive. Therefore, the leading coefficient of Λ_α is the same as that of $Q\,\Lambda_\gamma\,\Lambda_{\gamma\oplus\beta}$. This coefficient is one, by the inductive assumption. □

COROLLARY 8.5. *If $G \subset PSL_2\mathbb{C}$ is a Kleinian group which uniformizes a hyperbolic 2-bridge link then the traces (defined up to sign) of the elements of G are algebraic integers.*

PROOF. G is conjugate to the group

$$\left\langle \begin{bmatrix}1 & 1\\ 0 & 1\end{bmatrix}, \begin{bmatrix}1 & 0\\ -X_0 & 1\end{bmatrix}\right\rangle$$

where X_0 is a non-zero root of Λ_α for some $\alpha \in \mathbb{Q}_0$. The trace of the product of these two matrices is $2 - X_0$ and it follows that the traces of G lie in the ring $\mathbb{Z}[X_0]$. By Theorem 8.4, we know that this ring is made up entirely of algebraic integers. □

We conclude with three observations made from the data we've collected.

First, our computer experiments show that, except for $\Lambda_{1/4}$, $\Lambda_{p/q}$ has a non-trivial factorization over $\mathbb{Q}$ for every even q which is less than 300. In contrast, it is not difficult to find $\alpha \in \mathbb{Q}$ for which Λ_α is irreducible. For instance, Hoste and Shanahan prove in [**17**] that this is the case for each of the twist knots $L\left(\frac{n}{2n+1}\right)$.

In a similar vein, Theorem 3 of [**27**] states that Riley polynomials have no repeated factors. On the other hand, our experiments find many α's with even denominator less than 300 for which Λ_α has factors with multiplicity larger than one. In almost every such case, the factor with high multiplicity is the monomial X. Amongst the numbers we searched, we found exactly three exceptions to this. If $\alpha = 7/24$ then Λ_α is divisible by $(X^2-1)^2$ and if $\alpha \in \{41/264, 103/264\}$ then Λ_α is divisible by $(X^2-1)^3$. We remark that, in [**26**], Theorem 3 of [**27**] is used to show that there is no 2-bridge knot whose trace field contains $\mathbb{Q}(i)$ as a subfield.

Finally, Question 1.7 of [**22**] asks whether there is an $\alpha \in \mathbb{Q}$ with odd denominator and numerator not equal to one for which $\mathcal{T}_0(\alpha)$ is irreducible over $\mathbb{Q}$ but Λ_α factors non-trivially. They found no such examples amongst the double twist knots with crossing numbers at most 98. Since there are many α's with even denominator (links) for which $\mathcal{T}_0(\alpha)$ is irreducible, the observation above provides many examples with *even* denominators. However, if we look only at α's with odd denominators (knots), we find no examples up to denominator 200.

9. Lists

This section consists of lists of polynomials computed using the techniques of this paper.

$T_0(\alpha)$ with $\alpha \in \mathbb{Q} \cap \left[0, \frac{1}{2}\right]$ and denominator at most 18.

α	$T_0(\alpha)$
1/3	$X - 1$
1/4	$X - 2$
1/5	$X^2 - 3X + 1$
2/5	$XZ - Z - 1$
1/6	$(X - 1)(X - 3)$
1/7	$X^3 - 5X^2 + 6X - 1$
2/7	$X^2Z - 3XZ + 2Z - 1$
3/7	$X^2Z - XZ - 2X + 1$
1/8	$\left(X^2 - 4X + 2\right)(X - 2)$
3/8	$X^2Z - 2XZ - X + Z$
1/9	$\left(X^3 - 6X^2 + 9X - 1\right)(X - 1)$
2/9	$X^3Z - 5X^2Z + 7XZ - 2Z - 1$
4/9	$X^2Z^2 - XZ^2 - 3XZ + 2Z + 1$
1/10	$(X^2 - 3X + 1)(X^2 - 5X + 5)$
3/10	$X^3Z - 4X^2Z + 5XZ - 2X - 2Z + 3$
1/11	$X^5 - 9X^4 + 28X^3 - 35X^2 + 15X - 1$
2/11	$X^4Z - 7X^3Z + 16X^2Z - 13XZ + 3Z - 1$
3/11	$X^4Z - 5X^3Z + 8X^2Z - X^2 - 4XZ + X + 1$
4/11	$X^3Z^2 - 3X^2Z^2 - X^2Z + 3XZ^2 - Z^2 + Z + 1$
5/11	$X^3Z^2 - X^2Z^2 - 4X^2Z + 3XZ + 3X - 1$
1/12	$(X^2 - 4X + 1)(X - 1)(X - 2)(X - 3)$
5/12	$(XZ - Z - 2)(XZ - 1)(X - 1)$
1/13	$X^6 - 11X^5 + 45X^4 - 84X^3 + 70X^2 - 21X + 1$
2/13	$X^5Z - 9X^4Z + 29X^3Z - 40X^2Z + 22XZ - 3Z - 1$
3/13	$X^5Z - 7X^4Z + 17X^3Z - 16X^2Z - 2X^2 + 4XZ + 5X - 1$
4/13	$X^4Z^2 - 5X^3Z^2 + 9X^2Z^2 - 3X^2Z - 7XZ^2 + 8XZ + 2Z^2 - 5Z + 1$
5/13	$X^4Z^2 - 3X^3Z^2 - 2X^3Z + 3X^2Z^2 + 3X^2Z - XZ^2 + X^2 - XZ - X + 1$
6/13	$X^3Z^3 - X^2Z^3 - 5X^2Z^2 + 4XZ^2 + 6XZ - 3Z - 1$

1/14	$(X^3-5X^2+6X-1)(X^3-7X^2+14X-7)$
3/14	$X^5Z-8X^4Z+23X^3Z-28X^2Z-X^2+13XZ+2X-2Z+1$
5/14	$X^4Z^2-4X^3Z^2-X^3Z+6X^2Z^2-4XZ^2+3XZ+Z^2+2X-2Z-1$
1/15	$(X^4-9X^3+26X^2-24X+1)(X^2-3X+1)(X-1)$
2/15	$X^6Z-11X^5Z+46X^4Z-91X^3Z+86X^2Z-34XZ+4Z-1$
4/15	$(X^3Z-3X^2Z+2XZ-1)(X^2Z-4XZ+4Z-1)$
7/15	$X^4Z^3-X^3Z^3-6X^3Z^2+5X^2Z^2+10X^2Z-6XZ-4X+1$
1/16	$(X^4-8X^3+20X^2-16X+2)(X^2-4X+2)(X-2)$
3/16	$X^6Z-10X^5Z+38X^4Z-68X^3Z+58X^2Z-2X^2-22XZ+7X+3Z-4$
5/16	$X^5Z^2-6X^4Z^2+14X^3Z^2-4X^3Z-16X^2Z^2+15X^2Z+9XZ^2-18XZ-2Z^2+3X+7Z-4$
7/16	$(X^3Z^2-2X^2Z^2-3X^2Z+XZ^2+4XZ+X-Z)(XZ-2)$
1/17	$X^8-15X^7+91X^6-286X^5+495X^4-462X^3+210X^2-36X+1$
2/17	$X^7Z-13X^6Z+67X^5Z-174X^4Z+239X^3Z-166X^2Z+50XZ-4Z-1$
3/17	$X^7Z-11X^6Z+47X^5Z-98X^4Z+103X^3Z-X^3-51X^2Z+3X^2+9XZ-1$
4/17	$X^6Z^2-9X^5Z^2+31X^4Z^2-50X^3Z^2-3X^3Z+36X^2Z^2+14X^2Z-8XZ^2-18XZ+4Z+1$
5/17	$X^6Z^2-7X^5Z^2+19X^4Z^2-3X^4Z-25X^3Z^2+13X^3Z+16X^2Z^2-18X^2Z-4XZ^2+2X^2+8XZ-4X+1$
6/17	$X^5Z^3-5X^4Z^3-X^4Z^2+10X^3Z^3-10X^2Z^3+6X^2Z^2+5XZ^3+3X^2Z-8XZ^2-Z^3-3XZ+3Z^2-1$
7/17	$X^5Z^3-3X^4Z^3-4X^4Z^2+3X^3Z^3+9X^3Z^2-X^2Z^3+5X^3Z-6X^2Z^2-9X^2Z+XZ^2-2X^2+4XZ+4X-1$
8/17	$X^4Z^4-X^3Z^4-7X^3Z^3+6X^2Z^3+15X^2Z^2-10XZ^2-10XZ+4Z+1$
1/18	$(X^3-6X^2+9X-1)(X^3-6X^2+9X-3)(X-1)(X-3)$
5/18	$(X^5Z^2-7X^4Z^2+18X^3Z^2-2X^3Z-20X^2Z^2+7X^2Z+8XZ^2-5XZ+X-2Z-1)(X-1)$
7/18	$(X^4Z^3-3X^3Z^3-3X^3Z^2+3X^2Z^3+5X^2Z^2-XZ^3+3X^2Z-2XZ^2-3XZ-X+2Z+1)(X-1)$

Riley polynomials $\Lambda_\alpha(X)$ with $\alpha \in \mathbb{Q} \cap \left[0, \frac{1}{2}\right]$ and denominator at most 23.

1/3	$X-1$
1/4	$X-2$
1/5	X^2-3X+1
2/5	$-X^2+X-1$
1/6	$(X-1)(X-3)$
1/7	X^3-5X^2+6X-1
2/7	$-X^3+3X^2-2X-1$
3/7	$-X^3+X^2-2X+1$
1/8	$(X^2-4X+2)(X-2)$
3/8	$-(X^2-2X+2)X$

1/9	$(X^3-6X^2+9X-1)(X-1)$
2/9	$-X^4+5X^3-7X^2+2X-1$
4/9	$X^4-X^3+3X^2-2X+1$
1/10	$(X^2-3X+1)(X^2-5X+5)$
3/10	$-(X^2-X-1)(X^2-3X+3)$
1/11	$X^5-9X^4+28X^3-35X^2+15X-1$
2/11	$-X^5+7X^4-16X^3+13X^2-3X-1$
3/11	$-X^5+5X^4-8X^3+3X^2+X+1$
4/11	$X^5-3X^4+4X^3-X^2-X+1$
5/11	$X^5-X^4+4X^3-3X^2+3X-1$
1/12	$(X^2-4X+1)(X-1)(X-2)(X-3)$
5/12	$(X^2-X+2)(X^2+1)(X-1)$
1/13	$X^6-11X^5+45X^4-84X^3+70X^2-21X+1$
2/13	$-X^6+9X^5-29X^4+40X^3-22X^2+3X-1$
3/13	$-X^6+7X^5-17X^4+16X^3-6X^2+5X-1$
4/13	$X^6-5X^5+9X^4-4X^3-6X^2+5X+1$
5/13	$X^6-3X^5+5X^4-4X^3+2X^2-X+1$
6/13	$-X^6+X^5-5X^4+4X^3-6X^2+3X-1$
1/14	$(X^3-5X^2+6X-1)(X^3-7X^2+14X-7)$
3/14	$-(X^3-3X^2+2X-1)(X^3-5X^2+6X+1)$
5/14	$(X^3-X^2+1)(X^3-3X^2+4X-1)$
1/15	$(X^4-9X^3+26X^2-24X+1)(X^2-3X+1)(X-1)$
2/15	$-X^7+11X^6-46X^5+91X^4-86X^3+34X^2-4X-1$
4/15	$(X^4-3X^3+2X^2+1)(X^3-4X^2+4X+1)$
7/15	$-X^7+X^6-6X^5+5X^4-10X^3+6X^2-4X+1$
1/16	$(X^4-8X^3+20X^2-16X+2)(X^2-4X+2)(X-2)$
3/16	$-(X^4-6X^3+10X^2-2X-2)(X^3-4X^2+4X-2)$
5/16	$(X^4-4X^3+6X^2-2X-2)(X^3-2X^2+2)$
7/16	$-(X^4-2X^3+4X^2-4X+2)(X^2+2)X$
1/17	$X^8-15X^7+91X^6-286X^5+495X^4-462X^3+210X^2-36X+1$
2/17	$-X^8+13X^7-67X^6+174X^5-239X^4+166X^3-50X^2+4X-1$
3/17	$-X^8+11X^7-47X^6+98X^5-103X^4+50X^3-6X^2-1$
4/17	$X^8-9X^7+31X^6-50X^5+39X^4-22X^3+18X^2-4X+1$
5/17	$X^8-7X^7+19X^6-22X^5+3X^4+14X^3-6X^2-4X+1$
6/17	$-X^8+5X^7-11X^6+10X^5+X^4-10X^3+6X^2-1$
7/17	$-X^8+3X^7-7X^6+10X^5-11X^4+10X^3-6X^2+4X-1$
8/17	$X^8-X^7+7X^6-6X^5+15X^4-10X^3+10X^2-4X+1$

1/18	$(X^3-6X^2+9X-1)(X^3-6X^2+9X-3)(X-1)(X-3)$
5/18	$(X^4-3X^3+X^2+2X+1)(X^3-4X^2+5X-1)(X-1)$
7/18	$-(X^4-X^3+X^2+1)(X^3-2X^2+3X-1)(X-1)$
1/19	$X^9-17X^8+120X^7-455X^6+1001X^5-1287X^4+924X^3-330X^2+45X-1$
2/19	$-X^9+15X^8-92X^7+297X^6-541X^5+553X^4-296X^3+70X^2-5X-1$
3/19	$-X^9+13X^8-68X^7+183X^6-269X^5+211X^4-80X^3+18X^2-9X+1$
4/19	$X^9-11X^8+48X^7-105X^6+121X^5-73X^4+20X^3+6X^2-3X+1$
5/19	$X^9-9X^8+32X^7-55X^6+45X^5-19X^4+16X^3-10X^2-3X-1$
6/19	$-X^9+7X^8-20X^7+25X^6-X^5-31X^4+24X^3+6X^2-9X-1$
7/19	$-X^9+5X^8-12X^7+15X^6-9X^5-X^4+4X^3-2X^2-X+1$
8/19	$X^9-3X^8+8X^7-13X^6+17X^5-17X^4+12X^3-6X^2+X+1$
9/19	$X^9-X^8+8X^7-7X^6+21X^5-15X^4+20X^3-10X^2+5X-1$
1/20	$(X^4-8X^3+19X^2-12X+1)(X^2-3X+1)(X^2-5X+5)(X-2)$
3/20	$-(X^5-8X^4+21X^3-20X^2+7X-2)(X^4-6X^3+11X^2-6X-1)$
7/20	$-(X^5-4X^4+7X^3-4X^2-X+2)(X^4-2X^3+X^2+2X-1)$
9/20	$(X^4+3X^2+1)(X^3-X^2+3X-2)(X^2-X+1)$
1/21	$(X^6-13X^5+64X^4-146X^3+148X^2-48X+1)(X^3-5X^2+6X-1)(X-1)$
2/21	$-X^{10}+17X^9-121X^8+468X^7-1068X^6+1461X^5-1163X^4+496X^3-95X^2+5X-1$
4/21	$X^{10}-13X^9+69X^8-192X^7+300X^6-261X^5+107X^4+12X^3-29X^2+7X+1$
5/21	$X^{10}-11X^9+49X^8-112X^7+140X^6-107X^5+79X^4-56X^3+15X^2-7X+1$
8/21	$(X^6-3X^5+4X^4-2X^3+1)(X^4-2X^3+3X^2-X+1)$
10/21	$-X^{10}+X^9-9X^8+8X^7-28X^6+21X^5-35X^4+20X^3-15X^2+5X-1$
1/22	$(X^5+11X^4+44X^3+77X^2+55X+11)(X^5+9X^4+28X^3+35X^2+15X+1)$
3/22	$-(X^5+9X^4+28X^3+35X^2+17X+5)(X^5+7X^4+16X^3+13X^2+X-1)$
5/22	$(X^5+7X^4+16X^3+11X^2-X+3)(X^5+5X^4+8X^3+5X^2+3X+1)$
7/22	$-(X^5+5X^4+10X^3+7X^2-3X-5)(X^5+3X^4+2X^3-3X^2-3X+1)$
9/22	$(X^5+3X^4+6X^3+7X^2+5X+3)(X^5+X^4+2X^3+X^2+X+1)$
1/23	$-X^{11}-21X^{10}-190X^9-969X^8-3060X^7-6188X^6-8008X^5-6435X^4-3003X^3-715X^2-66X-1$
2/23	$X^{11}+19X^{10}+154X^9+695X^8+1912X^7+3300X^6+3544X^5+2269X^4+791X^3+125X^2+6X-1$
3/23	$X^{11}+17X^{10}+122X^9+481X^8+1136X^7+1644X^6+1432X^5+707X^4+171X^3+11X^2+2X+1$
4/23	$-X^{11}-15X^{10}-94X^9-319X^8-636X^7-756X^6-516X^5-181X^4-35X^3-17X^2-6X+1$
5/23	$-X^{11}-13X^{10}-70X^9-201X^8-332X^7-324X^6-196X^5-79X^4-15X^3-3X^2-2X-1$
6/23	$X^{11}+11X^{10}+50X^9+119X^8+156X^7+116X^6+72X^5+61X^4+23X^3-3X^2+6X-1$
7/23	$X^{11}+9X^{10}+34X^9+65X^8+52X^7-24X^6-76X^5-33X^4+23X^3+15X^2-2X+1$
8/23	$-X^{11}-7X^{10}-22X^9-35X^8-20X^7+24X^6+48X^5+23X^4-11X^3-13X^2-2X+1$
9/23	$-X^{11}-5X^{10}-14X^9-25X^8-32X^7-32X^6-28X^5-23X^4-15X^3-7X^2-2X-1$
10/23	$X^{11}+3X^{10}+10X^9+19X^8+32X^7+40X^6+40X^5+29X^4+15X^3+X^2-2X-1$
11/23	$X^{11}+X^{10}+10X^9+9X^8+36X^7+28X^6+56X^5+35X^4+35X^3+15X^2+6X+1$

Acknowledgement. Thanks to Kelly McKinnie for her numerous corrections and suggestions. We also thank the referee for several helpful comments

References

[1] H. Akiyoshi, M. Sakuma, M. Wada, and Y. Yamashita, *Ford domains of punctured torus groups and two-bridge knot groups*, Sūrikaisekikenkyūsho Kōkyūroku **1163** (2000), 67–77. Hyperbolic spaces and related topics, II (Japanese) (Kyoto, 1999). MR1799369

[2] H. Akiyoshi, M. Sakuma, M. Wada, and Y. Yamashita, *Punctured torus groups and 2-bridge knot groups. I*, Lecture Notes in Mathematics, vol. 1909, Springer, Berlin, 2007. MR2330319

[3] B. H. Bowditch, *Markoff triples and quasi-Fuchsian groups*, Proc. London Math. Soc. (3) **77** (1998), no. 3, 697–736, DOI 10.1112/S0024611598000604. MR1643429

[4] S. Boyer and X. Zhang, *On Culler-Shalen seminorms and Dehn filling*, Ann. of Math. (2) **148** (1998), no. 3, 737–801, DOI 10.2307/121031. MR1670053

[5] G. Burde, SU(2)-*representation spaces for two-bridge knot groups*, Math. Ann. **288** (1990), no. 1, 103–119, DOI 10.1007/BF01444524. MR1070927

[6] M. Chu, *Detecting essential surfaces as intersections in the character variety*, Algebr. Geom. Topol. **17** (2017), no. 5, 2893–2914, DOI 10.2140/agt.2017.17.2893. MR3704247

[7] M. Culler and P. B. Shalen, *Varieties of group representations and splittings of* 3*-manifolds*, Ann. of Math. (2) **117** (1983), no. 1, 109–146, DOI 10.2307/2006973. MR683804

[8] M. Culler, C. McA. Gordon, J. Luecke, and P. B. Shalen, *Dehn surgery on knots*, Ann. of Math. (2) **125** (1987), no. 2, 237–300, DOI 10.2307/1971311. MR881270

[9] D. B. A. Epstein and R. C. Penner, *Euclidean decompositions of noncompact hyperbolic manifolds*, J. Differential Geom. **27** (1988), no. 1, 67–80. MR918457

[10] F. González-Acuña and J. M. Montesinos-Amilibia, *On the character variety of group representations in* $\mathrm{SL}(2,\mathbf{C})$ *and* $\mathrm{PSL}(2,\mathbf{C})$, Math. Z. **214** (1993), no. 4, 627–652, DOI 10.1007/BF02572429. MR1248117

[11] C. McA. Gordon, *Riley's conjecture on* $\mathrm{SL}(2,\mathbb{R})$ *representations of 2-bridge knots*, J. Knot Theory Ramifications **26** (2017), no. 2, 1740003, 6, DOI 10.1142/S021821651740003X. MR3604485

[12] F. Guéritaud, *On canonical triangulations of once-punctured torus bundles and two-bridge link complements*, Geom. Topol. **10** (2006), 1239–1284, DOI 10.2140/gt.2006.10.1239. With an appendix by David Futer. MR2255497

[13] A. Hatcher and U. Oertel, *Boundary slopes for Montesinos knots*, Topology **28** (1989), no. 4, 453–480, DOI 10.1016/0040-9383(89)90005-0. MR1030987

[14] A. Hatcher and W. Thurston, *Incompressible surfaces in* 2*-bridge knot complements*, Invent. Math. **79** (1985), no. 2, 225–246, DOI 10.1007/BF01388971. MR778125

[15] Allen Hatcher. *Topology of Numbers.*

[16] H. M. Hilden, M. T. Lozano, and J. M. Montesinos-Amilibia, *On the character variety of group representations of a* 2*-bridge link* $p/3$ *into* $\mathrm{PSL}(2,\mathbf{C})$, Bol. Soc. Mat. Mexicana (2) **37** (1992), no. 1-2, 241–262. Papers in honor of José Adem (Spanish). MR1317576

[17] J. Hoste and P. D. Shanahan, *Trace fields of twist knots*, J. Knot Theory Ramifications **10** (2001), no. 4, 625–639, DOI 10.1142/S0218216501001049. MR1831680

[18] J. Hoste and P. D. Shanahan, *A formula for the A-polynomial of twist knots*, J. Knot Theory Ramifications **13** (2004), no. 2, 193–209, DOI 10.1142/S0218216504003081. MR2047468

[19] J. Hoste and P. D. Shanahan, *Commensurability classes of twist knots*, J. Knot Theory Ramifications **14** (2005), no. 1, 91–100, DOI 10.1142/S0218216505003737. MR2124555

[20] D. Lee and M. Sakuma, *Epimorphisms between 2-bridge link groups: homotopically trivial simple loops on 2-bridge spheres*, Proc. Lond. Math. Soc. (3) **104** (2012), no. 2, 359–386, DOI 10.1112/plms/pdr036. MR2880244

[21] D. D. Long and A. W. Reid, *Commensurability and the character variety*, Math. Res. Lett. **6** (1999), no. 5-6, 581–591, DOI 10.4310/MRL.1999.v6.n5.a11. MR1739217

[22] M. L. Macasieb, K. L. Petersen, and R. M. van Luijk, *On character varieties of two-bridge knot groups*, Proc. Lond. Math. Soc. (3) **103** (2011), no. 3, 473–507, DOI 10.1112/plms/pdr003. MR2827003

[23] D. Mumford, C. Series, and D. Wright, *Indra's pearls*, Cambridge University Press, Cambridge, 2015. The vision of Felix Klein; With cartoons by Larry Gonick; Paperback edition with corrections; For the 2002 edition see [MR1913879]. MR3558870

[24] T. Ohtsuki, R. Riley, and M. Sakuma, *Epimorphisms between 2-bridge link groups*, The Zieschang Gedenkschrift, Geom. Topol. Monogr., vol. 14, Geom. Topol. Publ., Coventry, 2008, pp. 417–450, DOI 10.2140/gtm.2008.14.417. MR2484712

[25] K. L. Petersen and A. T. Tran, *Character varieties of double twist links*, Algebr. Geom. Topol. **15** (2015), no. 6, 3569–3598, DOI 10.2140/agt.2015.15.3569. MR3450771

[26] A. W. Reid and G. S. Walsh, *Commensurability classes of 2-bridge knot complements*, Algebr. Geom. Topol. **8** (2008), no. 2, 1031–1057, DOI 10.2140/agt.2008.8.1031. MR2443107

[27] R. Riley, *Parabolic representations of knot groups. I*, Proc. London Math. Soc. (3) **24** (1972), 217–242, DOI 10.1112/plms/s3-24.2.217. MR300267

[28] R. Riley, *Nonabelian representations of 2-bridge knot groups*, Quart. J. Math. Oxford Ser. (2) **35** (1984), no. 138, 191–208, DOI 10.1093/qmath/35.2.191. MR745421

[29] Inc. SageMath. *CoCalc Collaborative Computation Online*, 2016. `https://cocalc.com/`.

[30] M. Sakuma and J. Weeks, *Examples of canonical decompositions of hyperbolic link complements*, Japan. J. Math. (N.S.) **21** (1995), no. 2, 393–439, DOI 10.4099/math1924.21.393. MR1364387

[31] C. Series, *The geometry of Markoff numbers*, Math. Intelligencer **7** (1985), no. 3, 20–29, DOI 10.1007/BF03025802. MR795536

[32] W. P. Thurston. The geometry and topology of 3-manifolds. mimeographed lecture notes, 1979.

Department of Mathematical Sciences, University of Montana
Email address: `eric.chesebro@mso.umt.edu`

Contemporary Mathematics
Volume **760**, 2020
https://doi.org/10.1090/conm/760/15285

Virtually special embeddings of integral Lorentzian lattices

Michelle Chu

ABSTRACT. The automorphism groups of integral Lorentzian lattices act by isometries on hyperbolic space with finite covolume. In the case of reflective integral lattices, the automorphism groups are commensurable to arithmetic hyperbolic reflection groups. However, for a fixed dimension, there is only finitely many reflective integral Lorentzian lattices, and these can only occur in small dimensions. The goal of this note is to construct embeddings of low-dimensional integral Lorentzian lattices into unimodular Lorentzian lattices associated to right-angled reflection groups. As an application, we construct many discrete groups of $\mathrm{Isom}(\mathbb{H}^n)$ for small n which are C-special in the sense of Haglund-Wise.

1. Introduction

Given a finite volume polyhedron P in hyperbolic space $\mathbb{H}^n$, let Γ be the group generated by the reflections on the sides of P. If the action of Γ tiles $\mathbb{H}^n$ without interiors of copies of P overlapping, we say that Γ is a *hyperbolic reflection group* and its fundamental polyhedron P is a *hyperbolic Coxeter polyhedron.* The quotient $\mathbb{H}^n/\Gamma$ is a finite-volume hyperbolic orbifold.

The theory of hyperbolic reflection groups provides many examples of finite volume hyperbolic orbifolds. However, in higher dimensions, these cease to exist [**Vin84, Kho86, Pro86**]. Another way to construct finite volume hyperbolic orbifolds in any dimension is as quotients of H^n by the automorphism groups $\mathrm{Aut}(L)$ of Lorentzian lattices L. These automorphism groups are examples of arithmetic groups of simplest type in $\mathrm{Isom}(\mathbb{H}^n)$. If the subgroup generated by reflections has finite index in $\mathrm{Aut}(L)$, we say the lattice L is *reflective.* Such a subgroup is an *arithmetic hyperbolic reflection group.*

In this note we construct embeddings of lattices into unimodular lattices of higher dimension. The Lorentzian unimodular lattices $\mathbf{I}_{n,1}$ are reflective for $2 \leq n \leq 19$ [**Vin72, KV78**]. Furthermore, for $2 \leq n \leq 8$, these are associated to reflection groups of hyperbolic right-angled polyhedra, which are geometric right-angled Coxeter groups [**PV05**]. Right-angled Coxeter groups, or RACGs, are particularly interesting because they have many nice properties which are inherited by their subgroups. For example, virtually embedding hyperbolic 3-manifold groups into RACGs has determined the virtual Haken and the virtual fibering conjectures for all finite volume hyperbolic 3-manifolds as well as LERFness of their fundamental

2010 *Mathematics Subject Classification.* Primary 11H56, 20F55; Secondary 22E40.
The author was supported by NSF Grant DMS-1803094.

groups [**Ago13, Wis11**]. In the sense of [**HW08**] we say a group is *C-special* if it embeds as a quasi-convex subgroup of a RACG.

We apply the lattice embeddings together with the explicit relationship between the unimodular lattices $\mathbf{I}_{n,1}$ and RACGs given by [**ERT12**] to construct many examples of C-special hyperbolic manifold groups in dimension 3 and 4. The following theorem extends and improves the results in [**Chu19**] (see also [**Chu18**]) and [**DMP18**, Theorem 2.6].

THEOREM 1.1. *Let Γ be an integral arithmetic group of simplest type in* $\mathrm{Isom}(\mathbb{H}^3)$ *or* $\mathrm{Isom}(\mathbb{H}^4)$. *Then $\Gamma_{(2)}$, the principal congruence subgroup of level 2, is compact C-special.*

For fixed dimensions, the indices of the principal congruence subgroups of level 2 contained in integral arithmetic group of simplest type are uniformly bounded. Theorem 1.1 extends [**Chu19**, Theorem 1.2, Proposition 5.3][**Chu18**, Theorem 3.1, Proposition 4.3, Proposition 5.3] and also improves on the bounds for the value of D found in [**DMP18**, Proposition 2.6] by removing the dependence on the discriminant. This gives a uniform bound for D which is independent of anything to get a strengthening of [**DMP18**, Theorem 2.2]. Prior to these results, Bergeron-Haglund-Wise showed that given an arithmetic group of simplest type in $\mathrm{O}^+(n,1)$, there exist some m such that the congruence subgroup of level m is special [**BHW11**], but Theorem 1.1 shows that $m = 2$ is enough for the cases included in Theorem 1.1.

As a consequence of Theorem 1.1 we can give a slight improvement to the results in [**Chu19**].

PROPOSITION 1.2. *Let d be a square-free positive integer and $\mathcal{O}_d$ the ring of integers in the quadratic imaginary field $\mathbb{Q}(\sqrt{-d})$. The principal congruence subgroup* $\mathrm{PSL}(2,\mathcal{O}_d)_{(2)}$ *embeds in a RACG and has index*

$$[\mathrm{PSL}(2,\mathcal{O}_d):\mathrm{PSL}(2,\mathcal{O}_d)_{(2)}] = \begin{cases} 48 & \text{if } d \equiv 1,2 \mod (4) \\ 60 & \text{if } d \equiv 3 \mod (8) \\ 36 & \text{if } d \equiv 7 \mod (8). \end{cases}$$

In particular, since the figure eight knot group intersects $\mathrm{PSL}(2,\mathcal{O}_3)_{(2)}$ in a subgroup of index 10, we also get the following corollary.

COROLLARY 1.3. *The figure eight knot complement has a special cover of degree 10.*

This note is organized as follows: In Section 2 we give the necessary preliminary background in integral lattices, their automorphism groups, and arithmetic groups of simplest type. In Section 3, inspired by the lattice gluings in [**All18**], we construct embeddings of integral lattices into unimodular lattices. In Section 4 we use these embeddings to prove Theorem 1.1. Finally, in Section 5 we give an explicit example.

2. Preliminaries

2.1. Integral lattices. A *lattice* L is a $\mathbb{Z}$-module equipped with a $\mathbb{Q}$-valued non-degenerate symmetric bilinear form $(\cdot,\cdot)$ on the vector space $V = L \otimes \mathbb{Q}$, called the *inner product*. L is called *Lorentzian* if its inner product has signature $+^n -^1$. The *norm* of a vector v is its inner product with itself (v,v). If the inner product of every pair of vectors in L is $\mathbb{Z}$-valued, L is called *integral*. In what follows, let L be an integral lattice unless otherwise noted.

The dual of L is the lattice $L^* = \{v \in L \otimes \mathbb{Q} : (v, L) \in \mathbb{Z}\}$. Notice that L is integral if and only if $L \subset L^*$. We define the *discriminant group* as $\Delta(L) = L^*/L$, a finite abelian group. We will refer to the minimal number of generators of $\Delta(L)$ as $\operatorname{rank}(\Delta(L))$.

The *determinant* of L, or $\det L$ is the determinant of an inner product matrix A_L, with respect to some $\mathbb{Z}$-basis of L. It is independent of the choice of $\mathbb{Z}$-basis and in fact $|\det L| = |\Delta(L)|$. If L' is a sublattice of L of index d then $\det L' = d^2 \cdot \det L$.

The $\mathbb{Z}$-valued inner product on L extends to a $\mathbb{Q}$-valued inner product on L^* and descends to a $\mathbb{Q}/\mathbb{Z}$-valued inner product on $\Delta(L)$.

An integral lattice L is called *strongly-square-free*, denoted by *SSF*, if the rank of $\Delta(L)$ is at most $\frac{1}{2}\dim(L)$ and every invariant factor of $\Delta(L)$ is square-free. In other words, $\Delta(L)$ is a direct product of at most $\frac{1}{2}\dim(L)$-many finite cyclic subgroups, each of square-free order. An integral lattice is called *unimodular* whenever $\Delta(L)$ is trivial.

Lattices may also be defined more generally over totally real number fields.

2.2. Automorphisms and arithmetic groups. Let k be a totally real number field with ring of integers $\mathcal{O}_k$ and let f be a quadratic form of signature $+^n-^1$ defined over k such that for every non-identity embedding $\sigma : k \hookrightarrow \mathbb{R}$, the form f^σ is positive definite. Let $\mathrm{O}(f;\mathbb{R})$ denote the orthogonal group that preserves f and $\mathrm{O}^+(f;\mathbb{R})$ its index-two subgroup which is time-orientation preserving. Then the group $\Gamma = \mathrm{O}^+(f;\mathcal{O}_k) = \mathrm{O}^+(f;\mathbb{R}) \cap \mathrm{GL}_{n+1}(\mathcal{O}_k)$ is a finite-covolume discrete subgroup of $\mathrm{O}^+(f;\mathbb{R})$ which is identified with $\mathrm{Isom}(\mathbb{H}^n)$ via the hyperboloid model. The field k is called the *field of definition* for Γ. Any discrete subgroup of $\mathrm{Isom}(\mathbb{H}^n)$ which is commensurable to some such $\mathrm{O}^+(f;\mathcal{O}_k)$ is called *arithmetic of simplest type*.

If L is an integral Lorentzian lattice, then its *automorphism group* is the group

$$\begin{aligned}\mathrm{Aut}(L) &= \{g \in \mathrm{GL}(V) | Lg = L \text{ and } f_L(xg, yg) = f_L(x,y) \text{ for all } x, y \in L\}\\ &= \{g \in \mathrm{GL}_{n+1}(\mathbb{Z}) | gA_Lg^{tr} = A_L\}\\ &= \mathrm{O}(f_L;\mathbb{Z}).\end{aligned}$$

We call $\mathrm{Aut}(L)$ and any finite index subgroup of it an *integral arithmetic group of simplest type.*

The principal congruence subgroup of level m in $\mathrm{Aut}(L)$ is the subgroup

$$\begin{aligned}\mathrm{Aut}(L)_{(m)} &= \{g \in \mathrm{GL}(V) | vg \equiv v \mod m \text{ for all } x \in L\}\\ &= \{g \in \mathrm{GL}_{n+1}(\mathbb{Z}) | gA_Lg^{tr} = A_L \text{ and } g \equiv I_{n+1} \mod m\}\\ &= \mathrm{O}(f_L;\mathbb{Z})_{(m)}\end{aligned}$$

where I_{n+1} is the $(n+1) \times (n+1)$ identity matrix.

If $n-1$ is not divisible by 8, there is, up to isomorphism over $\mathbb{Z}$, a unique unimodular Lorentzian lattice of signature $+^n-^1$ denoted $\mathbf{I}_{n,1}$. Let q_n be the standard Lorentzian quadratic form

$$q_n := -x_0^2 + x_1^2 + \cdots + x_n^2. \tag{2.1}$$

The unimodular lattice $\mathbf{I}_{n,1}$ has automorphism group $Aut(\mathbf{I}_{n,1}) = \mathrm{O}(q_n, \mathbb{Z})$ and is reflective for $n \leq 19$ [**Vin72, KV78**].

2.3. Invariants and existence of integral lattices. This section assumes familiarity with Conway-Sloane p-adic symbols [**CS99**, Chapter 15].

Over the p-adic integers, a form f associated to a $p-adic$ lattice L_p can be decomposed as a direct sum

$$f = f_1 \oplus p f_p \oplus p^2 f_{p^2} \oplus \cdots \oplus q f_q \oplus \ldots \tag{2.2}$$

where q is a p-power and f_q is a p-adic integral form with determinant prime to p. This Jordan decomposition is unique for odd primes.

For p odd, the p-adic symbol of f is the formal product of factors $q^{\epsilon_q n_q}$ with

$$\epsilon_q = \left(\frac{\det f_q}{p}\right) \text{ and } n_q = \dim f_q$$

where $\left(\frac{\alpha}{p}\right)$ denotes the Kronecker symbol.

For $p = 2$, the 2-adic symbol of f is the formal product of factors $q_{t_q}^{\epsilon_q n_q}$ or $q^{\epsilon_q n_q}$ where the former indicates f_q is of type I and the later indicates f_q is of type II and with

$$\epsilon_q = \left(\frac{\det f_q}{2}\right), \ n_q = \dim f_q \text{ and } t_q = \text{oddity}(f_q)$$

where the Kronecker symbol $\left(\frac{a}{2}\right)$ is $+1$ if $a \equiv \pm 1 \mod 8$ or -1 if $a \equiv \pm 3 \mod 8$.

Unfortunately, the 2-adic symbol is not unique, since a 2-adic form can have essentially different Jordan decompositions. However, Conway-Sloane define an abreviated 2-adic symbol using compartments and trains. Two abreviated 2-adic symbols represent the same form if and only they are related by sign walking see [**CS99**, Chapter 15, §7.5].

By [**CS99**, Theorem 11, Chapter 15], there exist an integral lattice L of determinant d having specified local forms L_p and signature $+^r-^s$ if and only if the determinant condition, the oddity formula, and the Jordan blocks conditions displayed below hold.

1. The determinant condition: for each p, the ϵ_q from the p-adic symbol satisfy

$$\prod \epsilon_q = \left(\frac{a}{p}\right) \tag{2.3}$$

where $\det(L) = p^\alpha a$.

2. The oddity formula:

$$\text{signature}(L) + \sum_{p \text{ odd}} p\text{–excess}(L_p) \equiv \text{oddity}(K_2) \mod 8 \tag{2.4}$$

where

$$\text{signature}(L) = r - s,$$

$$p\text{–excess}(L_p) \equiv \sum_q n_q(q-1) + 4 \cdot \#(\text{odd powers } q \text{ with } \epsilon_q = -1),$$

$$\text{and oddity}(L_2) = \sum t_q + 4 \cdot \#(\text{odd powers } q \text{ with } \epsilon_q = -1).$$

3. The Jordan blocks conditions: the 2-adic Jordan blocks satisfy the following

$$\text{if type II, } t_q \equiv 0 \mod 8 \tag{2.5}$$

$$\text{if } n_q = 1, \begin{cases} \epsilon_q = +1 \text{ then } t_q \equiv \pm 1 \mod 8 \\ \epsilon_q = -1 \text{ then } t_q \equiv \pm 3 \mod 8 \end{cases} \tag{2.6}$$

$$\text{if type I and } n_q = 2, \begin{cases} \epsilon_q = +1 \text{ then } t_q \equiv 0 \text{ or } \pm 2 \mod 8 \\ \epsilon_q = -1 \text{ then } t_q \equiv 4 \text{ or } \pm 2 \mod 8 \end{cases} \tag{2.7}$$

$$\text{and } t_q \equiv n_q \mod 2. \tag{2.8}$$

When working with the abbreviated 2-adic symbol, the Jordan blocks conditions on a compartment of total dimension at least 3 reduce to just one condition: the total oddity in the compartment has the same parity as its total dimension.

2.4. Some facts and observations. We state here some observations.

For odd p, the p-excess is always even.

If L is SSF, then since the rank of any invariant factor is square-free, the p-adic symbol for L_p will only contains terms for $q = 1$ and $q = p$. Furthermore, since the rank of $\Delta(L)$ is at most $\frac{1}{2}\dim(L)$, also $n_p \leq \frac{1}{2}\dim(L)$.

If we take L^{neg} to be as L with all inner products negated, its local forms will change as follows. If p is odd, then the p-adic symbol for L_p^{neg} is got from that of L_p by multiplying each superscript by $\left(\frac{(-1)^{n_q}}{p}\right)$. The 2-adic symbol for L_p^{neg} is got from that of L_p by negating each subscript.

If $p \nmid \det L$ or if $\left(\frac{-1}{p}\right) = 1$, then the p-excess of L_p and L_p^{neg} agree. If $p|L$ and $\left(\frac{-1}{p}\right) = 1$, then the p-excess of L_p and L_p^{neg} differ by $4 \mod 8$.

3. Lattice embeddings

The goal of this section is to prove the following proposition.

PROPOSITION 3.1. *Let L be an integral lattice of signature $+^r -^s$ and let $\delta = \operatorname{rank}(\Delta(L))$. Let $m = \max\{\delta + 1, 3\}$. Then L embeds in a unimodular lattice of signature $+^{r+m} -^s$.*

The proof will be separated into two cases depending on the parity of $\det(L)$. The main idea is to use a technique in [**All18**]. We will construct a lattice K of signature $+^{\delta+1} -^0$ with $\det(K) = (-1)^s \det(L)$ by specifying its local forms K_p, chosen such that there exist a group isomorphism $\phi : \Delta(L) \to \Delta(K)$ which negates norms and inner products. Gluing L to K along the graph of ϕ will then result in a unimodular lattice.

3.1. Case 1: d is odd. If $\det(L)$ is odd, the following lemma holds regardless of whether L is SSF.

LEMMA 3.2. *Let L be an integral lattice of signature $+^r -^s$ with $\det(L)$ odd and $\operatorname{rank}(\Delta(L)) = \delta$. Let $m = \max\{\delta + 1, 3\}$. Then L embeds in a unimodular lattice of signature $+^{r+m} -^s$.*

PROOF. Assume L is not unimodular and let $d := (-1)^s \det(L)$. Defined the local forms K_p as follows.

For odd $p \nmid d$, define K_p by $1^{(\frac{d}{p})m}$.

For odd $p|d$ with $d = p^\alpha a$, define K_p by the product of $q^{\epsilon_q n_q}$ where for $q = p^m > 1$, the term $q^{\epsilon_q n_q}$ matches that of L_p^{neg}, and where $1^{\epsilon_1 n_1}$ is chosen such that $\prod \epsilon_q = (\frac{a}{p})$ and $\sum n_q = m$.

Let $t \equiv m + \sum_{p \text{ odd}} p\text{--excess}(K_p)$. Since $\sum_{p \text{ odd}} p\text{--excess}(K_p)$ is even, t will have the parity of m. Define K_2 by $1_t^{(\frac{d}{2})m}$.

With these choices of local forms, all conditions Equations (2.3) to (2.8) are satisfied. So there exist an integral lattice K of signature $+^m-^0$ and determinant d with the prescribed local forms.

Now each local form K_p differs from L_p^{neg} by a unimodular factor. We have that $\Delta(K_p)$ and $\Delta(L_p^{neg})$ are isomorphic and correspond to the Sylow p-subgroups of $\Delta(K)$ and $\Delta(L^{neg})$. It follows that there exist a group isomorphism $\phi : \Delta(L) \to \Delta(K)$ which negates norms and inner products. Let $G = \{(x, \phi x)\}$ be the graph of ϕ. Then G is a totally isotropic subgroup of $\Delta(L \oplus K) = \Delta(L) \oplus \Delta(K)$, that is, the natural $\mathbb{Q}/\mathbb{Z}$-valued inner product on $\Delta(L \oplus K)$ vanishes on G.

Write $L \oplus_G K$ for the preimage of G in $(L \oplus K)^* = L^* \oplus K^*$. Since G is totally isotropic, $L \oplus_G K$ is an integral lattice containing $L \oplus K$ as a sublattice of index $|G|$ and therefore its determinant is given by

$$\det L \oplus_G K = \frac{\det L \oplus K}{|G|^2} = \frac{d^2}{(d)^2} = 1. \tag{3.1}$$

So $L \oplus_G K$ is a unimodular lattice containing L with orthogonal complement $L^\perp$ isomorphic to K. □

3.2. Case 2: d is even.

LEMMA 3.3. *Let L be an integral lattice of signature $+^r-^s$ with* $\det(L)$ *even and* $\operatorname{rank}(\Delta(L)) = \delta$. *Let $m = \max\{\delta + 1, 3\}$. Then L embeds in a unimodular lattice of signature $+^{r+m}-^s$.*

PROOF. Let $d := (-1)^s \det(L)$. Observe first that the SSF assumption guarantees that $\Delta(L_2)$ (the 2-Sylow subgroup of $\Delta(L)$) is $(\mathbb{Z}/2\mathbb{Z})^\alpha$ where $d = 2^\alpha a$ and $2 \nmid a$. This means that the natural $\mathbb{Q}_2/\mathbb{Z}_2$-valued norms and inner products in $\Delta(L_2)$ are in $\frac{1}{2}\mathbb{Z}_2$. In particular, negating norms and inner products in $\Delta(L_2)$ is trivial.

Defined the local forms K_p as follows.

For odd $p \nmid d$, define K_p by $1^{(\frac{d}{p})m}$.

For odd $p|d$ with $d = p^\alpha a$, define K_p by the product of $q^{\epsilon_q n_q}$ where for $q = p^m > 1$, the term $q^{\epsilon_q n_q}$ matches that of L_p^{neg}, and where $1^{\epsilon_1 n_1}$ is chosen such that $\prod \epsilon_q = (\frac{a}{p})$ and $\sum n_q = m$.

Let $t \equiv m + \sum_{p \text{ odd}} p\text{--excess}(K_p)$. Since $\sum_{p \text{ odd}} p\text{--excess}(K_p)$ is even, t will have the parity of m. Define K_2 by the reduced 2-adic symbol $[1^{(\frac{a}{2})(m-\alpha)}2^{+\alpha}]_t$ where $d = 2^\alpha a$.

With these choices of local forms, all conditions Equations (2.3) to (2.8) are satisfied. So there exist an integral lattice K of signature $+^m-^0$ and determinant d with the prescribed local forms.

Now each local form K_p for $p \neq 2$ differs from L_p^{neg} by a unimodular factor. We then have that for all p (including $p = 2$) $\Delta(K_p)$ and $\Delta(L_p^{neg})$ are isomorphic and correspond to the Sylow p-subgroups of $\Delta(K)$ and $\Delta(L^{neg})$. It follows that there exist a group isomorphism $\phi : \Delta(L) \to \Delta(K)$ which negates norms and inner products. Let $G = \{(x, \phi x)\}$ be the graph of ϕ. Then G is a totally isotropic subgroup of $\Delta(L \oplus K) = \Delta(L) \oplus \Delta(K)$.

The remaining follows exactly as in the proof of Theorem 3.2. □

4. Special subgroups

In this section we gather the necessary ingredients and prove Theorem 1.1 as a direct consequence of Theorem 4.1 and Theorem 4.2.

It can be shown that the automorphism group of a non-SSF lattice is always contained in the automorphism group of one which is SSF [**Wat62**, **Wat75**, **All12**]. For SSF lattices of dimension up to 5, the rank of their discriminant groups is at most 2. Therefore, the following corollary follows from Theorem 3.1 since $L \otimes \mathbb{R}$ is the orthogonal complement to the positive subgroups $K \otimes \mathbb{R}$ in the vector space $(L \oplus_G K) \otimes \mathbb{R}$.

COROLLARY 4.1. *If L is an integral lattice of signature $+^3-^1$ (resp. $+^4-^1$), then* $\mathrm{Aut}(L)$ *embeds as a geometrically finite subgroup of* $\mathrm{Aut}(\mathbf{I}_{6,1})$ *(resp.* $\mathrm{Aut}(\mathbf{I}_{7,1})$*) which preserves a copy of $\mathbb{H}^3$ (resp. $\mathbb{H}^4$) in $\mathbb{H}^6$ (resp $\mathbb{H}^7$).*

The following theorem of Everitt-Ratcliffe-Tschantz provides the explicit relationship between the automorphism groups of the unimodular lattices $\mathbf{I}_{n,1}$ for $n \leq 8$ and right-angled Coxeter groups. Recall that $q_n := -x_0^2 + x_1^2 + \cdots + x_n^2$ and $\mathrm{Aut}(\mathbf{I}_{n,1}) = \mathrm{O}(q_n, \mathbb{Z})$.

THEOREM 4.2 ([**ERT12**, Theorem 2.1]). *For $2 \leq n \leq 7$, $\mathrm{O}^+(q_n, \mathbb{Z})_{(2)}$ is a geometric RACG. It is the reflection group of an all-right hyperbolic polyhedron of dimension n. The group $\mathrm{O}^+(q_8, \mathbb{Z})_{(2)}$ contains a geometric RACG as a subgroup of index 2. This subgroup is the reflection group of an all-right hyperbolic polyhedron of dimension 8.*

LEMMA 4.3. *Let L be an integral lattice and K the corresponding integral lattice with $L \oplus_G K$ unimodular as constructed in the proof of Theorem 3.1. Then $\mathrm{Aut}(L)_{(2)}$ is contained in $\mathrm{Aut}(L \oplus_G K)_{(2)}$.*

PROOF. Recall that K was constructed so that $\Delta(L) \cong \Delta(K) \cong G$. An element $\gamma \in \mathrm{Aut}(L)$ is in $\mathrm{Aut}(L)_{(2)}$ if and only if for any $v \in L$, $v\gamma = v + 2v'$ where v' is some other element in L, or equivalently, $v\gamma \equiv v \mod 2$. Similarly, $\gamma \in \mathrm{Aut}(L \oplus_G K)$ if and only if for any $v \in L \oplus_G K$, $v\gamma \equiv v \mod 2$. If $v \in L \oplus K$ then clearly for $\gamma \in Aut(L)$, $v\gamma \equiv v \mod 2$.

For the remainder of the proof, set $\gamma \in Aut(L)_{(2)}$ and $u \in L \oplus_G K - L \oplus K$. Suppose m is the smallest positive integer so that $mu = u' \in L \oplus K$. Then $u'\gamma = u' + 2x$ where $x \in L \oplus K$. If v is some other element of $L \oplus K$ with $u' \perp v$, then $v\gamma = v + 2y$ for some $y \in L \oplus K$ and since $0 = (u', v) = (u'\gamma, v\gamma)$,

$$0 = (u'\gamma, v + 2y) = (u'\gamma, v) + (mu\gamma, 2y) = (u'\gamma, v) + 2m(u\gamma, y).$$

Thus $(u'\gamma, v) \equiv 0 \mod 2m$. Also,

$$(u'\gamma, u') = (u' + 2x, u') = (u', u') + (2x, mu) = (u', u') + 2m(x, u).$$

Thus $(u'\gamma, u') \equiv (u', u') \mod 2m$. Therefore $u'\gamma \equiv u' \mod 2m$ and since $u' = mu$ we have $u\gamma \equiv u \mod 2$. This shows $\gamma \in \mathrm{Aut}(L \oplus_G K)_{(2)}$. □

PROOF OF THEOREM 1.1. The proof follows from Theorem 4.1, Theorem 4.2, and Theorem 4.3 since the lattice L embeds as the orthogonal complement of the positive definite sublattice K and this induces the embedding of $\mathrm{Aut}(L)$ as a geometrically finite subgroup of $\mathrm{Aut}(L \oplus_G K)$. □

We remark here that the principal congruence subgroups of level 2 have uniformly bounded index in integral arithmetic subgroups of simplest type in $\mathrm{Isom}(\mathbb{H}^n)$. Indeed, any such Γ is contained in some $\mathrm{O}^+(f;\mathbb{Z}) \subset \mathrm{GL}_{n+1}(\mathbb{Z})$ and thus $[\Gamma : \Gamma_{(2)}] \leq |\mathrm{GL}_{n+1}(\mathbb{Z}/2\mathbb{Z})|$.

The Bianchi group $\mathrm{PSL}_2(\mathcal{O}_d)$ is contained in $\mathrm{O}^+(f_d;\mathbb{Z})$ where

$$f_d = \begin{cases} 2x_0x_1 + 2x_2^2 + 2dx_3^2 & \text{if } d \equiv 1,2 \mod 4 \\ 2x_0x_1 + 2x_2^2 + 2x_2x_3 + \frac{d+1}{2}x_3^2 & \text{if } d \equiv 3 \mod 4 \end{cases}$$

[**JM96**]. It is easy to see via the explicit embedding of $\mathrm{PSL}_2(\mathcal{O}_d)$ into $\mathrm{O}^+(f_d;\mathbb{Z})$ described in [**Chu19**, §3.1] that the image of $\mathrm{PSL}_2(\mathcal{O}_d)_{(2)}$ lands in $\mathrm{O}^+(f_d;\mathbb{Z})_{(2)}$. Therefore, we get that the principal congruence subgroup $\mathrm{PSL}(2,\mathcal{O}_d)_{(2)}$ embeds in a RACG. The index of this subgroup is well known to be

$$[\mathrm{PSL}(2,\mathcal{O}_d) : \mathrm{PSL}(2,\mathcal{O}_d)_{(2)}] = \begin{cases} 48 & \text{if } d \equiv 1,2 \mod (4) \\ 60 & \text{if } d \equiv 3 \mod (8) \\ 36 & \text{if } d \equiv 7 \mod (8), \end{cases}$$

(see for example [**Dic01**]), and thus, Theorem 1.2 follows.

5. An example

Consider the lattice L with basis $\{e_0, e_1, e_2, e_3\}$ and associated inner product matrix $A_L = \mathrm{diag}\{-7,1,1,1\}$. The dual L^* has $\mathbb{Z}$-basis $\{\frac{1}{7}e_0, e_1, e_2, e_3\}$ and the discriminant group $\Delta(L) \cong \mathbb{Z}/7\mathbb{Z}$ is generated by the image of the vector $\frac{1}{7}e_0$. We can take K to be the lattice with basis $\{e_4, e_5, e_6\}$ and associated inner product matrix$A_K = \mathrm{diag}\{7,1,1\}$. The dual K^* has $\mathbb{Z}$-basis $\{\frac{1}{7}e_4, e_5, e_6\}$ and the discriminant group $\Delta(K) \cong \mathbb{Z}/7\mathbb{Z}$ is generated by the image of the vector $\frac{1}{7}e_4$. Therefore $\Delta(L \oplus K) = \mathbb{Z}/7\mathbb{Z} \times \mathbb{Z}/7\mathbb{Z}$ and image of the vector $u = \frac{4}{7}e_0 + \frac{3}{7}e_4$ generates a totally isotropic subgroup G of $\Delta(L \oplus K)$ of order $\mathbb{Z}/7\mathbb{Z}$. Indeed, u has norm -1 and pairs integrally with $L \oplus K$. If we let $v = \frac{3}{7}e_0 + \frac{4}{7}e_4$ then $\{u, v, e_1, e_2, e_3, e_5, e_6\}$ defines an orthogonal $\mathbb{Z}$-basis for $L \oplus_G K$ with associated inner product matrix $\mathrm{diag}\{-1,1,1,1,1,1,1\}$. Therefore $L \oplus_G K = \mathbf{I}_{6,1}$.

Since $e_0 = 4u - 3v$, a copy of L is the sublattice with $\mathbb{Z}$-basis $\{4u-3v, e_1, e_2, e_3\}$ which is the orthogonal complement of a copy of K with $\mathbb{Z}$-basis $\{-3u+4v, e_5, e_6\}$. The change of basis matrix is

$$B = \begin{pmatrix} \frac{4}{7} & 0 & \frac{-3}{7} & \\ 0 & I_3 & 0 & \\ \frac{-3}{7} & 0 & \frac{4}{7} & \\ & & & I_2 \end{pmatrix} \tag{5.1}$$

where I_n denotes the $n \times n$ identity matrix. Therefore,

$$B \cdot \mathrm{diag}\{-7,1,1,1,7,1,1\} \cdot B^{tr} = \mathrm{diag}\{-1,1,1,1,1,1,1\}.$$

Let γ be in $\mathrm{Aut}(L)$. As a matrix with entries in $\mathbb{Z}$ preserving A_L, γ the first row $\begin{pmatrix} a & b & c & d \end{pmatrix}$ satisfies $a^2 \equiv 1 \mod 7$ and $b, c, d \equiv 1 \mod 7$. We extend γ in Aut_L

to γ in $\mathrm{Aut}(L \oplus K)$ as follows:

$$\gamma \mapsto \begin{cases} \begin{pmatrix} \gamma & \\ & I_3 \end{pmatrix} & \text{if } a \equiv 1 \mod 7 \\ \begin{pmatrix} \gamma & & \\ & -I_2 & \\ & & I_1 \end{pmatrix} & \text{if } a \equiv -1 \mod 7 \end{cases} \tag{5.2}$$

and such an integral matrix preserves integrality when conjugated by B, that is, $B \cdot \gamma \cdot B^{-1}$ is integral and preserves $\mathrm{diag}\{-1,1,1,1,1,1,1\}$. Indeed, if

$$\gamma_1 = \begin{pmatrix} 7a+1 & 7b & 7c & 7d & \\ e & f & g & h & \\ i & j & k & l & \\ m & n & o & p & \\ & & & & I_3 \end{pmatrix}, \gamma_2 = \begin{pmatrix} 7a-1 & 7b & 7c & 7d & & \\ e & f & g & h & & \\ i & j & k & l & & \\ m & n & o & p & & \\ & & & & -I_2 & \\ & & & & & I_1 \end{pmatrix}$$

then

$$B \cdot \gamma_1 \cdot B^{-1} = \begin{pmatrix} 16a+1 & 4b & 4c & 4d & 12a & 0 & 0 \\ 4e & f & g & h & 3e & 0 & 0 \\ 4i & j & k & l & 3i & 0 & 0 \\ 4m & n & o & p & 3m & 0 & 0 \\ -12a & -3b & -3c & -3d & 1-9a & 0 & 0 \\ 0 & 0 & 0 & 0 & 0 & 1 & 0 \\ 0 & 0 & 0 & 0 & 0 & 0 & 1 \end{pmatrix}$$

and

$$B \cdot \gamma_2 \cdot B^{-1} = \begin{pmatrix} 16a-1 & 4b & 4c & 4d & 12a & 0 & 0 \\ 4e & f & g & h & 3e & 0 & 0 \\ 4i & j & k & l & 3i & 0 & 0 \\ 4m & n & o & p & 3m & 0 & 0 \\ -12a & -3b & -3c & -3d & -9a-1 & 0 & 0 \\ 0 & 0 & 0 & 0 & 0 & -1 & 0 \\ 0 & 0 & 0 & 0 & 0 & 0 & 1 \end{pmatrix}.$$

Furthermore, If $\gamma \equiv I_7 \mod 2$ entrywise, so is $B \cdot \gamma \cdot B^{-1}$, and thus there is an inclusion of principal congruence subgroups of level 2: $\mathrm{Aut}(L)_{(2)} \subset \mathrm{Aut}(\mathbf{I}_{6,1})_{(2)}$.

Acknowledgments

The author thanks Daniel Allcock for introducing her to lattice gluing and encouraging this work.

References

[Ago13] I. Agol, *The virtual Haken conjecture*, Doc. Math. **18** (2013), 1045–1087, DOI 10.1016/j.procs.2013.05.269. With an appendix by Agol, Daniel Groves, and Jason Manning. MR3104553

[All12] D. Allcock, *The reflective Lorentzian lattices of rank 3*, Mem. Amer. Math. Soc. **220** (2012), no. 1033, x+108, DOI 10.1090/S0065-9266-2012-00648-4. MR3025355

[All18] D. Allcock, *Prenilpotent pairs in the E_{10} root lattice*, Math. Proc. Cambridge Philos. Soc. **164** (2018), no. 3, 473–483, DOI 10.1017/S0305004117000287. MR3784264

[BHW11] N. Bergeron, F. Haglund, and D. T. Wise, *Hyperplane sections in arithmetic hyperbolic manifolds*, J. Lond. Math. Soc. (2) **83** (2011), no. 2, 431–448, DOI 10.1112/jlms/jdq082. MR2776645

[Chu18] Michelle Chu, *Quantifying virtual properties of bianchi groups*, Phd dissertation, The University of Texas at Austin, 2018.

[Chu19] M. Chu, *Special subgroups of Bianchi groups*, Groups Geom. Dyn. **14** (2020), no. 1, 41–59, DOI 10.4171/ggd/533. MR4077653

[CS99] J. H. Conway and N. J. A. Sloane, *Sphere packings, lattices and groups*, 3rd ed., Grundlehren der Mathematischen Wissenschaften [Fundamental Principles of Mathematical Sciences], vol. 290, Springer-Verlag, New York, 1999. With additional contributions by E. Bannai, R. E. Borcherds, J. Leech, S. P. Norton, A. M. Odlyzko, R. A. Parker, L. Queen and B. B. Venkov. MR1662447

[DMP18] Jason DeBlois Miller Patel DeBlois, Nicholas Miller, and Priyam Patel, *Effective virtual and residual properties of some arithmetic hyperbolic 3-manifolds*, Trans. Amer. Math. Soc. **373** (2020), 8219–8257. DOI: https://doi.org/10.1090/tran/8190

[Dic01] St., *Literatur-Berichte: Niedere Zahlentheorie* (German), Monatsh. Math. Phys. **14** (1903), no. 1, A75–A77, DOI 10.1007/BF01707031. Von Prof. Dr. Paul Bachmann in Weimar. Erster Teil. (B. G. Teubners Sammlung von Lehrbüchern auf dem Gebiete der mathematischen Wissenschaften mit Einschlußihrer Anwendungen, Bd. X, 1.) B. G. Teubner, Leipzig, 1902. 8°. MR1547256

[ERT12] B. Everitt, J. G. Ratcliffe, and S. T. Tschantz, *Right-angled Coxeter polytopes, hyperbolic six-manifolds, and a problem of Siegel*, Math. Ann. **354** (2012), no. 3, 871–905, DOI 10.1007/s00208-011-0744-2. MR2983072

[HW08] F. Haglund and D. T. Wise, *Special cube complexes*, Geom. Funct. Anal. **17** (2008), no. 5, 1551–1620, DOI 10.1007/s00039-007-0629-4. MR2377497

[JM96] D. G. James and C. Maclachlan, *Fuchsian subgroups of Bianchi groups*, Trans. Amer. Math. Soc. **348** (1996), no. 5, 1989–2002, DOI 10.1090/S0002-9947-96-01606-6. MR1348863

[KV78] I. M. Kaplinskaja and È. B. Vinberg, *The groups $O_{18,1}(Z)$ and $O_{19,1}(Z)$* (Russian), Dokl. Akad. Nauk SSSR **238** (1978), no. 6, 1273–1275. MR0476640

[Kho86] A. G. Khovanskiĭ, *Hyperplane sections of polyhedra, toric varieties and discrete groups in Lobachevskiĭ space* (Russian), Funktsional. Anal. i Prilozhen. **20** (1986), no. 1, 50–61, 96. MR831049

[PV05] L. Potyagailo and E. Vinberg, *On right-angled reflection groups in hyperbolic spaces*, Comment. Math. Helv. **80** (2005), no. 1, 63–73, DOI 10.4171/CMH/4. MR2130566

[Pro86] M. N. Prokhorov, *Absence of discrete groups of reflections with a noncompact fundamental polyhedron of finite volume in a Lobachevskiĭ space of high dimension* (Russian), Izv. Akad. Nauk SSSR Ser. Mat. **50** (1986), no. 2, 413–424. MR842588

[Vin72] È. B. Vinberg, *The groups of units of certain quadratic forms* (Russian), Mat. Sb. (N.S.) **87(129)** (1972), 18–36. MR0295193

[Vin84] È. B. Vinberg, *Absence of crystallographic groups of reflections in Lobachevskiĭ spaces of large dimension* (Russian), Trudy Moskov. Mat. Obshch. **47** (1984), 68–102, 246. MR774946

[Wat62] G. L. Watson, *Transformations of a quadratic form which do not increase the class-number*, Proc. London Math. Soc. (3) **12** (1962), 577–587, DOI 10.1112/plms/s3-12.1.577. MR142512

[Wat75] G. L. Watson, *Transformations of a quadratic form which do not increase the class-number. II*, Acta Arith. **27** (1975), 171–189, DOI 10.4064/aa-27-1-171-189. MR364100

[Wis11] Daniel T. Wise, *The structure of groups with quasiconvex hierarchy*, 187 pages.

Department of Mathematics, Statisctics, and Computer Science, University of Illinois at Chicago, 851 S. Morgan St., Chicago, Illinois 60607-7045

Email address: michu@uic.edu

Contemporary Mathematics
Volume **760**, 2020
https://doi.org/10.1090/conm/760/15286

An exposition of the equivalence of Heegaard Floer homology and embedded contact homology

Vincent Colin, Paolo Ghiggini, and Ko Honda

Abstract. This article is a survey on the authors' proof of the isomorphism between Heegaard Floer homology and embedded contact homology appeared in [**4**–**7**].

1. Introduction

In the search for gluing formulas for gauge theoretic invariants of four-manifolds (Donaldson's invariants and Seiberg-Witten invariants) it was soon realised that the object which should be associated to a three-manifold is a Floer homology. The first to be constructed was *instanton Floer homology* in [**10**] by Floer himself, although only for integer homology spheres. On the other hand, the development of a Seiberg-Witten Floer homology languished for several years, until Kronheimer and Mrowka defined *monopole Floer homology* in [**29**].

In the meantime the deep connection between gauge theory and symplectic geometry became clear. For example, Atiyah proposed a conjectural reinterpretation of instanton Floer homology as a Lagrangian intersection Floer homology, now known as the "Atiyah-Floer conjecture", in [**1**] and Taubes proved in [**50**] that the Seiberg-Witten invariants of a symplectic four-manifold are equivalent to a count of J-holomorphic curves which, in most cases, are embedded. This result is usually called "SW=Gr".

Ozsváth and Szabó defined *Heegaard Floer homology* in [**39**, **40**] motivated by the Atiyah-Floer conjecture. Instead of the original definition, in this article we will use an equivalent one given by Lipshitz in [**36**]. The starting point of its construction is a *pointed Heegaard diagram* $(\Sigma, \boldsymbol{\alpha}, \boldsymbol{\beta}, z)$ which describes a three-manifold M. Here Σ is a Heegaard surface of genus $h > 0$ associated to some self-indexing Morse function with a unique maximum and a unique minimum, $\boldsymbol{\alpha}$ is the collection of the attaching circles for the index one critical points, $\boldsymbol{\beta}$ is the collection

---

2020 *Mathematics Subject Classification.* Primary 57K31; Secondary 57K33, 57R58.

Key words and phrases. Heegaard Floer homology, embedded contact homology, contact structure, open book decomposition.

The first and second authors were supported in part by ANR grant ANR-08-BLAN-0291-01 "Floer Power", ERC grant 278246 "GEODYCON" and ANR grant ANR-16-CE40-017 "Quantact".

The third author was supported in part by NSF Grants DMS-0805352, DMS-1105432, DMS-1406564, and DMS-1549147.

of the attaching circles for the index two critical points, and z is a basepoint in the complement of $\boldsymbol{\alpha}$ and $\boldsymbol{\beta}$. The Heegaard Floer complexes are generated by h-tuples of intersection points between the $\boldsymbol{\alpha}$-curves and the $\boldsymbol{\beta}$-curves and the differential counts certain J-holomorphic curves in $\mathbb{R} \times [0,1] \times \Sigma$ with boundary on $\mathbb{R} \times \{0\} \times \boldsymbol{\beta}$ and $\mathbb{R} \times \{1\} \times \boldsymbol{\alpha}$. We refer to Section 3 for an overview of Heegaard Floer homology in its cylindrical reformulation.

Hutchings, partially in collaboration with Taubes, defined *embedded contact homology* in [**18, 20, 21**] motivated by SW=Gr and by Eliashberg, Givental and Hofer's symplectic field theory [**9**]. The starting point for embedded contact homology is a contact form α on M. The contact form determines the *Reeb vector field* R by

$$\iota_R d\alpha = 0 \quad \text{and} \quad \alpha(R) = 1.$$

The embedded contact homology complex is generated by finite sets of simple Reeb orbits with finite multiplicities (called orbit sets) and its differential counts certain J-holomorphic curves in the symplectisation $(\mathbb{R} \times M, d(e^s\alpha))$. We refer to Section 4 for an overview of embedded contact homology.

Thus by 2009 we had three Floer homology theories for three-manifolds, each with its own strengths and weaknesses, which share many formal properties: among others a U-map, a contact invariant and a decomposition as direct sum of subgroups indexed by Spin^c structures in the case of monopole and Heegaard Floer homology and by first homology classes in the case of embeded contact homology. The relative advantage of monopole Floer homology is its link with geometry, that of embedded contact homology is its link with Reeb dynamics and that of Heegaard Floer homology is that it needs less sophisticated analytical tools than the previous two, which makes it relatively computable and easier to develop.

These three Floer homology theories were expected to be isomorphic according to some big picture, and the proofs of the isomorphisms, which appeared in the last ten years, span several hundred pages of very technical arguments. The isomorphism between monopole Floer homology and embedded contact homology is due to Taubes and appeared in [**43–47**]. The isomorphism between monopole Floer homology and Heegaard Floer homology is due to Kutluhan, Lee and Taubes and appeared in [**31–35**]. The isomorphism between Heegaard Floer homology and embedded contact homology, which appeared in [**4–7**], is the topic of this survey. On the other hand, the relation of these three theories with instanton Floer homology is still to be clarified.

These isomorphisms have already had various applications. The most striking ones are the proofs, for contact three manifolds, of the Weinstein conjecture by Taubes [**49**] and of the Arnold's chord conjecture by Hutchings and Taubes [**22, 23**].

The picture, however, is not yet complete for several reasons. First of all, the definitions of the isomorphisms require some choices, but it is not clear at what extent the result depends on them. Second, we do not know if composing two of the isomorphisms we obtain the third one. Of course this question will make sense only after the previous one has been answered. Finally, monopole Floer homology and Heegaard Floer homology are "almost" 3+1 topological quantum field theories, but it is not known if the isomorphisms commute with maps induced by cobordisms.

The isomorphism between Heegaard Floer homology and embedded contact homology is expressed in a more precise way by the following theorem.

THEOREM 1.1. *Let M be a closed three manifold and ξ a contact structure on M. Then there are isomorphisms*

$$\widehat{\Phi}_*: \widehat{HF}(-M) \to \widehat{ECH}(M),$$
$$\Phi^+_*: HF^+(-M) \to ECH(M)$$

such that the diagram

$$\begin{array}{ccccccccc} \dots \longrightarrow & \widehat{HF}(-M) & \longrightarrow & HF^+(-M) & \xrightarrow{U} & HF^+(-M) & \longrightarrow \dots \\ & \downarrow{\scriptstyle \widehat{\Phi}_*} & & \downarrow{\scriptstyle \Phi^+_*} & & \downarrow{\scriptstyle \Phi^+_*} & \\ \dots \longrightarrow & \widehat{ECH}(M) & \longrightarrow & ECH(M) & \xrightarrow{U} & ECH(M) & \longrightarrow \dots \end{array} \tag{1.1}$$

commutes. Moreover the isomorphisms map the contact class to the contact class and match the splitting according to $Spin^c$-structures in Heegaard Floer homology to the splitting according to first homology classes in embedded contact homology.

Vinicius Ramos in [**42**] proved that $\widehat{\Phi}_*$ and Φ^+_* respect also the grading by homotopy classes of plane fields which exists in both theories. For simplicity we will not talk about Spin^c-structures or gradings in this survey.

A natural setting for relating Heegaard Floer homology and embedded contact homology is that of *open book decompositions* (see Definition 3.9) because an open book decomposition determines both a Heegaard splitting and a contact structure. The Heegaard splitting is obtained by taking as Heegaard surface the union of two opposite pages. The contact structure is provided by a construction of Thurston and Winkelnkemper from [**51**].

Fix an open book decomposition $(S, \hbar)$ for M where S has genus $g > 0$ and connected boundary. The first step in the definition of the isomorphisms is to adapt the definitions of $\widehat{HF}(-M)$ and $\widehat{ECH}(M)$ to the open book decomposition $(S, \hbar)$. For Heegaard Floer homology this is achieved by pushing all interesting intersection points between the $\boldsymbol{\alpha}$- and $\boldsymbol{\beta}$-curves to one side of the Heegaard surface obtained from $(S, \hbar)$; see Subsection 3.3.

For embedded contact homology, this is achieved in [**4**] and reviewed in Subsections 4.3 and 4.4. There we introduce the group $\widehat{ECH}(N, \partial N)$ for the mapping torus N of $(S, \hbar)$ and prove that $\widehat{ECH}(M) \cong \widehat{ECH}(N, \partial N)$. This group is defined as a direct limit

$$\widehat{ECH}(N, \partial N) = \varinjlim ECH_i(N),$$

where i is the number of intersections of an orbit set in N with a fibre and the direct limit is taken with respect to maps $ECH_i(N) \to ECH_{j+1}(N)$ defined by increasing the multiplicity of an elliptic orbit on ∂N. This orbit can be regarded intuitively as a receptacle for the J-holomorphic curves in $\mathbb{R} \times M$ which intersect the cylinder over the binding.

Chain maps between Floer complexes are often defined by counting holomorphic curves in some symplectic cobordisms. Here, we use the open book $(S, \hbar)$ to build a symplectic cobordism W_+ from $[0,1] \times S$ to N and a symplectic cobordism X_+ from $[0,1] \times \Sigma$ to M. Here Σ is the union of two opposite pages of the open book decomposition, and therefore is a closed surface of genus $2g$. Our convention is that symplectic cobordisms have cylindrical ends and we say that they go *from* the convex (positive) end *to* the concave (negative) end.

Counting holomorphic curves in the cobordisms W_+ and X_+ with suitable asymptotics and Lagrangian boundary conditions gives maps $\Phi_*\colon \widehat{HF}(-M) \to ECH_{2g}(N)$ and $\Phi_*^+\colon HF^+(-M) \to ECH(M)$. After composing Φ_* with the natural map $ECH_{2g}(N) \to \widehat{ECH}(N, \partial N)$, we obtain the map $\widehat{\Phi}_*$. The maps Φ_* and $\widehat{\Phi}_*$ are defined in [**5**] (with a slightly different notation), while Φ_*^+ is defined in [**7**]. Both are reviewed in Section 5. Strictly speaking, in the construction of Φ we replace the embedded Floer homology groups of N with isomorphic periodic Floer homology groups; see Subsection 4.6.

In [**5**] we also define a map $\Psi_*\colon ECH_{2g}(N) \to \widehat{HF}(-M)$ by counting holomorphic curves in a symplectic cobordism $\overline{W}_-$ passing through a base point. The cobordism $\overline{W}_-$ is defined as a compactification of W_+ turned upside-down. The Lagrangian boundary condition on $\overline{W}_-$ is singular, and this leads to many more potential degenerations of holomorphic curves. For this reason the analysis of the J-holomorphic curves defining Ψ_* is longer and more difficult than the analysis of the J-holomorphic curves defining Φ_*. The construction of Ψ_* is reviewed in Section 6.

Then, in [**6**] we prove that Φ_* and Ψ_* are inverses of each other by composing the two cobordisms and degenerating them in a different way. The proof that $\Phi_* \circ \Psi_*$ and $\Psi_* \circ \Phi_*$ are isomorphisms is thus reduced to a computation of some relative Gromov-Taubes invariants. This step is briefly described in Section 7.

Finally we prove that the natural map $ECH_{2g}(N) \to \widehat{ECH}(N, \partial N)$ is an isomorphism by an argument based on stabilising the open book decomposition. This last step is described in Section 8. This finishes the proof that $\widehat{\Phi}_*$ is an isomorphism. A simple algebraic argument based on the commutativity of the diagram (1.1) and the properties of the map U in both theories shows that Φ_*^+ is also an isomorphism.

CONVENTION 1.2. Throughout the paper we will follow the convention that all three-manifolds are connected, oriented and compact. If X is a manifold with boundary, we will denote $\mathrm{int}(X) = X \setminus \partial X$.

We finish this introduction with a warning to the reader: the articles [**4**–**7**] are still under review at the time of writing this survey, and therefore the numbering of the references to those articles is likely to become obsolete.

2. Moduli spaces of J-holomorphic curves

In this section we review some generalities about moduli spaces of holomorphic curves with the goal of fixing notations and conventions. The reader should keep in mind that we are not trying to give a one-size-fits-all definition of the various moduli space we will use: each one will be defined in due course; here we introduce the terminology that we will use in their definitions.

Let $\mathcal{Y}$ be a compact, connected three-manifold. A *stable Hamiltonian structure* on $\mathcal{Y}$ is a pair (α, ω) where α is a 1-form, and ω is a closed 2-form which satisfy $\alpha \wedge \omega > 0$ and $\ker d\alpha \subset \ker \omega$. Stable Hamiltonian structures in this article will satisfy also one of the following conditions:

(i) $d\alpha = \omega$, or
(ii) $d\alpha = 0$.

Clearly when (i) is satisfied α is a contact form on $\mathcal{Y}$. Fibrations over S^1 are the main source of stable Hamiltonian structures satisfying (ii). In fact, if $\pi\colon \mathcal{Y} \to S^1$ is a locally trivial fibration with fibre $\mathcal{S}$, it is always possible to find a representative of the monodromy $\phi\colon \mathcal{S} \to \mathcal{S}$ preserving an area form ω, and therefore to regard ω as a 2-form on $\mathcal{Y}$. If dt is a length form on S^1 and we define $\alpha = \pi^* dt$, then (α, ω) will be called the stable Hamiltonian structure *induced by the fibration* $\pi\colon \mathcal{Y} \to S^1$. Stable Hamiltonian structures satisfying (ii) will arise, in the same way, also from fibrations $\pi\colon \mathcal{Y} \to I$ over a closed interval.

A stable Hamiltonian structure determines a Reeb vector field[1] R on $\mathcal{Y}$, which is the vector field defined by the equations

$$\begin{cases} \alpha(R) = 1, \\ \iota_R \omega = 0. \end{cases}$$

An almost complex structure J on $\mathbb{R} \times \mathcal{Y}$ is *compatible* with the stable Hamiltonian structure (α, ω), or with the contact form α when (i) is satisfied, if

- J is invariant under translations in the $\mathbb{R}$ direction,
- $J(\partial_s) = R$, where s denotes the coordinate on $\mathbb{R}$,
- $J(\xi) = \xi$, where $\xi = \ker \alpha$, and
- $\omega(\cdot, J\cdot)$ is an Euclidean metric on ξ.

Let $(\mathcal{X}, \Omega_{\mathcal{X}})$ be a symplectic four-manifold, possibly with boundary and cylindrical ends (in the latter case we also call $(\mathcal{X}, \Omega_{\mathcal{X}})$ a *symplectic cobordism*). The positive ends are identified with $(0, +\infty) \times \mathcal{Y}_+$ and the negative ends are identified with $(-\infty, 0) \times \mathcal{Y}_-$ where $\mathcal{Y}_+$ and $\mathcal{Y}_-$ are three-manifolds which are endowed with a stable Hamiltonian structure. On each end $\Omega_{\mathcal{X}} = d(e^s \alpha)$ if the stable Hamiltonian structure satisfies (i), or $\Omega_{\mathcal{X}} = ds \wedge \alpha + \omega$ if it satisfies (ii). The boundary of $\mathcal{X}$ has a vertical part which is foliated by symplectic submanifolds and a horizontal part which is convex. The vertical boundary of $\mathcal{X}$, if nonempty, is equipped with a Lagrangian submanifold $\mathcal{L}$ such that the intersection of $\mathcal{L}$ with an end of $\mathcal{X}$ is a half cylinder over a collection of properly embedded curves in $\partial\mathcal{Y}$ which are tangent to ξ. Those curves are called the boundary at infinity[2] of $\mathcal{L}$. The prototype is $\mathcal{X} = \mathbb{R} \times [0, 1] \times S$ with a product symplectic form $\Omega_{\mathcal{X}} = ds \wedge dt + \omega$, where S is a surface with boundary, endowed with the Lagrangian submanifold $\mathcal{L} = (\mathbb{R} \times \{0\} \times \mathbf{a}_0) \cup (\mathbb{R} \times \{1\} \times \mathbf{a}_1)$, where $\mathbf{a}_0$ and $\mathbf{a}_1$ are properly embedded one-dimensional submanifolds with boundary of S.

On $\mathcal{X}$ we choose an almost complex structure $J_{\mathcal{X}}$ which is compatible with $\Omega_{\mathcal{X}}$ and with the stable Hamiltonian structures on the ends. In order to have SFT compactness for $J_{\mathcal{X}}$-holomorphic curves, we assume that the vertical part of $\partial\mathcal{X}$ is foliated by $J_{\mathcal{X}}$-holomorphic submanifolds which form a barrier to $J_{\mathcal{X}}$-holomorphic curves and that the horizontal boundary is $J_{\mathcal{X}}$-convex, so that a maximum principle for $J_{\mathcal{X}}$-holomorphic curves holds. Note that in most cases $\mathcal{X}$ is the total space of a symplectic fibration $\mathcal{X} \xrightarrow{\varpi} \mathcal{B}$ and the condition on $J_{\mathcal{X}}$ at the vertical boundary is obtained by asking that ϖ should be holomorphic, at least near $\partial\mathcal{X}$. We often require more properties from the almost complex structures, but in the text we will recall only those which are relevant for a first reading.

[1] Also called *Hamiltonian vector field*, especially when α is not a contact form.

[2] This term will be used only in this section.

DEFINITION 2.1. A *$J_{\mathcal{X}}$-holomorphic curve* in $\mathcal{X}$ with boundary on $\mathcal{L}$ is a triple (F, j, u) such that (F, j) is a smooth Riemann surface, possibly with boundary and punctures, and $u\colon F \to \mathcal{X}$ is a proper map satisfying $du \circ j = J_{\mathcal{X}} \circ du$ and mapping each connected component of ∂F to a distinct connected component of $\mathcal{L}$. If F is connected, then (F, j, u) is called *irreducible*. If $F_0 \subset F$ is a connected component, the restriction $(F_0, j|_{F_0}, u|_{F_0})$ will be called an *irreducible component* of (F, j, u).

We will simply say a *holomorphic curve* when the almost complex structure is clear from the context. In Section 6 we will need to consider holomorphic curves with boundary on a singular Lagrangian $\mathcal{L}$. In that case we will assume that each connected component of ∂F is sent to a distinct connected component of the smooth part of $\mathcal{L}$.

When $\mathcal{X}$ is the total space of a symplectic fibration $\varpi\colon \mathcal{X} \to \mathcal{B}$ and $J_{\mathcal{X}}$ is compatible with the fibration, which means that ϖ is holomorphic, we say that a holomorphic curve (F, j, u) is a *degree d multisection* of $\mathcal{X}$ if $\varpi \circ u\colon F \to \mathcal{B}$ is a branched cover of degree d.

The punctures of F are divided into positive and negative punctures so that u maps a punctured neighbourhood of a positive puncture to a positive end of $\mathcal{X}$ and a punctured neighbourhood of a negative puncture to a negative end. Around each positive puncture we choose coordinates $(s, t) \in (0, +\infty) \times [0, 1]$ for a boundary puncture and $(s, t) \in (0, +\infty) \times S^1$ for an interior puncture. Around each negative puncture we choose similar coordinates with $(0, +\infty)$ replaced by $(-\infty, 0)$. If e is either a Reeb orbit in $\mathcal{Y}$ reparametrised so that it has period one, or a Reeb chord connecting two points of the boundary at infinity of $\mathcal{L}$, we say that u is *positively* (resp. *negatively*) *asymptotic* to e if, in the neighbourhood of a positive (resp. negative) puncture, $\lim_{s \to \pm\infty} u(s, t) = e(t)$ (with positive sign for positive punctures and negative sign for negative punctures). We call an *end* of a holomorphic curve the restriction of u which is mapped to an end of $\mathcal{X}$.

DEFINITION 2.2. If $\mathcal{X} = \mathbb{R} \times \mathcal{Y}$ and $J_{\mathcal{X}}$ is adapted to a stable Hamiltonian structure on $\mathcal{Y}$, a $J_{\mathcal{X}}$-holomorphic curve in $\mathcal{X}$ is either a *trivial cylinder* or *strip* if it parametrises $\mathbb{R} \times e$ where e is, respectively, a closed Reeb orbit or a Reeb chord. In a general symplectic manifold $\mathcal{X}$ we say that an end of a $J_{\mathcal{X}}$-holomorphic curve is *trivial* if it coincides with a portion of a trivial cylinder or strip.

We say that two holomorphic curves (F, j, u) and (F', j', u') are equivalent, and write $(F, j, u) \sim (F', j', u')$, if there exists an orientation-preserving diffeomorphism $\phi\colon F \to F'$ extending smoothly over the punctures such that $\phi_* j = j'$ and $u = u' \circ \phi$. Given a set $\mathbf{e}$ of chords and orbits in the various ends, the moduli space[3] $\mathcal{M}_{\mathcal{X}}(\mathbf{e})$ is the quotient by the equivalence relation $\sim$ of the space of $J_{\mathcal{X}}$-holomorphic curves in $\mathcal{X}$ with boundary on $\mathcal{L}$ which are asymptotic to the chords and orbits in $\mathbf{e}$. In the definition of $\mathcal{M}_{\mathcal{X}}(\mathbf{e})$ the topology of F is *not* fixed, and therefore different $J_{\mathcal{X}}$-holomorphic curves can have different genera, number of punctures and number of connected components. Moreover, multiple orbits are treated in the embedded contact homology way, which means that only their total multiplicity in $\mathbf{e}$ counts. For example, if $\mathbf{e}$ contains an orbit γ with multiplicity two, then

[3] In the next sections we will distinguish chords and orbits belonging to different ends of $\mathcal{X}$. Here, for simplicity, we don't.

holomorphic curves[4] in $\mathcal{M}_{\mathcal{X}}(\mathbf{e})$ can have either one end at a double cover of γ or two ends at γ. See [**18**] or Subsection 4.1 for more details.

Let $\mathcal{T}(F)$ be the Teichmüller space of complex structures on F and $\mathrm{Aut}(F, j)$ the group of automorphisms of the Riemann surface (F, j). To a $J_{\mathcal{X}}$-holomorphic curve (F, j, u) we associate a formal deformation complex

$$0 \to T_{id}\,\mathrm{Aut}(F, j) \xrightarrow{\mathfrak{L}} T_{[j]}\mathcal{T}(F) \oplus \Omega^0(F; u^*T\mathcal{X}) \xrightarrow{D_{j,u}} \Omega^{0,1}(S; u^*T\mathcal{X}) \to 0$$

where $D_{j,u}$ is the formal linearised Cauchy-Riemann operator where j is also considered as a variable, and $\mathfrak{L}$ is the formal linearisation of the action of $\mathrm{Aut}(F, j)$. If the ends of u are nondegenerate, a suitable Sobolev completion of the formal deformation complex is an elliptic complex. Let $\mathcal{H}^i$ (for $i = 0, 1, 2$) be its homology groups, which are finite dimensional because of the Fredholm property. We define the *Fredholm index* of a $J_{\mathcal{X}}$-holomorphic curve (F, j, u) as

$$\mathrm{ind}(u) = -\dim \mathcal{H}^0 + \dim \mathcal{H}^1 - \dim \mathcal{H}^2$$

If u is not constant (and it will never be constant in this article), then $\mathcal{H}^0 = 0$. We say that (F, j, u) is *regular* if the linearised operator $D_{j,u}$ is surjective in the appropriated Sobolev completion, i.e. if $\mathcal{H}^2 = 0$. If (F, j, u) is nonconstant and regular, then its equivalence class has a neighbourhood in the moduli spaces $\mathcal{M}_{\mathcal{X}}(\mathbf{e})$ that is diffeomorphic to a ball of dimension $\mathrm{ind}(u)$. This is all standard holomorphic curves theory, and the reader can find the details in the many expositions of the topic, e.g. [**53**, Lecture 7] and [**41**, Section 13a].

In the article we will more often use another index, which is specific to holomorphic curves in four-dimensional symplectic manifolds: the ECH-type index $I(u)$. For its definition we refer to the original articles ([**18, 19**], where it was first introduced for embedded contact homology, and [**5**–**7**] for its extension to the various symplectic cobordisms used in the proof of Theorem 1.1). Here we will only state its main properties:

(1) *homology invariance*: $I(u)$ depends only on the relative homology class defined by u (after a suitable compactification);
(2) *concatenation*: if two holomorphic curves (F_+, j_+, u_+) and (F_-, j_-, u_-) such that the positive ends of u_- match the negative ends of u_+ are glued along the matching ends to form a holomorphic curve (F, j, u), then $I(u) = I(u_-) + I(u_+)$; and
(3) *index inequality*: if $\delta(u)$ is an algebraic count of singularities of u as in [**38**, Appendix E], then

$$\mathrm{ind}(u) + 2\delta(u) \leq I(u). \tag{2.1}$$

Moreover there is equality if u has no end which is asymptotic to a closed Reeb orbit.

The index inequality was first proved by McDuff for closed holomorphic curves as an adjunction inequality (see [**38**, Appendix E] for a more modern exposition) and reinterpreted by Taubes as an index inequality (see [**48**]). For punctured holomorphic curves it was proved by Hutchings in [**18**] (see also [**19**]) in order to define embedded contact homology. The extension to other settings is straightforward and is done in [**5**].

[4] We will always call the elements of the moduli spaces "holomorphic curves" even if, strictly speaking, they are equivalence classes of holomorphic curves.

The index inequality has the following important consequence: if $I(u) < 2$ (and, in some cases, also if $I(u) = 2$) then u is embedded and $I(u) = \mathrm{ind}(u)$. Embeddedness in particular implies somewhere injectivity, and therefore for a generic almost complex structure $J_{\mathcal{X}}$ all holomorphic curves of low ECH-type index are regular by standard techniques; see in [**38**, Section 3.2]. A similar result holds for holomorphic curves of higher index satisfying constraints of sufficiently high codimension. For this reason in [**5**–**7**] we do not need to worry about the regularity issues which trouble many parts of symplectic topology.

Given a property $\bigstar$ (e.g. $I = 0$), we denote the subset of $J_{\mathcal{X}}$-holomorphic curves in $\mathcal{M}_{\mathcal{X}}(\mathbf{e})$ satisfying $\bigstar$ by $\mathcal{M}^{\bigstar}_{\mathcal{X}}(\mathbf{e})$. This convention will be used throughout the paper.

For sake of brevity, in the next sections we will always denote a holomorphic curve (F, j, u) simply by u. In view of the embeddedness of the holomorphic curves of low ECH-type index, we will also often identify a holomorphic curve with its image.

3. Heegaard Floer homology

3.1. A review of Heegaard Floer homology. In this section we briefly review the Heegaard Floer homology groups associated to a closed 3-manifold M. We will work with Lipshitz's "cylindrical reformulation" from [**36**] instead of with the original definition from [**39**].

Every closed three-manifold can be encoded by a *pointed Heegaard diagram* $(\Sigma, \boldsymbol{\alpha}, \boldsymbol{\beta}, z)$ associated to a Heegaard decomposition of M. Here Σ is a closed, oriented, connected surface of genus h which divides M into two handlebodies H_α and H_β, and $\boldsymbol{\alpha} = \{\alpha_1, \dots, \alpha_h\}$ and $\boldsymbol{\beta} = \{\beta_1, \dots, \beta_h\}$ are collections of h pairwise disjoint simple closed curves in Σ which bound discs in H_α and H_β respectively and are linearly independent in $H_1(\Sigma; \mathbb{Z})$. Finally $z \in \Sigma - \boldsymbol{\alpha} - \boldsymbol{\beta}$ is a base point, which is not needed to describe M, but is a crucial ingredient in the definition of the Heegaard Floer chain complexes.

After fixing an area form ω on Σ, we obtain a stable Hamiltonian structure (dt, ω) on $[0,1] \times \Sigma$ with Reeb vector field $R = \partial_t$. The submanifolds $L_{\boldsymbol{\alpha}} = \mathbb{R} \times \{1\} \times \boldsymbol{\alpha}$ and $L_{\boldsymbol{\beta}} = \mathbb{R} \times \{0\} \times \boldsymbol{\beta}$ of $\mathbb{R} \times [0,1] \times \Sigma$ are Lagrangian for the symplectic form $\Omega_X = ds \wedge dt + \omega$. We choose an almost complex structure J on X which is compatible with the stable Hamiltonian structure (dt, ω) and such that the surface $\mathbb{R} \times [0,1] \times \{z\}$ is holomorphic.

We denote by $\mathcal{S}_{\boldsymbol{\alpha},\boldsymbol{\beta}}$ the set of unordered h-tuples of intersection points $\mathbf{y} = \{y_1, \dots, y_h\}$ between α- and β-curves for which there is a permutation $\sigma \in \mathfrak{S}_h$ such that $y_i \in \alpha_i \cap \beta_{\sigma(i)}$. Given $\mathbf{y}_+, \mathbf{y}_- \in \mathcal{S}_{\alpha,\beta}$, we consider the moduli space[5] $\mathcal{M}_X(\mathbf{y}_+, \mathbf{y}_-)$ of J-holomorphic curves in X with boundary on $L_{\boldsymbol{\alpha}} \cup L_{\boldsymbol{\beta}}$ and asymptotic to chords $[0,1] \times \mathbf{y}_\pm$ for $s \to \pm\infty$ (*HF-curves* in the following). Translations in the $\mathbb{R}$ direction act on $\mathcal{M}_X(\mathbf{y}_+, \mathbf{y}_-)$ and we denote the quotient by $\widehat{\mathcal{M}}_X(\mathbf{y}_+, \mathbf{y}_-)$.

To a J-holomorphic curve $u \in \mathcal{M}_X(\mathbf{y}_+, \mathbf{y}_-)$ we associate two topological quantities: an "intersection number" $n_z(u)$, defined as the algebraic intersection of the image of u with $[-1,1] \times [0,1] \times \{z\}$ (see [**38**, Appendix E] for the intersection number between two J-holomorphic curves), and the ECH-type index[6] $I(u)$ ([**5**, Equation 4.5.4]), which in this context is a reformulation of Lipshitz's index formula

[5]The same moduli spaces are denoted by $\mathcal{M}^X_J(\mathbf{y}, \mathbf{y}')$ in [**5**].

[6]denoted by I_{HF} in [**5**].

[**36**, Corollary 4.3] (see also [**37**]). The index inequality is proved in [**5**, Theorem 4.5.13]. By positivity of intersection for J-holomorphic curves in dimension four, $n_z(u) \geq 0$. Moreover, for a generic almost complex structure, $I(u) \geq 0$ for each HF-curve u.

The chain complex $\widehat{CF}(\Sigma, \boldsymbol{\alpha}, \boldsymbol{\beta}, z)$ is freely generated by $\mathcal{S}_{\boldsymbol{\alpha},\boldsymbol{\beta}}$ as a vector space over $\mathbb{Z}/2\mathbb{Z}$. The differential is

$$\widehat{\partial}\mathbf{y}_+ = \sum_{\mathbf{y}_- \in \mathcal{S}_{\boldsymbol{\alpha},\boldsymbol{\beta}}} \#\widehat{\mathcal{M}}_X^{I=1, n_z=0}(\mathbf{y}_+, \mathbf{y}_-)\mathbf{y}_-.$$

The chain complex $CF^+(\Sigma, \boldsymbol{\alpha}, \boldsymbol{\beta}, z)$ is freely generated, as a vector space over $\mathbb{Z}/2\mathbb{Z}$, by the pairs $[\mathbf{y}, i]$ where $\mathbf{y} \in \mathcal{S}_{\boldsymbol{\alpha},\boldsymbol{\beta}}$ and $i \in \mathbb{N}$ (our convention is that $0 \in \mathbb{N}$). The differential is

$$\partial^+([\mathbf{y}_+, i]) = \sum_{\mathbf{y}_- \in \mathcal{S}_{\boldsymbol{\alpha},\boldsymbol{\beta}}} \sum_{0 \leq j \leq i} \#\widehat{\mathcal{M}}_X^{I=1, n_z=j}(\mathbf{y}_+, \mathbf{y}_-)[\mathbf{y}_-, i-j].$$

In order to have finite sums we need to assume *weak admissibility*, which can be rephrased by asking that the curves $\boldsymbol{\alpha}$ and $\boldsymbol{\beta}$ be exact Lagrangian submanifolds for some primitive of ω on $\Sigma \setminus \{z\}$. Weak admissibility can always be achieved up to isotopy of the α- and β-curves and deformation of ω. See [**39**, Section 4.2.2] or [**36**, Section 5].

The homologies of $\widehat{CF}(\Sigma, \boldsymbol{\alpha}, \boldsymbol{\beta}, z)$ and $CF^+(\Sigma, \boldsymbol{\alpha}, \boldsymbol{\beta}, z)$ are denoted by $\widehat{HF}(M)$ and $HF^+(M)$ respectively. The fact that they are well defined invariants of three-manifolds up to diffeomorphism was proved in [**39**] and reproved in [**36**] for the cylindrical reformulation we are using in this article. Naturality is addressed in [**28**].

The map $U\colon CF^+(\Sigma, \boldsymbol{\alpha}, \boldsymbol{\beta}, z) \to CF^+(\Sigma, \boldsymbol{\alpha}, \boldsymbol{\beta}, z)$ defined as

$$U([\mathbf{y}, i]) = \begin{cases} [\mathbf{y}, i-1] & \text{if } i \geq 1, \\ 0 & \text{if } i = 0 \end{cases}$$

is a chain map. The short exact sequence

$$0 \longrightarrow \widehat{CF}(\Sigma, \boldsymbol{\alpha}, \boldsymbol{\beta}, z) \longrightarrow CF^+(\Sigma, \boldsymbol{\alpha}, \boldsymbol{\beta}, z) \xrightarrow{U} CF^+(\Sigma, \boldsymbol{\alpha}, \boldsymbol{\beta}, z) \longrightarrow 0$$

induces the exact triangle

$$\begin{array}{ccc} HF^+(M) & \xrightarrow{\quad U \quad} & HF^+(M) \\ & \nwarrow \quad \swarrow & \\ & \widehat{HF}(M) & \end{array} \tag{3.1}$$

in homology. Heegaard Floer homology decomposes as a direct sum of groups indexed by Spin^c structures and the triangle (3.1) holds for every summand. For simplicity we will not discuss this decomposition.

3.2. A geometric interpretation of the U-map. This subsection is taken from [**7**, Section 3], but some of the arguments are slightly modified. In [**36, 39**], the U-map

$$U : CF^+(\Sigma, \boldsymbol{\alpha}, \boldsymbol{\beta}, z) \to CF^+(\Sigma, \boldsymbol{\alpha}, \boldsymbol{\beta}, z)$$

is defined algebraically as $U([\mathbf{y}, i]) = [\mathbf{y}, i-1]$, but we need to redefine it geometrically to be able to compare it with the U-map in ECH.

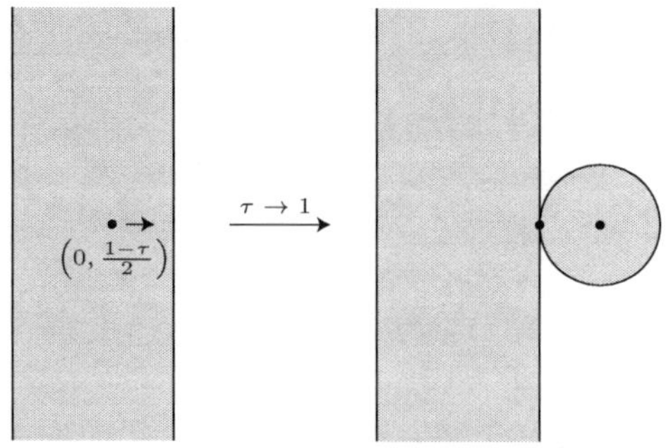

FIGURE 1. Moving the basepoint toward the boundary.

Let us fix[7] $\mathbf{z} = (0, \frac{1}{2}, z) \in X = \mathbb{R} \times [0,1] \times \Sigma$, and let $J^\Diamond$ be a generic small perturbation of J supported near $\mathbf{z}$ such that $\mathbb{R} \times [0,1] \times \{z\}$ remains holomorphic. This perturbation is needed so that $J^\Diamond$-holomorphic curves do not have a closed irreducible component passing through $\mathbf{z}$.

Given $\mathbf{y}_+, \mathbf{y}_- \in \mathcal{S}_{\boldsymbol{\alpha},\boldsymbol{\beta}}$ we denote by $\mathcal{M}_X(\mathbf{y}_+, \mathbf{y}_-; \mathbf{z})$ the moduli space of $J^\Diamond$-holomorphic curves in X with boundary on $L_{\boldsymbol{\alpha}} \cup L_{\boldsymbol{\beta}}$ which are positively asymptotic to $[0,1] \times \mathbf{y}_+$, negatively asymptotic to $[0,1] \times \mathbf{y}_-$ and pass through $\mathbf{z}$. Note that $n_z(u) \geq 1$ for $u \in \mathcal{M}_X(\mathbf{y}_+, \mathbf{y}_-; \mathbf{z})$.

DEFINITION 3.1. The *geometric U-map* with respect to the point $\mathbf{z}$ is the map

$$U_{\mathbf{z}}([\mathbf{y}_+, i]) = \sum_{\mathbf{y}_- \in \mathcal{S}_{\boldsymbol{\alpha},\boldsymbol{\beta}}} \sum_{j=1}^{i} \#\mathcal{M}_X^{I=2, n_z=j}(\mathbf{y}_+, \mathbf{y}_-; \mathbf{z})[\mathbf{y}_-, i-j].$$

Standard arguments in symplectic geometry based on the compactness and gluing results of [**36**] show that the geometric U-map

$$U_{\mathbf{z}} \colon CF^+(\Sigma, \boldsymbol{\alpha}, \boldsymbol{\beta}, z) \to CF^+(\Sigma, \boldsymbol{\alpha}, \boldsymbol{\beta}, z)$$

is a chain map. The main result of the subsection is the following.

THEOREM 3.2. *There exists a chain homotopy*

$$H : CF^+(\Sigma, \boldsymbol{\alpha}, \boldsymbol{\beta}, z) \to CF^+(\Sigma, \boldsymbol{\alpha}, \boldsymbol{\beta}, z)$$

such that

$$U_{\mathbf{z}} - U = H \circ \partial^+ + \partial^+ \circ H \tag{3.2}$$

and $H([\mathbf{y}, 0]) = 0$ *for all* $\mathbf{y} \in \mathcal{S}_{\boldsymbol{\alpha},\boldsymbol{\beta}}$.

PROOF. We will define a chain homotopy between $U_{\mathbf{z}}$ and U by moving $\mathbf{z}$ towards the boundary and identifying a count of holomorphic curves in the limit configuration with the map U. More precisely, for $\tau \in [0,1)$ we define $\mathbf{z}_\tau = \{0\} \times \{\frac{1-\tau}{2}\} \times \{z\}$, so that $\mathbf{z}_0 = \mathbf{z}$, and consider a generic smooth family of almost complex structures $J^\Diamond_\tau$, each obtained by a small perturbation of J supported in a neighbourhood of $\mathbf{z}_\tau$, such that $J^\Diamond_0 = J^\Diamond$ and $\mathbb{R} \times [0,1] \times \{z\}$ remains holomorphic for all τ.

Let $\mathcal{M}_X(\mathbf{y}_+, \mathbf{y}_-; \mathbf{z}_*)$ be the moduli space of $J^\Diamond_\tau$-holomorphic curves in X with boundary on $L_{\boldsymbol{\alpha}} \cup L_{\boldsymbol{\beta}}$ which are positively asymptotic to $[0,1] \times \mathbf{y}_+$, negatively

[7]What is called z and $\mathbf{z}$ here is called z^f and z respectively in [**7**].

asymptotic to $[0,1] \times \mathbf{y}_-$ and pass through $\mathbf{z}_\tau$ for some $\tau \in (0,1)$. The map H is defined as

$$H([\mathbf{y}_+, i]) = \sum_{\mathbf{y}_- \in \mathcal{S}_{\boldsymbol{\alpha},\boldsymbol{\beta}}} \sum_{j=1}^{i} \#\mathcal{M}_X^{I=1, n_z=j}(\mathbf{y}_+, \mathbf{y}_-; \mathbf{z}_*)[\mathbf{y}_-, i-j].$$

The structure of the boundary of the compactification of the moduli spaces $\mathcal{M}_X^{I=2}(\mathbf{y}_+, \mathbf{y}_-; \mathbf{z}_*)$ implies that H is a chain homotopy between $U_{\mathbf{z}}$ and a map defined by counting certain broken holomorphic curves in the limit $\tau \to 1$. To finish the proof we need to identify this map with U. This will be done in the rest of the section. □

In the limit $\tau \to 1$ we obtain $\widetilde{X} = \widetilde{B} \times \Sigma$ where $\widetilde{B} = (B \sqcup D)/\!\sim$ with $B = \mathbb{R} \times [0,1]$, $D = \{|x| \leq 1\} \subset \mathbb{C}$ and $\sim$ identifies $(0,0) \in B$ with $-1 \in D$. See Figure 1. For an appropriate choice of $J_\tau^\Diamond$, the limit almost complex structure is J on $B \times \Sigma$ and a small perturbation $J_D^\Diamond$ of a product almost complex structure J_D on $D \times \Sigma$ such that $J_D^\Diamond$ and J_D coincide outside of a small neighbourhood of $\mathbf{z}_1 = (0, z) \in D \times \Sigma$. From now on we write $\mathbf{z}_1 = \mathbf{z}$.

We define the ECH-type index I — in fact a relative version of Taubes's index from [**48**] in this case — for a homology class $A \in H_2(D \times \Sigma, \partial D \times \boldsymbol{\beta})$ which admits a representative F such that each component of ∂F is mapped to a distinct component of $\partial D \times \boldsymbol{\beta}$. Let (τ, τ') be the trivialisation of $T(D \times \Sigma)$ along $\partial D \times \boldsymbol{\beta}$ such that τ is induced by a radial and outward-pointing vector field along ∂D tangent to D, and τ' is induced by a vector field along $\boldsymbol{\beta}$ tangent to Σ and transverse to $\boldsymbol{\beta}$. Let $Q_{(\tau,\tau')}(A)$ be the intersection number between F and a push-off of F, where ∂F is pushed along τ'. We define[8]

$$I(A) = c_1(T(D \times \Sigma)|_A, (\tau, \tau')) + Q_{(\tau,\tau')}(A). \tag{3.3}$$

If u is a holomorphic curve in $D \times \Sigma$ with boundary on $\partial D \times \boldsymbol{\beta}$ representing a relative homology class A as above, then the index inequality gives

$$I(A) = \operatorname{ind}(u) + 2\delta(u). \tag{3.4}$$

In particular u is an embedding if and only if $\operatorname{ind}(u) = I(A)$. See [**5**, Theorem 4.5.13] for the proof of a similar result. Let $A_{k_0,k_1} = k_0[D \times \{pt\}] + k_1[\{pt\} \times \Sigma]$. If $k_0 \leq h$, where h is the genus of Σ, an easy computation gives

$$I(A_{k_0,k_1}) = k_0 + k_1(2 - 2h) + 2k_0k_1. \tag{3.5}$$

In particular, $I(A_{h,k_1}) = 2k_1 + h$.

We fix points[9] $y_i \in \beta_i$ for $i = 1, \ldots, h$ and denote by $\mathcal{M}_{D\times\Sigma}(A_{k_0,k_1}; \mathbf{z}, \mathbf{y})$ the moduli space of $J_D^\Diamond$-holomorphic curves in $D \times \Sigma$ with boundary on $\partial D \times \boldsymbol{\beta}$ representing the homology class A_{k_0,k_1} and passing through $(0, z)$ and k_0 points of the form $(-1, y_i)$.

REMARK 3.3. If $k_1 \leq 0$, a standard compactness argument shows that the moduli spaces $\mathcal{M}_{D\times\Sigma}(A_{k_0,k_1}; \mathbf{z}, \mathbf{y})$ are empty, provided that $J_D^\Diamond$ is close enough to a product almost complex structure.

[8]The formula in [**7**, Definition 3.2.1] has an extra term $\mu_{(\tau,\tau')}(\partial A)$, which vanishes here by our choice of trivialisation.

[9]These points are called w_i in [**7**].

LEMMA 3.4. *Let u_{τ_i}, for $\tau_i \to 1$ as $i \to \infty$, be a sequence of $J^{\Diamond}_{\tau_i}$-holomorphic curves in $\mathcal{M}^{I=2,n_z=j}_{X}(\mathbf{y}_+, \mathbf{y}_-; z_{\tau_i})$ for some fixed $\mathbf{y}_\pm \in \mathcal{S}_{\boldsymbol{\alpha},\boldsymbol{\beta}}$ and $j > 0$. Then, up to passing to a subsequence, u_{τ_i} converges to a pair of holomorphic curves (u_B, u_D) where u_B is a union of trivial strips $B \times \{\mathbf{y}\}$ in $B \times \Sigma$ over an intersection point $\mathbf{y} \in \mathcal{S}_{\boldsymbol{\alpha},\boldsymbol{\beta}}$ and $u_D \in \mathcal{M}_{D\times\Sigma}(A_{h,1}; \mathbf{z}, \mathbf{y})$. In particular $\mathbf{y}_+ = \mathbf{y}_- = \mathbf{y}$ and $j = 1$.*

PROOF. By Gromov compactness a subsequence of u_{τ_i} converges to a pair (u_B, u_D) where u_B is a J-holomorphic curve in $\mathcal{M}_X(\mathbf{y}_+, \mathbf{y}_-)$ and u_D is a $J^{\Diamond}_D$-holomorphic curve in $\mathcal{M}_{D\times\Sigma}(A_{h,k_1}; \mathbf{z}, \mathbf{y})$ for some $\mathbf{y} = (y_1, \dots, y_h)$. A simple computation shows that

$$I(u_B) + I(u_D) = 2 - h. \tag{3.6}$$

By Equation(3.5) we can rewrite Equation (3.6) as $I(u_B) + 2k_1 = 2$. Since $k_1 \geq 1$ by Remark 3.3 and $I(u_B) \geq 0$, this implies that $I(u_B) = 0$ and $k_1 = 1$; in particular u_B is a union of trivial strips and therefore $\mathbf{y}_\pm = \mathbf{y}$ and $j = 1$. □

The moduli spaces $\mathcal{M}_{D\times\Sigma}(A_{h,1}; \mathbf{z}, \mathbf{y})$ are zero-dimensional and consist of embedded $J^{\Diamond}_D$-holomorphic curves by Equations (3.4) and (3.5). We denote

$$G(h) = \#\mathcal{M}_{D\times\Sigma}(A_{h,1}; \mathbf{z}, \mathbf{y}).$$

Thus, by Lemma 3.4, H is a homotopy between $U_{\mathbf{z}}$ and $G(h)U$. In order to compute $G(h)$ we further degenerate $D \times \Sigma$. The first step is to degenerate D into $D \cup S^2$, where $0 \in D$ is identified with $\infty \in S^2 \cong \mathbb{C} \cup \{\infty\}$ and $\mathbf{z} = (0, z) \in S^2 \times \Sigma$. Let $J^{\Diamond}_S$ denote the limit almost complex structure on $S^2 \times \Sigma$, which we assume to be a small perturbation of a product almost complex structure J_S in a small neighbourhood of $\mathbf{z}$.

Holomorphic curves in $\mathcal{M}_{D\times\Sigma}(A_{h,1}; \mathbf{z}, \mathbf{y})$ degenerate into pairs of curves consisting in the trivial multisection $D \times \{\mathbf{y}\}$ in $D \times \Sigma$ and a $J^{\Diamond}_S$-holomorphic curve in $S^2 \times \Sigma$ representing the homology class $B_{h,1} = h[S^2] + [\Sigma]$ passing through $\mathbf{z} = (0, z)$ and $(\infty, y_1), \dots, (\infty, y_h)$. We denote the moduli space of such curves by $\mathcal{M}_{S^2\times\Sigma}(B_{h,1}; \mathbf{z}, \mathbf{y})$.

Holomorphic curves in $\mathcal{M}_{S^2\times\Sigma}(B_{h,1}; \mathbf{z}, \mathbf{y})$ can be reducible and only the irreducible component passing through $(0, z)$ has to be regular. We denote the subset of $\mathcal{M}_{S^2\times\Sigma}(B_{h,1}; \mathbf{z}, \mathbf{y})$ consisting of irreducible $J^{\Diamond}_S$-holomorphic curves by $\mathcal{M}^{irr}_{S^2\times\Sigma}(B_{h,1}; \mathbf{z}, \mathbf{y})$. Simple index considerations taking into account the homological and point constraints imply that the holomorphic curves in $\mathcal{M}_{S^2\times\Sigma}(B_{g,1}; \mathbf{z}, \mathbf{y}) \setminus \mathcal{M}^{irr}_{S^2\times\Sigma}(B_{g,1}; \mathbf{z}, \mathbf{y})$ consist of $\{\infty\} \times \Sigma$ with h spheres in the class $[S^2]$, one of which passes through $(0, z)$. However this irreducible configuration cannot appear in the limit of a sequence of curves in $\mathcal{M}_{D\times\Sigma}(A_{h,1}; \mathbf{z}, \mathbf{y})$ as D degenerates into $D \cup S^2$. This can be proved by a careful analysis of the limit in the SFT sense, but it can also be seen intuitively as follows: if $J^{\Diamond}_D$ is close enough to the product almost complex structure J_D, then the $J^{\Diamond}_D$-holomorphic curves in $\mathcal{M}_{D\times\Sigma}(A_{h,1}; \mathbf{z}, \mathbf{y})$ are C^0-close to $\{0\} \times \Sigma \cup D \times \{y_1\} \cup \ldots \cup D \times \{y_h\}$, and therefore the $J^{\Diamond}_S$-holomorphic component of the limit is C^0-close to $\{0\} \times \Sigma \cup S^2 \times \{y_1\} \cup \ldots \cup S^2 \times \{y_h\}$. Thus we have proved that

$$G(h) = \#\mathcal{M}^{irr}_{S^2\times\Sigma}(B_{h,1}; \mathbf{z}, \mathbf{y}).$$

The following lemma completes the proof of Theorem 3.2.

LEMMA 3.5. *$G(h) = 1$ for all $h \geq 1$.*

The proof will follow by combining the two lemmas below.

LEMMA 3.6. *For every $h \geq 1$ we have $G(h) = G(1)^h$.*

PROOF. In the proof we will degenerate Σ along $h-1$ separating curves in order to obtain a nodal surface $\widetilde{\Sigma}$ whose irreducible components are tori. We choose the curves so that each irreducible component contains exactly one of the points y_i. Since the basepoint $\mathbf{z}$ remains in one component, the almost complex structure on $S^2 \times \widetilde{\Sigma}$ is a product almost complex structure in all but one of the irreducible components of $S^2 \times \widetilde{\Sigma}$. Since a product almost complex structure is not generic enough, before degenerating Σ we need to modify $J_S^\diamondsuit$. To this aim we introduce a generic almost complex structure $J_S^\heartsuit$ on $D \times \Sigma$ among those making the projection $S^2 \times \Sigma \to \Sigma$ into a holomorphic map and keeping the section $\{\infty\} \times \Sigma$ holomorphic. By a standard continuation argument the cardinality of the set $\mathcal{M}^{irr}_{S^2\times\Sigma}(B_{h,1}; \mathbf{z}, \mathbf{y})$ is the same for $J_S^\diamondsuit$ and $J_S^\heartsuit$; from now on we will work with the latter almost complex structure.

As Σ degenerates towards $\widetilde{\Sigma}$, holomorphic curves in $\mathcal{M}^{irr}_{S^2\times\Sigma}(B_{1,h}; \mathbf{z}, \mathbf{y})$ degenerate into holomorphic curves in $S^2 \times \widetilde{\Sigma}$, with the same point constraints, which are irreducible in each irreducible component of $S^2 \times \widetilde{\Sigma}$, and such that two irreducible components meet at one point when two irreducible components of $S^2 \times \widetilde{\Sigma}$ meet. This shows that $G(h) = G(1)^h$. □

REMARK 3.7. The section $\{\infty\} \times \Sigma$ is not regular, and thus neither $J_S^\diamondsuit$ nor $J_S^\heartsuit$ are generic almost complex structures. What we are computing here is a simple instance of *relative Gromov-Witten invariant* in the sense of [**25**].

LEMMA 3.8. $G(1) = 1$.

PROOF. The lemma follows from [**38**, Example 8.6.12], but one can also argue more explicitly by degenerating $\Sigma = T^2$ into two spheres touching each other in two points, one of which contains z and the other one y_1. We call Σ_0 the component containing z and Σ_1 the component containing y_1. We also call 0 and ∞ the points where Σ_0 and Σ_1 meet.

As we degenerate Σ, the $J_S^\heartsuit$-holomorphic curves in $\mathcal{M}^{irr}_{S^2\times\Sigma}(B_{1,1}; \mathbf{z}, \mathbf{y})$ converge to pairs of holomorphic curves (u_0, u_1), where u_0 takes values in $S^2 \times \Sigma_0$ and passes through $(z, 0)$, and u_1 takes values in $S^2 \times \Sigma_1$ and passes through (y_1, ∞). Since $J_S^\heartsuit$ is close to a product almost complex structure, $J_S^\heartsuit$-holomorphic curves in $\mathcal{M}^{irr}_{S^2\times\Sigma}(B_{1,1}; \mathbf{z}, \mathbf{y})$ are C^0-close to $\{0\} \times \Sigma \cup S^2 \times \{y_1\}$. This implies that the curve u_0 represents the homology class $[S^2] + [\Sigma_0]$, while the curve u_1 represents the homology class $[\Sigma_0]$. Moreover the images of u_0 and u_1 match at $S^2 \times \{0, \infty\}$. The image of u_1 is a small perturbation of the graph of a degree zero holomorphic map $\Sigma_1 \to S^2$ and the image of u_0 is a small perturbation of the graph of a degree one holomorphic map $\Sigma_0 \to S^2$. Then elementary complex analysis implies that there is a unique choice for u_1, while the choice for u_0 becomes unique once the intersection of its image with $S^2 \times \{0, \infty\}$ is fixed. This implies that $G(1) = 1$. □

3.3. Adapting $\widehat{HF}$ to an open book decomposition. Let M be a closed 3-manifold, $B \subset M$ a link, S a compact, oriented, connected surface with nonempty boundary, and $\hbar\colon S \to S$ an orientation-preserving diffeomorphism such that $\hbar|_{\partial S} = id$.

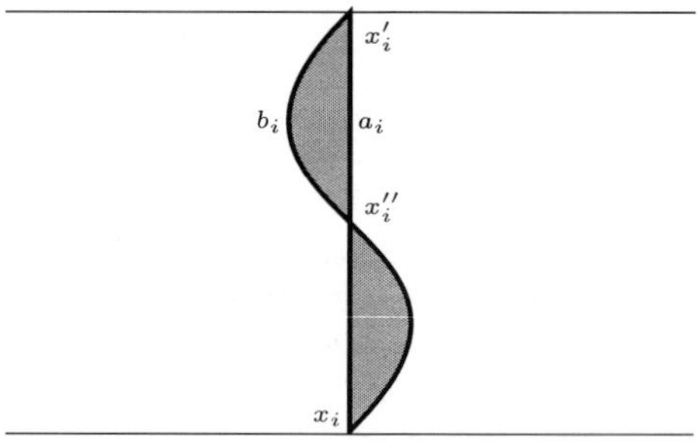

FIGURE 2. A portion of $-S_1$. The shaded regions are the "thin strips" D_i and D_i' which connect x_i'' to x_i or x_i'.

DEFINITION 3.9. An *open book decomposition* of M with *binding* B, *page* S and *monodromy* $\hslash$ is a locally trivial fibration $\pi\colon M\setminus B \to S^1$ with fibre $\mathrm{int}(S)$ and monodromy $\hslash|_{\mathrm{int}(S)}$ such that the closure of any fibre is a surface with boundary whose boundary is B. The pair $(S,\hslash)$ is called an *abstract open book decomposition*.

We will call a page not only the abstract surface S, but also all surfaces $S_t = \overline{\pi^{-1}(t)}$ for $t \in S^1$.

In this section we explain how to associate a pointed Heegaard diagram to an open book decomposition and compute $\widehat{HF}(-M)$ from the page and the monodromy, using a construction from [**16**]. From now on we assume that ∂S is connected and S has genus g.

We identify $S^1 \cong [0,2]/0 \sim 2$. An open book decomposition with page S and monodromy $\hslash$ gives rise to a Heegaard decomposition $M = H_{\boldsymbol{\alpha}} \cup H_{\boldsymbol{\beta}}$, where $H_{\boldsymbol{\alpha}} = \overline{\pi^{-1}([0,1])}$, $H_{\boldsymbol{\beta}} = \overline{\pi^{-1}([1,2])}$, and the Heegaard surface $\Sigma = S_1 \cup -S_0$ is the union of two pages glued along the binding.

A *basis of arcs* for S is a collection of $2g$ pairwise disjoint properly embedded arcs $\mathbf{a} = \{a_1,\dots,a_{2g}\}$ in S such that $S\setminus \mathbf{a}$ is a connected polygon. Starting from a basis $\mathbf{a}$ for S, we can construct α- and β-curves for Σ as follows:

$$\boldsymbol{\alpha} = (\mathbf{a}\times\{1\})\cup(\mathbf{a}\times\{0\}) \quad\text{and}\quad \boldsymbol{\beta} = (\mathbf{b}\times\{1\})\cup(\hslash(\mathbf{a})\times\{0\}).$$

Here $\mathbf{b}$ is a small deformation of $\mathbf{a}$ relative to its endpoints, so that each pair a_i and b_i intersects each other transversely at three points: two of the intersections are their endpoints x_i and x_i' on ∂S and the third intersection is an interior point x_i''; see Figure 2.[10] We assume that $\hslash$ is chosen so that the α- and β-curves are smooth and intersect transversely. These conditions can be easily achieved by an isotopy of $\hslash$ relative to the boundary.

All the intersection points of $\boldsymbol{\alpha}$ and $\boldsymbol{\beta}$ lie in S_0, with the exception of the points $x_i'' = x_i''\times\{1\}$. We place the basepoint z on S_1, away from the "thin strips" D_i and D_i', $i=1,\dots,2g$, given in Figure 2. The positioning of z prevents holomorphic curves involved in the differential for $\widehat{CF}(-\Sigma,\boldsymbol{\alpha},\boldsymbol{\beta},z)$ — besides the ones corresponding to the thin strips — from entering $-S_1$. Hence all of the nontrivial holomorphic curve information is concentrated on S_0. The homology of $\widehat{CF}(-\Sigma,\boldsymbol{\alpha},\boldsymbol{\beta},z)$ is isomorphic to $\widehat{HF}(-M)$ since we reversed the orientation of Σ.

[10][**5**, Figure 1] is flipped with rispect to Figure 2 because it shows S_1 instead of $-S_1$.

Let $\mathcal{S}_{\mathbf{a},\hbar(\mathbf{a})} \subset \mathcal{S}_{\boldsymbol{\alpha},\boldsymbol{\beta}}$ consist of the $2g$-tuples of intersection points $\mathbf{y}$ all of whose components are in S_0. We define[11] $\overline{CF}(S, \mathbf{a}, \hbar(\mathbf{a}))$ as the chain complex generated by $\mathcal{S}_{\mathbf{a},\hbar(\mathbf{a})}$ and whose differential counts HF-curves in $\mathbb{R} \times [0,1] \times S$. Then we define $\widehat{CF}(S, \mathbf{a}, \hbar(\mathbf{a}))$ as the quotient of $\overline{CF}(S, \mathbf{a}, \hbar(\mathbf{a}))$ modulo the identifications $\{x_i\} \cup \mathbf{y}_0 \sim \{x_i'\} \cup \mathbf{y}_0$ for all $(2g-1)$-tuples of chords $\mathbf{y}_0$. The quotient inherits a differential because no holomorphic curve in $\mathbb{R} \times [0,1] \times S$ can have a positive end at one of the chords over x_i, x_i' unless it has an irreducible component which is a trivial strip over that chord by [**5**, Claim 4.9.2], and therefore the elements of the form $\{x_i\} \cup \mathbf{y}_0 + \{x_i'\} \cup \mathbf{y}_0$ generate a subcomplex.

The identifications above are the algebraic consequence of the thin strips D_i and D_i'. The following proposition was proved in [**5**, Theorem 4.9.4]:

PROPOSITION 3.10. $\widehat{HF}(S, \mathbf{a}, \hbar(\mathbf{a}), z) \cong \widehat{HF}(-\Sigma, \boldsymbol{\alpha}, \boldsymbol{\beta}, z)$.[12]

REMARK 3.11. $\mathbf{x} = \{x_1, \dots, x_{2g}\}$ is a cycle and its homology class $[\mathbf{x}] \in \widehat{HF}(-M)$ is the contact invariant of the contact structure supported by the open book decomposition; see [**16**].

4. Embedded contact homology

4.1. A review of embedded contact homology. In this section we briefly review the embedded contact homology groups associated to a closed 3-manifold M. Embedded contact homology was defined by Hutchings [**18**], partially in collaboration with Taubes [**20, 21**], and is intimately connected with the dynamics of a Reeb vector field.

Let α be a contact form on M. We assume that every closed Reeb orbit of α is *nondegenerate*, which means that its linearised first return map does not have 1 as an eigenvalue. Contact forms satisfying this property are called *nondegenerate*. The linearised first return map is a symplectic transformation, and therefore its eigenvalues are $\{\lambda, \lambda^{-1}\}$, where λ is either real or in the unit circle. A closed Reeb orbit is:

- *hyperbolic* if the eigenvalues of its linearised first return map are real, or
- *elliptic* if they lie on the unit circle.

The chain complex $ECC(M, \alpha)$ is generated by finite sets $\boldsymbol{\gamma} = \{(\gamma_i, m_i)\}$, called *orbit sets*, where:

- γ_i is a simple closed Reeb orbit,
- m_i is a positive integer, and
- if γ_i is a hyperbolic orbit, then $m_i = 1$.

We will denote the set of orbit sets by $\mathcal{O}$. An orbit set $\boldsymbol{\gamma}$ will also be written multiplicatively as $\prod \gamma_i^{m_i}$, with the convention that $\gamma_i^2 = 0$ whenever γ_i is hyperbolic. The empty orbit set $\varnothing$ will be written multiplicatively as 1.

We choose an almost complex structure J on $\mathbb{R} \times M$ compatible with α. Let $\boldsymbol{\gamma}_+ = \{(\gamma_i^+, m_i^+)\}$ and $\boldsymbol{\gamma}_- = \{(\gamma_i^-, m_i^-)\}$ be orbit sets. We denote by $\mathcal{M}_{\mathbb{R}\times M}(\boldsymbol{\gamma}_+, \boldsymbol{\gamma}_-)$ the moduli space of J-holomorphic curves in $\mathbb{R} \times M$ which are positively asymptotic to covers of the closed Reeb orbits γ_i^+ with total multiplicity m_i^+ as $s \to +\infty$ and negatively asymptotic to covers of the closed Reeb orbits γ_i^- with total multiplicity

[11] The same chain complex is called $\widehat{CF}'(S, \mathbf{a}, \hbar(\mathbf{a}))$ in [**5**].

[12] We consider $\widehat{HF}(-\Sigma, \boldsymbol{\alpha}, \boldsymbol{\beta}, z)$ here and $\widehat{HF}(\Sigma, \boldsymbol{\beta}, \boldsymbol{\alpha}, z)$ in [**5**]. The two groups are isomorphic.

m_i^- as $s \to -\infty$. Moreover, we consider equivalent two J-holomorphic curves which differ only by (branched) covers of trivial cylinders with the same multiplicity. Such curves are called *connectors*. Translations in the $\mathbb{R}$ direction acts on the moduli spaces; we denote the quotient by $\widehat{\mathcal{M}}_{\mathbb{R}\times M}(\gamma_+, \gamma_-)$.

The ECH index for J-holomorphic curve in $\mathbb{R} \times M$ is defined in [**18**, Definition 1.5]. The following lemma is a consequence of the index inequality proved in [**19**, Theorem 4.15].

LEMMA 4.1 ([**20**, Proposition 7.15]). *Let J be a generic almost complex structure on $\mathbb{R} \times M$ compatible with α. Then:*

(1) *A J-holomorphic curve u with $I(u) = 0$ is a union of branched covers of trivial cylinders over simple closed Reeb orbits (i.e. a connector).*
(2) *A J-holomorphic curve u with $I(u) = 1$ (respectively $I(u) = 2$) is a disjoint union of a connector and an embedded J-holomorphic curve u' with $I(u') = \operatorname{ind}(u') = 1$ (respectively $I(u') = \operatorname{ind}(u') = 2$).*

The ends of a J-holomorphic curve u in $\mathcal{M}_{\mathbb{R}\times M}(\gamma_+, \gamma_-)$ without connector components determine partitions of the multiplicities of the elliptic orbits in γ_+ and γ_-. It turns out that, when $I(u) = 1$ or $I(u) = 2$, these partitions must coincide with preferred partitions called the *outgoing* and *incoming* partitions for positive and negative ends, respectively. The incoming and outgoing partitions can be computed from the dynamics of the linearised Reeb flow. For their definition see [**18**, Section 4.1] or [**19**, Definition 4.14]. For the relation between these partitions and the ECH index see [**19**, Theorem 4.15], for example. In this article we will not need the precise definition of those partitions, except for the following fact, which is a direct consequence of [**19**, Definition 4.14].

LEMMA 4.2. *Let γ be a simple elliptic orbit and suppose that its linearised Reeb flow is conjugated to a rotation by an angle $2\pi\theta$. If $0 < \theta < \frac{1}{m}$, then the incoming partition of (γ, m) is (m) and the outgoing partition is $(\underbrace{1, \dots, 1}_{m})$. On the other hand, if $-\frac{1}{m} < \theta < 0$, then the incoming partition of (γ, m) is $(\underbrace{1, \dots, 1}_{m})$ and the outgoing partition is (m).*

The differential on $ECC(M, \alpha)$ is defined as

$$\partial \gamma_+ = \sum_{\gamma_- \in \mathcal{O}} \#\widehat{\mathcal{M}}^{I=1}_{\mathbb{R}\times M}(\gamma_+, \gamma_-)\gamma_-.$$

The map ∂ was shown to satisfy $\partial^2 = 0$ by Hutchings and Taubes in [**20, 21**]. The homology of $ECC(M, \alpha)$ is the *embedded contact homology* group $ECH(M)$. Independence of the choice of the contact form α, the contact structure ξ, and the compatible almost complex structure J was shown by Taubes in [**43–47**] by proving an isomorphism between embedded contact homology and monopole Floer homology.

REMARK 4.3. The moduli spaces $\mathcal{M}(\gamma_+, \gamma_-)$ can be nonempty only if γ_+ and γ_- define the same homology class. Thus $ECC(M, \alpha)$ decomposes as a direct sum of subcomplexes indexed by classes in $H_1(M; \mathbb{Z})$. This decomposition is analogous to the decomposition of the Heegaard Floer (or monopole) chain complexes according to Spin^c-structures. The induced decomposition in $ECH(M)$ is independent of all choices, except for the indexing by elements in $H_1(M;\mathbb{Z})$ which depends on the contact structure only through its Euler class.

Given a point $\mathbf{z} \in \mathbb{R} \times M$, we denote by $\mathcal{M}(\gamma_+, \gamma_-; \mathbf{z})$ the moduli space of J-holomorphic curves asymptotic to $\gamma_\pm$ and passing through $\mathbf{z}$ where, as before, we identify curves which differ by a connector component. For a generic point $\mathbf{z}$, the map[13] $U_{\mathbf{z}}\colon ECC(M, \alpha) \to ECC(M, \alpha)$ is defined as

$$U_{\mathbf{z}}\gamma_+ = \sum_{\gamma_- \in \mathcal{O}} \#\mathcal{M}^{I=2}_{\mathbb{R}\times M}(\gamma_+, \gamma_-; \mathbf{z})\gamma_-.$$

The same techniques used to show that $\partial^2 = 0$ also show that U is a chain map; see [**24**, Section 2.5] for more details. Then $\widehat{ECC}(M, \alpha)$ is defined as the cone of $U_{\mathbf{z}}$ and $\widehat{ECH}(M)$ is its homology. As before, the work of Taubes in [**43**] and the following articles shows that $\widehat{ECH}(M)$ is invariant of all choices.

4.2. Morse-Bott theory for embedded contact homology. In several steps of the proof of Theorem 1.1 we will use Morse-Bott techniques extensively. Here we give a brief description of those techniques and refer the reader to [**2**] and [**3**] for a general treatment and to [**4**, Section 4] for one which is more adapted to embedded contact homology. For the purpose of this paper, a contact form is *Morse-Bott* if every closed orbit of its Reeb vector field is either isolated and nondegenerate, or belongs to an S^1-family and is nondegenerate in the normal direction.[14] We denote a Morse-Bott S^1-family of simple closed Reeb orbits by $\mathcal{N}$ and the Morse-Bott torus corresponding to $\mathcal{N}$ by $T_{\mathcal{N}} = \cup_{x \in \mathcal{N}} x$.

Let $\{v_1, v_2\}$ be an oriented basis for ξ at some point $p \in T_{\mathcal{N}}$ so that v_1 is transverse to $T_{\mathcal{N}}$ and v_2 is tangent to $T_{\mathcal{N}}$. The linearised first return map of the Reeb flow on ξ_p is given, in the basis $\{v_1, v_2\}$, by the matrix $\begin{pmatrix} 1 & 0 \\ a & 1 \end{pmatrix}$ with $a \neq 0$.

DEFINITION 4.4. $T_{\mathcal{N}}$ is called a *positive* Morse-Bott torus if $a > 0$ and a *negative* Morse-Bott torus if $a < 0$.

A Morse-Bott contact form α can be perturbed into nondegenerate forms α_ϵ, for $\epsilon > 0$ sufficiently small, which depend on the choice of a Morse function on each Morse-Bott S^1-family. The close Reeb orbits of the forms α_ϵ will be the nondegenerate closed Reeb orbits of α together with the closed Reeb orbits in the Morse-Bott tori corresponding to the critical points of the Morse functions on the Morse-Bott families. We will sweep under the carpet the actual construction of the contact forms α_ϵ and the fact that they are nondegenerate only up to orbits of some action (i.e. period) L_ϵ, with $L_\epsilon \to +\infty$ as $\epsilon \to 0$, so that a limiting procedure is involved in computing $ECH(M)$ from a Morse-Bott contact form.

If we choose a family of almost complex structures J_ϵ compatible with α_ϵ and converging to an almost complex structure J compatible with α as $\epsilon \to 0$, a sequence of J_ϵ-holomorphic curves converges to a "cascade" of J-holomorphic curves some of whose ends are connected to negative gradient trajectories in the Morse-Bott families. The negative gradient trajectories which can appear are of three types:

- flow-lines between critical points,

[13]This map is called U in [**4**] and U' in [**7**].

[14]In general, there is also the case where the Reeb orbits come in two-dimensional families; however this will not occur here.

- semi-infinite trajectories between a critical point and an end of a holomorphic piece of the cascade, and
- finite trajectories joining a positive and a negative end of different holomorphic pieces.

See [**3**, Section 11.2] for a precise statement.

The converse is more delicate: a cascade of J-holomorphic curves whose ends are joined by negative gradient flow trajectories in general cannot be deformed to a J_ϵ-holomorphic curve without using some abstract perturbation theory which has not been rigorously established yet. The main problem is that a family of simply covered J_ϵ-holomorphic curves — which are thus regular for a generic J_ϵ — could degenerate into a cascade in which some of the holomorphic curves are multiply covered and could even have negative Fredholm index.

We will therefore restrict our attention to a special class of Morse-Bott cascades in which finite negative gradient trajectories are not allowed and at most one of the holomorphic pieces is not a cover of a trivial cylinder. Those cascades are called *very nice Morse-Bott buildings* in [**4**]. Let $\gamma_\pm = \{(\gamma_i^\pm, m_i^\pm)\}$ be orbit sets where each $\gamma_i^\pm$ is either a nondegenerate closed Reeb orbit of α or a closed Reeb orbit corresponding to a critical point of the Morse function on a Morse-Bott family. We denote by[15] $\mathcal{M}^{MB}_{\mathbb{R}\times M}(\gamma_+,\gamma_-)$ the moduli space of very nice Morse-Bott buildings in $\mathbb{R}\times M$ with positive ends at γ_+ and negative ends at γ_-. We denote by $\widehat{\mathcal{M}}^{MB}_{\mathbb{R}\times M}(\gamma_+,\gamma_-)$ the quotient by translations and by $\mathcal{M}^{MB}_{\mathbb{R}\times M}(\gamma_+,\gamma_-;\mathbf{z})$ the subspace of those cascades passing through a generic point $\mathbf{z}\in\mathbb{R}\times M$. In the Appendix to [**4**] we prove the following.

THEOREM 4.5. *For a generic almost complex structure J compatible with α and ϵ sufficiently small there are bijections*

$$\widehat{\mathcal{M}}^{MB,I=1}_{\mathbb{R}\times M}(\gamma_+,\gamma_-)\cong\widehat{\mathcal{M}}^{I=1}_{\mathbb{R}\times M}(\gamma_+,\gamma_-)\quad\textit{and}$$
$$\mathcal{M}^{MB,I=2}_{\mathbb{R}\times M}(\gamma_+,\gamma_-;\mathbf{z})\cong\mathcal{M}^{I=2}_{\mathbb{R}\times M}(\gamma_+,\gamma_-;\mathbf{z})$$

where the moduli spaces on the left-hand side are defined using the almost complex structure J and the moduli spaces on the right-hand side are defined using a perturbed almost complex structure J_ϵ.

4.3. Embedded contact homology of manifolds with torus boundary. In this subsection we define several flavours of embedded contact homology for manifolds with torus boundary; see [**4**, Section 7] for more details. Let N be a three-manifold with $\partial N\cong T^2$ and let α be a contact form on N such that ∂N is foliated by Reeb orbits. We assume that the foliation is linear for some choice of coordinates in ∂N. Then, we say that α is *rational* if the Reeb orbits on ∂N are closed, and *irrational* if they are dense. If α is rational, we assume that ∂N is a Morse-Bott torus. We choose a Morse function with unique maximum and minimum on the corresponding Morse-Bott S^1-family. After perturbing α using this Morse function, the Morse-Bott family of orbits foliating ∂N is replaced by a pair of orbits: one elliptic called e and one hyperbolic called h. To perform the perturbation it is convenient to enlarge N slightly; we will ignore this technical point. If ∂N is a positive Morse-Bott torus, then e comes from the maximum and

[15]In [**4**] the same moduli spaces are denoted by $\mathcal{M}^{MB,vn}(\gamma_+,\gamma_-)$ because they consist of very nice Morse-Bott buildings.

h from the minimum, while if ∂N is negative, then h comes from the maximum and e from the minimum. In the following definitions we will assume that ∂N is a negative Morse-Bott torus when α is rational. The similar definitions in the positive case are left to the reader.

DEFINITION 4.6. The chain complex $ECC(\text{int}(N), \alpha)$ is generated by orbit sets contained in the interior of N and the differential is defined from J-holomorphic curves in $\mathbb{R} \times \text{int}(N)$.

$ECC(\text{int}(N), \alpha)$ is a chain complex because the foliation by Reeb orbits in ∂N prevents J-holomorphic curves with asymptotics in $\text{int}(N)$ from touching ∂N by positivity of intersection. (Alternatively, we can say that $\mathbb{R} \times \partial N$ is a Levi-flat[16] surface.)

DEFINITION 4.7. The chain complex $ECC(N, \alpha)$ coincides with $ECC(\text{int}(N), \alpha)$ if α is irrational, and is generated by orbit sets built from closed Reeb orbits in $\text{int}(N)$ plus e and h on ∂N. The differential counts very nice Morse-Bott buildings contained in $\mathbb{R} \times N$.

The proof that $ECC(N, \alpha)$ is a chain complex when α is rational consists in showing that a family of very nice Morse-Bott buildings in $\mathbb{R} \times N$ cannot break into a non very nice Morse-Bott building; see [**4**, Lemma 7.12]. This is a consequence of the fact that, by the trapping lemma [**4**, Lemma 5.3.2], positive ends of J-holomorphic curves in $\mathbb{R} \times N$ asymptotic to the closed Reeb orbits foliating ∂N must be trivial.

DEFINITION 4.8. The chain complex $ECC^\flat(N, \alpha)$ is generated by orbit sets built from closed Reeb orbits in $\text{int}(N)$ plus e and the chain complex $ECC^\sharp(N, \alpha)$ is generated by orbit sets built from closed Reeb orbits in $\text{int}(N)$ plus h.

Both $ECC^\flat(N, \alpha)$ and $ECC^\sharp(N, \alpha)$ are subcomplexes of $ECC(N, \alpha)$ because e and h can appear only at the negative end of a very nice Morse-Bott building, except for connectors and two negative gradient flow trajectories from h to e in their Morse-Bott family.

Invariance was addressed only for $ECH(N, \alpha)$ when α is an irrational contact form; for the other flavours it will not be needed.

PROPOSITION 4.9 ([**4**, Proposition 7.2.1]). *Let α_1 and α_2 be contact forms on N which agree on ∂N to first order (and in particular the Reeb vector fields and the characteristic foliations of α_1 and α_2 on ∂N are equal) and define contact structures $\xi_i = \ker \alpha_i$ which are isotopic relative to the boundary. If ∂N is foliated by Reeb orbits of irrational slope, then there is an isomorphism*

$$ECH(N, \alpha_1) \cong ECH(N, \alpha_2).$$

The strategy of the proof is to extend (N, α_i), $i = 1, 2$, to closed contact manifolds so that the closed Reeb orbits not contained in N have much larger action than those contained in N, and to use the action properties of the continuation maps for embedded contact homology of closed three-manifolds.

When ∂N is a negative Morse-Bott torus, we define two further versions of embedded contact homology for (N, α) which are, in some sense, embedded contact

[16]Named after Eugenio Elia Levi, an Italian mathematician who was killed in action during WWI. His brother Beppo Levi, also a mathematician, was forced into exile by the fascist racial laws of 1938.

homology groups relative to the boundary. We recall that $ECC(N,\alpha)$ can be seen as a commutative algebra generated by simple closed Reeb orbits with the relation that $\gamma^2=0$ if γ is hyperbolic. Let $\langle e-1\rangle$ be the ideal generated by $e-1$. Even if the differential does not respect the multiplicative structure, this ideal is a subcomplex because only connectors can have e at a positive end.

DEFINITION 4.10. We define

$$ECC(N,\partial N,\alpha)=ECC^\flat(N,\alpha)/\langle e-1\rangle \quad \text{and}$$
$$\widehat{ECC}(N,\partial N,\alpha)=ECC(N,\alpha)/\langle e-1\rangle.$$

We denote the differential in $ECC(N,\partial N,\alpha)$ by ∂_{rel} and in $\widehat{ECC}(N,\partial N,\alpha)$ by $\widehat{\partial}_{rel}$. Factoring out h, we can decompose $\widehat{\partial}_{rel}$ as follows. If γ is an orbit set not containing h, we can write

$$\widehat{\partial}_{rel}(\gamma)=\partial_{rel}(\gamma)+hU_{rel}(\gamma)$$
$$\widehat{\partial}_{rel}(h\gamma)=h\widehat{\partial}_{rel}(\gamma).$$

From $(\widehat{\partial}_{rel})^2=0$ we deduce that the map

$$U_{rel}\colon ECC(N,\partial N,\alpha)\to ECC(N,\partial N,\alpha)$$

is a chain map. Its name is motivated by the fact that it plays the role of the map U in embedded contact homology relative to the boundary.

Given $a\in H_1(N;\mathbb{Z})$, let $ECC_a(N,\alpha)$ be the subcomplex of $ECC(N,\alpha)$ generated by orbit sets in the homology class a. The same notation is used for the sharp ($\sharp$) and flat ($\flat$) flavours. From the fact that only connectors can have e at a positive end, it is easy to see that the map

$$ECC_a(N,\alpha)\to ECC_{a+[e]}(N,\alpha)$$
$$\gamma\mapsto e\gamma$$

is a chain map.

If $\overline{a}\in H_1(N;\mathbb{Z})/\langle[e]\rangle$, let $ECC_{\overline{a}}(N,\partial N,\alpha)$ be the subcomplex of $ECC(N,\partial N,\alpha)$ generated by orbit sets in the class $\overline{a}$. Note that $H_1(N;\mathbb{Z})/\langle[e]\rangle$ is the first homology group of the closed manifold obtained by Dehn filling N along the slope of e. The following lemma is almost immediate.

LEMMA 4.11. *Suppose $[e]\neq 0$ in $H_1(N;\mathbb{Z})$. If $a\in H_1(N;\mathbb{Z})$ and $\overline{a}$ is its image in $H_1(N;\mathbb{Z})/\langle[e]\rangle$, then*

$$ECH_{\overline{a}}(N,\partial N,\alpha)=\varinjlim\left\{ECH^\flat_{a+j[e]}(N,\alpha)\right\}_{j\in\mathbb{N}}$$
$$\widehat{ECH}_{\overline{a}}(N,\partial N,\alpha)=\varinjlim\left\{ECH_{a+j[e]}(N,\alpha)\right\}_{j\in\mathbb{N}}.$$

The relation between embedded contact homology of N relative to the boundary and embedded contact homology of the Dehn filling of N along the slope of e is the main result of [**4**]. The following theorem is a generalisation of [**4**, Theorem 1.1.1], expressed in a slightly different language.

THEOREM 4.12. *Let N be a three-manifold with torus boundary and let α be a contact form on N such that ∂N is a negative Morse-Bott torus for the Reeb vector field. Let moreover M be the Dehn filling of N with respect of the slope of e. If* *there is a cohomology class $\varphi \in H^1(N;\mathbb{Z})$ such that $\varphi([\gamma]) \geq 0$ for all closed Reeb orbits γ and $\varphi([e]) > 0$, then there exist isomorphisms*

$$\sigma_*\colon ECH(N, \partial N, \alpha) \to ECH(M),$$
$$\widehat{\sigma}_*\colon \widehat{ECH}(N, \partial N, \alpha) \to \widehat{ECH}(M)$$

such that the diagram

$$\begin{array}{ccccccccc}
\cdots \longrightarrow & \widehat{ECH}(N, \partial N) & \longrightarrow & ECH(N, \partial N) & \xrightarrow{U_{rel}} & ECH(N, \partial N) & \longrightarrow \cdots \\
& \downarrow \widehat{\sigma}_* & & \downarrow \sigma_* & & \downarrow \sigma_* & \\
\cdots \longrightarrow & \widehat{ECH}(M) & \longrightarrow & ECH(M) & \xrightarrow{U} & ECH(M) & \longrightarrow \cdots
\end{array}$$

commutes. Moreover, the maps σ_* and $\widehat{\sigma}$ are compatible with the direct sum decompositions according to homology classes.

The proof of Theorem 4.12 will be sketched in the next subsection. A heuristic reason why it is expected to hold is the following. Let α_δ, $0 < \delta \ll 1$, be a family of contact forms on $M = N \cup V$ such that, on the solid torus $V \cong D^2 \times (\mathbb{R}/\mathbb{Z})$ with cylindrical coordinates (r, θ, z), their Reeb vector fields R_δ are tangent to the concentric tori $\{r = const\}$ and have constant slope $\frac{1}{\delta}$ (away from the core). As we send $\delta \to 0$, the Conley-Zehnder index of the core goes to $+\infty$ and we should therefore be able to ignore it. At the same time, one expects that $I_{ECH} = 1$ holomorphic curves that cross the core when $\delta > 0$ are in one-to-one correspondence with holomorphic curves which have e at the negative end when $\delta = 0$. This is the reason for identifying $e = 1$. Similarly, if we place the marked point z on the core and h correspondingly, then $I_{ECH} = 2$ holomorphic curves that pass through z should be in one-to-one correspondence with $I_{ECH} = 1$ holomorphic curves which have h at a negative end when $\delta = 0$.

4.4. How to prove Theorem 4.12.

4.4.1. *A special contact form on M.* Let α be a contact form on N which is nondegenerate in $\mathrm{int}(N)$ and has a negative Morse-Bott torus of closed Reeb orbits at ∂N. To keep the discussion closer to [**4**] we assume that there is a properly embedded surface $S \subset N$ such that $[\partial S] \cdot [e] = 1$ and the Reeb vector field of α is positively transverse to S. This ensures the existence of a cohomology class $\varphi \in H^1(N;\mathbb{Z})$ such that $\varphi(\gamma) \geq 0$ and $\varphi([e]) = 1$.

We decompose $M = N \cup (T^2 \times [1, 2]) \cup V$ where V is a solid torus whose meridian is attached to the slope of e. We identify $T^2 \times \{1\}$ with ∂N and $T^2 \times \{2\}$ with ∂V. The region $T^2 \times (1, 2)$ will be called the *no man's land.*[17] We choose a contact form α_M on M such that:

(1) α_M restricts to α on N, so that, in particular, ∂N is a negative Morse-Bott torus,
(2) the closed Reeb orbits in the no man's land have arbitrarily large action and foliate the tori $T^2 \times \{c\}$ for $c \in (1, 2)$,
(3) $\alpha_V = \alpha_M|_V$ is nondegenerate in $\mathrm{int}(V)$ and ∂V is a positive Morse-Bott torus,
(4) every closed Reeb orbit in $\mathrm{int}(V)$ is transverse to a meridional disc of V,

[17] The name was suggested to us by a visit to the remains of Berlin's wall.

(5) there is an exhaustion of concentric solid tori $V_0 \subset V_1 \subset \ldots \subset \text{int}(V)$ such that ∂V_i is linearly foliated by Reeb orbits of irrational slope r_i with $r_i \to +\infty$ as $i \to +\infty$ (for some choice of coordinates in which e has slope $+\infty$) and all closed Reeb orbits in $V \setminus V_i$ have slope larger that r_i.

On the Morse-Bott S^1-family of closed Reeb orbits corresponding to ∂V we choose a Morse function with a unique minimum and a unique maximum and denote by e' the (elliptic) orbit corresponding to the maximum and h' the (hyperbolic) orbit corresponding to the minimum.

Condition (2) on α_M implies that the closed Reeb orbits in the no man's land can effectively be ignored in computing $ECH(M)$. In order to prove this rigorously we need a direct limit argument, and therefore we need a family of contact forms α_M for which the closed Reeb orbits in the no man's land have larger and larger action. This direct limit is somewhat involved because it must be repeated at each step of the proof of Theorem 4.12, and therefore we prefer to ignore it. For this reason, from now on, every statement in this section will hold after an unexpressed direct limit. See [**4**, Sections 9.2 and 9.3] for the detailed construction of the contact forms.

After ignoring the closed Reeb orbits in the no man's land, $ECC(M, \alpha_M)$ becomes isomorphic to $ECC(V, \alpha_V) \otimes ECC(N, \alpha)$ with a differential ∂_M which is, however, not the usual differential of a tensor product of complexes.

4.4.2. *A filtration.* The next step is to introduce a filtration[18] $\mathcal{F}$ on the chain complex $(ECC(V, \alpha_V) \otimes ECC(N, \alpha_N), \partial_M)$ to simplify the differential. To construct this filtration, we choose the generator $\eta \in H^1(V; \mathbb{Z})$ which evaluates positively on the closed Reeb orbits in $\text{int}(V)$ and define $\mathcal{F}^p$ as the subspace generated by orbit sets $\boldsymbol{\gamma} \otimes \boldsymbol{\Gamma}$ with $\eta([\boldsymbol{\gamma}]) \leq p$. Note that $\mathcal{F}^p = \{0\}$ for $p < 0$ and $\mathcal{F}^0$ is generated by orbit sets of the form $\boldsymbol{\gamma} \otimes \boldsymbol{\Gamma}$ where $\boldsymbol{\gamma}$ contains only orbits in ∂V.

LEMMA 4.13 ([**4**, Corollary 9.4.2]). *The differential ∂_M preserves the vector spaces $\mathcal{F}^p$ for all p.*

SKETCH OF PROOF. Let u be an embedded holomorphic curve in $\mathbb{R} \times M$ between $\boldsymbol{\gamma}_+ \otimes \boldsymbol{\Gamma}_+$ and $\boldsymbol{\gamma}_- \otimes \boldsymbol{\Gamma}_-$. We denote by $[u] \in H_2(M, \bigcup \gamma_i^+ \cup \bigcup \gamma_j^-)$ the relative homology class determined by the projection of u to M. If $\boldsymbol{\gamma}_+$ and $\boldsymbol{\gamma}_-$ do not contain orbits in ∂V, then it is easy to see that

$$\eta([\boldsymbol{\gamma}_+]) - \eta([\boldsymbol{\gamma}_-]) = [u] \cdot [e'] \geq 0$$

by positivity of the intersections between the image of u and the holomorphic cylinder $\mathbb{R} \times e'$. If $\boldsymbol{\gamma}_+$ or $\boldsymbol{\gamma}_-$ contain orbits in ∂V the proof is more subtle because one has to use positivity of intersection with the nonclosed Reeb orbits on the tori ∂V_i. □

The next step is to describe the differential induced in the graded complex or, in fancy words, in the zero page of the spectral sequence associated to the filtration. For that we use Morse-Bott techniques, which are justified by the following lemma.

LEMMA 4.14 (See [**4**, Corollary 9.5.2]). *A Morse-Bott building u between $\boldsymbol{\gamma}_+ \otimes \boldsymbol{\Gamma}_+$ and $\boldsymbol{\gamma}_- \otimes \boldsymbol{\Gamma}_-$ with $I(u) = 1$ and $\eta(\boldsymbol{\gamma}_+) = \eta(\boldsymbol{\gamma}_-)$ is very nice. Moreover its projection to M is contained either in N or in V or in the no man's land.*

[18] The idea of this filtration was suggested to us by Michael Hutchings.

PROOF. If $I(u) = 1$ and u is not very nice, it must have two irreducible components joined by a Morse trajectory of finite length: the condition $I(u) = 1$ rules out the possibility that u is not very nice because of two irreducible components which are not connectors and are not linked by a Morse trajectory, since those two components could be translated independently in the $\mathbb{R}$-direction. Then u must have either a positive end at an orbit of ∂N or a negative end at an orbit of ∂V. In either case, positivity of intersection with nearby Reeb orbits implies that u approaches that orbit with a framing different from the framing induced by the Morse-Bott torus, and therefore a topological argument forces $\eta(\gamma_+) > \eta(\gamma_-)$. For the details see [**4**, Lemma 9.5.1 and Corollary 9.5.2]. The last claim holds because the tori in the no man's land foliated by Reeb orbits form a barrier for the holomorphic curve by the blocking lemma [**4**, Lemma 5.2.3], which is a consequence of positivity of intersection between the projection of u to M and the Reeb vector field. For the details of this last argument see the proof of [**4**, Lemma 9.5.3]. □

The holomorphic curves in the symplectisation of the no man's land can be described using a finite energy foliation first constructed by Wendl in [**52**]; see [**4**, Section 8.4]: there is a foliation of $\mathbb{R} \times T^2 \times [1,2]$ by a cylinders with positive end at the closed Reeb orbits in ∂V and negative end at the closed Reeb orbits in ∂N. After a Morse-Bott perturbation, this foliation gives two $I = 1$ holomorphic curves in $\mathbb{R} \times T^2 \times [1,2]$: one cylinder from e' to h and one cylinder from h' to e. Moreover, the foliation obstructs the existence of any other embedded holomorphic curve in $\mathbb{R} \times T^2 \times [1,2]$. The same holds for Morse-Bott buildings u with $I(u) = 2$ and constrained to pass through a generic point in $\mathbb{R} \times M$.

Given two orbit sets $\boldsymbol{\gamma}' = \prod \gamma_i^{m_i'}$ and $\boldsymbol{\gamma} = \prod \gamma_i^{m_i}$ (in multiplicative notation), we set $\boldsymbol{\gamma}/\boldsymbol{\gamma}' = \prod \gamma_i^{m_i - m_i'}$ if $m_i' \leq m_i$ for all i; otherwise we set $\boldsymbol{\gamma}/\boldsymbol{\gamma}' = 0$. We denote $E_0^p = \mathcal{F}^p/\mathcal{F}^{p+1}$; then $E_0 = \bigoplus E_0^p$ is a chain complex with differential ∂_0 induced by ∂_M. The differential ∂_0 respects the direct sum decomposition by construction, and we denote by E_1^p the homology of E_0^p.

As a vector space we can still identify $E_0 \cong ECC(V, \alpha_V) \otimes ECC(N, \alpha)$. More precisely, if we denote the subcomplex of $ECC(V, \alpha_V)$ generated by orbit sets $\boldsymbol{\gamma}$ such that $\eta([\boldsymbol{\gamma}]) = p$ by $ECC_p(V, \alpha_V)$, then $E_0^p \cong ECC_p(V, \alpha_V) \otimes ECC(N, \alpha)$.

From Lemma 4.14 and the description of the $I = 1$ holomorphic curves in the symplectisation of the no man's land, it follows that the differential ∂_0 is

$$\partial_0(\boldsymbol{\gamma} \otimes \boldsymbol{\Gamma}) = (\partial_V \boldsymbol{\gamma}) \otimes \boldsymbol{\Gamma} + (\boldsymbol{\gamma}/e') \otimes h\boldsymbol{\Gamma} + (\boldsymbol{\gamma}/h') \otimes e\boldsymbol{\Gamma} + \boldsymbol{\gamma} \otimes (\partial_N \boldsymbol{\Gamma}), \tag{4.1}$$

where ∂_N and ∂_V denote the differential of $ECC(N, \alpha)$ and $ECC(V, \alpha_V)$ respectively. See [**4**, Lemma 9.5.3].

4.4.3. *Computation of E_1.* By factoring out the terms h' and h, we can write the differentials ∂_V and ∂_N as:

$$\begin{cases} \partial_V \boldsymbol{\gamma} = \partial_V^\flat \boldsymbol{\gamma} \\ \partial_V(h'\boldsymbol{\gamma}) = h'\partial_V^\flat \boldsymbol{\gamma} + \partial_V'(h'\boldsymbol{\gamma}) \end{cases} \qquad \begin{cases} \partial_N \boldsymbol{\Gamma} = \partial_N^\flat \boldsymbol{\Gamma} + h\partial_N' \boldsymbol{\Gamma} \\ \partial_N(h\boldsymbol{\Gamma}) = h\partial_N^\flat \boldsymbol{\Gamma} \end{cases} \tag{4.2}$$

where $\boldsymbol{\gamma} \in ECC^\flat(V, \alpha_V)$, $\boldsymbol{\Gamma} \in ECC^\flat(N, \alpha)$, $\partial_V^\flat$ and $\partial_N^\flat$ are the differentials for the chain complexes $ECC^\flat(V, \alpha_V)$ and $ECC^\flat(N, \alpha)$ respectively, and the terms $\partial_V'(h'\boldsymbol{\gamma})$ and $\partial_N' \boldsymbol{\Gamma}$ do not contain h'.

If we write

$$C_{k,k'}^p = (h')^{k'} ECC_p^\flat(V, \alpha) \otimes h^k ECC^\flat(N, \alpha)$$

with $(k, k') \in \{0,1\}^2$, then we can describe the differential ∂_0 on E_0^p by the following diagram

$$
\begin{array}{ccc}
C^p_{0,1} & \xrightarrow{1\otimes h\partial'_N + \cdot/e'\otimes h} & C^p_{1,1} \\
\Big\downarrow{\scriptstyle \partial'_V\otimes 1+\cdot/h'\otimes e} & & \Big\downarrow{\scriptstyle \partial'_V\otimes 1+\cdot/h'\otimes e} \\
C^p_{0,0} & \xrightarrow{1\otimes h\partial'_N + \cdot/e'\otimes h} & C^p_{1,0}
\end{array}
\tag{4.3}
$$

where each $C^p_{k,k'}$ carries the internal differential $\partial^\flat_V \otimes 1 + 1 \otimes \partial^\flat_N$. Thus we can filter each E_0^p by $k'-k$ and the first term of the corresponding spectral sequence (i.e. the homology of the graded complex associated to the filtration) is $ECH^\flat_p(V, \alpha_V) \otimes ECH^\flat(N, \alpha)$.

We make a digression into the embedded contact homology of the solid torus.

LEMMA 4.15 ([**4**, Lemma 8.1.2(4)]). *$ECH^\flat(V, \alpha_V) \cong \mathbb{Z}/2\mathbb{Z}[e']$.*

SKETCH OF PROOF. We start by computing $ECH(\mathrm{int}(V), \alpha_V)$. We recall the exhaustion $V_0 \subset V_i \subset \ldots \subset \mathrm{int}(V)$ by concentric solid tori. Each ∂V_i is linearly foliated by Reeb orbits with irrational slope. Positivity of intersection with the Reeb flow implies that holomorphic curves with ends in V_i are contained in V_i and thus, by [**4**, Lemma 8.3.1], we have

$$ECH(\mathrm{int}(V), \alpha_V) \cong \varinjlim ECH(V_i, \alpha_V|_{V_i}).$$

Since $\alpha_V|_{V_i}$ is irrational, Proposition 4.9 applies, and therefore we can compute $ECH(V_i, \alpha_V|_{V_i})$ using a different contact form with a unique simple closed Reeb orbit; see [**4**, Section 8.2]. For those contact forms the ECH index I can be lifted to an absolute index on closed orbits, which is moreover preserved by continuation maps; see [**4**, Lemma 8.2.2 and Lemma 8.2.3]. A direct computation shows that the ECH index of elements in $ECC(V_i, \alpha_V|_{V_i})$, besides the class of the empty set, grows with i. Hence in the direct limit only the empty set survives, and therefore we have $ECH(\mathrm{int}(V), \alpha_V) \cong \mathbb{Z}/2\mathbb{Z}$. See [**4**, Proposition 8.3.2].

In order to compute $ECH^\flat(V, \alpha_V)$ we restrict the filtration $\mathcal{F}$ induced by $\eta \in H^1(V;\mathbb{Z})$ to $ECC^\flat(V, \alpha_V)$. The induced spectral sequence collapses at $E^1 = \mathbb{Z}/2\mathbb{Z}[e'] \otimes ECH(\mathrm{int}(V), \alpha_V) \cong \mathbb{Z}/2\mathbb{Z}[e']$. See [**4**, Section 8.5]. □

This implies that $ECH^\flat(V, \alpha_V) \otimes ECH^\flat(N, \alpha) = 0$ for $p > 0$ and therefore $E_0^p = 0$ for $p > 0$. Thus the spectral sequence induced by $\mathcal{F}$ collapses at the E_1 term and the inclusion $\mathcal{F}^0 \subset ECC(M, \alpha_M)$ induces an isomorphism $E_1^0 = H(\mathcal{F}^0) \cong ECH(M)$.

Now we compute E_1^0. Given a finite set of simple closed Reeb orbits $\gamma_1, \ldots, \gamma_n$ we define $\mathcal{R}[\gamma_1, \ldots, \gamma_n]$ as the polynomial algebra generated by those orbits over $\mathbb{Z}/2\mathbb{Z}$ with the relation that $\gamma^2 = 0$ if γ is hyperbolic. Using this notation, $ECC_0(V, \alpha_V) = \mathcal{R}[e', h']$. There are only three holomorphic curves in V with $I = 1$ and asymptotic to e' or h': two cylinders from e' to h' coming from gradient flow trajectories in the Morse-Bott family corresponding to ∂V and one holomorphic plane positively asymptotic to h'. See [**4**, Proposition 8.4.4 and Proposition 8.4.5] for the construction of this plane. Thus $\partial_V((e')^i) = 0$ and $\partial_V((e')^i h') = (e')^i$, so we can write $E_0^0 \cong \mathcal{R}[e', h'] \otimes ECC(N, \alpha)$ with differential

$$\partial_0(\gamma \otimes \mathbf{\Gamma}) = \gamma \otimes (\partial_N \mathbf{\Gamma}) + (\gamma/h') \otimes (1+e)\mathbf{\Gamma} + (\gamma/e') \otimes h\mathbf{\Gamma}. \tag{4.4}$$

A simple algebraic argument now implies that $E_0^1 \cong ECH(N, \partial N, \alpha)$. In fact $\gamma \otimes \mathbf{\Gamma}$ is a cycle only if $\gamma = 1$ and $\partial_N \gamma = 0$; the third term of Equation (4.4) implies that, if $h\gamma$ is a cycle, then it is a boundary, and the second term implies that $e\mathbf{\Gamma} = \mathbf{\Gamma}$ in homology.

4.4.4. *Definition of σ.* The argument of the previous paragraphs gives a non-canonical isomorphism $ECH(M, \alpha_M) \cong ECH(N, \partial N, \alpha)$. Now we show that this isomorphism is in fact induced by a geometrically meaningful chain map. We define the complex

$$ECC^{\natural}(N, \alpha) = \mathcal{R}[h'] \otimes ECC^{\flat}(N, \alpha)$$

with differential

$$\partial^{\natural}(\gamma \otimes \mathbf{\Gamma}) = \gamma \otimes \partial_N^{\flat}(\mathbf{\Gamma}) + \gamma/h' \otimes (1 + e)\mathbf{\Gamma}. \tag{4.5}$$

It is easy to show that $ECH^{\natural}(N, \alpha) \cong ECH(N, \partial N, \alpha)$.

We define

$$\sigma(\gamma \otimes \mathbf{\Gamma}) = \sum_{i=0}^{\infty} (e')^i \gamma \otimes (\partial_N')^i \mathbf{\Gamma}. \tag{4.6}$$

The sum is well defined because $(\partial_N')^{k+1}(\mathbf{\Gamma}) = 0$ if $\varphi([\mathbf{\Gamma}]) = k$, where φ is a class in $H^1(N; \mathbb{Z})$ such that $\varphi([e]) = 1$.

Since $\gamma = (h')^j$ for $j = 0, 1$, the image of σ is contained in $\mathcal{F}^0$, which is the first nonzero group of the filtration, and therefore to show that σ is a chain map it is enough to verify that $\sigma \circ \partial^{\natural} = \partial_0 \circ \sigma$. This is an easy verification using Equation (4.1) and the fact that ∂_N' commutes with $\partial_N^{\flat}$ and the multiplication by e. Finally one can prove without much effort by algebraic considerations similar to those of the previous paragraph that the map σ induces an isomorphism in homology.

4.4.5. *The U map and the hat version.* We recall that the U map in $ECC(M, \alpha_M)$ counts $I = 2$ holomorphic curves in $\mathbb{R} \times M$ passing through a generic base point $z \in \mathbb{R} \times M$. In order to simplify the computation, we put the base point in the symplectisation of the no man's land $\mathbb{R} \times T^2 \times (1, 2)$. The following lemma is proved by the same techniques we used to study the differential.

LEMMA 4.16 (See [**4**, Lemma 9.9.3]). *The map U preserves the filtration $\mathcal{F}$. In the lowest filtration level $\mathcal{F}^0$, it is given by*

$$U(\gamma \otimes \mathbf{\Gamma}) = \gamma/e' \otimes \mathbf{\Gamma}. \tag{4.7}$$

The unique holomorphic curve contributing to U in the lowest filtration level is a cylinder from e' to e belonging to the finite energy foliation of $\mathbb{R} \times T^2 \times [1, 2]$ which was already used to understand the differential ∂_0. This foliation is described in [**4**, Section 8.4].

On the other hand, we define the map $U^{\natural} : ECC^{\natural}(N, \alpha) \to ECC^{\natural}(N, \alpha)$ by

$$U^{\natural}(\gamma \otimes \mathbf{\Gamma}) = \gamma \otimes \partial_N' \mathbf{\Gamma}. \tag{4.8}$$

By comparing Equation (4.8) with Equations (4.5), (4.6) and (4.7), we obtain that $U^{\natural}$ is a chain map and that the diagram

$$\begin{array}{ccc} ECC^{\natural}(N, \alpha) & \xrightarrow{U^{\natural}} & ECC^{\natural}(N, \alpha) \\ \sigma \downarrow & & \downarrow \sigma \\ ECC(M, \alpha_M) & \xrightarrow{U} & ECC(M, \alpha_M) \end{array}$$

commutes. The map $U^\natural$ corresponds to U_{rel} under the isomorphism $ECH^\natural(N,\alpha) \cong ECH(N,\partial N,\alpha)$, and therefore we have established part of Theorem 4.12.

We also define the chain complex

$$\widehat{ECC}^\natural(N,\alpha) = \mathcal{R}[h'] \otimes ECC(N,\alpha)$$

with differential

$$\widehat{\partial}^\natural(\gamma \otimes \mathbf{\Gamma}) = \gamma \otimes \partial_N \mathbf{\Gamma} + \gamma/h' \otimes (1+e)\mathbf{\Gamma}.$$

The decomposition of the differential ∂_N described in Equation (4.2) implies that $\widehat{ECC}^\natural(N,\alpha)$ is the cone of the map $U^\natural$. Moreover one can easily prove that $\widehat{ECH}^\natural(N,\alpha) \cong \widehat{ECH}(N,\partial N,\alpha)$, and therefore the remaining part of Theorem 4.12 follows from the naturality of the mapping cone construction.

4.5. Sutured embedded contact homology. Sutured embedded contact homology is another version of embedded contact homology for manifolds with boundary which was defined in [**8**]. Sutured manifolds are three-manifolds with decorated boundary which were introduced by Gabai in [**11**] for the study of foliations. Convex surfaces were introduced by Giroux in [**13**] as a tool for studying contact structures. The parallel between the theory of sutured manifolds and convex surfaces has been explored since [**15**].

In this subsection we relate sutured embedded contact homology to the versions of embedded contact homology introduced in Subsection 4.3. First, we recall the definition of a family of sutured manifolds which is less general than those considered by Gabai but closer to the theory of convex surfaces.

DEFINITION 4.17. A *balanced sutured manifold* (M,Γ) is a three-manifold with boundary M decorated by an embedded one-dimensional oriented submanifold $\Gamma \subset \partial M$ such that:

- each connected component of M intersects ∂M,
- each connected component of ∂M intersects Γ,
- if $N(\Gamma)$ denotes a closed tubular neighbourhood of Γ in ∂M, then
$$\partial M \setminus \mathrm{int}(N(\Gamma)) = R_+(\Gamma) \sqcup R_-(\Gamma),$$
- if we orient $R_+(\Gamma)$ with the orientation of ∂M and $R_-(\Gamma)$ with the opposite orientation, then the orientation of Γ coincides with the orientation of $\partial R_+(\Gamma)$ and is opposite to the orientation of $R_-(\Gamma)$, and finally
- $\chi(R_+(\Gamma)) = \chi(R_-(\Gamma))$.

REMARK 4.18. If (M,ξ) is a contact three-manifold with convex boundary and Γ is the dividing set ∂M, then (M,Γ) is a balanced sutured manifold by [**13**].

It is often useful to think of a sutured manifold as a manifold with corners, where $R_+(\Gamma) \cup R_-(\Gamma)$ is the "horizontal boundary", $N(\Gamma)$ is the "vertical boundary" and $\partial R_+(\Gamma) \cup \partial R_-(\Gamma) = \partial N(\Gamma)$ are the corners. The following definition is best visualised in this point of view.

DEFINITION 4.19 (See [**8**, Definition 2.8]). A contact form α is *adapted* to the balanced sutured manifold (M,Γ) if $\alpha|_{R_\pm(\Gamma)}$ are Liouville forms on $R_\pm(\Gamma)$ and the Reeb vector field is transverse and outward pointing on $R_+(\Gamma)$, transverse and inward pointing at $R_-(\Gamma)$ and tangent to $N(\Gamma)$, so that $N(\Gamma)$ is foliated by Reeb trajectories going from $R_-(\Gamma)$ to $R_+(\Gamma)$.

The simplest example of a sutured manifold with an adapted contact form is the following: we take a compact, oriented surface with boundary P and we define $M = P \times [-1,1]$ and $\Gamma = \partial P \times \{0\}$. We identify $R_\pm(\Gamma) = P \times \{\pm 1\}$ and $N(\Gamma) = \partial P \times [-1,1]$. The contact form is $\alpha = \lambda + dt$, where λ is a Liouville form on P. However many more interesting examples exist: in fact, contact sutured manifolds are a fairly general concept, as the following proposition shows.

PROPOSITION 4.20 (See [**8**, Lemma 4.1]). *Let (M, ξ) be a contact manifold with convex boundary and no closed connected component, and let Γ be the dividing set of ∂M. Then ξ admits a contact form which is adapted to the balanced sutured manifold (M, Γ).*

The interest of sutured contact manifolds is that they form a fairly large class of contact manifolds for which SFT and ECH compactness hold: see [**8**, Corollary 5.19 and Corollary 5.21]. In [**8**] we considered sutured contact manifolds of any dimension, and that level of generality introduced extra complications in the proof of compactness. In dimension three, however, the story is fairly simple: all we need to show is that a J-holomorphic curve in $\mathbb{R} \times M$ which is asymptotic to closed Reeb orbits of α cannot touch $\mathbb{R} \times \partial M$ if J belongs to a suitable class of almost complex structures. Roughly speaking, we define a function t on a neighbourhood of ∂M by integrating the Reeb vector field, and a function τ in a neighbourhood on $N(\Gamma)$ such that, in that neighbourhood, we can write $\alpha = dt + e^\tau d\theta$ with θ a coordinate on Γ. We say that an almost complex structure J on $\mathbb{R} \times M$ is *tailored* to (M, Γ, α) if it is compatible with α and moreover t is harmonic and τ is subharmonic with respect to the Laplacian induced by J; see [**8**, Section 5.1, Section 5.2 and Section 5.3]. Thus, if u is a J-holomorphic curve which is asymptotic to closed Reeb orbits, the maximum principle applied to $t \circ u$ and $\tau \circ u$ is enough to show that u cannot intersect the region where t and τ are defined. It is easy to see that the space of tailored almost complex structures is nonempty and contractible.

All other aspects of the theory of J-holomorphic curves (i.e. regularity, Fredholm theory, transversality, gluing, etc.) are semi-local in nature, and therefore do not change if the ambient manifold is not closed. Thus, we can define the *sutured embedded contact homology* complex $ECC(M, \Gamma, \alpha)$ as the vector spaces over $\mathbb{Z}/2\mathbb{Z}$ generated by orbit sets in M with differential counting J-holomorphic curves in $\mathbb{R} \times M$ with ECH index $I = 1$ for a tailored almost complex structure J. These complexes were defined in [**8**] and were inspired by sutured Floer homology, which was defined by Juhász in the context of Heegaard Floer homology in [**26**]. We conjectured the following.

CONJECTURE 4.21 ([**8**, Conjecture 1.5]). *Let (M, Γ, α) be a sutured contact manifold. Then there is an isomorphism*

$$ECH(M, \Gamma, \alpha) \cong SFH(-M, -\Gamma),$$

which moreover respects the decomposition into first homology classes on the left-hand side and relative $Spin^c$-structures on the right-hand side.

As an evidence to this conjecture, we constructed gluing maps which are formally analogous to Juhász's map for sutured decompositions from [**27**] and Honda, Kazez and Matić's map for gluing along convex surfaces from [**17**]: see [**8**, Theorem 1.9 and Theorem 1.10]. The proof is currently a work in progress of the authors with Gilberto Spano.

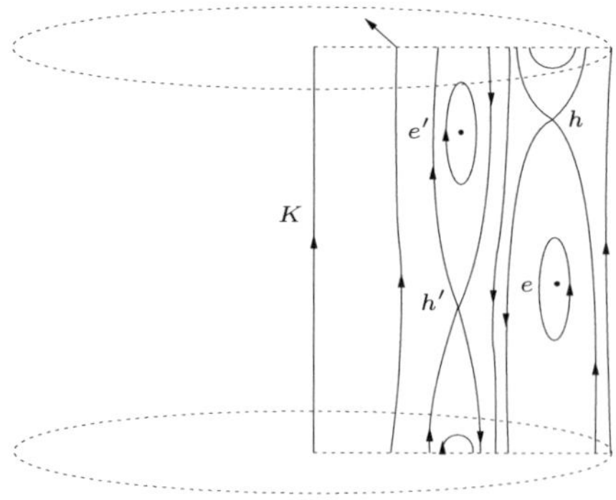

FIGURE 3. The first return map of the Reeb vector field of α_M on a longitudinal section of a neighbourhood of K. The top and the bottom are identified.

The sutured embedded contact homology groups are independent of the contact form and the almost complex structure in the following sense.

THEOREM 4.22 ([**4**, Theorem 10.2.2]). *Let α_1 and α_2 be contact forms adapted to the balanced sutured manifold (M,Γ) and let J_1 and J_2 be almost complex structures on $\mathbb{R}\times M$ such that J_i is tailored to (M,Γ,α_i) for $i=1,2$. If $\xi_1=\ker\alpha_1$ and $\xi_2=\ker\alpha_2$ are isotopic through contact structures making ∂M convex with dividing set Γ, then*

$$ECH(M,\Gamma,\alpha_1)\cong ECH(M,\Gamma,\alpha_2),$$

where the first group is defined using J_1 and the second using J_2. Moreover this isomorphism preserves the decomposition of the sutured embedded contact homology groups as direct sums of subgroups indexed by homology classes in $H_1(M)$.

The proof is similar to the proof of Proposition 4.9. This result has been proved independently, but with similar techniques, by Kutluhan, Sivek and Taubes [**30**], who also prove naturality of the isomorphism. In light of Conjecture 4.21 we expect that the sutured contact homology groups should be independent also of the contact structure, but at the moment we are not able to find a direct proof of this more general invariance.

Two more basic but important examples of sutured manifolds are the following:

- given a closed three-manifold M and an embedded codimension zero ball $B\subset M$, we define the balanced sutured manifold $(M(1),\Gamma_1)$, where $M(1)=M\setminus \mathrm{int}(B)$ and Γ_1 is a connected curve in $\partial M(1)$;
- given a closed three-manifold M and a knot $K\subset M$, we define a balanced sutured manifold $(M(K),\Gamma_K)$, where $M(K)$ is the complement of an open tubular neighbourhood of K and Γ_K consists of two parallel copies of the meridian of K with opposite orientations.

Sutured embedded contact homology for these two sutured manifolds is related to the embedded contact homology groups defined in the previous sections as follows.

THEOREM 4.23 ([**4**, Theorem 10.3.1 and Theorem 10.3.2]). *Let M be a closed three-manifold, $K\subset M$ a knot and N the complement of an open tubular neighbourhood of K in M. Then there exist a contact form on N as in Subsection 4.3*

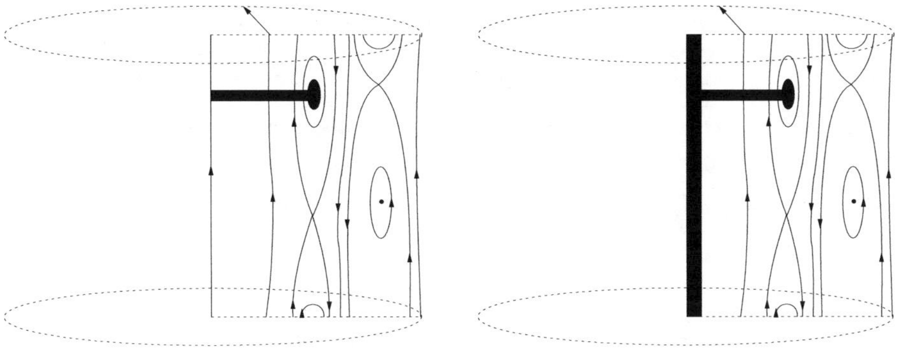

FIGURE 4. The ball B and the neighbourhood of K are obtained by rotating the shaded regions on the left and on the right, respectively, around the vertical axis. The top and the bottom are identified.

and adapted contact forms α_1 on $(M(1), \Gamma_1)$ and α_K on $(M(K), \Gamma_K)$ such that

$$ECH(M(1), \Gamma_1, \alpha_1) \cong \widehat{ECH}(N, \partial N, \alpha) \quad \textit{and}$$
$$ECH(M(K), \Gamma_K, \alpha_K) \cong ECH^\sharp(N, \alpha).$$

SKETCH OF PROOF. Let α_M be a contact form on M as in Subsection 4.4. The first return map of the Reeb vector field of α_M in a neighbourhood of K containing ∂N and after the Morse-Bott perturbation is depicted in Figure 3. The sutured manifold $(M(1), \Gamma_1)$ is identified to the subset of M described in the left side of Figure 4. The adapted contact forms α_1 is the restrictions of α_M to $M(1)$. However, the contact form α'_K obtained by restricting α_M to $M(K)$ is not adapted to the suture because its restriction to $R_\pm(\Gamma_K)$ is not a Liouville form. This can be seen from the fact that the contact structure is negatively transverse to the component of $\partial R_\pm(\Gamma_K)$ which is closer to K. The issue can be corrected by adding a layer to that component in which the Reeb vector field remains constant and the contact structure rotates until it satisfies the sutured condition. The new sutured manifold is diffeomorphic to $(M(K), \Gamma_K)$ and the contact form α_K we obtain is adapted to $(M(K), \Gamma_K)$.

Both $ECC(M(1), \Gamma_1, \alpha_1)$ and $ECC(M(K), \Gamma_K, \alpha_K)$ are generated by orbit sets containing closed Reeb orbits in N and h'; moreover the holomorphic plane in $\mathbb{R} \times M$ which is positively asymptotic to h' does not contribute to the differential of $ECC(M(K), \Gamma_K, \alpha_K)$. At this point, the result follows from algebraic manipulations similar to those in the proof of Theorem 4.12, but easier. □

Combining Theorem 4.12 with Theorem 4.23 we obtain the following corollary.

COROLLARY 4.24. *$ECH(M(1), \Gamma_1, \alpha_1) \cong \widehat{ECH}(M, \alpha_M)$.*

Combining Conjecture 4.21 with Theorem 4.23 and the relation between sutured Floer homology and knot Floer homology, we obtain the following conjecture.

CONJECTURE 4.25. *If $K \subset M$ is a null-homologous knot of genus g, N is the complement of an open tubular neighbourhood of K and α is a contact form on N*

for which ∂N is a negative Morse-Bott torus foliated by meridians of K, then

$$ECH_i^\sharp(N,\alpha) \cong \widehat{HFK}(M,K,i-g),$$

where $ECH_i^\sharp(N,\alpha)$ denotes the homology of the subcomplex of $ECH^\sharp(N,\alpha)$ generated by orbit sets in N with total linking number i with K.

Both the Alexander grading in knot Floer homology and the linking number depend on the choice of a Seifert surface if $H_2(M;\mathbb{Z}) \neq 0$. In the conjecture we assume that we have made the same choice in both cases.

4.6. Periodic Floer homology and open books. We recall from Giroux [**14**] that a contact form is supported by an open book decomposition if its Reeb vector field is tangent to the binding and positively transverse to the interior of the pages. In the same article he also sketched an equivalence between contact structures up to isotopy and open book decompositions up to positive stabilisation, but here we will need only the easy part of the equivalence for our purposes: namely the existence of contact forms supported by open book decomposition and the isotopy between contact structures supported by open book decompositions related by positive stabilisations.

Let $(S,\hbar)$ be an abstract open book decomposition for M. We assume, without loss of generality, that ∂S is connected, and identify the mapping torus of $\hbar$, denoted by N, with the complement of an open tubular neighbourhood of the binding.

ASSUMPTION 4.26. On $\hbar$ we assume the following:

- $\hbar^*\beta - \beta$ is exact for some Liouville form β on S, and
- $\hbar(y,\theta) = (y,\theta - y)$ for coordinates $(y,\theta) \in (-\varepsilon, 0] \times \mathbb{R}/\mathbb{Z}$ in a collar of ∂S.

By a Moser's trick argument one can always find a representative of the monodromy satisfying the conditions above (see [**4**, Lemma 9.3.2]) and, by a refinement of Thurston and Winkelnkemper's construction from [**51**], one can construct a contact form α on M which is supported by the open book decomposition $(S,\hbar)$, and such that the first return map of the Reeb flow in N is $\hbar$ and ∂N is a negative Morse-Bott torus; see [**4**, Lemma 9.3.3] and [**5**, Lemma 2.12]. Thus the Reeb flow of α on N satisfies the conditions of Subsections 4.3 and 4.4 permitting to define $ECH(N,\partial N,\alpha)$ and prove the isomorphism with $ECH(M,\alpha)$.

We define

$$ECH_i(N,\alpha) = \bigoplus_{a\cdot[S]=i} ECH_a(N,\alpha),$$

where $a \in H_1(N)$ and $[S] \in H_2(N,\partial N)$ is the relative homology class of a page of the open book decomposition. Thus, by Lemma 4.11,

$$\widehat{ECH}(N,\partial N,\alpha) = \varinjlim ECH_i(N,\alpha).$$

In the definition of the isomorphism between $\widehat{HF}(-M)$ and $\widehat{ECH}(M)$ it will be useful to pass to periodic Floer homology groups.

There is a family of stable Hamiltonian structures $(\alpha_\varsigma,\omega_\varsigma)_{\varsigma\in[0,1]}$ such that $\omega_\varsigma = d\alpha_\varsigma$ for $\varsigma \in (0,1]$ (i.e. α_ς is a contact form for $\varsigma \neq 0$), $\alpha_1 = \alpha$ and (α_0,ω_0) is the stable Hamiltonian structure induced by the fibration $\pi\colon N \to S^1$. See [**5**, Section 3.1]. Let J_ς be a smooth family of almost complex structures on $\mathbb{R}\times N$ compatible with $(\alpha_\varsigma,\omega_\varsigma)$.

The *Periodic Floer homology* complexes $PFC_i(N, \alpha_0, \omega)$ are defined in the same way as the embedded contact homology complexes $ECC_i(N, \alpha)$: the different names have only a historical motivation. We can also define all decorated version of periodic Floer homology as in Subsection 4.3. By comparing regular J_0-holomorphic curves with J_ς-holomorphic curves with $\varsigma > 0$ sufficiently small, we obtain the following result.

LEMMA 4.27 (See [**5**, Theorem 3.6.1]). *For every $i \in \mathbb{N}$ there exists $\varsigma_i \in (0,1]$ such that, for all $\varsigma \in (0, \varsigma_i]$, there is an isomorphism of chain complexes*

$$PFC_i(N, \alpha_0, \omega_0) \cong ECC_i(N, \alpha_\varsigma).$$

This is not yet enough for taking direct limits because ς_i becomes smaller as i increases and might tend to zero.[19] However, arguing with more care, we can prove the following.

LEMMA 4.28. *For α and (α_0, ω_0) as above, there is an isomorphism*

$$\varinjlim PFH_i(N, \alpha_0, \omega_0) = \widehat{PFH}(N, \partial N, \alpha_0, \omega_0) \cong \widehat{ECH}(N, \partial N, \alpha).$$

PROOF. We need to define continuation maps $ECC_i(N, \alpha) \to ECC_i(N, \alpha_{\varsigma_i})$ for every $i \in \mathbb{N}$ such that the diagram

$$\begin{array}{ccccc} ECC_i(N,\alpha) & \longrightarrow & ECC_i(N,\alpha_{\varsigma_i}) & = & PFC_i(N,\alpha_0,\omega_0) \\ \downarrow{\scriptstyle \cdot e} & & & & \downarrow{\scriptstyle \cdot e} \\ ECC_{i+1}(N,\alpha) & \longrightarrow & ECC_{i+1}(N,\alpha_{\varsigma_{i+1}}) & = & PFC_i(N,\alpha_0,\omega_0) \end{array}$$

commutes. To define the continuation maps $ECC_i(N, \alpha) \to ECC_i(N, \alpha_{\varsigma_i})$, we slightly enlarge N to N' so that $\partial N'$ is foliated by Reeb orbits of irrational slope and no new closed Reeb orbit intersecting a page less than $i + 2$ times is created, and invoke Proposition 4.9. The continuation maps are supported on holomorphic curves which are contained in $\mathbb{R} \times M$ endowed with a symplectic form which interpolates between α and α_{ς_i}. See [**4**, Theorem 3.1.2] for the definition of "supported on holomorphic curves" and an overview on continuation maps in ECH and [**23**] for the full story. By the trapping lemma [**4**, Lemma 5.3.2], curves with a positive end at e must contain a trivial cylinder over e, and therefore the commutativity of the diagram follows. □

5. The open-closed maps Φ and Φ^+

In this section we define maps from Heegaard Floer homology to embedded contact homology. As it often happens in symplectic geometry, these maps will be defined by counting holomorphic curves in symplectic cobordisms. Here the symplectic cobordisms will come from an open book decomposition $(S, \mathfrak{h})$ of M.

5.1. The maps Φ and $\widehat{\Phi}_*$.

Let B_+ be the unit disc with one puncture in the interior and one puncture on the boundary, which we identify biholomorphically with the subset of the cylinder $\mathbb{R} \times \mathbb{R}/2\mathbb{Z}$ obtained by rounding the corners of $\mathbb{R} \times \mathbb{R}/2\mathbb{Z} \setminus (2, \infty) \times (1, 2)$. We define a fibration

$$\pi_{W_+} : W_+ \to B_+$$

[19] This issue was pointed out to us by Thomas Brown.

with fibre S and monodromy $\hbar$, and equip the total space with the symplectic form $\Omega_+ = ds \wedge dt + \omega$, where (s,t) are coordinates on $B_+ \subset \mathbb{R} \times \mathbb{R}/2\mathbb{Z}$ and $\omega = d\beta$ is an area form on S which is preserved by $\hbar$. We can view (W_+, Ω_+) as a symplectic cobordism (with boundary) between $[0,1] \times S$ and N. We refer to [**5**, Section 5] for more details about this construction.

Let $\mathbf{a} = \{a_1, \dots, a_{2g}\}$ be a basis of arcs of S. We define a Lagrangian submanifold $L_{\mathbf{a}} \subset \partial W$ as the trace of the parallel transport of $\{3\} \times \{1\} \times \mathbf{a}$ along ∂B_+ with respect to the symplectic connection defined as the Ω_+-orthogonal to the tangent spaces of the fibres. Then,

$$L_{\mathbf{a}} \cap [3, +\infty) \times \{1\} \times S = [3, +\infty) \times \{1\} \times \mathbf{a}$$
$$L_{\mathbf{a}} \cap [3, +\infty) \times \{0\} \times S = [3, +\infty) \times \{0\} \times \hbar(\mathbf{a}).$$

In the following $\widehat{\mathcal{O}}_k$ will denote the set of orbit sets in N with total intersection number k with a fibre. We fix an almost complex structure J_+ on W_+ which is compatible with Ω_+ and with the stable Hamiltonian structures at the ends. Given a Heegaard Floer generator $\mathbf{y} \in \mathcal{S}_{\mathbf{a},\hbar(\mathbf{a})}$ and a periodic Floer homology generator $\gamma \in \widehat{\mathcal{O}}_{2g}$, we denote by $\mathcal{M}_{W_+}(\mathbf{y}, \gamma)$ the moduli space of J_+-holomorphic curves in W_+ with boundary on $L_{\mathbf{a}}$ which are positively asymptotic to $\mathbf{y}$ and negatively asymptotic to γ. Holomorphic curves in $\mathcal{M}_{W_+}(\mathbf{y}, \gamma)$ are somewhere injective because distinct connected components of the boundary are on distinct connected components of $L_{\mathbf{a}}$. Thus, for generic J_+, the moduli spaces $\mathcal{M}_{W_+}(\mathbf{y}, \gamma)$ are (disjoint unions of) smooth manifolds of the dimension predicted by the Fredholm index.

The ECH-type index I for J_+-holomorphic curves in $\mathcal{M}_{W_+}(\mathbf{y}, \gamma)$ is defined in [**5**, Definition 5.6.5]. By the index inequality [**5**, Theorem 5.6.9], J_+-holomorphic curves u with $I(u) = 0$ are embedded, have Fredholm index $\operatorname{ind}(u) = 0$, and therefore are isolated. We define

$$\overline{\Phi} \colon \overline{CF}(S, \mathbf{a}, \hbar(\mathbf{a})) \to PFC_{2g}(N, \alpha_0, \omega)$$

such that, on each $\mathbf{y} \in \mathcal{S}_{\mathbf{a},\hbar(\mathbf{a})}$, it is given by

$$\overline{\Phi}(\mathbf{y}) = \sum_{\gamma \in \widehat{\mathcal{O}}_{2g}} \#\mathcal{M}^{I=0}_{W_+}(\mathbf{y}, \gamma)\gamma, \tag{5.1}$$

Compactness and gluing for holomorphic curves in Heegaard Floer homology and embedded contact homology extend to holomorphic curves in W_+. Therefore the usual argument on the ends of one-dimensional moduli spaces proves that $\overline{\Phi}$ is a chain map; see [**5**, Proposition 6.2.2].

LEMMA 5.1 ([**5**, Theorem 6.2.4]). *The map* $\overline{\Phi}$ *induces a chain map*

$$\Phi \colon \widehat{CF}(S, \mathbf{a}, \hbar(\mathbf{a})) \to PFC_{2g}(N, \alpha_0, \omega)$$

and moreover $\Phi(\mathbf{x}) = e^{2g}$, *where* $\mathbf{x}$ *represents the contact class.*

SKETCH OF PROOF. The statement is a consequence of the following fact, proved in [**5**, Lemma 6.2.3]: a J_+-holomorphic curve in W_+ with boundary on $L_{\mathbf{a}}$ which is positively asymptotic to a chord over an intersection point x on ∂S (i.e. $x = x_i$ or $x = x_i'$ for some i) has an irreducible component consisting of a trivial section $B_+ \times \{x\}$ followed by a gradient flow trajectory connecting the orbit over x to e in the Morse-Bott family associated to ∂N. In fact a holomorphic curve with a nonconstant positive end at x cannot be contained in N. □

Finally we define the map $\widehat{\Phi}_*\colon \widehat{HF}(-M) \to \widehat{ECH}(M)$ of Theorem 1.1 as the composition of $\Phi_*\colon \widehat{HF}(S, \mathbf{a}, h(\mathbf{a})) \to ECH_{2g}(N, \alpha_0, \omega)$ with the inclusion

$$PFC_{2g}(N, \alpha_0, \omega) \to \widehat{PFC}(N, \partial N, \alpha_0, \omega) \cong \widehat{ECH}(N, \partial N, \alpha)$$

and the map $\widehat{\sigma}_*\colon \widehat{ECH}(N, \partial N, \alpha) \to \widehat{ECH}(M)$ of Theorem 4.12.

REMARK 5.2. We could have defined the map Φ using the contact form α instead of the stable Hamiltonian structure (α_0, ω) at the negative end of W_+. However, working with the stable Hamiltonian structure induced by the fibration will be useful in the construction of the inverse map Ψ in Section 6.

5.2. The map Φ^+. The map Φ^+ will be defined by counting holomorphic curves in a symplectic cobordism with boundary (X_+, Ω_{X_+}) from $[0,1] \times \Sigma$ to M. The symplectic form Ω_{X_+} will not be exact, and therefore the count will require some extra care.

5.2.1. *The symplectic cobordism* (X_+, Ω_{X_+}). We describe a simplified construction of (X_+, Ω_{X_+}) and refer the reader to [**7**, Section 4] for the actual details, with the warning that the notations will not correspond exactly.

Let $(\Sigma, \boldsymbol{\alpha}, \boldsymbol{\beta}, z)$ be the pointed Heegaard diagram for $-M$ associated to the open book decomposition (S, h) as in Subsection 3.3.[20] In particular $\Sigma = S_0 \cup -S_1$. For technical reasons we need to assume that S has genus at least two. We also fix an area form ω^+ on Σ, whose properties will be specified in the construction of X_+.

We decompose $M = N \cup (T^2 \times [1,2]) \cup V$ as in Subsection 4.4 and choose a contact form λ on M which is supported by the open book decomposition (S, h) and moreover satisfies the following properties:

(1) $\lambda|_N = \alpha$ (in particular ∂N is a negative Morse-Bott torus),
(2) ∂V is a positive Morse-Bott torus,
(3) Reeb orbits in $\mathrm{int}(V)$ and in the no man's land $T^2 \times (1,2)$ are long[21], and
(4) α is close to the stable Hamiltonian structure (α_0, ω) induced by the fibration $N \to S^1$ with fibre S, so that $ECC_{2g}(N, \lambda)$ is isomorphic to $PFC_{2g}(N, \lambda_0, d\lambda)$.

At least at a first approximation, the symplectic cobordism (X_+, Ω_+) will satisfy the following properties:

(1) the positive end is
$$X_+^> = [3, +\infty) \times [0,1] \times \Sigma, \quad \Omega_{X_+}|_{X_+^>} = ds \wedge dt + \omega^+;$$
(2) the negative end is
$$X_+^< = (-\infty, -1] \times M, \quad \Omega_{X_+}|_{X_+^<} = d(e^s \lambda);$$
(3) there is a symplectic surface $S_z \subset X_+$ such that
$$S_z \cap X_+^> = [3, +\infty) \times [0,1] \times \{z\} \quad \text{and} \quad S_z \cap X_+^< = \emptyset;$$
(4) $\Omega_{X_+}|_{X_+ \setminus S_z} = d\Theta$ for some one-form Θ such that $\Theta|_{X_+^<} = e^s\lambda$;

---

[20]What we call Σ here was called $-\Sigma$ in [**7**].
[21]What "long" means will be made clear in Lemma 5.5.

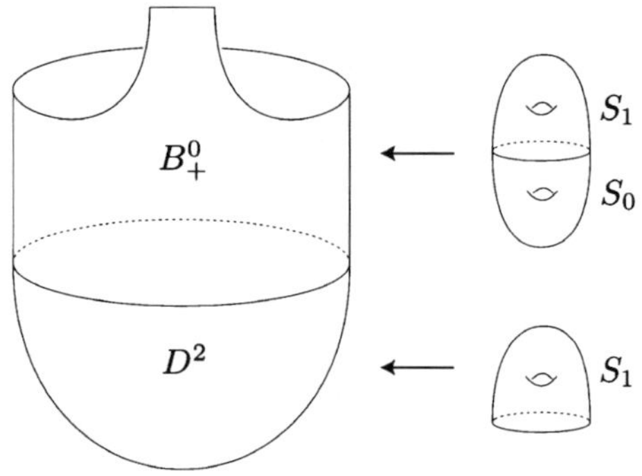

FIGURE 5. Schematic diagram for $X^0_+ \cup X^1_+$ which indicates the fibres over each subsurface.

(5) there is a Lagrangian submanifold $L_{\boldsymbol{\alpha}} \subset \partial X_+$ such that

$$X^{\geq}_+ \cap L_{\boldsymbol{\alpha}} = ([3,+\infty) \times \{0\} \times \boldsymbol{\beta}) \cup ([3,+\infty) \times \{1\} \times \boldsymbol{\alpha})$$

and Θ is exact on $L_{\boldsymbol{\alpha}}$; and

(6) there is an embedding of W_+ into X_+ such that $W_+ \cap S_z = \emptyset$,

$$X^{\geq}_+ \cap W_+ = [3,+\infty) \times [0,1] \times S_0, \quad X^{\leq}_+ \cap W_+ = \mathbb{R} \times N,$$

and the restriction of Ω_{X_+} to W_+ is close to Ω_+ (see the proof of Lemma 5.5).

We refer to [**7**, Lemma 4.1.1] for a more precise description of (X_+, Ω_{X_+}). We will give here only a sketch of the construction; the details, which are quite delicate, can be found in [**7**, Section 4.1].

We identify $S_0 \cong S$ and extend $\hbar\colon S_0 \to S_0$ to a diffeomorphism $\hbar^+ \colon \Sigma \to \Sigma$ which, on S_1, restricts to a perturbation of the identity by a small Hamiltonian isotopy. We require that $\boldsymbol{\beta} = \hbar^+(\boldsymbol{\alpha})$. Let $B^0_+ = B_+ \cap [0,+\infty) \times S^1$; i.e. we cut the negative end out of B_+. We define X^0_+ as the bundle over B^0_+ with fibre Σ and monodromy $\hbar^+$. If ω^+ is an area form on Σ preserved by $\hbar^+$, then we define $\Omega_{X^0_+} = ds \wedge dt + \omega^+$. The component of the boundary of X^0_+ over $\{0\} \times S^1$ is a three-manifold N^+ obtained as mapping torus of $(\Sigma, \hbar^+)$. It contain the mapping torus N of $(S, \hbar)$ as a closed submanifold of codimension zero.

We recall that in the following construction the orientation of S_1 will be the opposite of the orientation as a page of the open book decomposition $(S, \hbar)$ of M. We define $X^1_+ = D^2 \times S_1$ and $\Omega_{X^1_+} = \omega_{D^2} + \omega^+|_{S_1}$ where, of course, ω_{D^2} is the standard symplectic form on D^2. Then we glue X^0_+ and X^1_+ by identifying $\partial D^1 \times S_1 \subset \partial X^1_+$ with $N^+ \setminus \mathrm{int}(N)$, which is the mapping torus of $(S_1, \hbar^+|_{S_1})$. We can glue the symplectic forms $\Omega_{X^0_+}$ and $\Omega_{X^1_+}$ because $\hbar^+|_{S_1}$ is Hamiltonian isotopic to the identity. The gluing of X^0_+ and X^1_+ is schematically depicted in Figure 5. Thus $X^0_+ \cup X^1_+$ is a manifold with boundary and (concave) corners and the compact boundary component of $X^0_+ \cup X^1_+$ is homeomorphic to M. We smooth the corner and modify the symplectic form near the compact boundary component so that, in a collar, it looks like a piece of the symplectisation of (M, λ). Finally we obtain (X_+, Ω_{X_+}) by completing $X^0_+ \cup X^1_+$ with the negative half-symplectisation $((-\infty, 0] \times M, d(e^r\lambda))$ of (M, λ).

We assume that the base point $z \in S_1$ is a fixed point for $\hbar^+$; then we define the symplectic surface

$$S_z = (B^0_+ \cup D^2) \times \{z\}.$$

Finally, we define a Lagrangian submanifold $L_{\boldsymbol{\alpha}} \subset \partial X_+$ such that

$$X^{\geq}_+ \cap L_{\boldsymbol{\alpha}} = ([3, +\infty) \times \{0\} \times \boldsymbol{\beta}) \cup ([3, +\infty) \times \{1\} \times \boldsymbol{\alpha}).$$

Note that $\partial X_+ \subset X^0_+$, which is the portion of X_+ which has the structure of a symplectic surface bundle, and therefore has a symplectic connection. As it was the case for $L_{\mathbf{a}} \subset \partial W_+$, we define $L_{\boldsymbol{\alpha}}$ as the trace of the parallel transport of $\{3\} \times \{1\} \times \boldsymbol{\alpha}$ along ∂B_+ with respect to this symplectic connection.

5.2.2. *Definition of* Φ^+. We fix a generic almost complex structure J^+ on X_+ which is compatible with Ω_{X_+} and the stable Hamiltonian structures at the ends, and makes S_z holomorphic. Given a $2g$-tuple of intersection points $\mathbf{y} \in \mathcal{S}_{\boldsymbol{\alpha}, \boldsymbol{\beta}}$ and an orbit set $\boldsymbol{\gamma} \in \mathcal{O}$ (i.e. contained in M) we denote by $\mathcal{M}_{X_+}(\mathbf{y}, \boldsymbol{\gamma})$ the moduli space of J^+-holomorphic curves in X_+ with boundary on $L_{\boldsymbol{\alpha}}$ which are positively asymptotic to $[0,1] \times \mathbf{y}$ and negative asymptotic to $\boldsymbol{\gamma}$. To a curve $u \in \mathcal{M}_{X_+}(\mathbf{y}, \boldsymbol{\gamma})$ we associate an ECH-type index $I(u)$ and the intersection number $\mathcal{F}(u)$ between the image of u and the surface S_z. By positivity of intersection, $\mathcal{F}(u) \geq 0$.

We define the map $\Phi^+ \colon CF^+(\Sigma, \boldsymbol{\alpha}, \boldsymbol{\beta}, z) \to ECC(M, \lambda)$ as

$$\Phi^+([\mathbf{y}, i]) = \sum_{\boldsymbol{\gamma} \in \mathcal{O}} \sum_{j=0}^{i} \# \mathcal{M}^{I=0, \mathcal{F}=j}_{X_+}(\mathbf{y}, \boldsymbol{\gamma}) \boldsymbol{\gamma}. \tag{5.2}$$

THEOREM 5.3. *Φ^+ is a well defined chain map.*

This follows from the usual analysis of moduli spaces. However, the symplectic form Ω_{X_+} is not exact, and therefore some extra care is needed. For example, fixing the intersection number $\mathcal{F}$ is an important ingredient for proving compactness of the moduli spaces. Moreover, J^+-holomorphic curves with no positive ends can exist, and we have to exclude them from the count. This is done by showing that they have large ECH-type index; see [**7**, Lemma 5.6.2 and Lemma 5.6.3]. The assumption on the genus of S is used there.

We need to show that the maps $\widehat{\Phi}_*$ and Φ^+_* fit into the commutative diagram (1.1). In [**7**, Section 6] this is done by comparing the mapping cones of U in Heegaard Floer homology and embedded contact homology. For simplicity we prefer to ignore algebraic complications and illustrate only the geometric ideas of [**7**, Section 6]. Thus, instead to work with cones, we will sketch only the proofs of commutativity of the two squares displayed in (1.1). The next theorem shows commutativity of the right-hand square.

THEOREM 5.4 ([**7**, Theorem 5.9.1]). *The map Φ^+ commutes up to homotopy with the maps U on Heegaard Floer homology and embedded contact homology.*

PROOF. The maps U are defined by counting $I = 2$ holomorphic curves passing through a generic point: in embedded contact homology by definition and in Heegaard Floer homology by Theorem 3.2. We define a chain homotopy between $\Phi^+ \circ U$ and $U \circ \Phi^+$ by moving a base point from the positive end of X_+ to the negative end, and counting $I = 1$ holomorphic curves in X_+ passing through the moving point at some isolated instant. □

Now we prove commutativity of the left-hand square.

LEMMA 5.5 ([**7**, Lemmas 5.8.1 and 5.8.2]). *If* $\mathbf{y} \in \mathcal{S}_{\mathbf{a},\hbar(\mathbf{a})}$ *and* $u \in \mathcal{M}_{X_+}^{\mathcal{F}=0}(\mathbf{y},\gamma)$, *then* $\gamma \in \widehat{\mathcal{O}}_{2g}$ *(i.e. it is built from closed Reeb orbits in* N*) and the image of* u *is contained in* W_+.

PROOF. Since $\mathcal{F}(u) = 0$, the image of u is contained in $X_+ \setminus S_z$, where Ω_{X_+} is exact. Then an energy estimate shows that γ is built from closed Reeb orbits in N plus e' and h' because orbits in $\operatorname{int}(V)$ and in the no man's land are long. Then, using positivity of intersection with the J^+-holomorphic surfaces $C_\theta = B_+ \times \{\theta\} \subset W_+$, where $\theta \in \partial S$, we obtain that γ is built from closed orbits in N and the image of u is contained in W_+. □

If $\Omega_{X_+}|_{W_+}$ is sufficiently close to Ω_+, not only we can identify $ECC_{2g}(N,\alpha)$ and $PFC_{2g}(N,\alpha_0,\omega)$, but we can also choose J^+ on X_+ and J_+ on W_+ so that there is a bijection between J^+-holomorphic curves contained in W_+ and J_+-holomorphic curves. Thus Lemma 5.5 implies that there is a commutative diagram

$$\begin{array}{ccc} \overline{CF}(S,\mathbf{a},\hbar(\mathbf{a})) & \longrightarrow & CF^+(\Sigma,\boldsymbol{\alpha},\boldsymbol{\beta},z) \\ \Big\downarrow \overline{\Phi} & & \Big\downarrow \Phi^+ \\ ECC_{2g}(N) & \longrightarrow & ECC(M) \end{array}$$

where the top arrow is the inclusion $\mathbf{y} \mapsto [\mathbf{y},0]$. In homology, both this map and $\overline{\Phi}$ factor through $\widehat{HF}(S,\mathbf{a},\hbar(\mathbf{a}))$, and this proves the commutativity of the left-hand square.

The next theorem is proved by an algebraic trick which generalises Ozsváth and Szabó's observation that $\widehat{HF}(M)$ is trivial if and only if $HF^+(M)$ is trivial[22] in [**40**, Proposition 2.1].

PROPOSITION 5.6 (See [**7**, Theorem 6.1.5]). Φ^+_* *is an isomorphism if and only if* $\widehat{\Phi}_*$ *is an isomorphism.*

6. The closed-open map Ψ

In this section we will define a closed-open map

$$\Psi\colon PFC_{2g}(N,\omega,\alpha_0) \to \widehat{CF}(S,\mathbf{a},\hbar(\mathbf{a})).$$

First, we observe that the naive map that we would obtain by turning the cobordism W_+ "upside-down" does not work: if we stack W_+ on top of the upside-down cobordism W_-, the map we obtain is not homotopic to the identity because holomorphic curves from an Heegaard Floer generator to itself have negative index. A similar phenomenon happens in [**12**], where the composition of the open-closed map from Hochschild homology to symplectic cohomology with the closed-open map from symplectic cohomology to Hochschild cohomology is a "Poincaré duality" rather than the identity. In our case, to amend the naive map, we compactify the cobordism W_- and count holomorphic curves in the compactified cobordism. However the price of this operation is to work with holomorphic curves with boundary conditions on a *singular* Lagrangian *with boundary*, which introduces an incredible amount of complication.

---

[22] $\widehat{HF}(M)$ and $HF^+(M)$ are never trivial, but subgroups associated to some Spin^c structures can be trivial.

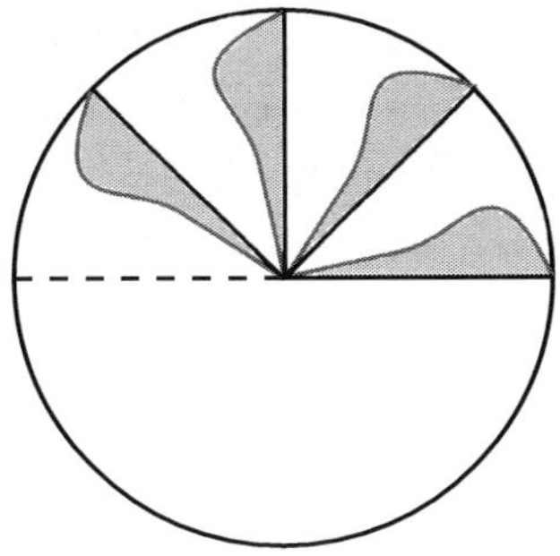

FIGURE 6. The extended arcs and the action of $\overline{h}_m$ on D. The black arcs are portions of the $\overline{a}_i$ and the red ones are portions of the $\overline{h}(\overline{a}_i)$. The gray regions are the bigons covered by the *thin strips* of Definition 6.2.

6.1. Definition of Ψ. Let D be a disc. On D we consider polar coordinates (ρ, ϕ) with $\rho \in [0, 1]$ and write z_∞ for the origin. Let $\overline{S}$ be the capped-off surface obtained by gluing D to S along their boundary. For every $m \gg 0$ (i.e. large enough so that all conditions we impose on it make sense) we define a diffeomorphism $\overline{h}_m : \overline{S} \to \overline{S}$ such that

$$\overline{h}_m|_S = h \quad \text{and} \quad \overline{h}_m|_D(\rho, \phi) = (\rho, \phi + \nu_m(\rho))$$

where $\nu_m(\rho) = \frac{2\pi}{m}$ for $\rho \leq \frac{1}{2}$, $\nu_m(1) = 0$ and $\nu_m(\rho) \geq 0$ for $\rho \in [0, 1]$. We also assume that $\overline{h}_m$ converges to a diffeomorphism $\overline{h}_\infty$ for $m \to \infty$. For the complete list of properties of ν_m see [**5**, Subsection 5.1.2] and for a graphical description of the action of $\overline{h}_m$ on D see Figure 6. Most of the time we will write $\overline{h}$ for $\overline{h}_m$ and assume $m \gg 0$. Similarly, every object depending on $\overline{h}_m$ may or may not have an index m depending on the context.

We extend the arcs a_i to arcs $\overline{a}_i$ by straight lines in D, so that they all meet at z_∞. We assume, among other things, that the angles at z_∞ between two consecutive arcs is an integer multiple of $\frac{2\pi}{m}$ larger than $\frac{4\pi g}{m}$ and that, as $m \to \infty$, the portions of the arcs in $D_{1/2} = \{\rho \leq 1/2\}$ converge to the segment $\{\phi = 0\}$. This last property will be used in the proof of Theorem 6.4. See [**5**, Subsection 5.2.2] for the complete description of the arcs $\overline{a}_i$.

Let B_- be the unit disc with one puncture in the interior and one puncture on the boundary, which we identify biholomorphically with the subset of the cylinder $\mathbb{R} \times \mathbb{R}/2\mathbb{Z}$ obtained by rounding the corners of $(\mathbb{R} \times \mathbb{R}/2\mathbb{Z}) \setminus ((-\infty, -2) \times (1, 2))$. We define a fibration

$$\pi_{\overline{W}_-} : \overline{W}_- \to B_-$$

with fibre $\overline{S}$ and monodromy $\overline{h}$, and equip the total space with the symplectic form $\overline{\Omega}_- = ds \wedge dt + \overline{\omega}$, where (s, t) are coordinates on $B_- \subset \mathbb{R} \times \mathbb{R}/2\mathbb{Z}$ and $\overline{\omega}$ is an area form on $\overline{S}$ which is preserved by $\overline{h}$ and extends ω. We can view $(\overline{W}_-, \overline{\Omega}_-)$ as a symplectic cobordism (with boundary) between $[0, 1] \times \overline{S}$ and $\overline{N}$, where the latter is the mapping torus of $(\overline{S}, \overline{h})$. We refer to [**5**, Section 5] for more details about this construction. Note that $\overline{N}$ carries a stable Hamiltonian structure $(\overline{\alpha}_0, \overline{\omega})$ which is induced by the fibration $\overline{\pi} : \overline{N} \to S^1$. Its only closed Reeb orbit in $\overline{N} \setminus N$ intersecting a fibre at most $2g$ times is the orbit δ_0 corresponding to z_∞.

In $\overline{W}_-$ there is the subset W_- which is the total space of the fibration over B_- with fibre S and monodromy $\mathcal{h}$. Since $\overline{\mathcal{h}}|_D$ is isotopic to the identity, there is an identification $\overline{W}_- \setminus \mathrm{int}(W_-) \cong B_- \times D$ which induces a map[23]

$$\pi_D \colon \overline{W}_- \setminus \mathrm{int}(W_-) \to D.$$

We will denote similar maps for $\mathbb{R} \times [0,1] \times \overline{S}$ and $\mathbb{R} \times \overline{N}$ by the same symbol.

We define an immersed Lagrangian submanifold with boundary $L_{\overline{\mathbf{a}}}$ by parallel transport of $\overline{\mathbf{a}} = (\overline{a}_1, \ldots, \overline{a}_{2g})$ placed at the fibre over $(-3, 1)$. Then,

$$L_{\overline{\mathbf{a}}} \cap (-\infty, -3] \times \{1\} \times \overline{S} = (-\infty, -3] \times \{1\} \times \overline{\mathbf{a}}$$
$$L_{\overline{\mathbf{a}}} \cap (-\infty, -3] \times \{0\} \times \overline{S} = (-\infty, -3] \times \{0\} \times \overline{\mathcal{h}}(\overline{\mathbf{a}}).$$

The manifold $L_{\overline{\mathbf{a}}}$ is embedded in the interior and has a clean self-intersection at the boundary.

We fix an almost complex structure J_- on $\overline{W}_-$ which is generic among those whose restriction to $\overline{W}_- \setminus \mathrm{int}(W_-)$ is the pull back of a split almost complex structure on $B_- \times D$ so that, in particular, the map π_D is holomorphic.

We call $\sigma_\infty^- = B_- \times \{z_\infty\} \subset \overline{W}_-$ the *section at infinity*. By our choice of almost complex structure, σ_∞^- is holomorphic, regular (i.e. its linearised Cauchy-Riemann operator is surjective) and $\mathrm{ind}(\sigma_\infty^-) = 0$; see [**5**, Lemma 5.8.9].

We denote by $\overline{\mathcal{O}}_{2g}$ the set of orbit sets built from closed Reeb orbits in $\overline{N}$ with total intersection number $2g$ with a fibre and by $\overline{\mathcal{S}}_{\mathbf{a},\mathcal{h}(\mathbf{a})}$ the set of $2g$-tuples of intersection points between $\overline{\mathbf{a}}$ and $\overline{\mathcal{h}}(\overline{\mathbf{a}})$. A subtlety in the definition of $\overline{\mathcal{S}}_{\mathbf{a},\mathcal{h}(\mathbf{a})}$ is that z_∞ can be repeated, unlike all other intersection points, because it can be seen as an intersection point between different pairs of arcs. We denote by $\mathcal{M}_{\overline{W}_-}(\gamma, \mathbf{y})$ the moduli space of J_--holomorphic curves in $\overline{W}_-$ with boundary on $L_{\overline{\mathbf{a}}}$ which are positively asymptotic to $\gamma \in \overline{\mathcal{O}}_{2g}$ and negatively asymptotic to $\mathbf{y} \in \overline{\mathcal{S}}_{\mathbf{a},\mathcal{h}(\mathbf{a})}$. We also fix the basepoint $\mathfrak{m} = \{(0, \frac{3}{2})\} \times \{z_\infty\} \in \overline{W}_-$ and denote the set of J_--holomorphic curves in $\mathcal{M}_{\overline{W}_-}(\gamma, \mathbf{y})$ which pass through $\mathfrak{m}$ by $\mathcal{M}_{\overline{W}_-}(\gamma, \mathbf{y}; \mathfrak{m})$.

Let $z_\infty^\dagger \in \overline{S}$ be a point near z_∞ in the complement of the arcs $\overline{\mathbf{a}}$ and $\overline{\mathcal{h}}(\overline{\mathbf{a}})$. We denote by $(\sigma_\infty^-)^\dagger$ the J_--holomorphic multisection which intersects each fibre in the orbit of $z_\infty^\dagger$ under the monodromy $\overline{\mathcal{h}}$. We recall that $\overline{\mathcal{h}}$ has order $m \gg 0$ near z_∞ and therefore $(\sigma_\infty^-)^\dagger$ intersects each fibre m times. Given a J_--holomorphic map $u \in \mathcal{M}_{\overline{W}_-}(\gamma, \mathbf{y})$, we define $n_*(u)$ as the algebraic intersection number between $(\sigma_\infty^-)^\dagger$ and the image of u. The orbit of $z_\infty^\dagger$ gives rise to holomorphic multisections and to intersection numbers, still denoted by n_*, also in $\mathbb{R} \times \overline{N}$ and $\mathbb{R} \times [0,1] \times \overline{S}$. Since both $(\sigma_\infty^-)^\dagger$ and u are J_--holomorphic, $n_*(u) \geq 0$ by positivity of intersection. To u we associate also an ECH-type index $I(u)$; see [**5**, Definition 5.6.6 and Subsection 5.7.7]. If u has no ends at δ_0 or at the chord over z_∞, then $I(u) \geq 0$ for a generic J_- by the index inequality [**5**, Theorem 5.6.9]. Moreover, $I(\sigma_\infty^-) = 0$.

We define a map $\overline{\Psi} \colon ECC_{2g}(N, \alpha_0, \omega) \to \overline{CF}(S, \mathbf{a}, \mathcal{h}(\mathbf{a}))$ by

$$\overline{\Psi}(\gamma) = \sum_{\mathbf{y} \in \mathcal{S}_{\mathbf{a},\mathcal{h}(\mathbf{a})}} \#\mathcal{M}_{\overline{W}_-}^{I=2, n_*=m}(\gamma, \mathbf{y}; \mathfrak{m})\mathbf{y}$$

(where $\gamma \in \widehat{\mathcal{O}}_{2g}$) and, finally, the map $\Psi \colon ECC_{2g}(N, \alpha_0, \omega) \to \widehat{CF}(S, \mathbf{a}, \mathcal{h}(\mathbf{a}))$ by composing $\overline{\Psi}$ with the projection $\overline{CF}(S, \mathbf{a}, \mathcal{h}(\mathbf{a})) \to \widehat{CF}(S, \mathbf{a}, \mathcal{h}(\mathbf{a}))$.

---

[23]This map is called $\overline{\pi}_{D^2}$ in [**5**].

By a standard argument involving the index inequality we can prove that the moduli spaces $\mathcal{M}_{\overline{W}_-}^{I=2,n_*=m}(\gamma, \mathbf{y}; \mathfrak{m})$ are zero-dimensional manifolds. However, showing that they are compact, and even more showing that Ψ is a chain map, occupy a large portion of [**5**] because of the unusual structure of the Lagrangian boundary condition $L_{\mathbf{a}}$. In the next subsection we will give an outline of the argument.

6.2. Ψ is a chain map. To prove that Ψ is a chain map we analyse the boundary of the compactification of the one-dimensional moduli spaces $\mathcal{M}_{\overline{W}_-}^{I=3,n_*=m}(\gamma, \mathbf{y}; \mathfrak{m})$. The first step is to verify that SFT compactness holds.

THEOREM 6.1 (See [**5**, Section 7.3]). *Let u_n be a sequence of holomorphic curves in $\mathcal{M}_{\overline{W}_-}^{I=i,n_*=m}(\gamma, \mathbf{y}; \mathfrak{m})$. Then, up to taking a subsequence, u_n converges in the SFT sense to a building $u_\infty = (u_\infty^{-h}, \dots, u_\infty^0, \dots, u_\infty^l)$, where u_∞^k is one of the following:*

- *a holomorphic curve in $\overline{W}_-$ with boundary on $L_{\overline{\mathbf{a}}}$ and passing through $\mathfrak{m}$ for $k = 0$,*
- *a holomorphic curve in $\mathbb{R} \times \overline{N}$ for $k > 0$, or*
- *a holomorphic curve in $\mathbb{R} \times [0,1] \times \overline{S}$ with boundary on $(\mathbb{R} \times \{1\} \times \overline{\mathbf{a}}) \cup \mathbb{R} \times (\{0\} \times \overline{h}(\overline{\mathbf{a}}))$ for $k < 0$.*

Moreover u_∞^l has positive ends at γ, u_∞^{-h} has negative ends at $\mathbf{y}$ and the positive ends of u_∞^k match with the negative ends of u_∞^{k+1} for $k = -h, \dots, l-1$. Finally, $I(u_\infty^{-h}) + \dots + I(u_\infty^l) = i$ and $n_*(u_\infty^{-h}) + \dots + n_*(u_\infty^l) = m$.

Despite the unusual situation, the proof of Theorem 6.1 follows the proof of the usual SFT compactness theorem in [**3**] very closely. See [**5**, Section 7.3] for the details. Even if $\gamma \in \widehat{\mathcal{O}}_{2g}$ and $\mathbf{y} \in \mathcal{S}_{\mathbf{a},h(\mathbf{a})}$, the holomorphic curves u_∞^k can have one of the following features, which we will call "bad degenerations":

(i) ends at δ_0 (for $k \geq 0$) or at the chord over z_∞ (for $k \leq 0$),
(ii) boundary points at the boundary[24] of $L_{\overline{\mathbf{a}}}$, or
(iii) closed irreducible components.

Any bad degeneration appearing in the compactification of $\mathcal{M}_{\overline{W}_-}^{I=3,n_*=m}(\gamma, \mathbf{y}; \mathfrak{m})$ gives rise to a boundary component which contributes neither to $\partial \circ \overline{\Psi}$ nor $\overline{\Psi} \circ \partial$. In fact, it will turn out that $\overline{\Psi}$ is not a chain map, but we will be able to exclude enough bad degenerations and show that the contribution of the remaining ones is in the kernel of the projection from $\overline{CF}(S, \mathbf{a}, h(\mathbf{a}))$ to $\widehat{CF}(S, \mathbf{a}, h(\mathbf{a}))$.

In [**5**, Section 7.5] we use topological tools — the intersection numbers n_* and the ECH-type index — to prove that bad degenerations can follow only a limited number of patterns, which are described in [**5**, Theorem 7.6.1], and then we exclude all patterns but one by studying, via a rescaling argument inspired by [**25**], how the sequence u_n approaches the limit. For sake of simplicity we wil describe only the main ideas of the proof.

DEFINITION 6.2. A *thin strip* is a holomorphic curve u in $\mathbb{R} \times [0,1] \times D$ such that $\pi_D \circ u$ covers the small bigon in D between an arc in $\overline{a}_i$ and its image under $\overline{h}$. See Figure 6.

A thin strip is positively asymptotic to the chord over z_∞ and negatively asymptotic to a chord over one among the intersection points x_i or x_i'. Moreover, if

[24] This type of bad degeneration is called "boundary points at z_∞" in [**5**].

u is a thin strip, then it is regular and satisfies $I(u) = 1$ and $n_*(u) = 1$. The next proposition summarises some of the results of [**5**, Section 7.5].

PROPOSITION 6.3. *Let u_n be a sequence of J_--holomorphic curves in the moduli space $\mathcal{M}_{\overline{W}_-}^{n_*=m}(\gamma, \mathbf{y}; \mathfrak{m})$ converging to $u_\infty = (u_\infty^{-h}, \ldots, u_\infty^0, \ldots, u_\infty^l)$. Then the following hold:*

(1) *there is at most one nontrivial negative end (i.e. which does not belong to a trivial cylinder) at a cover of δ_0 in all u_∞^k with $k > 0$;*
(2) *if one of the holomorphic curves u_∞^k has a negative end at a cover of δ_0, then u_∞^0 has an irreducible component which is a cover of the section at infinity σ_∞^-;*
(3) *if some u_∞^k with $k < 0$ has a positive end at the chord over z_∞, then some u_∞^k with $k > 0$ has a negative end at a cover of δ_0;*
(4) *any nontrivial irreducible component of u_∞^k (with $k < 0$) with a positive end at the chord over z_∞ is a thin strip;*
(5) *no u_∞^k has a boundary point at the boundary of $L_{\overline{\mathbf{a}}}$; and*
(6) *no u_∞^k has a closed irreducible component.*

SKETCH OF PROOF. The main tool is the analysis of the contributions of ends at covers of δ_0, ends at the chord over z_∞ and boundary points at the boundary of $L_{\overline{\mathbf{a}}}$ to $n_*(u_\infty)$. By [**5**, Lemma 7.4.1], a nontrivial positive end at a p-fold cover of δ_0 contributes at least p to $n_*(u_\infty)$, while a nontrivial negative end contributes at least $m - p$. By [**5**, Lemma 7.4.2] a nontrivial positive end at the chord over z_∞ contributes 1 to $n_*(u_\infty)$ if it belongs to a thin strip, and at least $2g$ otherwise, while a nontrivial negative end contributes at least $2g$. Finally, by [**5**, Lemma 7.4.4] a boundary point at the boundary of $L_{\overline{\mathbf{a}}}$ contributes at least $2g$. All these claims are easy consequences of positivity of intersection between holomorphic curves. As an example, we will compute the contribution of a boundary point at the boundary of $L_{\overline{\mathbf{a}}}$ in u_∞^0. The composition $\pi_D \circ u_\infty^0$ is defined and holomorphic in a neighbourhood of such a point. Since nonconstant holomorphic functions are open, its image covers a region between two arcs in $\overline{\mathbf{a}}$ in a neighbourhood of z_∞. Thus, if we take $z_\infty^\dagger$ close enough to z_∞, the section $(\sigma_\infty^-)^\dagger$ will intersect the image of u_∞^0 at least $2g$ times near the boundary point at the boundary of $L_{\overline{\mathbf{a}}}$.

Suppose that there are two nontrivial negative ends at δ_0 with multiplicity p' and p'' respectively. Then they contribute $2m - p' - p''$ to $n_*(u_\infty)$. Since $p' + p'' \leq 2g \ll m$ and $n_*(u_\infty) = m$, we have a contradiction. This proves (1).

The map $\pi_D \circ u_\infty^0$ is holomorphic and nonconstant, and therefore open, in a neighbourhood of the point mapped to $\mathfrak{m}$. Thus, by unique continuation of holomorphic curves, a neighbourhood of that point contributes m to $n_*(u_\infty)$, unless the irreducible component of u_∞^0 passing through $\mathfrak{m}$ is a cover of the section at infinity σ_∞^-. This proves (2).

A nontrivial negative end at a chord over z_∞ contributes at least $2g$ to $n_*(u_\infty)$. Thus, by the argument above, if there is a positive end at a chord over z_∞, an irreducible component of u_∞^0 is a cover of σ_∞^0, and therefore, for some $k > 0$, the holomorphic curve u_∞^k has a nontrivial negative end at a cover of δ_0. This proves (3).

A nontrivial positive end at a chord over z_∞ contributes at least $2g$ to $n_*(u_\infty)$ unless it belongs to a thin strip. By (3), there is a nontrivial negative end at a cover of δ_0 which contributes $m - p$ with $p < 2g$. This is a contradiction, and thus (4) is proved.

Either there is a nontrivial negative end at a cover of δ_0, which contributes $m-p$ to $n_*(u_\infty)$, or the irreducible component of u_∞^0 passing through $\mathfrak{m}$ is not a cover of σ_∞^- and therefore contributes m to $n_*(u_\infty)$. A boundary point at the boundary of $L_{\overline{\mathbf{a}}}$ contributes at least $2g$, which is a contradiction in both cases because $p < 2g$. This proves (5).

A closed irreducible component of u_∞ is homologous to a branched cover of a fibre $\overline{S}$. Its Fredholm index, computed by the Riemann-Roch formula, is negative, and therefore such a component cannot exist for a generic almost complex structure J_-.[25] This proves (6). □

Now suppose $\sum_{i=-h}^{l} I(u_\infty^k) = 3$. By [**5**, Lemma 7.5.5] $I(u_\infty^0) \geq 0$ and $I(u_\infty^k) \geq 1$ for $i \neq 0$, which implies that we can have buildings with at most four levels in the boundary of the compactification of $\mathcal{M}_{\overline{W}_-}^{I=3,n_*=m}(\boldsymbol{\gamma}, \mathbf{y}; \mathfrak{m})$. Moreover $I(u_\infty^0) = 2$ if no irreducible component of u_∞^0 is a cover of σ_∞^- because passing through $\mathfrak{m}$ is a codimension-two constraint. Therefore, if u_∞ is a two-level building, then $I(u_\infty^0) = 2$ and u_∞ contributes to either $\overline{\Psi} \circ \partial$ or $\partial \circ \overline{\Psi}$. On the other hand, bad degenerations necessarily appear in buildings with three or four levels and have an irreducible component which is a cover of the section at infinity σ_∞^-.

A holomorphic building u_∞ in the compactification of $\mathcal{M}_{\overline{W}_-}^{I=3,n_*=m}(\boldsymbol{\gamma}, \mathbf{y}; \mathfrak{m})$ with a bad degeneration is called of type (A) if $I(u_\infty^1) = 2$ and of type (B)[26] if $I(u_\infty^1) = 1$. If u_∞ is of type (A), then $u_\infty = (u_\infty^{-1}, u_\infty^0, u_\infty^1)$, where u_∞^0 contains σ_∞^- as an irreducible component and u_∞^{-1} is a thin strip. Buildings of type (B) can be ignored as a consequence of the next theorem. We will use a limiting argument for $m \to \infty$, and therefore we will have manifolds $\overline{W}_{-,m}$ constructed from $\hbar_m$, almost complex structures $J_{-,m}$ and so on. However, all manifolds $\overline{W}_{-,m}$ are diffeomorphic, so we fix a model $\overline{W}_-$ and regard all $J_{-,m}$ as almost complex structures on it.

THEOREM 6.4 ([**5**, Theorem 7.10.1]). *If m is large enough, buildings of type (B) do not appear in the boundary of the compactification of $\mathcal{M}_{(\overline{W}_-,J_{-,m})}^{I=3,n_*=m}(\boldsymbol{\gamma}, \mathbf{y}; \mathfrak{m})$ for any $\boldsymbol{\gamma} \in \widehat{\mathcal{O}}_{2g}$ and $\mathbf{y} \in \mathcal{S}_{\mathbf{a},\hbar(\mathbf{a})}$.*

The proof of this theorem occupies the subsections from 7.7 to 7.10 of [**5**]; here we will sketch only the main ideas. Assume that there is a sequence $m_k \to \infty$ such that, for any k, there is a sequence of J_{-,m_k}-holomorphic curves $u_{k,j}$ in $\mathcal{M}_{(\overline{W}_-,J_{-,m_k})}^{I=3,n_*=m_k}(\boldsymbol{\gamma}, \mathbf{y}; \mathfrak{m})$ converging to a building $u_{k,\infty}$ with $I(u_{k,\infty}^1) = 1$. We know from Proposition 6.3 that there is only one nontrivial negative end in $u_{k,\infty}$ at a cover of δ_0 for every k and, moreover, it must belong to an irreducible component of index one. For simplicity, we assume that the end at δ_0 is simple; the case of a nontrivial cover is slightly more complicated, but does not need any new idea. To simplify the situation further, we assume that every $u_{k,\infty}$ contains a level $u_{k,\infty}^1$ with $I(u_{k,\infty}^1) = 1$ from an orbit set $\boldsymbol{\gamma}_+$ to an orbit set $\delta_0\boldsymbol{\gamma}_-$, a level $u_{k,\infty}^0$ which contains σ_∞^- as an irreducible component and has other components contained in

[25]In [**5**] we use an ECH index computation to show that a building u_∞ with a closed component has $I > 3$.

[26]Buildings of type (A) correspond to buildings of type (1) and buildings of type (B) correspond to buildings of type (2) to (6) in [**5**, Theorem 7.6.1].

W_-, and a level $u^{-1}_{k,\infty}$ which is a thin strip from the chord over z_∞ to a chord over one of the intersection points x_i or x'_i. There may be one more level, but it will not be relevant to our argument.

The end of $u^1_{k,\infty}$ at δ_0 (which, we recall, is parametrised by $(s,t) \in (-\infty, 0] \times \mathbb{R}/2\mathbb{Z}$) satisfies the Fourier expansion

$$\pi_D \circ u^1_{k,\infty}(s,t) = e^{\pi(1-\frac{1}{m_k})s} f_k(t) + o(e^{\pi(1-\frac{1}{m_k})s}) \tag{6.1}$$

where $f_k(t) = c_k e^{\pi i t}$, with $c_k \in \mathbb{C}$, is the *asymptotic eigenfunction* of the end. See [**5**, Lemma 7.7.3]. For a generic choice of almost complex structure, and up to taking a subsequence, the coefficients c_k converge to $c \neq 0$ by Lemma 7.7.6 and Lemma 7.7.9 in [**5**]. If we translate s, we multiply c_k by a positive real number, and therefore we can assume that $|c| = 1$. Since the moduli spaces of holomorphic curves in $\mathbb{R} \times \overline{N}$ with $I = 1$ and fixed asymptotics are finite up to translations, the set $\mathcal{C}$ of possible limit values c is finite.

DEFINITION 6.5 (See [**5**, Definition 7.7.10][27]). A *bad radial ray* in $\mathbb{C}$ is a half-line $-ic\mathbb{R}_+$ for $c \in \mathcal{C}$.

We observe that $-i = e^{\pi \frac{3}{2} t}$ and, as we will see later, it is no coincidence that the projection of $\mathfrak{m}$ to B_- is $(0, \frac{3}{2})$. As explained in [**5**, Remark 7.7.11] we can assume that $-1\mathbb{R}_+$ is not a bad radial ray.

After extracting a diagonal subsequence $u_{k,j(k)}$, we find sequences $R^\pm_k \to +\infty$ for $k \to +\infty$ and holomorphic maps $\tilde{u}_k \colon B_- \cap ([-R^-_k, R^+_k] \times \mathbb{R}/2\mathbb{Z}) \to \overline{W}_-$ (called *truncations* in [**5**, Subsection 7.8.1]) which parametrise the portions of $u_{k,j(k)}$ contained in a neighbourhood of the section at infinity σ^-_∞. We define holomorphic functions $\widetilde{w}_k = \pi_D \circ \widetilde{u}_k$ and, after fixing a compact neighbourhood K of $(0, \frac{3}{2})$ in B_-, we define constants $C_k = \|\widetilde{w}_k\|_{L^\infty(K)}$ and holomorphic functions $w_k = \dfrac{\widetilde{w}_k}{C_k}$. If the diagonal subsequence and the truncations are chosen carefully (i.e. as in [**5**, Lemma 7.8.4]), the functions w_k converge to a holomorphic function $w_\infty \colon B_- \to \mathbb{C}$ such that:

- w_∞ has a unique and simple zero at $(0, \frac{3}{2})$,
- $w_\infty(\partial B_-) \subset \mathbb{R}_+$,
- $\lim_{s \to +\infty} \frac{w_\infty(s,t)}{|w_\infty(s,t)|} = c e^{i\pi t}$ for some $c \in \mathcal{C}$, and
- $\lim_{s \to -\infty} w_\infty(s,t) \in \mathbb{R}_+$.

See [**5**, Theorem 7.8.15].

The map $(s,t) \mapsto (s, 1-t)$ defines an anti-holomorphic involution of B_- fixing $(0, \frac{3}{2})$ — remember that $t \in \mathbb{R}/2\mathbb{Z}$. The function

$$f(s,t) = w_\infty(s,t)/\overline{w_\infty(s,1-t)}$$

is thus holomorphic, bounded, and has real boundary conditions. Moreover, it satisfies $\lim_{s\to-\infty} f(s,t) = 1$ and therefore, by standard one-variable complex analysis, $f(s,t) = 1$ for all $(s,t) \in B_-$, so $w_\infty(s,t) = \overline{w_\infty(s,1-t)}$. This implies that $w_\infty(s, \frac{3}{2}) \in \mathbb{R}$ for all $s \in [-2, +\infty) = B_- \cap \left(\mathbb{R} \times \{\frac{3}{2}\}\right)$. Since $(0, \frac{3}{2})$ is the unique zero of w_∞ and $w_\infty(\partial B_+) \subset \mathbb{R}_+$, we have $-ic \in -1\mathbb{R}_+$. This is a contradiction because $-1\mathbb{R}_+$ is not a bad radial ray, and therefore we have shown that, for k

[27] [**5**, Definition 7.7.10] differs from the definition we give here in the notation and for the fact that, in [**5**], we consider also the case of ends to multiple covers of δ_0.

sufficiently large, no building of type (B) with a simple end at δ_0 can be in the boundary of the moduli space $\mathcal{M}^{I=3,n_*=m_k}_{(W_-,J_{-,m_k})}(\gamma,\mathbf{y};\mathfrak{m})$.

If the nontrivial end of each $u_{k,\infty}$ at a cover of δ_0 is not simple, the limit must be taken in the SFT sense and yields a building $w_\infty = (w_\infty^{-a},\dots,w_\infty^b)$ of holomorphic functions $w_\infty^h\colon \widetilde{F}_\infty^h \to \mathbb{C}$ where $\widetilde{F}_\infty^h$ is a branched cover of $\mathbb{R}\times[0,1]$ for $h<0$, of B_- for $h=0$ and of $\mathbb{R}\times S^1$ for $h>0$. See [**5**, Subsection 7.8.7]. Then, we apply the involution argument component by component and obtain a similar contradiction. This ends the proof of Theorem 6.4.

From now on we will assume that m is large enough so that Theorem 6.4 holds. In order to prove that Ψ is a chain map, it remains to prove the following.

THEOREM 6.6 (Reformulation of [**5**, Theorem 7.2.2]). *If $\gamma \in \widehat{\mathcal{O}}_{2g}$ is an orbit set and $\mathbf{y},\mathbf{y}' \in \mathcal{S}_{\mathbf{a},\hslash(\mathbf{a})}$ are $2g$-tuples of intersection points which differ only by replacing one intersection point x_i with x_i' for some i, the numbers of boundary points corresponding to buildings of type (A) in the compactifications of $\mathcal{M}^{I=3,n_*=m}(\gamma,\mathbf{y};\mathfrak{m})$ and $\mathcal{M}^{I=3,n_*=m}(\gamma,\mathbf{y}';\mathfrak{m})$ are equal.*

Let $\mathbf{y}_\infty \in \overline{\mathcal{S}}_{\mathbf{a},\hslash(\mathbf{a})}$ be the $2g$-tuple of intersection points obtained by replacing x_i or x_i' in $\mathbf{y}$ or $\mathbf{y}'$ with z_∞. We fix $\gamma' \in \widehat{\mathcal{O}}_{2g-1}$ and, for $r \gg 0$, define

$$\mathfrak{P} = [r,+\infty)^2\times\widehat{\mathcal{M}}^{I=1,n_*=1}_{\mathbb{R}\times[0,1]\times\overline{S}}(\mathbf{y}_\infty,\mathbf{y})\times\mathcal{M}^{I=0,n_*=0}_{\overline{W}_-}(\delta_0\gamma',\mathbf{y}_\infty)\times\widehat{\mathcal{M}}^{I=2,n_*=m-1}_{\mathbb{R}\times\overline{N}}(\gamma,\delta_0\gamma').$$

The moduli spaces in the definition of $\mathfrak{P}$ have dimension 0, 0 and 1 respectively, and therefore $\mathfrak{P}$ has dimension 3.

For the rest of the section we will identify the quotient of a moduli space by the $\mathbb{R}$ action with a slices for that action (i.e the moduli spaces $\widehat{\mathcal{M}}$ with a slice in the corresponding moduli spaces $\mathcal{M}$). Gluing gives us an embedding

$$G\colon \mathfrak{P} \to \mathcal{M}^{I=3,n_*=m}_{\overline{W}_-}(\gamma,\mathbf{y}).$$

Strictly speaking, we should allow the glued curve to have boundary on the Lagrangian submanifolds associated to the extension of the arcs $\overline{a}_i$ to the other side of z_∞; these are the *extended moduli spaces* of [**5**, Definition 5.7.24]. However, [**5**, Claim 7.12.12] shows that this is a technical complication which can be ignored at a first reading.

To prove Theorem 6.6 we need to understand when the image of G belongs to $\mathcal{M}^{I=3,n_*=m}_{\overline{W}_-}(\gamma,\mathbf{y};\mathfrak{m})$ and show that, essentially, it will only depend on how holomorphic curves in $\widehat{\mathcal{M}}^{I=2,n_*=m-1}_{\mathbb{R}\times\overline{N}}(\gamma,\delta_0\gamma')$ approach δ_0.

To a sequence $(T_{-,n},T_{+,n},u^{-1},u^0,u^1) \in \mathfrak{P}$ with $T_{\pm,n} \to +\infty$ we associate a holomorphic function $w\colon B_- \to \mathbb{C}$ by the truncate-and-rescale technique used in the proof of Theorem 6.4 applied to the sequence $G(T_{-,n},T_{+,n},u^{-1},u^0,u^1)$. We can see this function as a sort of normal vector to (u^{-1},u^0,u^1) in the compactification of $\mathcal{M}^{I=3,n_*=m}_{\overline{W}_-}(\gamma,\mathbf{y})$ along which the sequence $G(T_{-,n},T_{+,n},u^{-1},u^0,u^1)$ approaches (u^{-1},u^0,u^1).

The function w satisfies the following properties:

- $w(s,t) \in \mathbb{R}^+\cdot e^{i\eta(t)}$ for all $(s,t)\in\partial B_-$, where $\eta\colon[0,1]\to\mathbb{R}$ is a decreasing function with $\eta(0)=\frac{\pi}{m}$ and $\eta(1)=0$;
- $\lim_{s\to-\infty}\left|w(s,t)-c_1e^{-\frac{\pi}{m}(s+it-i)}\right| < \infty$ for some $c_1\in\mathbb{R}^+$; and
- $\lim_{s\to+\infty}\left|w(s,t)-c_2e^{\pi(s+it)}\right| < \infty$ for some $c_2\in\mathbb{C}^\times$.

Functions of this sort form a non-empty open cone $\mathcal{N}$ inside a three-dimensional real vector space, are determined by the constants c_1 and c_2 and have a unique zero. See [**5**, Subsection 7.12.1] for the proof of all these properties (and more). Moreover, the zero of a holomorphic function w associated to a sequence $(T_{-,n}, T_{+,n}, u^{-1}, u^0, u^1)$ is the limit of the projections to B_- of the intersection points between the image of $G(T_{-,n}, T_{+,n}, u^{-1}, u^0, u^1)$ and the section at infinity σ_∞^-.

The careful reader has surely observed that the truncate-and-rescale technique in the proof of Theorem 6.4 produced functions which were bounded for $s \to -\infty$, while here they have a pole of order $\frac{\pi}{m}$. The reason is that there we took a limit for $m \to +\infty$ which made the pole disappear; on the other hand, here we work with a large, but fixed, value of m.

We fix $r_0 > r$ and denote $\partial\mathfrak{P}_{(r_0)} = \{T_+ = r_0\} \subset \mathfrak{P}$; thus $\partial\mathfrak{P}_{(r_0)}$ is two-dimensional. If r_0 is generic and sufficiently large, $G(\partial\mathfrak{P}_{(r_0)})$ intersects transversely every branch of the moduli space $\mathcal{M}^{I=3,n_*=m}_{\overline{W}_-}(\gamma, \mathbf{y}; \mathfrak{m})$ converging to a building (u^{-1}, u^0, u^{+1}) with $u^{-1} \in \widehat{\mathcal{M}}^{I=1,n_*=1}_{\mathbb{R}\times[0,1]\times\overline{S}}(\mathbf{y}_\infty, \mathbf{y})$, $u_0 \in \mathcal{M}^{I=0,n_*=0}_{\overline{W}_-}(\delta_0\gamma', \mathbf{y}_\infty)$ and $u_1 \in \widehat{\mathcal{M}}^{I=2,n_*=m-1}_{\mathbb{R}\times\overline{N}}(\gamma, \delta_0\gamma')$.

We define $\Upsilon' \colon \partial\mathfrak{P}_{(r_0)} \to B_-$ such that $\Upsilon'(T_-, r_0, u^{-1}, u^0, u^{+1})$ is the projection to B_- of the intersection point between the image of $G(T_-, r_0, u^{-1}, u^0, u^{+1})$ and σ_∞^-. We define a second map $\Upsilon'' \colon \partial\mathfrak{P}_{(r_0)} \to B_-$ such that $\Upsilon''(T_-, r_0, u^{-1}, u^0, u^{+1})$ is the (unique) zero of a holomorphic function w whose constants c_1 and c_2 are determined by the asymptotic eigenfunctions of $u^{\pm 1}$. Both functions are proper, and therefore their degrees are well defined.

For r_0 large enough, the holomorphic curves in $G(\partial\mathfrak{P}_{(r_0)})$ which intersect σ_∞^- near $\mathfrak{m}$ are close to breaking into the building (u^{-1}, u^0, u^{+1}) by [**5**, Lemma 7.12.15], so Υ' and Υ'' are C^0 close by the truncate-and-rescale argument defining w and therefore they have the same degree. Since $(0, \frac{3}{2})$ is a regular value for Υ' and Υ'' depends only on the asymptotic eigenfunctions of the ends of $u^{\pm 1}$, this implies that the cardinality of the set

$$G(\partial\mathfrak{P}_{(r_0)}) \cap \mathcal{M}^{I=3,n_*=m}_{\overline{W}_-}(\gamma, \mathbf{y}; \mathfrak{m})$$

depends only on the asymptotic eigenfunctions of the ends of $u^{\pm 1}$. Now recall that u^{-1} is a thin strip. The asymptotic eigenfunction of its positive end, which can be easily computed explicitly — see [**5**, Subsection 7.7.1], depends only on the local behavior of the arcs $\overline{a}_i$ and $\overline{h}(\overline{a}_i)$, which is the same for both ends of the arc. Thus the same construction with $\mathbf{y}'$ in the place of $\mathbf{y}$ gives the same result. Thus we have proved the following theorem.

THEOREM 6.7. *Ψ is a chain map.*

7. Homotopies

In this section we prove the following.

THEOREM 7.1. *$\Phi_* \colon \widehat{HF}(S, \mathbf{a}, h(\mathbf{a})) \to PFH_{2g}(N, \alpha_0, \omega)$ is an isomorphism.*

In order to prove Theorem 7.1 we define a chain homotopy

$$H \colon \widehat{CF}(S, \mathbf{a}, h(\mathbf{a})) \to \widehat{CF}(S, \mathbf{a}, h(\mathbf{a}))$$

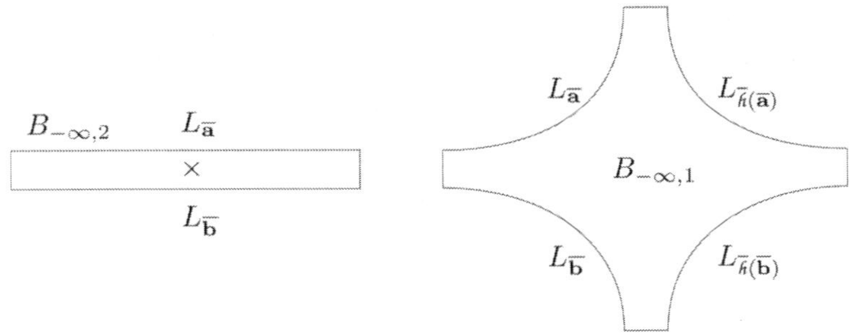

FIGURE 7. The surface $B_{-\infty,2}$ to the left and $B_{-\infty,1}$ to the right. The $\times$ represents the point $\mathfrak{b}_{-\infty}$ and the labels on the boundary are the Lagrangian submanifolds.

between $\Psi \circ \Phi$ and a quasi-isomorphism Θ, and a chain homotopy

$$K: PFC_{2g}(N, \alpha_0, \omega) \to PFC_{2g}(N, \alpha_0, \omega)$$

between $\Phi \circ \Psi$ and the identity by counting isolated holomorphic curves in one-parameter families of cobordisms. The definition of these maps and the proof that they have the required properties occupies the largest part of [**6**].

We define the fibration $\pi_{\overline{W}_+}: \overline{W}_+ \to B_+$ with fibre $\overline{S}$ and monodromy $\overline{\hbar}$. In $\overline{W}_+$ we consider the singular Lagrangian submanifold $L_{\mathbf{a}} \subset \partial\overline{W}_+$ defined as the trace of the parallel transport of a copy of $\overline{\mathbf{a}}$ on $\pi_{\overline{W}_+}^{-1}(3,1)$ along ∂B_+. Given $\mathbf{y} \in \mathcal{S}_{\mathbf{a},\hbar(\mathbf{a})}$ and $\gamma \in \widehat{\mathcal{O}}_{2g}$, we denote by $\mathcal{M}_{\overline{W}_+}(\mathbf{y}, \gamma)$ the moduli space of holomorphic curves in $\overline{W}_+$ with boundary on $L_{\overline{\mathbf{a}}}$ which are positively asymptotic to the chords over $\mathbf{y}$ and negatively asymptotic to γ. We define the multisection $(\sigma_\infty^+)^\dagger$ in analogy with $(\sigma_\infty^-)^\dagger$ in $\overline{W}_-$ and denote the induced intersection number by n_*. If we choose the almost complex structure on $\overline{W}_+$ so that $(\sigma_\infty^+)^\dagger$ is holomorphic, by [**5**, Lemma 5.4.9] we have

$$\mathcal{M}_{\overline{W}_+}^{n_*=0}(\mathbf{y}, \gamma) = \mathcal{M}_{W_+}(\mathbf{y}, \gamma). \tag{7.1}$$

7.1. Homotopy for $\Psi \circ \Phi$. For $\tau \in \mathbb{R}$ we define B_τ as a smoothing of

$$\mathbb{R} \times \mathbb{R}/2\mathbb{Z} \setminus ((e^\tau, +\infty) \times (1,2) \cup (-\infty, -e^\tau) \times (1,2)),$$

which can also be seen, up to a biholomorphism, as an annulus with a puncture on each boundary component. We choose a smooth family of points[28]

$$\mathfrak{b}_\tau \in (-e^\tau, e^\tau) \times \left\{\frac{3}{2}\right\} \subset B_\tau$$

such that $(B_\tau, \mathfrak{b}_\tau)$ converges (in the SFT sense; see [**3**, Section 3.4]) to $(B_{\pm\infty}, \mathfrak{b}_{\pm\infty})$, as $\tau \to \pm\infty$, where:

- $B_{+\infty} = B_+ \sqcup B_-$ (with B_+ on top of B_-) and $\mathfrak{b}_{+\infty} = (0, \frac{3}{2}) \in B_-$, and
- $B_{-\infty} = B_{-\infty,1} \sqcup B_{-\infty,2}$ where $B_{-\infty,1}$ is obtained from $\{-2 \leq s \leq 2\} \cup \{0 \leq t \leq 1\} \subset \mathbb{R}^2$ by smoothing the corners (and therefore is biholomorphic to $D^2 \setminus \{\pm 1, \pm i\}$), $B_{-\infty,2} = [-2,2] \times \mathbb{R}$ and $\mathfrak{b}_{-\infty} \in (0,0) \in [-2,2] \times \mathbb{R}$. See Figure 7.

[28]These points are denoted by $\overline{\mathfrak{m}}^b(\tau)$ in [**6**].

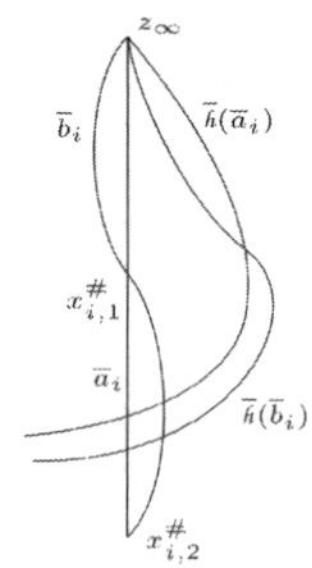

FIGURE 8. The arcs $\overline{a}_i$, $\overline{b}_i$, $\overline{h}(\overline{a}_i)$, and $\overline{h}(\overline{b}_i)$ near z_∞.

Every $(B_\tau, \mathfrak{b}_\tau)$ admits an anti-holomorphic involution which contains $\mathfrak{b}_\tau$ in its fixed point set. This property is used in a symmetry argument similar to that of Subsection 6.2.

For each $\tau \in \mathbb{R}$, we define a fibration

$$\pi_{\overline{W}_\tau} : \overline{W}_\tau \to B_\tau$$

with fibre $\overline{S}$ and monodromy $\overline{h}$ (depending on an integer $m \gg 0$ which will be suppressed from the notation; the reader should keep in mind that all statements hold for m large enough). For $\tau = \pm\infty$ they extend to fibrations over the limit surfaces. The total space $\overline{W}_\tau$ (for $\tau \in \mathbb{R} \cup \{\pm\infty\}$) admits a symplectic form $\Omega_{\overline{W}_\tau}$ such that all fibres are symplectic.

Let $\overline{\mathbf{b}} = (\overline{b}_1, \dots, \overline{b}_{2g})$ be small Hamiltonian perturbations of $\overline{\mathbf{a}}$ (depending on m) such that

- $\overline{a}_i \cap \overline{b}_i = \{z_\infty, x^{\#}_{i,1}, x^{\#}_{i,2}, x^{\#}_{i,3}\}$, where $x^{\#}_{i,2} \in \mathrm{int}(S)$ and $x^{\#}_{i,1}, x^{\#}_{i,3} \in \overline{S} \setminus S$,
- $\overline{b}_i$, near z_∞, is obtained from $\overline{a}_i$ by a small clockwise rotation, and
- $\overline{b}_i$ approaches $\overline{a}_i$ sufficiently fast as $m \to \infty$.

See Figure 8. We denote $\partial_+ B_\tau = \partial B_\tau \cap [0, +\infty) \times \mathbb{R}/2\mathbb{Z}$ and $\partial_- B_\tau = \partial B_\tau \cap (-\infty, 0] \times \mathbb{R}/2\mathbb{Z}$. On $\partial_+ B_\tau$ we consider the singular Lagrangian submanifold $L_{\overline{\mathbf{a}}}$ obtained by parallel transporting $\overline{\mathbf{a}}$ and on $\partial_- B_\tau$ we consider the singular Lagrangian submanifold $L_{\overline{\mathbf{b}}}$ obtained by parallel transporting $\overline{\mathbf{b}}$, so that

$$\begin{aligned}
L_{\mathbf{a}} \cap [2e^\tau, +\infty) \times \{1\} \times \overline{S} &= [2e^\tau, +\infty) \times \{1\} \times \overline{\mathbf{a}}, \\
L_{\mathbf{a}} \cap [2e^\tau, +\infty) \times \{0\} \times \overline{S} &= [2e^\tau, +\infty) \times \{0\} \times \overline{h}(\overline{\mathbf{a}}), \\
L_{\overline{\mathbf{b}}} \cap (-\infty, -2e^\tau] \times \{1\} \times \overline{S} &= (-\infty, -2e^\tau] \times \{1\} \times \overline{\mathbf{b}}, \\
L_{\overline{\mathbf{b}}} \cap (-\infty, -2e^\tau] \times \{0\} \times \overline{S} &= (-\infty, -2e^\tau] \times \{0\} \times \overline{h}(\overline{\mathbf{b}}).
\end{aligned}$$

The Lagrangian submanifolds $L_{\overline{\mathbf{a}}}$ and $L_{\overline{\mathbf{b}}}$ induce Lagrangian submanifolds in $\partial \overline{W}_{\pm\infty}$ in an obvious way. See Figure 7 for the Lagrangian submanifolds in $\partial \overline{W}_{-\infty}$.

Let $W_\tau \subset \overline{W}_\tau$ be the total space of a fibration with fibre S and monodromy h. We choose a generic family of almost complex structures J_τ on $\overline{W}_\tau$ which restrict to the pull back of a split almost complex structure on $\overline{W}_\tau \setminus W_\tau \cong B_\tau \times D$. We also fix a family of marked points $\mathfrak{m}_\tau = (\mathfrak{b}_\tau, z_\infty) \in \overline{W}_\tau$.

Given $\mathbf{y}_+ \in \mathcal{S}_{\mathbf{a}, h(\mathbf{a})}$ and $\mathbf{y}_- \in \mathcal{S}_{\mathbf{b}, h(\mathbf{b})}$, we denote by $\mathcal{M}_{\overline{W}_*}(\mathbf{y}_+, \mathbf{y}_-; \mathfrak{m}_*)$ the moduli space of pairs (τ, u), where $\tau \in \mathbb{R}$ and u is a J_τ-holomorphic curves in $\overline{W}_\tau$ with boundary on $L_{\mathbf{a}} \cup L_{\mathbf{b}}$ which is positively asymptotic to the chords over

$\mathbf{y}_+$, negatively asymptotic to the chords $\mathbf{y}_-$ and passes through $\mathfrak{m}_\tau$. To any such u we associate the ECH-type index $I(u)$ satisfying the index inequality — see [**6**, Subsection 3.2.6] — and the intersection number $n_*(u)$, which is the algebraic intersection between u and the J_τ-holomorphic multisection $(\sigma_\infty^\tau)^\dagger$ defined by a point $z_\infty^\dagger \in \overline{S}$ close to z_∞ and not on the arcs $\overline{\mathbf{a}}$ and $\overline{\mathbf{b}}$; see [**6**, Notation 3.2.9].

DEFINITION 7.2. We define[29] the map $\overline{H}\colon \overline{CF}(S, \mathbf{a}, \hbar(\mathbf{a})) \to \overline{CF}(S, \mathbf{b}, \hbar(\mathbf{b}))$ by

$$\overline{H}(\mathbf{y}_+) = \sum_{\mathbf{y}_- \in \mathcal{S}_{\mathbf{b},\hbar(\mathbf{b})}} \#\mathcal{M}_{\overline{W}_*}^{I=1, n_*(u)=m}(\mathbf{y}_+, \mathbf{y}_-; \mathfrak{m}_*)\mathbf{y}_-. \tag{7.2}$$

Let $\mathcal{S}^\sharp$ be the set of unordered $2g$-tuple of intersection points $x_{1,j(1)}^\sharp, \dots, x_{2g,j(2g)}^\sharp$ between $\overline{\mathbf{a}}$ and $\overline{\mathbf{b}}$. In particular no intersection point is equal to z_∞. Given $\mathbf{y}_+ \in \mathcal{S}_{\mathbf{a},\hbar(\mathbf{a})}$, $\mathbf{y}_- \in \mathcal{S}_{\mathbf{b},\hbar(\mathbf{b})}$ and $\mathbf{x}_\pm \in \mathcal{S}^\sharp$, we will denote the moduli spaces of holomorphic curves in $\overline{W}_{-\infty,1}$ with boundary conditions $L_{\overline{\mathbf{a}}} \cup L_{\overline{\mathbf{b}}} \cup L_{\overline{\hbar}(\overline{\mathbf{b}})} \cup L_{\overline{\hbar}(\overline{\mathbf{a}})}$ and asymptotic to the chords over $\mathbf{y}_+$, $\mathbf{x}_+$, $\mathbf{y}_-$ and $\overline{\hbar}(\mathbf{x}_-)$ at the four ends of $\overline{W}_{-\infty,1}$ (in counterclockwise order starting from the top) by $\mathcal{M}_{\overline{W}_{-\infty,1}}(\mathbf{y}_+, \mathbf{x}_+, \mathbf{y}_-, \overline{\hbar}(\mathbf{x}_-))$. Similarly, we denote the moduli space of holomorphic curves in $W_{-\infty,2}$ with boundary conditions $L_{\overline{\mathbf{a}}} \cup L_{\overline{\mathbf{b}}}$ which are asymptotic to the chords over $\mathbf{x}_\pm^\sharp$ for $t \to \pm\infty$ and pass through $\mathfrak{m}_{-\infty} = (0,0)$ by $\mathcal{M}_{\overline{W}_{-\infty,2}}(\mathbf{x}_+^\sharp, \mathbf{x}_-^\sharp; \mathfrak{m}_{-\infty})$. In the following definition we abbreviate

$$\mathcal{M}_1(\mathbf{y}_+, \mathbf{x}_+^\sharp, \mathbf{y}_-, \overline{\hbar}(\mathbf{x}_-^\sharp)) = \mathcal{M}_{\overline{W}_{-\infty,1}}^{I=0,n_*=0}(\mathbf{y}_+, \mathbf{x}_+, \mathbf{y}_-, \overline{\hbar}(\mathbf{x}_-)) \quad \text{and}$$

$$\mathcal{M}_2(\mathbf{x}_+^\sharp, \mathbf{x}_-^\sharp) = \mathcal{M}_{\overline{W}_{-\infty,2}}^{I=2,n_*=m}(\mathbf{x}_+^\sharp, \mathbf{x}_-^\sharp; \mathfrak{m}_{-\infty})$$

DEFINITION 7.3. We define the map $\overline{\Theta}\colon \overline{CF}(S, \mathbf{a}, \hbar(\mathbf{a})) \to \overline{CF}(S, \mathbf{b}, \hbar(\mathbf{b}))$ as

$$\overline{\Theta}(\mathbf{y}_+) = \sum_{\mathbf{y}_- \in \mathcal{S}_{\mathbf{b},\hbar(\mathbf{b})}} \sum_{\mathbf{x}_\pm^\sharp \in \mathcal{S}^\sharp} \#\left(\mathcal{M}_1(\mathbf{y}_+, \mathbf{x}_+^\sharp, \mathbf{y}_-, \overline{\hbar}(\mathbf{x}_-^\sharp)) \times \mathcal{M}_2(\mathbf{x}_+^\sharp, \mathbf{x}_-^\sharp)\right)\mathbf{y}_-. \tag{7.3}$$

THEOREM 7.4. *The maps $\overline{H}$ and $\overline{\Theta}$ satisfy the relation*

$$\overline{\partial} \circ \overline{H} + \overline{H} \circ \overline{\partial} = \overline{\Psi} \circ \overline{\Phi} + \overline{\Theta} + \overline{V} \tag{7.4}$$

for a map $\overline{V}$ whose image is generated by elements of the form $\{x_i\} \cup \mathbf{y}' + \{x_i'\} \cup \mathbf{y}'$ where x_i and x_i' here denote the intersection points of b_i ad $\hbar(b_i)$ on ∂S and $\mathbf{y}'$ is a $(2g-1)$-tuple of intersection points between the remaining arcs.

Theorem 7.4 is proved by analysing the degenerations of the moduli spaces $\mathcal{M}_{\overline{W}_*}^{I=2,n_*=m}(\mathbf{y}_+, \mathbf{y}_-; \mathfrak{m}_*)$. Degenerations which occur at some $\tau \in \mathbb{R}$ contribute to the left-hand side, degenerations which occur at $\tau \to +\infty$ contribute to $\overline{\Psi} \circ \overline{\Phi} + \overline{V}$ and degenerations which occur at $\tau \to -\infty$ contribute to $\overline{\Theta}$. For $\tau \to +\infty$, the cobordisms $\overline{W}_\tau$ degenerate into the juxtaposition of $\overline{W}_+$ on top of $\overline{W}_-$ with the basepoint $\mathfrak{m}_{+\infty} \in \overline{W}_-$. Sequences of holomorphic curves in $\mathcal{M}_{\overline{W}_*}^{I=2,n_*=m}(\mathbf{y}_+, \mathbf{y}_-; \mathfrak{m}_*)$ degenerate, for $\tau \to \infty$, either to two-level buildings in

$$\mathcal{M}_{\overline{W}_+}^{I=0,n_*=0}(\mathbf{y}_+, \gamma) \times \mathcal{M}_{\overline{W}_-}^{I=2,n_*=m}(\gamma, \mathbf{y}; \mathfrak{m}_{+\infty})$$

[29] For sake of exposition we define here a simplified homotopy map. As explained later, the actual homotopy map must take into account the chord over z_∞.

for some $\gamma \in \widehat{\mathcal{O}}_{2g}$ or to three-level buildings involving a holomorphic curve in $\overline{W}_+$ with a negative and at δ_0, the section at infinity σ_∞^- in $\overline{W}_-$, and a thin strip in $\mathbb{R} \times [0,1] \times \overline{S}$. The two-level degenerations contribute to $\overline{\Psi} \circ \overline{\Phi}$ by Equation (7.1), while the three-level degenerations contribute to $\overline{V}$.

The main tools we use in the analysis of bad degenerations are, as for proving that Ψ is a chain map, the intersection numbers n_*, the ECH-type index I and truncate-and-rescale arguments to analyse how a degenerating sequence approaches its limit. However, the cases to treat are many more than in the previous section and their analysis is more difficult.

THEOREM 7.5. *The map $\overline{\Theta}$ induces a quasi-isomorphism*

$$\Theta \colon \widehat{CF}(S, \mathbf{a}, h(\mathbf{a})) \to \widehat{CF}(S, \mathbf{b}, h(\mathbf{b})).$$

Before proving Theorem 7.5 we explain how, together with Equation (7.4), it proves that $\Psi_* \circ \Phi_*$ is an isomorphism. The missing step is to prove that $\overline{H}$ induces a map $H \colon \widehat{CF}(S, \mathbf{a}, h(\mathbf{a})) \to \widehat{CF}(S, \mathbf{b}, h(\mathbf{b}))$. Unfortunately we were not able to prove directly that $\overline{H}(\{x_i\} \cup \mathbf{y}') = \overline{H}(\{x_i'\} \cup \mathbf{y}')$ if $\mathbf{y}'$ is a $(2g-1)$-tuple of intersection points between the remaining arcs, and thus we extend all maps in Equation (7.4) to a chain complex generated by $2g$-tuples of intersection points which can contain z_∞ and whose homology is $\widehat{HF}(S, \mathbf{a}, h(\mathbf{a}))$. This, of course, brings in more bad degenerations to control.

Now we sketch the main steps of the proof of Theorem 7.5. We consider the two subsets $\mathcal{S}^\sharp_{odd}$ and $\mathcal{S}^\sharp_{even}$ of $\mathcal{S}^\sharp$: the first consists of $2g$-tuple $\{x^\sharp_{1,j(1)}, \dots, x^\sharp_{2g,j(2g)}\}$ where $j(i) \in \{1,3\}$, and the second of the unique element $\{x^\sharp_{1,2}, \dots, x^\sharp_{2g,2}\}$. By [6, Lemma 3.2.15], the moduli spaces $\mathcal{M}^{I=2,n_*=m}_{W_{-\infty,2}}(x^\sharp_+, x^\sharp_-; \mathfrak{m}_{-\infty})$ is nonempty only when $x^\sharp_- \in \mathcal{S}^\sharp_{odd}$ and $x^\sharp_+ \in \mathcal{S}^\sharp_{even}$.

For any $x^\sharp_- \in \mathcal{S}^\sharp_{odd}$ and $x^\sharp_+ \in \mathcal{S}^\sharp_{even}$, the number of elements in the moduli space $\mathcal{M}^{I=2,n_*=m}_{W_{-\infty,2}}(x^\sharp_+, x^\sharp_-; \mathfrak{m}_{-\infty})$ is equal to the relative Gromov-Taubes invariant G_3 defined as the number of embedded holomorphic curves in $D^2 \times \overline{S}$ with boundary on $\partial D^2 \times \mathbf{a}$, representing the relative homology class $[\overline{S}] + 2g[D^2]$ and passing through $(0, z_\infty), (x_1, 1), \dots, (x_{2g}, 1), (x_{2g+1}, -1), \dots, (x_{4g}, -1)$ with $x_i, x_{i+2g} \in \overline{a}_i$. See [6, Subsection 2.4.1].

By [6, Theorem 2.4.2] $G_3 = 1$; the proof is similar to the computation of the relative Gromov-Taubes invariant in Subsection 3.2. Thus we can rewrite

$$\overline{\Theta}(\mathbf{y}_+) = \sum_{\mathbf{y}_- \in \mathcal{S}_{\mathbf{b}, h(\mathbf{b})}} \sum_{\mathbf{x}^\sharp_+ \in \mathcal{S}^\sharp_{odd}} \sum_{x^\sharp_- \in \mathcal{S}^\sharp_{even}} \#\mathcal{M}^{I=0,n_*=0}_{\overline{W}_{-\infty,1}}(\mathbf{y}_+, \mathbf{x}^\sharp_+, \mathbf{y}_-, \overline{h}(\mathbf{x}^\sharp_-))\mathbf{y}_-.$$

This sum now has a fairly standard interpretation in Heegaard Floer theory as composition of two triangle maps which are know to give an isomorphism in homology: see [6, Lemma 3.3.13]. We can also argue as follows: any $2g$-tuple of intersection points $\mathbf{y} \in \mathcal{S}_{\mathbf{a}, h(\mathbf{a})}$ has a closest $2g$-tuple of intersection points $\mathbf{y}_{\in} \mathcal{S}_{\mathbf{a}, h(\mathbf{a})}$. There is a unique holomorphic curve in $W_{-\infty,1}$ of small energy with $I = 0$, $n_* = 0$ and ends at the chords over $\mathbf{y}_+$, $\mathbf{x}^\sharp_+ \in \mathcal{S}^\sharp_{even}$, $\mathbf{y}_-$ and $\mathbf{x}_- \in \mathcal{S}^\sharp_{odd}$: it is a union of $2g$ sections which project to the fibre on the fishtail-shaped quadrilateral in Figure 8.

REMARK 7.6. It might be surprising, at a first uncareful look, that those curves have index zero. The reason is that the concave angle gives a holomorphic map from a disc with four punctures on the boundary to $\overline{S}$ covering the quadrilateral for any conformal structure of the punctured disc, which depends on the cross-ratio of the four punctures. However, we are not looking just for maps into $\overline{S}$ but for multisections of $\overline{W}_{-\infty,1} \to B_{-\infty,1}$, so we also have a holomorphic map from the four-punctured disc to $B_{-\infty,1}$. The conformal structure of $B_{-\infty,1}$ is fixed, and that fixes the conformal structure of the four-punctured disc.

Thus the "low energy part" of $\overline{\Theta}$ is an isomorphism, and therefore $\overline{\Theta}$ is a quasi-isomorphism by standard homological algebra. If a component of $\mathbf{y}_+$ is either x_i or x_i', a direct inspection shows that every element in the moduli space $\mathcal{M}^{I=0,n_*=0}_{\overline{W}_{-\infty,1}}(\mathbf{y}_+, \mathbf{x}^\sharp_+, \mathbf{y}_-, \overline{h}(\mathbf{x}^\sharp_-))$ must have the small energy quadrilateral starting at the chord over x_i or x_i' as an irreducible component. This implies that $\overline{\Theta}(\{x_i\}\cup\mathbf{y}') = \overline{\Theta}(\{x_i'\}\cup\mathbf{y}')$ for all $\mathbf{y}'$, and therefore Θ induces a well defined map $\Theta\colon \widehat{CF}(S, \mathbf{a}, h(\mathbf{a})) \to \widehat{CF}(S, \mathbf{a}, h(\mathbf{a}))$ which is still a quasi-isomorphism.

7.2. Homotopy for $\Phi \circ \Psi$. In this section B_τ will denote a family of cylinders with a disc removed which, for $\tau \to \pm\infty$, converges to $B_{\pm\infty}$ such that:

- $B_{+\infty} = B_- \sqcup B_+$ (with B_- on top of B_+), and
- $B_{-\infty} = B_{-\infty,1} \vee B_{-\infty,2}$ where $B_{-\infty,1} = \mathbb{R} \times \mathbb{R}/2\mathbb{Z}$, $B_{-\infty,2} = D^2$, and the two components are attached by identifying $(0, \frac{3}{2}) \in \mathbb{R} \times \mathbb{R}/2\mathbb{Z}$ with $0 \in D^2$.

In B_τ we choose a marked point $\mathfrak{b}_\tau$ (smoothly in τ) such that $\lim_{\tau\to\pm\infty} \mathfrak{b}_\tau = \mathfrak{b}_{\pm\infty}$ where $\mathfrak{b}_{+\infty} = (0, \frac{3}{2}) \in B_-$ and $\mathfrak{b}_{-\infty} = \frac{i}{2} \in D^2$. We will also assume that every $(B_\tau, \mathfrak{b}_\tau)$ admits an anti-holomorphic involution and that $\mathfrak{b}_\tau$ is contained in its fixed point set. This property is used in a symmetry argument similar to that of Subsection 6.2.

DEFINITION 7.7. The Reeb vector field of N satisfies *Property* $(\dagger\dagger)_i$ if every simple elliptic orbit γ intersecting a fibre at most i times has first return map which is conjugated to a rotation by a sufficiently small negative angle so that the incoming partition of (γ, j), for $j = 1, \dots, i$, is $\underbrace{(1, \dots, 1)}_{j \text{ times}}$ and the outgoing partition is (j).

By [**5**, Section 2.5] it is possible to assume Property $(\dagger\dagger)_i$, for any i, after a C^1-small modification of the stable Hamiltonian structure supported on arbitrarily small neighbourhoods of the elliptic orbits intersecting a fibre at most i times. Note that we can see Property $(\dagger\dagger)_i$ as a property of the monodromy h. In this subsection we will always assume Property $(\dagger\dagger)_{2g}$.

For each $\tau \in \mathbb{R}$, we define a fibration

$$\pi_{\overline{W}_\tau} : \overline{W}_\tau \to B_\tau$$

with fibre $\overline{S}$ and monodromy $\overline{h}$ depending on an integer $m \gg 0$ which will be suppressed from the notation. As in the previous subsection, the reader should keep in mind that all statements hold for m large enough. For $\tau = \pm\infty$ those fibrations extend to fibrations over the limit surfaces. The total space $\overline{W}_\tau$ (for $\tau \in \mathbb{R} \cup \{\pm\infty\}$) admits a symplectic form $\Omega_{\overline{W}_\tau}$ such that all fibres are symplectic. On $\partial\overline{W}_\tau$, for $\tau \in \mathbb{R} \cup \{\pm\infty\}$, we consider the Lagrangian submanifold $L_{\overline{\mathbf{a}}}$ defined as the trace of the symplectic parallel transport along ∂B_τ applied to a copy of

$\overline{\mathbf{a}}$ on the fibre over a point of ∂B_τ. We also fixed the family of marked points $\mathfrak{m}_\tau = (\mathfrak{b}_\tau, z_\infty) \in \overline{W}_\tau$.

Let $W_\tau \subset \overline{W}_\tau$ be the total space of the fibration over B_τ with fibre S and monodromy $\hbar$. We choose a generic family of almost complex structures J_τ on $\overline{W}_\tau$ among those which restrict to the pull back of a split almost complex structure on $\overline{W}_\tau \setminus W_\tau \cong B_\tau \times D$. Since $\overline{W}_{-\infty,1} = \mathbb{R} \times \overline{N}$, we also require that $J_{-\infty,1}$ be compatible with the stable Hamiltonian structure on $\overline{N}$.

Given orbit sets $\boldsymbol{\gamma}_\pm \in \widehat{\mathcal{O}}_{2g}$, we denote by $\mathcal{M}_{\overline{W}_*}(\boldsymbol{\gamma}_+, \boldsymbol{\gamma}_-; \mathfrak{m}_*)$ the moduli space of pairs (u, τ) where $\tau \in \mathbb{R}$ and u is a J_τ-holomorphic curves in $\overline{W}_\tau$ with boundary on $L_{\overline{\mathbf{a}}}$ which is positively asymptotic to $\boldsymbol{\gamma}_+$, negatively asymptotic to $\boldsymbol{\gamma}_-$ and passes through $\mathfrak{m}_\tau$. To each u in this moduli space we associate the ECH-type index $I(u)$ satisfying the index inequality — see [**6**, Subsection 4.2.6] — and the intersection number $n_*(u)$, which is the algebraic intersection between u and the J_τ-holomorphic multisection $(\sigma_\infty^\tau)^\dagger$ defined by a point $z_\infty^\dagger \in \overline{S}$ close to z_∞ and not on the arcs $\overline{\mathbf{a}}$.

Definition 7.8. We define the map $K\colon PFC_{2g}(N, \alpha_0, \omega) \to PFC_{2g}(N, \alpha_0, \omega)$ by

$$K(\boldsymbol{\gamma}_+) = \sum_{\boldsymbol{\gamma}_- \in \widehat{\mathcal{O}}_{2g}} \#\mathcal{M}^{I=1, n_*=m}_{\overline{W}_*}(\boldsymbol{\gamma}_+, \boldsymbol{\gamma}_-; \mathfrak{m}_*)\boldsymbol{\gamma}_-. \tag{7.5}$$

We say that a holomorphic u curve in $\overline{W}_{-\infty,1}$ passes through $\zeta \in \pi^{-1}_{\overline{W}_{-\infty,1}}(0, \frac{3}{2})$ with multiplicity r if the multiplicity of ζ as intersection point between u and $\pi^{-1}_{\overline{W}_{-\infty,1}}(0, \frac{3}{2})$ is r. To make sense of the multiplicity of the intersection we must assume that $\pi^{-1}_{\overline{W}_{-\infty,1}}(0, \frac{3}{2})$ is holomorphic. In fact, in [**5**] and [**6**] we assume that most almost complex structures are compatible with the various fibrations.

We trivialise the fibration

$$\pi_{\overline{W}_{-\infty,2}} \colon \overline{W}_{-\infty,2} \to B_{-\infty,2} = D^2$$

and obtain a commutative diagram

$$\begin{array}{ccc} \overline{W}_{-\infty,2} & \xrightarrow{\ \cong\ } & D^2 \times \overline{S} \\ & {\scriptstyle \pi_{\overline{W}_{-\infty,2}}} \searrow \quad \swarrow & \\ & D^2. & \end{array}$$

We pull back the identification $\pi^{-1}_{\overline{W}_{-\infty,2}}(0) \cong \overline{S}$ given by the diagram above to an identification $\pi^{-1}_{\overline{W}_{-\infty,1}}(0, \frac{3}{2}) \cong \overline{S}$ via the gluing $\pi^{-1}_{\overline{W}_{-\infty,1}}(0, \frac{3}{2}) \cong \pi^{-1}_{\overline{W}_{-\infty,2}}(0)$. Thus a set of "points with multiplicities" $\mathfrak{z} = \{(\zeta_1, r_1), \dots, (\zeta_l, r_l)\}$ with $\zeta_i \in \pi^{-1}_{\overline{W}_{-\infty,1}}(0, \frac{3}{2})$ with $r_i > 0$ and $r_1 + \ldots + r_l = 2g$ can be seen as an element $\mathfrak{z} \in \mathrm{Sym}^{2g}(\overline{S})$.

Given $\mathfrak{z} = \{(\zeta_1, r_1), \dots, (\zeta_l, r_l)\} \in \mathrm{Sym}^{2g}(\overline{S})$, we denote by $\mathcal{M}_{\overline{W}_{-\infty,1}}(\boldsymbol{\gamma}_+, \boldsymbol{\gamma}_-; \mathfrak{z})$ the moduli space of holomorphic curves in $\overline{W}_{-\infty,1} = \mathbb{R} \times \overline{N}$ which are positively asymptotic to $\boldsymbol{\gamma}_+$ and negatively asymptotic to $\boldsymbol{\gamma}_-$ for $\boldsymbol{\gamma}_\pm \in \widehat{\mathcal{O}}_{2g}$ and pass through ζ_i with multiplicity r_i for each $i = 1, \dots, l$.

Similarly, we denote by $\mathcal{M}_{\overline{W}_{-\infty,2}}(\mathfrak{m}, \mathfrak{z})$ the moduli space of holomorphic curves in $\overline{W}_{-\infty,2} = D^2 \times \overline{S}$ with boundary on $L_{\overline{\mathbf{a}}} = \partial D^2 \times \overline{\mathbf{a}}$, representing the relative homology class $[\overline{S}] + 2g[D^2]$ and passing through $\mathfrak{m}_{-\infty} = (\frac{i}{2}, z_\infty)$ and $\mathfrak{z}$.

DEFINITION 7.9. We define the map $\Xi\colon ECC_{2g}(N) \to ECC_{2g}(N)$ as

$$\Xi(\gamma_-) = \sum_{\gamma_- \in \widehat{\mathcal{O}}_{2g}} \sum_{\mathfrak{z} \in \mathrm{Sym}^{2g}(\overline{S})} \# \left(\mathcal{M}^{I=0,n^*=0}_{\overline{W}_{-\infty,1}}(\gamma_+, \gamma_-; \mathfrak{z}) \times \mathcal{M}_{\overline{W}_{-\infty,2}}(\mathfrak{m}, \mathfrak{z}) \right).$$

The sum is finite because the moduli spaces $\mathcal{M}^{I=0,n^*=0}_{\overline{W}_{-\infty,1}}(\gamma_+, \gamma_-; \mathfrak{z})$ are nonempty only for finitely many $\mathfrak{z} \in \mathrm{Sym}^{2g}(\overline{S})$.

The following theorem is proved by analysing the degenerations of the moduli spaces $\mathcal{M}^{I=2,n_*=m}_{\overline{W}_\tau}(\gamma_+, \gamma_-; \mathfrak{m})$. To control the bad degenerations and show that either they cannot occur or they comes in pairs we use a combination of intersection theory, index computations, truncate-and-rescale and symmetry arguments like in Subsection 6.2.

THEOREM 7.10. *The maps K and Ξ satisfy the relation*

$$\partial \circ K + K \circ \partial = \Phi \circ \Psi + \Xi. \tag{7.6}$$

Degenerations of $\mathcal{M}^{I=2,n_*=m}_{\overline{W}_\tau}(\gamma_+, \gamma_-; \mathfrak{m})$ which occur at some $\tau \in \mathbb{R}$ contribute to the left-hand side, degenerations which occur at $\tau \to +\infty$ contribute to $\Phi \circ \Psi$ because $\overline{W}_{+\infty}$ is the juxtaposition of $\overline{W}_-$ on top of $\overline{W}_+$ and the basepoint $\mathfrak{m}_{+\infty}$ is in $\overline{W}_-$, as in the previous subsection, and degenerations which occur at $\tau \to +\infty$ contribute to Ξ.

Property $(\dagger\dagger)_{2g}$ is used in the proof Theorem 7.10 to simplify the gluing of the broken holomorphic curves appearing in the definition of Ξ to holomorphic curves in $\mathcal{M}^{I=2,n_*=m}_{\overline{W}_\tau}(\gamma, \gamma; \mathfrak{m})$ when there is an elliptic orbit with multiplicity larger that one in γ; see the proof of [**6**, Lemma 4.3.4]. We take $\gamma_+ = \gamma_- = \gamma$ because holomorphic curves in $\mathcal{M}^{I=0,n_*=0}_{\overline{W}_{-\infty.1}}(\gamma_+, \gamma_-)$ are necessarily branched covers of trivial cylinders because we have chosen $J_{-\infty,i}$ compatible with the stable Hamiltonian structure.

To exemplify the use of Property $(\dagger\dagger)_{2g}$ suppose for simplicity that we want to glue (v_1, v_2) where $v_1 \in \mathcal{M}^{I=0,n^*=0}_{\overline{W}_{-\infty,1}}(\gamma^{2g}, \gamma^{2g}; \mathfrak{z})$ is a degree $2g$ branched cover of the trivial cylinder over some elliptic orbit γ. Then $\mathfrak{z} = \{(\zeta, 2g)\}$ with $\zeta = \gamma(\frac{3}{2})$. Recall that, in the definition of the moduli spaces for embedded contact homology, branched covers of trivial cylinders with the same degree are identified, but when we glue we must choose a representative. The ends of the curve after gluing must satisfy the incoming and outgoing partitions, which means that we must choose a representative of v_1 satisfying the same conditions. The form of the partitions of γ implies that we can choose v_1 to be a branched cover of γ with a unique branch point at ζ of order $2g - 1$.

To prove that $\Phi_* \circ \Psi_*$ is an isomorphism, it remains to compute Ξ.

THEOREM 7.11. *Ξ is the identity.*

To prove the theorem, we first compute the relative Gromov-Taubes invariant G_2 defined as the number of elements in the moduli spaces $\mathcal{M}_{\overline{W}_{-\infty,2}}(\mathfrak{m}_{-\infty}, \mathfrak{z})$; see [**6**, Section 2.3]. To do so, we degenerate $B_{-\infty,2} \cong D^2$ into a copy of D^2 attached to a copy of S^2 along a point so that $\mathfrak{m}_{-\infty}$ remains in S^2. Holomorphic curves in $\mathcal{M}_{\overline{W}_{-\infty,2}}(\mathfrak{m}_{-\infty}, \mathfrak{z})$ degenerate to constant sections of $D^2 \times \overline{S}$ passing through the points in $\mathfrak{z}$ and holomorphic curves in $S^2 \times \overline{S}$ in the homology class $[\overline{S}] \times 2g[S^2]$ passing through $\mathfrak{m}_{-\infty}$ and $\mathfrak{z}$. We denote the moduli space of the latter holomorphic curves by $\mathcal{M}_{S^2\times\overline{S}}(\mathfrak{m}_{-\infty},\mathfrak{z})$. Thus $\#\mathcal{M}_{\overline{W}_{-\infty,2}}(\mathfrak{m}_{-\infty},\mathfrak{z}) = \#\mathcal{M}_{S^2\times\overline{S}}(\mathfrak{m}_{-\infty},\mathfrak{z})$. By [**6**, Theorem 2.3.3] $\#\mathcal{M}_{S^2\times\overline{S}}(\mathfrak{m}_{-\infty},\mathfrak{z}) = 1$ for any $\mathfrak{z}\in \mathrm{Sym}^{2g}(\overline{S})$. The proof is similar to the computation of the relative Gromov-Taubes invariant in Subsection 3.2. Then

$$\Xi(\gamma_+) = \sum_{\gamma_-\in\widehat{\mathcal{O}}_{2g}} \#\mathcal{M}^{I=0,n^*=0}_{\overline{W}_{-\infty,1}}(\gamma_+,\gamma_+)\gamma_-.$$

Since, for a suitable choice of almost complex structure, $\mathcal{M}^{I=0,n^*=0}_{\overline{W}_{-\infty,1}}(\gamma_+,\gamma_+)$ consists of trivial cylinders, Ξ is the identity map. This proves Theorem 7.11 and therefore concludes the proof of Theorem 7.1.

8. Stabilisation

In Section 7 we proved that $\Phi_*\colon \widehat{HF}(S,\mathbf{a},h(\mathbf{a}))\to PFH_{2g}(N,\alpha_0,\omega)$ is an isomorphism if h satisfies Property $(\dagger\dagger)_{2g}$. In view of Lemma 4.28, to prove that $\widehat{\Phi}_*$ is an isomorphism it remains to prove that $PFH_{2g}(N,\alpha_0,\omega)$ is isomorphic to $\widehat{PFH}(N,\partial N,\alpha_0,\omega)$; this will be done by a stabilisation argument.

In this section it would be desirable to assume Property $(\dagger\dagger)_i$ for all i, but this is not possible. Nevertheless, in practice, we can pretend it is for the following reason.

Lemma 8.1 (See [**6**, Lemma 5.5.2]). *There is a sequence of stable Hamiltonian structures (α_0^i,ω^i) with Reeb vector field R_i satisfying Property $(\dagger\dagger)_i$ such that*

- *R_i and R_{i+1} coincide outside of a small neighbourhood of the orbits of R_i intersecting a fibre at most $i+1$ times, and*
- *the continuation maps*

$$\mathfrak{K}_i\colon PFC_i(N,\alpha_0^i,\omega^i)\to PFC_i(N,\alpha_0^{i+1},\omega^{i+1})$$

satisfy $\mathfrak{K}_i(\gamma)=\gamma$ for every generator of $PFC_i(N,\alpha_0^i,\omega^i)$.

This implies that

$$\varinjlim PFH_i(N,\alpha_0^i,\omega^i) = \widehat{PFH}(N,\partial N,\alpha_0^0,\omega^0).$$

Proposition 8.2. *If h satisfies Property $(\dagger\dagger)_{2g+2}$, then the map*

$$\mathfrak{j}\colon PFC_{2g}(N)\to PFC_{2g+2}(N),\quad \mathfrak{j}(\gamma)=e^2\gamma$$

induces an isomorphism in homology.

We recall that $\mathfrak{j}$ is a chain map because no holomorphic curve in $\mathbb{R}\times N$ can have a nontrivial positive end at a cover of e.

Let (T,h_T) be the open book decomposition of S^3 such that T is a torus with one boundary component and h_T is isotopic, relative to the boundary, to the product of right-handed Dehn twists along two curves intersecting exactly at one point. In particular, the binding is the right-handed trefoil knot. Let N_T be the mapping torus of (T,h_T). We also assume that $h_T^*\beta_T-\beta_T$ is exact for some Liouville form β_T on T and that the Reeb vector field of the stable Hamiltonian structure on N_T induced by the fibration is nondegenerate and satisfies Property $(\dagger\dagger)_{2g+2}$ on $\mathrm{int}(N_T)$, while ∂N_T is a negative Morse-Bott torus.

Next, we take a pair of pants P with $\partial P=\partial_1P\sqcup\partial_2P\sqcup\partial_3P$, we fix an area form on P and choose a function $h\colon P\to\mathbb{R}$ which is sufficiently close to 1 in the C^∞ topology and such that its negative gradient flow is as shown in Figure 9. We

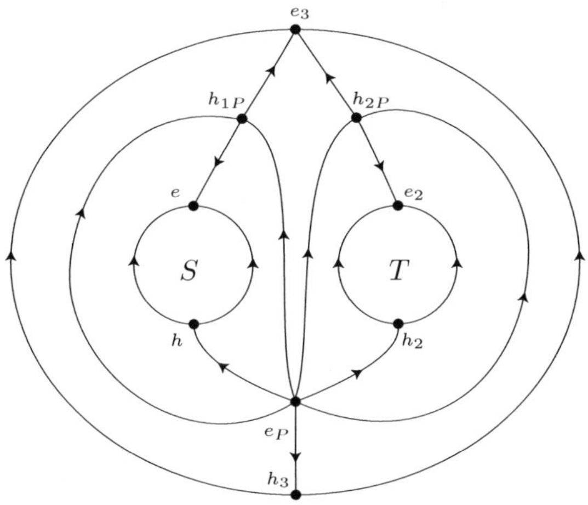

FIGURE 9. The pair of pants P. The negative gradient trajectories of the Morse function f are shown. The critical points e and h belong to the boundary component $\partial_1 P$. The critical points e_i and h_i, for $i = 2, 3$, belong to the boundary component $\partial_i P$.

denote the time 1 flow of f by $\hbar_P$ and the mapping torus of $(P, \hbar_P)$ by N_P. Finally, we define $\widetilde{S} = P \cup S \cup T$, where $\partial_1 P$ is glued to ∂S and $\partial_2 P$ to ∂T. We define also $\widetilde{\hbar}\colon \widetilde{S} \to \widetilde{S}$ such that $\widetilde{\hbar}|_S = \hbar$, $\widetilde{\hbar}_T = \hbar_T$ and $\widetilde{\hbar}|_P = \hbar_P$. Then $(\widetilde{S}, \widetilde{\hbar})$ is an abstract open book decomposition of M with page of genus $2g+2$ and we denote the mapping torus of $(\widetilde{S}, \widetilde{\hbar})$ by $\widetilde{N}$.

Let $\{a_{2g+1}, a_{2g+2}\}$ be a basis of arcs for T and, for $i = 2g+1$ or $2g+2$, let $\{x_i, x_i'\}$ be the intersection points between a_i and $\hbar_T(a_i)$ on ∂N_T. On $\widetilde{S}$ we consider a basis of arcs $\widetilde{\mathbf{a}} = \{\widetilde{a}_1, \dots, \widetilde{a}_{2g+2}\}$ where $\widetilde{a}_i$, for $i = 1, \dots, 2g+2$, is obtained by extending a_i up to $\partial \widetilde{S} = \partial_3 S$. The intersection points between $\widetilde{a}_i$ and $\widetilde{\hbar}(\widetilde{a}_i)$ in $\widetilde{S}$ will be denoted by $\widetilde{x}_i$ and $\widetilde{x}_i'$. Observe that the boundary behavior of the monodromy $\hbar_P$ forces one intersection point in $\mathrm{int}(P)$ between $\widetilde{a}_i$ and $\widetilde{\hbar}(\widetilde{a}_i)$ for each segment of $\widetilde{a}_i \setminus a_i$.

We define the stabilisation map

$$\mathfrak{S}\colon \widehat{CF}(S, \mathbf{a}, \hbar(\mathbf{a})) \to \widehat{CF}(\widetilde{S}, \widetilde{\mathbf{a}}, \widetilde{\hbar}(\widetilde{\mathbf{a}})), \quad \mathfrak{S}(\mathbf{y}) = \mathbf{y} \cup \{\widetilde{x}_{2g+1}, \widetilde{x}_{2g+2}\}.$$

Similarly, we define the stabilisation map

$$\mathfrak{T}\colon PFC_{2g}(N) \to PFC_{2g+2}(\widetilde{N}), \quad \mathfrak{T}(\gamma) = e_3^2 \gamma.$$

It is easy to see that $\mathfrak{S}$ and $\mathfrak{T}$ are chain maps.

LEMMA 8.3. *$\mathfrak{S}$ induces an isomorphism in homology.*

PROOF. One can check that

$$\widehat{HF}(\widetilde{S}, \widetilde{\mathbf{a}}, \widetilde{\hbar}(\widetilde{\mathbf{a}})) \cong \widehat{HF}(S, \mathbf{a}, \hbar(\mathbf{a})) \otimes \widehat{HF}(T, \mathbf{a}_T, \hbar_T(\mathbf{a}_T))$$

and $\widehat{HF}(T, \mathbf{a}_T, \hbar_T(\mathbf{a}_T))$ is generated by the class of $\{\widetilde{x}_{2g+1}, \widetilde{x}_{2g+2}\}$. Since $\mathbf{y} \cup \{\widetilde{x}_{2g+1}, \widetilde{x}_{2g+2}\}$ is homologous to $\mathbf{y} \cup \{x_{2g+1}, x_{2g+2}\}$ because of the intersection points between $\widetilde{a}_i$ and $\widetilde{\hbar}(\widetilde{a}_i)$ contained in $\mathrm{int}(P)$, the lemma follows. □

LEMMA 8.4. *$\mathfrak{T}$ induces an isomorphism in homology.*

PROOF. By an argument similar to the proof of Lemma [**6**, Lemma 5.2.1],

$$\Phi_{(\widetilde{S},\widetilde{h})}(\mathbf{y} \cup \{\widetilde{x}_{2g+1}, \widetilde{x}_{2g+1}\}) = e_3^2 \Phi_{(S,h)}(\mathbf{y}),$$

and therefore there is a commutative diagram

$$\begin{array}{ccc} \widehat{CF}(S, \mathbf{a}, h(\mathbf{a})) & \xrightarrow{\mathfrak{S}} & \widehat{CF}(\widetilde{S}, \widetilde{\mathbf{a}}, \widetilde{h}(\widetilde{\mathbf{a}})) \\ \Phi_{(S,h)} \downarrow & & \downarrow \Phi_{(\widetilde{S},\widetilde{h})} \\ PFC_{2g}(N) & \xrightarrow{\mathfrak{T}} & PFC_{2g+2}(\widetilde{N}). \end{array}$$

Then Theorem 7.1 and Lemma 8.3 imply that $\mathfrak{S}$ induces an isomorphism in homology. □

Let $\iota \colon PFC_{2g+2}(N) \to PFC_{2g+2}(\widetilde{N})$ be the inclusion of complexes induced by the inclusion $N \subset \widetilde{N}$.

LEMMA 8.5. *The inclusion ι is a chain map and induces an injective map $\iota_* \colon PFH_{2g+2}(N) \to PFH_{2g+2}(\widetilde{N})$.*

PROOF. Both claims are a consequence of [**6**, Lemma 5.2.1], stating that each irreducible component of a holomorphic curve in $\mathbb{R} \times \widetilde{N}$ is contained in one of $\mathbb{R} \times N$, $\mathbb{R} \times N_T$ or $\mathbb{R} \times N_P$. This is a consequence of the blocking lemma [**4**, Lemma 5.2.3], the trapping lemma [**4**, Lemma 5.3.2] and Property $(\dagger\dagger)_{2g+2}$, which help us control the behavior of the curves at the ends, and therefore their relative homology classes. □

PROOF OF PROPOSITION 8.2. We claim that the diagram

$$\begin{array}{ccc} PFH_{2g}(N) & \xrightarrow{\mathfrak{T}_*} & PFH_{2g+2}(\widetilde{N}) \\ \mathfrak{j}_* \downarrow & \nearrow \iota_* & \\ PFH_{2g+2}(N) & & \end{array}$$

commutes. In fact, by Morse-Bott theory, there are holomorphic cylinders from h_{1P} to e and e_3 coming from the unstable manifolds of h_{1P} for the negative gradient flow of f (see Figure 9 and [**6**, Lemma 5.2.2]) and therefore, if $\boldsymbol{\gamma}$ is a cycle in $PFC_{2g}(N)$, we have $\partial(h_{1P}(e+e_3)\boldsymbol{\gamma}) = e^2\boldsymbol{\gamma} + e_3^2\boldsymbol{\gamma}$.

The claim implies that $\iota_* \circ \mathfrak{j}_*$ is an isomorphism, and therefore ι_* is surjective. Thus it is invertible by Lemma 8.5, and this shows that $\mathfrak{j}_* = \iota_*^{-1} \circ \mathfrak{T}_*$ is invertible. □

In view of Lemma 8.1 we can repeat the stabilisation argument arbitrarily many times and conclude that the natural map $PFH(N)_{2g} \to \widehat{PFH}(N, \partial N)$ is an isomorphism. This implies that $\widehat{\Phi}_*$ is an isomorphism and therefore finishes the proof of Theorem 1.1.

Acknowledgments

We are indebted to Michael Hutchings for many helpful conversations. We also thank Denis Auroux, Thomas Brown, Tobias Ekholm, Dusa McDuff, Ivan Smith, Jean-Yves Welschinger and Chris Wendl for illuminating exchanges. Finally we are grateful to the organisers of the Boyerfest for giving us the opportunity of writing this survey. Since the beginning of the project VC and PG have been partially supported by ANR grant ANR-08-BLAN-0291-01 "Floer Power", ERC grant 278246 "GEODYCON" and ANR grant ANR- 16-CE40-017 "Quantact". KH was supported by NSF Grants DMS-0805352, DMS-1105432, DMS-1406564, and DMS-1549147. An important part of the work leading to [**4**–**7**] was carried out when PG and KH were visiting MSRI in Spring 2010; we are grateful to that institution for the hospitality.

References

[1] M. Atiyah, *New invariants of* 3- *and* 4-*dimensional manifolds*, The mathematical heritage of Hermann Weyl (Durham, NC, 1987), Proc. Sympos. Pure Math., vol. 48, Amer. Math. Soc., Providence, RI, 1988, pp. 285–299, DOI 10.1090/pspum/048/974342. MR974342

[2] F. Bourgeois, *A Morse-Bott approach to contact homology*, ProQuest LLC, Ann Arbor, MI, 2002. Thesis (Ph.D.)–Stanford University. MR2703292

[3] F. Bourgeois, Y. Eliashberg, H. Hofer, K. Wysocki, and E. Zehnder, *Compactness results in symplectic field theory*, Geom. Topol. **7** (2003), 799–888, DOI 10.2140/gt.2003.7.799. MR2026549

[4] V. Colin, P. Ghiggini and K. Honda, *Embedded contact homology and open book decompositions*, preprint, `arXiv:1008.2734`.

[5] V. Colin, P. Ghiggini, and K. Honda, *Equivalence of Heegaard Floer homology and embedded contact homology via open book decompositions*, Proc. Natl. Acad. Sci. USA **108** (2011), no. 20, 8100–8105, DOI 10.1073/pnas.1018734108. MR2806645

[6] V. Colin, P. Ghiggini, and K. Honda, *Equivalence of Heegaard Floer homology and embedded contact homology via open book decompositions*, Proc. Natl. Acad. Sci. USA **108** (2011), no. 20, 8100–8105, DOI 10.1073/pnas.1018734108. MR2806645

[7] V. Colin, P. Ghiggini and K. Honda, *The equivalence of Heegaard Floer homology and embedded contact homology III: from hat to plus*, preprint, `arXiv:1208.1526`.

[8] V. Colin, P. Ghiggini, K. Honda, and M. Hutchings, *Sutures and contact homology I*, Geom. Topol. **15** (2011), no. 3, 1749–1842, DOI 10.2140/gt.2011.15.1749. MR2851076

[9] Y. Eliashberg, A. Givental, and H. Hofer, *Introduction to symplectic field theory*, Geom. Funct. Anal. **Special Volume** (2000), 560–673, DOI 10.1007/978-3-0346-0425-3_4. GAFA 2000 (Tel Aviv, 1999). MR1826267

[10] A. Floer, *An instanton invariant for* 3-*manifolds*, Comm. Math. Phys. **118** (1988), no. 2, 215–240. MR956166

[11] D. Gabai, *Foliations and the topology of* 3-*manifolds*, J. Differential Geom. **18** (1983), no. 3, 445–503. MR723813

[12] S. Ganatra, *Symplectic Cohomology and Duality for the Wrapped Fukaya Category*, ProQuest LLC, Ann Arbor, MI, 2012. Thesis (Ph.D.)–Massachusetts Institute of Technology. MR3121862

[13] E. Giroux, *Convexité en topologie de contact* (French), Comment. Math. Helv. **66** (1991), no. 4, 637–677, DOI 10.1007/BF02566670. MR1129802

[14] E. Giroux, *Géométrie de contact: de la dimension trois vers les dimensions supérieures* (French, with French summary), Proceedings of the International Congress of Mathematicians, Vol. II (Beijing, 2002), Higher Ed. Press, Beijing, 2002, pp. 405–414. MR1957051

[15] K. Honda, W. H. Kazez, and G. Matić, *Tight contact structures and taut foliations*, Geom. Topol. **4** (2000), 219–242, DOI 10.2140/gt.2000.4.219. MR1780749

[16] K. Honda, W. H. Kazez, and G. Matić, *On the contact class in Heegaard Floer homology*, J. Differential Geom. **83** (2009), no. 2, 289–311. MR2577470

[17] K. Honda, W. Kazez and G. Matić, *Contact structures, sutured Floer homology and TQFT*, preprint, `arXiv:0807.2431`.

[18] M. Hutchings, *An index inequality for embedded pseudoholomorphic curves in symplectizations*, J. Eur. Math. Soc. (JEMS) **4** (2002), no. 4, 313–361, DOI 10.1007/s100970100041. MR1941088

[19] M. Hutchings, *The embedded contact homology index revisited*, New perspectives and challenges in symplectic field theory, CRM Proc. Lecture Notes, vol. 49, Amer. Math. Soc., Providence, RI, 2009, pp. 263–297. MR2555941

[20] M. Hutchings and C. H. Taubes, *Gluing pseudoholomorphic curves along branched covered cylinders. I*, J. Symplectic Geom. **5** (2007), no. 1, 43–137. MR2371184

[21] M. Hutchings and C. H. Taubes, *Gluing pseudoholomorphic curves along branched covered cylinders. II*, J. Symplectic Geom. **7** (2009), no. 1, 29–133. MR2491716

[22] M. Hutchings and C. H. Taubes, *Proof of the Arnold chord conjecture in three dimensions 1*, Math. Res. Lett. **18** (2011), no. 2, 295–313, DOI 10.4310/MRL.2011.v18.n2.a8. MR2784673

[23] M. Hutchings and C. H. Taubes, *Proof of the Arnold chord conjecture in three dimensions, II*, Geom. Topol. **17** (2013), no. 5, 2601–2688, DOI 10.2140/gt.2013.17.2601. MR3190296

[24] M. Hutchings and C. H. Taubes, *The Weinstein conjecture for stable Hamiltonian structures*, Geom. Topol. **13** (2009), no. 2, 901–941, DOI 10.2140/gt.2009.13.901. MR2470966

[25] E.-N. Ionel and T. H. Parker, *Relative Gromov-Witten invariants*, Ann. of Math. (2) **157** (2003), no. 1, 45–96, DOI 10.4007/annals.2003.157.45. MR1954264

[26] A. Juhász, *Holomorphic discs and sutured manifolds*, Algebr. Geom. Topol. **6** (2006), 1429–1457, DOI 10.2140/agt.2006.6.1429. MR2253454

[27] A. Juhász, *Floer homology and surface decompositions*, Geom. Topol. **12** (2008), no. 1, 299–350, DOI 10.2140/gt.2008.12.299. MR2390347

[28] A. Juhász, D. Thurston, and I. Zemke, *Naturality and mapping class groups in Heegaard Floer homology*, preprint `arXiv:1210.4996`.

[29] P. Kronheimer and T. Mrowka, *Monopoles and three-manifolds*, New Mathematical Monographs, vol. 10, Cambridge University Press, Cambridge, 2007. MR2388043

[30] Ç. Kutluhan, S. Sivek and C. H. Taubes, *Sutured ECH is a natural invariant*, preprint, `arXiv:1312.3600`.

[31] Ç. Kutluhan, Y.-J. Lee and C. H. Taubes, *$HF = HM$ I: Heegaard Floer homology and Seiberg–Witten Floer homology*, preprint, `arXiv:1007.1979`.

[32] Ç. Kutluhan, Y.-J. Lee and C. H. Taubes, *$HF = HM$ II: Reeb orbits and holomorphic curves for the ech/Heegaard-Floer correspondence*, preprint, `arXiv:1008.1595`.

[33] Ç. Kutluhan, Y.-J. Lee and C. H. Taubes, *$HF = HM$ III: Holomorphic curves and the differential for the ech/Heegaard Floer correspondence*, preprint, `arXiv:1010.3456` .

[34] Ç. Kutluhan, Y.-J. Lee and C. H. Taubes, *$HF = HM$ IV: The Seiberg-Witten Floer homology and ech correspondence*, preprint, `arXiv:1107.2297`.

[35] Ç. Kutluhan, Y.-J. Lee and C. H. Taubes, *$HF = HM$ V: Seiberg-Witten-Floer homology and handle addition*, preprint, `arXiv:1204.0115`.

[36] R. Lipshitz, *A cylindrical reformulation of Heegaard Floer homology*, Geom. Topol. **10** (2006), 955–1096, DOI 10.2140/gt.2006.10.955. [Paging previously given as 955–1097]. MR2240908

[37] R. Lipshitz, *Correction to the article: A cylindrical reformulation of Heegaard Floer homology [MR2240908]*, Geom. Topol. **18** (2014), no. 1, 17–30, DOI 10.2140/gt.2014.18.17. MR3158771

[38] D. McDuff and D. Salamon, *J-holomorphic curves and symplectic topology*, American Mathematical Society Colloquium Publications, vol. 52, American Mathematical Society, Providence, RI, 2004. MR2045629

[39] P. Ozsváth and Z. Szabó, *Holomorphic disks and topological invariants for closed three-manifolds*, Ann. of Math. (2) **159** (2004), no. 3, 1027–1158, DOI 10.4007/annals.2004.159.1027. MR2113019

[40] P. Ozsváth and Z. Szabó, *Holomorphic disks and three-manifold invariants: properties and applications*, Ann. of Math. (2) **159** (2004), no. 3, 1159–1245, DOI 10.4007/annals.2004.159.1159. MR2113020

[41] P. Seidel, *Fukaya categories and Picard-Lefschetz theory*, Zurich Lectures in Advanced Mathematics, European Mathematical Society (EMS), Zürich, 2008. MR2441780

[42] V. G. B. Ramos, *Absolute gradings on ECH and Heegaard Floer homology*, Quantum Topol. **9** (2018), no. 2, 207–228, DOI 10.4171/QT/107. MR3812797

[43] C. H. Taubes, *Embedded contact homology and Seiberg-Witten Floer cohomology I*, Geom. Topol. **14** (2010), no. 5, 2497–2581, DOI 10.2140/gt.2010.14.2497. MR2746723

[44] C. H. Taubes, *Embedded contact homology and Seiberg-Witten Floer cohomology II*, Geom. Topol. **14** (2010), no. 5, 2583–2720, DOI 10.2140/gt.2010.14.2583. MR2746724

[45] C. H. Taubes, *Embedded contact homology and Seiberg-Witten Floer cohomology III*, Geom. Topol. **14** (2010), no. 5, 2721–2817, DOI 10.2140/gt.2010.14.2721. MR2746725

[46] C. H. Taubes, *Embedded contact homology and Seiberg-Witten Floer cohomology IV*, Geom. Topol. **14** (2010), no. 5, 2819–2960, DOI 10.2140/gt.2010.14.2819. MR2746726

[47] C. H. Taubes, *Embedded contact homology and Seiberg-Witten Floer cohomology V*, Geom. Topol. **14** (2010), no. 5, 2961–3000, DOI 10.2140/gt.2010.14.2961. MR2746727

[48] C. H. Taubes, *Counting pseudo-holomorphic submanifolds in dimension 4 [MR1438194 (97k:58029)]*, Seiberg Witten and Gromov invariants for symplectic 4-manifolds, First Int. Press Lect. Ser., vol. 2, Int. Press, Somerville, MA, 2000, pp. 99–161. MR1798138

[49] C. H. Taubes, *The Seiberg-Witten equations and the Weinstein conjecture*, Geom. Topol. **11** (2007), 2117–2202, DOI 10.2140/gt.2007.11.2117. MR2350473

[50] C. H. Taubes, *Seiberg Witten and Gromov invariants for symplectic 4-manifolds*, First International Press Lecture Series, vol. 2, International Press, Somerville, MA, 2000. Edited by Richard Wentworth. MR1798809

[51] W. P. Thurston and H. E. Winkelnkemper, *On the existence of contact forms*, Proc. Amer. Math. Soc. **52** (1975), 345–347, DOI 10.2307/2040160. MR375366

[52] C. Wendl, *Finite energy foliations on overtwisted contact manifolds*, Geom. Topol. **12** (2008), no. 1, 531–616, DOI 10.2140/gt.2008.12.531. MR2390353

[53] C. Wendl, *Lectures on Symplectic Field Theory*, `arXiv: 1612.01009`.

Laboratoire de Mathématiques Jean Leray, Université de Nantes, F-44322 Nantes
Email address: `vincent.colin@univ-nantes.fr`

Laboratoire de Mathématiques Jean Leray, Université de Nantes, F-44322 Nantes
Email address: `paolo.ghiggini@univ-nantes.fr`

University of California, Los Angeles, Los Angeles, California 90095
Email address: `honda@math.ucla.edu`

Contemporary Mathematics
Volume **760**, 2020
https://doi.org/10.1090/conm/760/15287

Degenerations of representations and thin triangles

Daryl Cooper

ABSTRACT. There is a compactification of the space of representations of a finitely generated group into the groups of isometries of all spaces with Δ-thin triangles. The ideal points are actions on $\mathbb{R}$-trees. It is a geometric reformulation and extension of the Culler-Morgan-Shalen theory concerning limits of representations into $\mathrm{SL}_2\,\mathbb{C}$ and more generally $\mathrm{SO}(n,1)$.

This paper was written and circulated in the early 90's, but never published. Much has happened since then, but we have not attempted to update it.

Let G be a finitely generated group. A map $L : G \to \mathbb{R}$ is called a *length function*. The space of length functions is $\mathrm{LF}(G) = \mathbb{R}^G$ with the product topology. This is compactified by the space, $\mathbb{P}\mathrm{LF}(G) = \mathbb{P}(\mathbb{R}^G)$, of projectivized length functions.

Suppose $\mathcal{H}$ is a geodesic space with Δ-thin triangles and isometry group $\mathrm{Isom}(\mathcal{H})$. A homomorphism $\rho : G \to \mathrm{Isom}(\mathcal{H})$ determines a length function $L(\rho) : G \longrightarrow \mathbb{R}$

$$L(\rho)(g) = \inf\{d_{\mathcal{H}}(x, \rho(g)x) \ : \ x \in \mathcal{H}\}$$

The collection of all such ρ is denoted $\mathrm{Hom}_\Delta(G)$, and there is a map

$$L : \mathrm{Hom}_\Delta(G) \longrightarrow \mathrm{LF}(G)$$

The image of L has compact closure, and limit points in $\mathbb{P}\mathrm{LF}(G)$ are projective classes of length functions of certain actions of G on $\mathbb{R}$-trees. This is because a limit of rescaled Δ-thin spaces is an $\mathbb{R}$-tree.

This may be viewed as a generalization of the compactification of spaces of characters of representations into the Lie groups $\mathrm{SO}(n,1)$ used by Culler, Morgan and Shalen [**4, 6, 7, 9**] by actions on $\mathbb{R}$-trees. This idea was used by Bestvina [**1**] to provide new proofs of some results of Morgan and Shalen [**9**]. In particular he applied a geometric method to obtain this compactification for the subspace of discrete faithful representations into $\mathrm{O}(n,1)$. This paper removes the hypotheses of discrete and faithful, and applies the technique to a wider class of groups than $\mathrm{O}(n,1)$.

Next we give sufficient conditions on a length function to ensure it is the length function of an action of G on a *simplicial* tree. This is used to recover some results of Culler and Shalen [**6**] on compactifications of curves of characters by actions on simplicial trees. The standard approach to this uses the the Tits–Bass–Serre approach [**12**] to trees via $\mathrm{SL}_2\,\mathcal{F}$, where $\mathcal{F}$ is a field with a discrete rank-1

2010 *Mathematics Subject Classification.* 57M50, 57M07, 57M25, 20E08.
Research supported in part by the NSF circa 1992.

valuation. We have reduced the use of valuations to showing that a certain $\mathbb{R}$-tree has a minimal invariant subtree that is simplicial.

In what follows $\mathcal{H}$ is a geodesic space with thin triangles. The starting point in section 1 is the observation that if the vertices of a finite connected graph in $\mathcal{H}$ are moved, so that the diameter of the graph goes to infinity, and if the induced metric on the graph is rescaled to keep the diameter bounded, then the metric on the graph converges to a pseudo-metric for which the associated metric space is a tree.

Let K be the complete graph with vertex set G. Given a homomorphism $\rho : G \longrightarrow \mathrm{Isom}(\mathcal{H})$ there is a G-equivariant map of $f : K \to \mathcal{H}$. Each edge maps linearly. The map is determined by the choice of a single point $x \in \mathcal{H}$ that is a modified version of the *center* of ρ introduced by Bestvina [**1**]. In some sense ρ *operates* near x.

Given a sequence $\rho_n : G \to \mathrm{Isom}(\mathcal{H}_n)$ of homomorphisms for which the translation length of some element goes to infinity, there is a subsequence such that the rescaled pullback metrics on K converge to an $\mathbb{R}$-tree, Γ, on which G acts by isometries. This leads to the compactification result. We then show that, if the translation length of every element of G acting on Γ is an integer, and if the action is irreducible, then Γ has an invariant *simplicial* subtree. This is done in section 2.

Section 3 discusses how spaces of representations can be compactified by mapping the representation space into the space of length functions on the group and compactifying this latter space projectively. This is how Thurston compactified Teichmuller space. The argument up to this point is geometric and combinatorial.

In section 4, we specialize to the action of $\mathrm{SL}_2\,\mathbb{C}$ by isometries on $\mathbb{H}^3$. Valuations make their appearance to show that, if the ρ_n all lie on an *algebraic curve* $C \subset \mathrm{Hom}(G, \mathrm{SL}_2\,\mathbb{C})$ then, after rescaling the limit $\mathbb{R}$-tree, the translation length of every element is an integer. The idea is that the translation length of an element $g \in G$ is approximately $\log(1 + |p_g|)$ where $p_g = \mathrm{trace}(\rho g)$ is a function of $\rho \in C$.

If $p(z)$ is a polynomial of degree d in *one variable* z, then $\log(1 + |p(z)|) \sim d \log |z|$ for z large. For each end ϵ of C there is a valuation ν_ϵ on the function field of C, that generalizes the degree of a polynomial. It can be scaled to take integer values. If $f : C \to \mathbb{C}$ is an algebraic function and $z \in C$ moves out into ϵ, then $f(z)$ behaves like a rational function of one variable, with a degree of growth given by $\nu_\epsilon(f)$. This culminates in the main theorem (4.4).

These ideas were surely known to Culler, Morgan and Shalen, and motivated the search for a more algebraic formulation. However we believe the geometric point of view deserves to be more widely known.

1. Rescaling graphs in negatively curved spaces

The results of this section apply to geodesic spaces satisfying the *thin triangles* property introduced by Gromov in [**8**]. The main result of this section is (1.4) which states that the limit of certain rescaled metric spaces is an $\mathbb{R}$-tree.

Suppose (X, d) is a metric space. The *length* of a path $\gamma : [a, b] \longrightarrow X$ is defined to be the supremum over finite dissections of $[a, b]$ of the sum of the distances between endpoints of the dissecting subintervals. A *path-metric* is a metric with the property that the distance between two points is the infimum of the lengths of paths connecting the points. Thus a Riemannian manifold is a path-metric space.

A *metric simplicial tree* is a path-metric space that is also a simplicial tree. An $\mathbb{R}$-*tree* is a path-metric space, Γ, with the property that given any two points $a, b \in \Gamma$ there is a unique arc γ in Γ with endpoints a, b and γ is isometric to an interval in $\mathbb{R}$, see **[5]** or **[13]**. Thus every metric simplicial tree is an $\mathbb{R}$-tree.

Suppose X is a path metric space. A *geodesic* is an arc in X with the property that the length of the arc equals the distance in X between its endpoints. In a simply-connected complete Riemannian manifold of non-positive curvature this is equivalent to the usual definition of geodesic. A path-metric space in which every pair of points is contained in a geodesic is called a *geodesic space.* Given two points $a, b \in X$ then $[a, b]$ denotes some choice of geodesic with endpoints a, b. A *degenerate geodesic* $[a, a]$ consists of a single point.

A *triangle* in a geodesic space consists of three geodesics $[a, b], [b, c], [c, a]$ called the *sides* or *edges* of the triangle. Given $\Delta > 0$ a triangle is called Δ-*thin* if each side is contained in the Δ-neighborhood of the union of the other two sides. A geodesic space is called *negatively curved* or Δ-*hyperbolic* of *Gromov hyperbolic* if there is $\Delta > 0$ such that every triangle is Δ-thin.

It is clear that simplicial trees, and more generally $\mathbb{R}$-trees, have Δ-thin triangles for all $\Delta > 0$. The term *hyperbolic space* refers to $\mathbb{H}^n$, which is the complete simply connected Riemannian n-manifold with constant sectional curvature -1. It is well known that triangles in $\mathbb{H}^n$ are Δ-thin with $\Delta = \log\left(1 + \sqrt{2}\right)$.

PROPOSITION 1.1. *Suppose $(\mathcal{H}, d_{\mathcal{H}})$ is a geodesic space with Δ-thin triangles, and P is a polygon in $\mathcal{H}$ that has n geodesic sides. Then every side of P lies within a distance $(n-2)\Delta$ of the union of the other sides of P*

PROOF. The polygon P may be triangulated using $n - 2$ triangles by coning from a vertex of P, and the result follows from the thin triangles property. □

From now on we will assume that $(\mathcal{H}, d_{\mathcal{H}})$ is a geodesic space with Δ-thin triangles. Often we will write $d(x, y)$, instead of $d_{\mathcal{H}}(x, y)$. Let K be a *connected graph,* in other words a connected 1-dimensional simplicial complex. A map $f : K \longrightarrow \mathcal{H}$ is called a *straight map* if, for each edge e of K, the restriction of f to e is *linear* in the following sense. Since e is a 1-simplex, we may identify e with $[0, 1]$ by an affine map, and *linear* means $d_{\mathcal{H}}(f(0), f(t)) = t \cdot d_{\mathcal{H}}(f(0), f(1))$.

A *pseudo-metric* satisfies the same conditions as a metric, except that the distance between two distinct points might be zero. There is a pseudo-metric, $f^* d_{\mathcal{H}}$, on K defined by pulling back the metric on $\mathcal{H}$ that is given by

$$f^* d_{\mathcal{H}}(x, y) = d_{\mathcal{H}}(fx, fy)$$

A pseudo-metric space (K, d) determines an equivalence relation on K given by $x \sim y$ if $d(x, y) = 0$. Then d induces a metric, also denoted d, on $K/\sim$. We call $(K/\sim, d)$ the metric space *associated* to (K, d). Clearly f induces an isometry of $(K/\sim, d)$ with the subset $f(K)$ of $\mathcal{H}$. However the associated metric for $f^* d_{\mathcal{H}}$ is usually not a path-metric because there may be no geodesic in $f(K)$ that connects some pair of points in $f(K)$.

PROPOSITION 1.2. *Given $\Delta > 0$, and a finite connected graph K, suppose that $f_n : K \longrightarrow \mathcal{H}_n$ is a sequence of straight maps into geodesic spaces $\mathcal{H}_n$ with Δ-thin triangles. Suppose that $\lambda_n \to 0$, and $\lambda_n \cdot f_n^* d_{\mathcal{H}_n}$ converges pointwise on K to a pseudo-metric $\overline{d}$. Then the metric space associated to $(K, \overline{d})$ is a metric simplicial tree.*

PROOF. Let $(\Gamma, \overline{d})$ be the metric space associated to $\overline{d}$. Since f_n is linear on each edge it is clear that subset of Γ corresponding to an edge of K is an interval, possibly with zero length. Hence γ is the union of finitely many metric intervals.

The proof proceeds by induction on the number of edges of K. It suffices to prove the result for $K^+ = K \cup \beta$ where β is an edge, and $K \cap \beta$ is one or both endpoints of β. Let $\beta_- \subset \beta$ consist of all points $x \in \beta$ with the property that there is $y \in K$ and $\overline{d}(x, y) = 0$. We claim that β_- is a closed subinterval of β. For notational simplicity we assume that all the spaces $\mathcal{H}_n$ are the same.

To prove the claim, suppose $x', y' \in \beta_-$. There are $x, y \in K$ with $\overline{d}(x, x') = 0$ with $\overline{d}(y, y') = 0$. Since K is connected, we may choose an arc ℓ in K that contains x and y. Then ℓ is the union of some finite number, m, of intervals each contained in some edge of K. Thus $f_n(\ell)$ is a polygonal path in $\mathcal{H}$ with endpoints $f_n x$ and $f_n y$ that is the union of m geodesic segments, where m does not depend on n.

Let γ_n be the geodesic segment in $\mathcal{H}$ with endpoints $f_n x$ and $f_n y$. Every point of γ_n lies within a distance $D = (m-1)\Delta$ of $f_n(\ell)$. This follows from (1.1) applied to the polygon $\gamma_n \cup f_n(\ell)$. Now $f_n \beta_-$ is a geodesic arc in $\mathcal{H}$ with endpoints $f_n x'$ and $f_n y'$. Since $\overline{d}(x, x') = 0$ it follows that $\lambda_n d_{\mathcal{H}}(f_n x', f_n x) \to 0$ as $n \to \infty$. Similarly $\lambda_n d_{\mathcal{H}}(f_n y', f_n y) \to 0$.

Suppose $[a_1, a_2]$ and $[b_1, b_2]$ are geodesic arcs in $\mathcal{H}$ and $d_{\mathcal{H}}(a_i, b_i) \leq M$ for $i = 1, 2$. By applying (1.1) to the quadrilateral $[a_1, a_2] \cup [a_2, b_2] \cup [b_1, b_2] \cup [b_2, a_1]$ it follows $[a_1, a_2]$ is contained in a neighborhood of size $M + 2\Delta$ of $[b_1, b_2]$.

We apply this to the arcs γ_n and $f_n(\beta_-)$. It follows that given z' on β_- there is z_n on γ_n with $d_{\mathcal{H}}(f_n z', z_n) \leq M + 2\Delta$. Since the point z_n on γ_n is within a $d_{\mathcal{H}}$-distance D of some point on $f_n(\ell)$ it follows that $d_n(z', \ell) \leq M + 2\Delta + D$. Thus $\overline{d}(z', \ell) = \lim \lambda_n d_n(z', \ell) = 0$. This proves that β_- is an interval. It is easy to show that β_- is closed, establishing the claim.

Since β is attached to K at one endpoint, if $\beta - \beta_-$ is not empty, then it is an interval. If $\beta_- = \beta$ then it is clear that the tree formed by adding β to K is the same as the tree formed from just K. Let $\gamma = [a', b']$ be the closure of $\beta - \beta_-$. One of the endpoints, say a', of γ is a $\overline{d}$-distance of zero from some point, say a, of K. Thus $\overline{d}(a, a') = 0$. The topological space $\Gamma \equiv (K \cup \gamma)/\sim$ is the identification space formed from the union $K/\sim$ with γ by identifying a' with a. Therefore the result is homeomorphic to a simplicial tree. It remains to show that the metric induced by $\overline{d}$ makes Γ into a metric tree.

Let $x \in \gamma$ and $y \in K$ be arbitrary points then there are arcs $[y, a] \subset K/\sim$ and $[a', x] \subset \gamma$. The union of these arcs, $[y, a] \cup [a', x]$, is the unique arc in γ from x to y. The proof is completed by the claim that $\overline{d}(y, x) = \overline{d}(y, a) + \overline{d}(a, x)$.

Suppose that, given $\mu > 0$ that for all sufficiently large n that there is a point, z, on $[f_n x, f_n y]$ with $d_{\mathcal{H}}(f_n a, z) \leq \mu \lambda_n^{-1} + 2\Delta$. Then

$$\begin{aligned}
&| \, d_n(x, y) - (d_n(x, a) + d_n(a, y)) \, | \\
&\leq | \, d_{\mathcal{H}}(f_n x, z) + d_{\mathcal{H}}(z, f_n y) - d_{\mathcal{H}}(f_n x, f_n a) - d_{\mathcal{H}}(f_n a, f_n y) \, | \\
&\leq | \, d_{\mathcal{H}}(f_n x, z) - d_{\mathcal{H}}(f_n x, f_n a) + d_{\mathcal{H}}(z, f_n y) - d_{\mathcal{H}}(f_n a, f_n y) \, | \\
&\leq | \, d_{\mathcal{H}}(z, f_n a) + d_{\mathcal{H}}(z, f_n a) \, | \\
&\leq 2(\mu \lambda_n^{-1} + 2\Delta)
\end{aligned}$$

Multipying both sides by λ_n and using that $\lambda_n \to 0$, the claim follows in this case.

We consider the thin triangle with sides $[f_nx, f_na]$, $[f_nx, f_na']$ and $[f_na, f_na']$. Since $\overline{d}(a,a') = 0$ it follows that given $\epsilon > 0$ for sufficiently large n we have $d_n(a,a') \le \epsilon\lambda_n^{-1}$. Hence for all points q_n on $[f_nx, f_na]$

$$d_{\mathcal{H}}(q_n, [f_nx, f_na']) \le \epsilon\lambda_n^{-1} + \Delta$$

We may assume that there is some $\mu > 0$ such that for arbitrarily large n that $d_{\mathcal{H}}(f_na, [f_nx, f_ny]) > \mu\lambda_n^{-1}+2\Delta$. Define p_n to be the point on $[y,a]$ with $d_n(p_n, a) = \mu\lambda_n^{-1}$ and q_n to be the point on $[f_nx, f_na]$ with $d_{\mathcal{H}}(q_n, f_na) = \mu\lambda_n^{-1}$. Now:

$$\begin{aligned} d_{\mathcal{H}}(f_np_n, [f_ny, f_nx]) &\ge d_{\mathcal{H}}(f_na, [f_ny, f_nx]) - d_{\mathcal{H}}(f_na, f_np_n) \\ &\ge \mu\lambda_n^{-1} + 2\Delta - \mu\lambda_n^{-1} \\ &= 2\Delta \end{aligned}$$

Then thin triangles applied to the triangle with vertices f_na, f_nx, f_y gives

$$d_{\mathcal{H}}(f_np_n, [f_na, f_nx]) \le \Delta$$

Let r_n be a point on $[f_na, f_nx]$ with $d_{\mathcal{H}}(f_np_n, r_n) \le \Delta$. Then

$$\mid d_{\mathcal{H}}(f_na, r_n) - d_{\mathcal{H}}(f_na, f_np_n) \mid \le \Delta$$

Since r_n and q_n are both on $[f_na, f_nx]$ and $d_n(a, p_n) = d_{\mathcal{H}}(f_na, q_n)$ it follows that $d_{\mathcal{H}}(r_n, q_n) \le \Delta$. Thus

$$d_{\mathcal{H}}(f_np_n, q_n) \le d_{\mathcal{H}}(f_np_n, r_n) + d_{\mathcal{H}}(r_n, q_n) \le 2\Delta$$

Combining this with the bound on $d_{\mathcal{H}}(q_n, [f_nx, f_na'])$ gives

$$d_{\mathcal{H}}(f_np_n, [f_nx, f_na']) \le d_{\mathcal{H}}(f_np_n, q_n) + d_{\mathcal{H}}(q_n, [f_nx, f_na']) \le 2\Delta + \epsilon\lambda_n^{-1} + \Delta$$

As $n \to \infty$ there is a subsequence of p_n which converges to a point p in $[a,y]$. We get $\overline{d}(p, [x,a']) \le \epsilon$. Since $\epsilon > 0$ is arbitrary this gives $\overline{d}(p, [x,a']) = 0$. Now $\overline{d}(p,a) = \mu > 0$ so $p \ne a$ but this contradicts that a is the only point on $[a,y]$ zero distance from $[x,a']$. This proves the claim □

LEMMA 1.3. *Suppose that (Γ, d) is a metric space which is the union of an increasing sequence of subspaces T_n, each of which is a metric simplicial tree. Then Γ is an $\mathbb{R}$-tree*

PROOF. Suppose that x, y are two distinct points in Γ then there is some T_m which contains both of them. There is a unique arc $[x,y]$ in T_m. Suppose that there is a different arc α in Γ with the same endpoints. The intersection of α with $[x,y]$ is closed. We may assume that $[x,y]$ intersects α only in the points x, y. For otherwise we may replace $[x,y]$ by a suitable subarc.

Define a nearest-point projection

$$\pi : \Gamma \longrightarrow [x,y]$$

as follows. Given a point z in Γ there is an $n > m$ such that T_n contains z. Define πz to be the unique point on $[x,y]$ which is closest to z in T_n. Since the trees T_n are an increasing sequence, it is clear that this definition of πz is independent of n. We claim that π is continuous. For if $d_\Gamma(z,z') < \epsilon$ then if n is chosen large enough that T_n contains both z and z' then $d_{T_n}(z,z') < \epsilon$. But the restriction of π to the simplicial tree T_n clearly does not increase distance thus $d_{T_n}(\pi z, \pi z') < \epsilon$. This proves the claim.

We claim that $\pi\alpha = x$. Let β be the maximal subarc of α containing x such that $\pi\beta = x$. Since $\pi x = x$ it follows that β is not empty. Since π is continuous it follows that β is closed. Writing $\beta = [x, z]$ choose $n > m$ with T_n containing z. Assuming $z \neq x, y$ then $d_{T_n}(z, \pi z) = d_{T_n}(z, [x, y]) = \delta > 0$. Now $\pi y = y$ so $\beta \neq \alpha$ thus we may choose z' a point in $\alpha - \beta$ with $\pi z' \neq x$. Now choose $n > m$ such that T_n contains both z, z'. By choice of z' we have that $[z, z'] = [z, \pi z] \cup [\pi z, \pi z'] \cup [\pi z', z']$. Therefore $d_{T_n}(z, z') \geq d_{T_n}(z, \pi z) = \delta$. Since z' can be arbitrarily close to z this is a contradiction. Hence $z = x$.

It follows that there is a unique arc in Γ between any pair of distinct points. Since there is some T_n containing both these points, this arc is metrized as an interval. Hence Γ is an $\mathbb{R}$-tree □

Combining these two results together gives:

THEOREM 1.4. *Suppose that $\Delta > 0$, and $\mathcal{H}_n$ is a sequence of geodesic spaces with Δ-thin triangles, and K is a connected graph with countable many edges. Suppose that $f_n : K \to \mathcal{H}_n$ is a sequence of straight maps.*

Suppose that $\lambda_n \to 0$, and $\lambda_n \cdot f_n^* d_{\mathcal{H}_n}$ converges pointwise on K to a pseudo-metric $\overline{d}$. Then the metric space, Γ, associated to $(K, \overline{d})$ is an $\mathbb{R}$-tree

2. Limiting actions on $\mathbb{R}$-trees

In this section G is a finitely generated group, and we study actions of G by isometries on a geodesic space $\mathcal{H}$ with Δ-thin triangles. A sequence of homomorphism (*representations*) $\rho_n : G \longrightarrow \text{Isom}(\mathcal{H})$ *blows up* if there is $\alpha \in G$ such that $\lim_{n\to\infty} t(\rho_n \alpha) = \infty$, see [**3**]. If this case, after rescaling, there is a limiting action of G on an $\mathbb{R}$-tree Γ. With extra hypotheses, Γ contains a G-invariant simplicial subtree.

Suppose that (X, d) is a metric space and τ is an isometry of this space then the *translation length* of τ is

$$t(\tau) = \inf\{d(x, \tau x) : \ x \in X \ \}$$

This defines a map $t : \text{Isom}(X) \longrightarrow \mathbb{R}$.

Let $K \equiv K(G)$ be the complete graph with G as vertex set. This is a graph with one edge between every pair of vertices. Choose a finite generating set S for G. Given x in $\mathcal{H}$, and a representation $\rho : G \longrightarrow \text{Isom}(\mathcal{H})$, define

$$r_S(x) = \max\{d(x, (\rho\alpha)x) : \ \alpha \in S \ \}$$

An *approximate center* of ρ is a choice of x in $\mathcal{H}$ for which $r_S(x) \leq r_S(y) + 1$ for all $y \in \mathcal{H}$, compare [**1**].

For example, if G is the fundamental group of a compact convex hyperbolic surface with boundary, the developing map identifies the basepoint of the universal cover with some point in the hyperbolic plane. This is an approximate center. Intuitively the representation *operates near here*. It is used to overcome the problems associated with changing a representation by conjugacy.

Choose an approximate center x_n for ρ_n and define $\iota_n : K \longrightarrow \mathcal{H}$ on the vertices of $K = K(G)$ by:

$$\iota_n(\alpha) = (\rho_n \alpha) x_n$$

Then extend ι_n linearly over each edge of K so that ι_n is a straight map. Let $d_n = \iota_n^* d_{\mathcal{H}}$ be the pullback pseudo-metric on K. Now G acts on the vertices of K by multiplication on the left. Extend this action linearly over the edges of K to get an action $\sigma : G \longrightarrow \text{Isom}(K, d_n)$ by isometries on the pseudo-metric space (K, d_n) It is clear that ι_n is a G-equivariant isometry from K onto its image.

Define the *rescaling factor*

$$\lambda_n = (\max\{1, d_n(1_G, \beta) : \ \beta \in S\})^{-1}$$

and the *rescaled pseudo-metric* by $\tilde{d}_n = \lambda_n d_n$. Since ρ_n is blowing up it is easy to see that $\lambda_n \to 0$.

PROPOSITION 2.1. *Suppose that G is a finitely generated group and $\rho_n : G \to$ Isom$(\mathcal{H})$ is a sequence of representations which blows up. Let $\tilde{d}_n$ be as above. Then there is a subsequence of $\tilde{d}_n$ that converges pointwise to a pseudo-metric $\overline{d}$ on K. Furthermore, the metric space associated to $(K, \overline{d})$ is an $\mathbb{R}$-tree*

PROOF. Define K_S to be the subgraph of K which is the Cayley graph corresponding to the generating set S. Thus for each $s \in S$ there is an edge in K_S connecting g to gs. This condition is preserved by multiplication on the left, hence K_S is preserved by G. Thus K_S is a connected graph on which G acts.

The rescaling factor was chosen so that the $\tilde{d}_n$-length of every edge in K_S adjacent to the identity is at most 1. Since G acts transitively on vertices, and by $\tilde{d}_n$-isometries, it follows that the $\tilde{d}_n$-length of every edge in K_S is at most 1. Hence, for every pair of vertices x, y in K_S we have that $\tilde{d}_n(x, y)$ is bounded by the number of edges in a path from x to y. Since $\tilde{d}_n$ metrizes every edge of K as an interval, it follows that for every pair of points x, y in K there is a bound on $\tilde{d}_n(x, y)$

Choose a countable dense subset, D, of K. For x, y in D we have that $\tilde{d}_n(x, y)$ is bounded independently of n. Hence there is a subsequence, n_i, so that for all x, y in D then $\tilde{d}_{n_i}(x, y)$ converges. We claim for all x, x' in K that $\tilde{d}_{n_i}(x, x')$ converges. The pointwise-limit is clearly a pseudo-metric, $\tilde{d}$, on K. In what follows we assume the sequence has been replaced by the subsequence.

To prove the claim, consider an edge e of K, then the map ι_n is linear on e. The limit pseudo-metric restricted to e metrizes e as a closed interval. Given a point, x, on e we may choose x_1, x_2 in $D \cap e$ with x between them and with the limiting distance between x_1 and x_2 very small so that for all n sufficiently large $\tilde{d}_n(x, x_1) < \epsilon$. Given another point x' on some other edge e' we may similarly choose x'_1, x'_2 with limiting distance between them ϵ'. Then for n sufficiently large:

$$\begin{aligned}|\tilde{d}_n(x, x') - \tilde{d}_n(x_1, x'_1)| &\le \tilde{d}_n(x, x_1) + \tilde{d}_n(x'_1, x')\\ &\le \tilde{d}_n(x_2, x_1) + \tilde{d}_n(x'_1, x'_2)\\ &\le 2\epsilon\end{aligned}$$

Since $\tilde{d}_n(x_1, x'_1)$ converges it follows that $\tilde{d}_n(x, x')$ converges as $n \to \infty$. It follows from (1.4) that the metric space associated to $(K, \overline{d})$ is an $\mathbb{R}$-tree □

We now describe isometries of $\mathbb{R}$ trees, see **[5]** for a more complete treatment. The *core* of isometry, τ, of an $\mathbb{R}$-tree Γ is the set of points in Γ which are moved the minimal distance by τ :

$$C(\tau) = \{\ x \in \Gamma \ : \ d_\Gamma(x, \tau x) = t(\tau)\ \}$$

It is clear that this a closed convex set (ie $x, y \in$ core(τ) implies $[x, y]$ is in core(τ)) which is invariant under τ. An isometry, τ, of Γ is called *elliptic* if τ has a fixed point. It is called *hyperbolic* if $t(\tau) > 0$ and the core of τ is isometric to a line. This line is called the *axis* of τ.

LEMMA 2.2. *Every isometry of an $\mathbb{R}$-tree is either elliptic or hyperbolic.*

PROOF. Choose a point x in Γ and consider the intersection of $[x, \tau x]$ with $\tau^{-1}([x, \tau x])$. This is a geodesic $[x, y]$. If $y = \tau y$ then τ is elliptic. Otherwise, from the definition of y it follows that the intersection of $[y, \tau y]$ with $\tau^{-1}[y, \tau y]$ is the single point y, because any other point of intersection would also give an extra point of intersection of $[x, \tau x]$ with $\tau^{-1}([x, \tau x])$. Thus $\ell = \bigcup_n \tau^n[y, \tau y]$ is isometric to $\mathbb{R}$ and invariant under τ. Now $[x, \tau x] = [x, y] \cup [y, \tau y] \cup [\tau y, \tau x]$ and so $d(x, \tau x) \geq d(y, \tau y)$ with equality if and only if x is on ℓ. Thus the points moved the minimal distance by τ are exactly those points on ℓ. Hence τ has axis ℓ and is thus hyperbolic $\square$

LEMMA 2.3. *Suppose that α, β are isometries of an $\mathbb{R}$-tree and that* $\mathrm{core}(\alpha)$ *is disjoint from* $\mathrm{core}(\beta)$. *Then* $t(\alpha\beta) = t(\alpha) + t(\beta) + 2d(\mathrm{core}(\alpha), \mathrm{core}(\beta))$

PROOF. Let a be a point in $\mathrm{core}(\alpha)$ and b a point in $\mathrm{core}(\beta)$ such that $d(a, b)$ is minimal. To see that such points exist, choose points a' in $\mathrm{core}(\alpha)$ and b' in $\mathrm{core}(\beta)$. Define a by $[a', a] = [a', b'] \cap \mathrm{core}(\alpha)$ and b by $[b, b'] = [a', b'] \cap \mathrm{core}(\beta)$. Given any arc, γ, with one endpoint in $\mathrm{core}(\alpha)$ and the other endpoint in $\mathrm{core}(\beta)$ then γ contains $[a, b]$. For otherwise, since cores are convex we can join to γ an arc in $\mathrm{core}(\alpha)$ connecting a to γ and another arc in $\mathrm{core}(\beta)$ connecting b to γ. Observe that the arcs joined to γ meet $[a, b]$ only at a or b. The union of these three arcs is a path from a to b which contains a subset which is an embedded arc from a to b. But $[a, b]$ is the unique arc in the $\mathbb{R}$-tree with these endpoints. Hence γ contains $[a, b]$ as asserted. Hence $\mathrm{core}(\gamma) \geq d(a, b)$. Thus a, b have the claimed property.

Define

$$\ell = [\alpha(b), \beta^{-1}(b)] = \alpha([b, a]) \cup [\alpha(a), a] \cup [a, b] \cup [b, \beta^{-1}(b)]$$

Observe that $[\alpha(a), a]$ is contained in the $\mathrm{core}(\alpha)$ and $[b, \beta^{-1}(b)]$ is contained in the $\mathrm{core}(\beta)$. Also $[a, b]$ intersects $\mathrm{core}(\alpha) \cup \mathrm{core}(\beta)$ only at a, b. It follows from this that the four arcs on the right hand side only intersect at endpoints. Since $a \in \mathrm{core}(\alpha)$ we have $t(\alpha) = d(a, \alpha(a))$ and similarly $t(b) = d(\beta(b), b) = d(b, \beta^{-1}(b))$. Thus

$$\begin{aligned}
\mathrm{core}(\ell) &= d(\alpha(b), \alpha(a)) + d(\alpha(a), a) + d(a, b) + d(b, \beta^{-1}(b)) \\
&= d(\alpha(b), \alpha(a)) + t(a) + d(a, b) + t(b) \\
&= t(a) + 2d(a, b) + t(b) \\
&= t(a) + 2d(\mathrm{core}(\alpha), \mathrm{core}(\beta)) + t(b).
\end{aligned}$$

We claim that $\ell \cap \alpha\beta(\ell) = \alpha(b)$, from which it follows that ℓ is contained in the axis of $\alpha\beta$ and is a fundamental domain for $\alpha\beta$. The computation of $\mathrm{core}(\ell)$ above then completes the proof.

Now $[b, \beta^{-1}(b)]$ is contained in $\mathrm{core}(\beta)$. Since $\mathrm{core}(\beta)$ is invariant under β it follows that $\beta([b, \beta^{-1}(b)])$ is contained in $\mathrm{core}(\beta)$. Since $[b, a] \cap \mathrm{core}(\beta) = b$ it follows that $[b, a] \cap \beta([b, \beta^{-1}(b)]) = b$. If the claim is false then $\ell \cap \alpha\beta(\ell)$ is an interval of positive length containing $\alpha(b)$. Now $\alpha(b)$ is the start of ℓ and $\alpha([b, a])$ is a subinterval at the start of ℓ. Therefore this interval at the start of ℓ intersects the interval $\alpha\beta([b, \beta^{-1}(b)])$ at the end of $\alpha\beta(\ell)$ in an interval of positive length. Thus $\alpha([b, a]) \cap \alpha\beta([b, \beta^{-1}(b)])$ has positive length. But this implies $[b, a] \cap \beta([b, \beta^{-1}(b)]) = b$ has positive length, a contradiction $\square$

An action of G on an $\mathbb{R}$-tree Γ is *minimal* if there is no proper subtree that is preserved by G. The action is *irreducible* if there are two elements a, b of G which are both hyperbolic and whose translation axes are disjoint. A *global fixed point* is a point in Γ that is fixed by every element of G.

It can be shown that *irreducible* is equivalent to the statement that there is no point in $\Gamma \cup \partial\Gamma$ which is fixed by all of G. Here, $\partial\Gamma$ is the *boundary* of the $\mathbb{R}$-tree Γ. This boundary can be described as either the set of ends of Γ, or as the Gromov-boundary of Γ, but we will not need this. If two hyperbolics α, β, have axes whose intersection is compact, then for large n the axes of α and $\beta^n\alpha\beta^{-n}$ are disjoint.

LEMMA 2.4. *Suppose G is a finitely generated group of isometries of an $\mathbb{R}$-tree Γ. If every element of G is elliptic, then there is a global fixed point.*

PROOF. If α and β are elliptic then

$$t(\alpha\beta) = 2d(\text{core}(\alpha), \text{core}(\beta))$$

Thus if $\alpha\beta$ is also elliptic then core(α) intersects core(β). It follows that the intersection of the cores of a finite number of elements which generate the group is non-empty and each point in the intersection is fixed by every element in the group □

It is clear that the action of G on K discussed above induces an action $\sigma : G \longrightarrow \text{Isom}(\Gamma)$ on Γ by isometries.

ADDENDUM 2.5. *Assume the hypotheses of* (2.1). *Then G acts isometrically on the $\mathbb{R}$-tree Γ with no global fixed point. Moreover the action contains a hyperbolic.*

PROOF. Suppose that G fixes the point x in Γ. Then given $\epsilon > 0$, for n sufficiently large, there is a point x_n' in $(K, \tilde{d}_n)$ with image x in Γ such that x_n' is moved a $\tilde{d}_n$-distance at most ϵ by all elements of S. Taking $\epsilon = 1/2$ this means that for all sufficiently large n and all $\alpha \in S$ that:

$$d_{\mathcal{H}}(\iota_n x_n', \alpha\iota_n x_n') \leq \lambda_n/2$$

Thus $r_S(\iota_n x_n') \leq \lambda_n/2$ which contradicts the definition of λ_n and of the center of ρ_n. The conclusion that some element of G acts as a hyperbolic follows from lemma (2.4). □

LEMMA 2.6. *Suppose that G acts by isometries on an $\mathbb{R}$-tree Γ. Assume that some element of G is hyperbolic. Let Γ_- be the intersection of all connected subsets of Γ containing the axis of every hyperbolic. Then Γ_- is the unique G-invariant minimal subtree of Γ.*

PROOF. Let g be a hyperbolic element of G with axis ℓ. Then $t(g) > 0$. Suppose that Γ' is a G-invariant subtree of Γ then the restriction of g to Γ' is an isometry with translation length in Γ' at least as large as $t(g)$. By (2.2) it follows that $g|\Gamma'$ is a hyperbolic therefore it has an axis, ℓ', which is invariant under $g|\Gamma'$. It is easy to check that the only line preserved by a hyperbolic is its axis. Hence $\ell' = \ell$. Hence every invariant subtree contains the axis of every hyperbolic. Therefore Γ' contains Γ_-. Also Γ_- is clearly G-invariant and connected therefore an $\mathbb{R}$-tree □

PROPOSITION 2.7. *Suppose G be a finitely generated group which acts by isometries on an $\mathbb{R}$-tree Γ. Suppose the action and is minimal, and irreducible, and that the translation length of every element of G is an integer. Then Γ is a simplicial tree.*

PROOF. We call a point v in Γ a *vertex* if $\Gamma - v$ has more than 2 components. We will show that the distance between vertices in Γ is half of an integer. An inductive argument now implies that Γ is a simplicial tree.

Given two hyperbolic isometries a, b in G define d to be the distance between their axes. If $d > 0$ then by (2.3) we have $t(ab) = t(a) + t(b) + 2d$. Since all translation lengths are assumed to be integers this implies that d is a half integer.

Given two distinct vertices u_1, u_2 in Γ we claim that, for $i = 1, 2$, there are hyperbolics τ_i with axis ℓ_i such that $\ell_i \cap [u_1, u_2] = u_i$. Then $[u_1, u_2]$ is the shortest arc between ℓ_1 and ℓ_2 and therefore by the previous remark $d(u_1, u_2)$ is half an integer.

Let P be the component of $\Gamma - interior[u_1, u_2]$ containing u_1. Since u_1 is a vertex it follows that $P - u_1$ is not connected. Let Q, R be the closures of two distinct components of $P - u_1$. Since the action of G is irreducible, there are hyperbolics α, β with disjoint axes. Our subclaim is that there is a hyperbolic, τ_Q, with axis, ℓ_Q, contained in $Q - u_1$. Let A be the axis of α and B the axis of β. If the orbit of A under G is disjoint from Q then there is an invariant subtree given by taking the intersection of all subtrees containing the orbit of A. This subtree does not contain Q contradicting minimality. Thus we may assume that A intersects Q. Then $A \cap Q$ is an infinite half-line, since the axis A can only enter or leave Q at the point u_1. We may replace α by α^{-1} if necessary to ensure that $\alpha(A \cap Q) \subset A \cap Q$. The axis of $\alpha^n \beta \alpha^{-n}$ is $\alpha^n(B)$. For n large $\alpha^n(B)$ contains points in Q and does not contain u_1 therefore it is contained in $Q - u_1$. This proves the subclaim. Similarly there is a hyperbolic, τ_R, with axis, ℓ_R, contained in $R - u_1$.

Suppose that δ is an arc connecting ℓ_Q to ℓ_R. Then $interior(\delta)$ contains u_1 because $Q \cap R = u_1$. Furthermore δ is contained in $Q \cup R$. Hence δ intersects $[u_1, u_2]$ only at u_1. Choose δ to be the shortest arc connecting the axes of τ_Q and τ_R. The proof of (2.3) shows that the axis, ℓ_1, of the hyperbolic $\tau_Q \tau_R$ contains δ. Therefore ℓ_1 intersects $[u_1, u_2]$ in the single point u_1. Otherwise the intersection is an interval which contains a subinterval of positive length in δ and this is impossible. Similarly one obtains a hyperbolic with axis intersecting $[u_1, u_2]$ in the single point u_2. This proves the claim □

3. Length functions and compactifications

In this section we show how spaces of representations can be compactified by mapping the representation space into the space of length functions on the group and compactifying this latter space projectively.

Using the ideas in the previous section, any action, ρ, of a group G by isometries on a geodesic space $\mathcal{H}$ and any choice of a point, x, in $\mathcal{H}$ gives an equivariant straight map

$$\iota : K(G) \longrightarrow \mathcal{H}$$

Let d be the pseudo-metric on $K(G)$ obtained by pull-back of the $d_{\mathcal{H}}$ metric. This gives an action, σ, of G on $K(G)$ by d-isometries. Given g in G the translation length $t(\sigma g)$ is at least as large as the translation length $t(\rho g)$. It may be larger, for example if $\rho(g)$ is hyperbolic and no point on the axis of $\rho(g)$ in $\mathcal{H}$ is in image(ι). However there is a bound on the difference in the translation lengths of ρg and σg which depends only on the thin triangles constant Δ. We now prove this.

LEMMA 3.1. *Suppose that $\mathcal{H}$ has Δ-thin triangles and suppose that τ is an isometry of $\mathcal{H}$ and x is any point in $\mathcal{H}$. Then there is a point y in $[x, \tau x]$ such that*

$$|d_{\mathcal{H}}(y, \tau y) - t(\tau)| \leq 28\Delta$$

PROOF. If $d(x, \tau x) = t(\tau)$ then $y = x$ works. Otherwise choose z in $\mathcal{H}$ such that

$$d(z, \tau z) \ < \ \min\left(d(x, \tau x), t(\tau) + \Delta\right)$$

Every point in the orbit $\{\tau^n z : n \in \mathbb{Z}\}$ of z under the group generated by τ is moved by τ the same distance that z is moved. Thus we may assume that z is chosen in this orbit so that it is almost a closest point in the orbit to x in the sense that $d(z, x) < d(\tau^n z, x) + \Delta$ for all n.

The first case is that there are points y in $[x, \tau x]$ and w in $[z, \tau z]$ such that $d(y, w) \leq 2\Delta$. Now

$$d(w, \tau w) \leq d(w, \tau z) + d(\tau z, \tau w) = d(w, \tau z) + d(z, w) = d(z, \tau z)$$

Also

$$d(y, \tau y) \leq d(y, w) + d(w, \tau w) + d(\tau w, \tau y) = 2d(y, w) + d(w, \tau w)$$

Combining these gives

$$d(y, \tau y) \leq 4\Delta + d(z, \tau z) \leq 4\Delta + t(\tau) + \Delta$$

This proves that this y works in this case.

Next we prove that if the first case does not hold then $d(z, \tau z) \leq 20\Delta$. Assume the contrary. Let a, b be the points on $[z, \tau z]$ with $d(z, a) = d(b, \tau z) = 5\Delta$. Let a' be the point on $[x, z]$ with $d(z, a') = 5\Delta$ then we claim $d(a, a') \leq 4\Delta$. In the thin quadrilateral with vertices $x, \tau x, \tau z, z$ we have that every point on $[z, \tau z]$ is within a distance of 2Δ of the union of the two sides $[x, z]$ and $[\tau x, \tau z]$. This is because the first case does not hold thus no point on $[z, \tau z]$ is within 2Δ of $[x, \tau x]$. In particular there is some point, v say, in $[z, x]$ or $[\tau z, \tau x]$ with $d(a, v) \leq 2\Delta$. First suppose that v is in $[\tau z, \tau x]$ then:

$$\begin{aligned} d(v, \tau z) &\geq d(a, \tau z) - d(a, v) \\ &= d(z, \tau z) - d(z, a) - d(a, v) \\ &\geq 20\Delta - 5\Delta - 2\Delta \\ &= 13\Delta \end{aligned}$$

Hence

$$\begin{aligned} d(\tau^{-1} z, x) &= d(z, \tau x) \\ &\leq d(z, a) + d(a, \tau x) \\ &\leq 5\Delta + d(a, v) + d(v, \tau x) \\ &\leq 5\Delta + 2\Delta + (d(\tau z, \tau x) - d(v, \tau z)) \\ &\leq 7\Delta + d(z, x) - 13\Delta \\ &= d(z, x) - 6\Delta. \end{aligned}$$

This implies that $d(z, x) > d(\tau^{-1} z, x) + \Delta$ which contradicts the choice of z, proving the subclaim that v is in $[z, x]$. Since $d(a, z) = 5\Delta$ and $d(a, v) \leq 2\Delta$ it follows that $3\Delta \leq d(v, z) \leq 7\Delta$. Since v and a' are both on $[z, x]$ and $d(a', z) = 5\Delta$ it follows that $d(v, a') \leq 2\Delta$. Hence $d(a, a') \leq 4\Delta$ as claimed. Now define $b' = \tau a'$ thus

b' is the point on $[\tau x, \tau z]$ with $d(\tau z, b') = 5\Delta$. A similar argument shows that $d(b, b') \leq 4\Delta$. It now follows that

$$\begin{aligned} d(a', \tau a') = d(a', b') &\leq d(a', a) + d(a, b) + d(b, b') \\ &\leq 4\Delta + (d(z, \tau z) - d(z, a) - d(\tau z, b)) + 4\Delta \\ &\leq 8\Delta + (d(z, \tau z) - 10\Delta) \\ &\leq d(z, \tau z) - 2\Delta \\ &< (t(\tau) + \Delta) - 2\Delta \\ &< t(\tau) \end{aligned}$$

It is impossible that τ moves a' less than $t(\tau)$ thus $d(z, \tau z) \leq 20\Delta$. It remains to show y exists with this hypothesis.

Since the first case does not hold it follows that every point on $[x, \tau x]$ is within 2Δ of the union of $[x, z]$ and $[\tau x, \tau z]$. Hence there is some point y on $[x, \tau x]$ which is within a distance of 2Δ of a point c on $[x, z]$ and also of a point e on $[\tau x, \tau z]$. This is proved by an easy continuity argument. Now

$$\begin{aligned} \mid d(e, \tau z) - d(c, z) \mid &\leq d(e, c) + d(\tau z, z) \\ &\leq d(e, y) + d(y, c) + 20\Delta \\ &\leq 24\Delta. \end{aligned}$$

This choice of y works because:

$$\begin{aligned} d(y, \tau y) &\leq d(y, e) + d(e, \tau c) + d(\tau c, \tau y) \\ &= d(y, e) + d(e, \tau c) + d(c, y) \\ &\leq 2\Delta + d(e, \tau c) + 2\Delta \\ &= 4\Delta + |d(e, \tau z) - d(\tau c, \tau z)| \\ &= 4\Delta + |d(e, \tau z) - d(c, z)| \\ &\leq 4\Delta + 24\Delta. \end{aligned}$$

Observe that we use that $e, \tau z, \tau c$ all are on $[\tau x, \tau z]$ to obtain the second equality. □

COROLLARY 3.2. *Suppose that G is a group and K is the complete graph with vertex set G. Suppose that $\mathcal{H}$ is a geodesic space with Δ-thin triangles. The action of G on itself by left multiplication extends to an action, σ, on K.*

Suppose that $\rho : G \longrightarrow \mathrm{Isom}(\mathcal{H})$, and $\iota : K \longrightarrow \mathcal{H}$ is a ρ-equivariant straight map. Then σ preserves $d = \iota^* d_{\mathcal{H}}$. Moreover, for all g in G

$$|t(\sigma g) - t(\rho g)| \leq 28\Delta$$

PROOF. Given g in G since ι is an isometry onto its image it follows that $t(\rho g) \leq t(\sigma g)$. Consider the edge $[1, g]$ in K. Then $\iota([1, g]) = [x, \rho(g).x]$ where $x = \iota(1)$. The lemma (3.1) gives a point $y = \iota(z)$ such that

$$t(\sigma g) \leq d_K(z, \sigma(g).z) = d_{\mathcal{H}}(y, \rho(g).y) \leq t(\rho g) + 28\Delta$$

□

A *length function* on a group G is a function $L : G \longrightarrow \mathbb{R}$. The set of length functions on G is denoted $\mathrm{LF}(G)$ and equals $\prod_{g \in G} \mathbb{R}$ with the product topology.

Thus a sequence L_n in $\mathrm{LF}(G)$ converges to L if for all g in G we have that $L_n(g)$ converges to $L(g)$.

It is clear that conjugate isometries of X have the same translation length, thus some authors regard a length function as a real valued map from *conjugacy classes.* Given a representation $\rho : G \longrightarrow \mathrm{Isom}(X)$ we define the *length function* $L_\rho : G \longrightarrow \mathbb{R}$ associated to ρ by

$$L_\rho(g) = t(\rho g)$$

for g in G. This in turn defines a map

$$L : \mathrm{Hom}(G, \mathrm{Isom}(X)) \longrightarrow \mathrm{LF}(G)$$

by $L(\rho) = L_\rho$. We will consider the cases that X is $\mathcal{H}$ or an $\mathbb{R}$-tree.

In the previous section we introduced rescaling factors to obtain limiting $\mathbb{R}$-trees. Thus we wish to discuss convergence of length functions in a projective sense. Note that $\mathbb{R} - 0$ acts on $\mathrm{LF}(G)$ by multiplication. The space of *projectivized length functions* on a group G, denoted $\mathbb{P}\mathrm{LF}(G)$ is the set of projective equivalence classes of non-zero length functions

$$\mathbb{P}\mathrm{LF}(G) \equiv \frac{\mathrm{LF}(G) - 0}{\mathbb{R} - 0}$$

This is given the quotient topology of the subspace topology on $\mathrm{LF}(G) - 0$. Thus a sequence of length functions L_n define projective classes which converge to the projective class of a length function L if there are $\lambda_n \neq 0$ such that for all g in G

$$\lim \ \lambda_n L_n(g) \ = \ L(g)$$

We assume here that none of these length functions is identically zero.

The following is the main result of this section:

PROPOSITION 3.3. *Suppose that G is a finitely generated group and $\mathcal{H}_n$ is a geodesic space with Δ-thin triangles. Suppose that $\rho_n : G \longrightarrow \mathrm{Isom}(\mathcal{H}_n)$ is a sequence of representations which blows up. Let K be the complete graph with vertex set G. Let x_n be an approximate center of ρ_n and $\iota_n : K \longrightarrow \mathcal{H}_n$ an equivariant straight map such that $\iota_n(g) = \rho_n(g)x_n$, and set $d_n = \iota_n^* d_{\mathcal{H}_n}$.*

Then there is a subsequence, still denoted $\{\rho_n\}$, and $\lambda_n \to 0$ so that $\lambda_n \cdot d_n$ converges pointwise to a pseudo-metric $\overline{d}$ on K. Furthermore, the metric space associated to $(K, \overline{d})$ is an $\mathbb{R}$-tree Γ.

The action of G on K covers an action $\sigma : G \longrightarrow \mathrm{Isom}(\Gamma)$ and some element of $\sigma(G)$ is hyperbolic. Furthermore $\lim_{n\to\infty} L(\rho_n) = L(\sigma)$ in $\mathbb{P}\mathrm{LF}(G)$.

Suppose, in addition, that σ is irreducible and that $L(\sigma) : G \to \mathbb{Z}$. Then the minimal G-invariant subtree of Γ is a simplicial tree on which G acts with the same length function as σ

PROOF. The existence of the $\mathbb{R}$-tree Γ comes from (2.1) and (2.5) gives the action, σ, of G on Γ and guarantees that for some element g of G that σg is hyperbolic. Thus $t(\sigma g) > 0$ so $L(\sigma) \neq 0$ defines a (non-zero !) projective length function. Now (3.2) implies that $L(\rho_n) \to L(\sigma)$ in $\mathbb{P}\mathrm{LF}(G)$. To see this, let

$$\sigma_n : G \longrightarrow \mathrm{Isom}(K, \lambda_n d_n)$$

be the action action of G on K, with K regarded as a pseudo-metric space using the rescaled pseudo-metric $\lambda_n d_n$. Then $|t(\sigma_n g) - \lambda_n t(\rho_n g)| \leq \lambda_n 28\Delta$. Now $\lambda_n 28\Delta \to 0$ and so $L(\sigma_n) - \lambda_n L(\rho_n) \to 0$ in $\mathrm{LF}(G)$. Now $L(\sigma_n) \to L(\sigma)$ in $\mathrm{LF}(G)$ hence in

$\mathbb{P}\mathrm{LF}(G)$. Also $\lambda_n L(\rho_n)$ has the same limit in $\mathbb{P}\mathrm{LF}(G)$ as $L(\rho_n)$. This proves the claim. If, in addition, σ is irreducible and $L(\sigma)$ is integer valued, then (2.7) supplies the G-invariant simplicial sub-tree □

We can embed $\mathrm{LF}(G)$ into

$$\overline{\mathrm{LF}(G)} \equiv \mathrm{LF}(G) \cup \mathbb{P}\mathrm{LF}(G)$$

topologized so that a sequence $L_n \in \mathrm{LF}(G)$ converges to $[L] \in \mathbb{P}\mathrm{LF}(G)$ if and only if there are $\lambda_n \to 0$ such that $\lambda_n L_n \to L$ in $\mathrm{LF}(G)$. It was shown in [**6**] that if a sequence of representations of G into $\mathrm{SL}_2(\mathbb{C})$ does not blow up then there is a subsequence which, after suitable conjugacy, converges to a representation of G into $\mathrm{SL}_2\,\mathbb{C}$. See [**3**] corollary (2.1) for a geometric proof of this fact. Then (3.3) gives:

COROLLARY 3.4. *Suppose G is a finitely generated group, and $\Delta > 0$ and $\mathrm{Hom}_\Delta(G)$ is the set of all homomorphisms of G into $\mathrm{Isom}(X)$ for all geodesic spaces X with Δ-thin triangles. Then the image of $L : \mathrm{Hom}_\Delta(G) \longrightarrow \mathrm{LF}(G)$ has compact closure in $\overline{\mathrm{LF}(G)}$. Every point in $\overline{\mathrm{LF}(G)} \cap \mathbb{P}\mathrm{LF}(G)$ is the projectivized length function of an action of G on an $\mathbb{R}$-tree which has no global fixed point*

PROOF. Since G is countable, every sequence in $\mathrm{Hom}_\Delta(G)$ has a subsequence whose length functions converge in $\overline{\mathrm{LF}(G)}$. It follows from (3.3) that limit points in $\mathbb{P}\mathrm{LF}(G)$ are projectivized length functions of actions on $\mathbb{R}$-trees. □

Specializing (3.3) to this situation, if the sequence blows up, then there is a subsequence which converges, in $\mathbb{P}\mathrm{LF}(G)$, to an action of G on an $\mathbb{R}$-tree:

THEOREM 3.5. *Suppose that G is a finitely generated group and $\rho_n : G \longrightarrow \mathrm{SL}_2(\mathbb{C})$ is a sequence of representations. Then either there is a subsequence which converges in $\mathrm{Hom}(G, \mathrm{SL}_2\,\mathbb{C})$, or the sequence blows up and there is an $\mathbb{R}$-tree Γ and an action $\sigma : G \longrightarrow \mathrm{Isom}(\Gamma)$ with the following properties.*

- *There is no point of Γ fixed by G; equivalently $L(\sigma)$ is not identically zero.*
- *There is a subsequence ρ_{n_i} such that $\lim L(\rho_{n_i}) = L(\sigma)$ in $\mathbb{P}\mathrm{LF}(G)$.*

If, in addition, $L(\sigma)$ is integer valued, and σ is irreducible, then Γ may be chosen to be a simplicial tree.

4. Length functions and valuations

In this section we show that given a curve $C \subset \mathrm{Hom}(G, \mathrm{SL}_2\,\mathbb{C})$ satisfying certain mild hypotheses, and given an end ϵ of C, then the rescaled length functions of representations approaching ϵ converge to an integer valued function. This is the translation length function of a certain irreducible action on a simplicial tree.

The translation length of an element $A \in \mathrm{SL}_2\,\mathbb{C}$ with large trace is approximately twice the logarithm of the absolute value of its trace. Now trace is a polynomial function on the representation variety, and thus we are concerned with the logarithm of a polynomial when the variables are large. The logarithm of a polynomial of one complex variable is approximately the degree of the polynomial times the logarithm of the variable when the variable is large.

For a polynomial p of several independent variables the asymptotic behaviour of $\log|p|$ may be rather complicated. However the restriction of $\log|p|$ to a *complex curve* behaves in this respect like a polynomial function of one variable. The notion of the degree of a polynomial of one variable is replaced by evaluating a certain *valuation* on the polynomial. This valuation gives the rate of growth of the polynomial on the curve as one goes to infinity along the curve towards a particular topological end of the curve.

We will now give a geometric interpretation of the valuations we need. Let C be a complex curve in $\mathbb{C}^n$ and let $\pi : \mathbb{C}^n \longrightarrow \mathbb{C}$ be a linear projection such that $z = (\pi|C)$ is proper. It is not hard to see that such a projection always exists. Then C is a smooth 2-dimensional manifold except at finitely many points (see [**10**]) and the derivative of z is non-zero except at finitely many points. Let d be the degree (as a map) of z, which is the same as the number of pre-images of a regular value since z is complex differentiable.

PROPOSITION 4.1. *Let C be a complex affine algebraic curve in $\mathbb{C}^n$. Then there is an integer $d > 0$ with the following property. Let $y : \mathbb{C}^n \longrightarrow \mathbb{C}$ be a polynomial and let $\gamma : [0, \infty) \longrightarrow C$ be a proper arc. Then* $\lim_{t\to\infty} \log|y(\gamma t)|/\log|z(\gamma t)| = b/a$ *for some integers a, b with $|a| \leq d$.*

PROOF. The closure of the image of C in $\mathbb{C}^2$ under the polynomial map (y, z) is a complex curve K, (see [**10**]). A curve in $\mathbb{C}^2$ is the zero set of a polynomial, thus K is the zero set of a polynomial $p(y, z)$. The *Newton polygon*, Newt(p), of p is the convex hull of the finite set of points in the plane:

$$\{(m, n) : \text{ the coefficient of } \ y^m z^n \ \text{ in } \ p(y, z) \text{ is not zero } \}$$

Since $z = (\pi|C)$ is proper, for t large $|z(\gamma t)|$ is large. Thus, in looking at the order of magnitude of a term in $p(y(\gamma t), z(\gamma t))$ with t large, we may ignore the modulus of the coefficients in p.

The order of magnitude of $y^m z^n$ is $m \ \log|y| + n \ \log|z| \equiv \theta(m, n)$ which is a linear function of (m, n). Thus for a given large t the order of magnitude of the term $a.y^m z^n$ in $p(y, z)$ is approximately given by the linear map $\theta(m, n)$. Now $p(y(\gamma t), z(\gamma t)) = 0$ hence there cannot be a single term of $p(y, z)$ which is an order of magnitude larger than all other terms in $p(y, z)$. The terms of approximately largest magnitude in $p(y, z)$ nearly lie along a level set for the linear function θ. This implies that the terms of largest order of magnitude are those terms lying along some edge, e, of Newt(p), and that this edge is almost contained in a level set of θ.

As $t \to \infty$ the level sets of θ become more nearly parallel to the edge e. We claim that for all t sufficiently large, the edge e is the independent of t. For otherwise, as t increases and the edge e changes to a new edge e' and there is a value of t when the terms along both edge e and edge e' are the same order of magnitude, and this is impossible.

Let $p_e(y, z)$ be the *edge polynomial* consisting of the sum of those terms of p which lie along e. Now $p_e(y, z) = y^r z^s q(y^a z^b)$ where a, b are coprime integers such that e has slope b/a, and q is a polynomial of one variable with $q(0) \neq 0$. Since the terms in p_e are larger than any other terms in p it follows that $\lim_{t\to\infty} q(y^a(\gamma t) z^b(\gamma t)) = 0$. Hence $\lim_{t\to\infty} y^a z^b = \eta$ where η is a (necessarily non-zero) root of q. Now $\log|y|, \log|z| \to \pm\infty$ along γ so that

$$\lim_{t\to\infty} \log|y(\gamma t)|/\log|z(\gamma t)| = -b/a$$

For fixed large z the equation $q(y^a z^b) = 0$ has $|a|$ distinct solutions, and thus $p(y, z) = 0$ also has $|a|$ distinct solutions. But since the degree of z is d we have $|a| \leq d$ as claimed □

Thus γ determines a map $D_\gamma : \{rational\ functions\ on\ \mathbb{C}^n\} \longrightarrow \mathbb{Q}$ by

$$D_\gamma(y) = \lim_{t\to\infty} \log|y(\gamma t)|/\log|z(\gamma t)|$$

Take the smallest integer p such that $V_\gamma(y) = pD_\gamma(y)$ is an integer for all polynomials y. Observe that such p exists and that $p \leq d!$ by the above. We claim that changing the arc γ by a proper homotopy does not change V_γ. Suppose that γ' is properly homotopic to γ. Then the argument in (4.1) shows that there is an edge e for γ containing the terms of largest order of magnitude for large t, and $D_\gamma(y)$ is determined by this edge. Similarly there is an edge e' for γ'. The proper homotopy gives a 1-parameter family, γ_s, of proper arcs with $\gamma_0 = \gamma$ and $\gamma_1 = \gamma'$. Thus we get a 1-parameter family of edges e_s for these arcs. Now all the edges in this family are the same, by the same reasoning in (4.1) that the edge e was indepenent of t. This proves the claim.

Thus the map $V_\epsilon \equiv V_\gamma$ depends only on the end, ϵ, of C to which γ converges. Now ends of C correspond to ideal points of the smooth model $\tilde{C}$ of the projectivization of C and all we have done is to construct the (negative of the) valuation associated to an ideal point of $\tilde{C}$. Note that our sign convention means that $V_\epsilon(y) > 0$ iff $\lim_{t\to\infty}|y(\gamma t)| = \infty$. We say that a function $f(t)$ *blows up* as $t \to \infty$ if $\lim_{t\to\infty}|f(t)| = \infty$. Thus a rational function values positive if it blows up as you go out to the end of the curve in question.

Let G be a finitely generated group and $\mathrm{Hom}(G, \mathrm{SL}_2\,\mathbb{C})$ the affine algebraic set of representations $\rho : G \longrightarrow \mathrm{SL}_2\,\mathbb{C}$, following common usage we will call this the *representation variety*. Let C be a curve in $\mathrm{Hom}(G, \mathrm{SL}_2\,\mathbb{C})$ and ϵ an end of C such that there is some element $\alpha \in G$ for which $\mathrm{trace}(\rho\alpha) \to \infty$ as $\rho \to \epsilon$. Let V_ϵ be the associated valuation. It is an observation of Culler and Shalen that that if one works with a curve of characters instead of representations, then every end has this property eg. by [**3**] Corollary 2.1.

We next show that for a representation near an end, ϵ, of C the translation lengths of elements of G are approximately certain integer multiples of a large parameter. The integer assigned to an element α of G is the twice the valuation associated to ϵ of the trace of α. The formula for translation length of an element A of $\mathrm{SL}_2\,\mathbb{C}$ is

$$t(A) = 2\cosh^{-1}(|\,\mathrm{trace}(A)|/2)$$

For $\mathrm{trace}(A)$ large this is approximately $2\log|\,\mathrm{trace}(A)|$. Our primary concern is with elements of G whose translation length goes to infinity as one goes out along C towards ϵ. Such elements act on the limiting tree with non-zero translation length given by this integer.

This is why, following Culler and Shalen, we introduce the function

$$f_\alpha(\rho) = (\mathrm{trace}\ \rho\alpha)^2 - 4$$

then

$$t(\rho\alpha)/\log|f_\alpha(\rho)| \approx 1$$

when $\mathrm{trace}(\rho\alpha)$ is large. If $\mathrm{trace}(A) \approx \pm 2$ this measures how fast the translation length goes to zero.

Let ρ_n be a sequence converging to an end ϵ of C. Using the above approximation to $t(\rho_n\alpha)$ and proposition (4.1) there are λ_n such that for all α in G

$$\lim_{n\to\infty} \lambda_n t(\rho_n\alpha) = \max\{V_\epsilon(f_\alpha), 0\}$$

The corresponding formula of Morgan and Shalen [**9**] Proposition [II.3.15] has a negative sign because our V is the negative of the valuation that they use. By (3.5) there is a subsequence which converges to an action $\sigma : G \longrightarrow Aut(\Gamma)$ on an $\mathbb{R}$-tree Γ. Hence

PROPOSITION 4.2. *Let G be a finitely generated group and let C be a an affine algebraic curve of representations of G into $\mathrm{SL}_2\,\mathbb{C}$. Suppose that ϵ is an end of C and that ρ_n is a sequence in C which blows up and converges to ϵ. Define a length function on G by $p(g) = \max\{V_\epsilon(f_g), 0\}$ for each g in G. Then there is an action, σ, of G on an $\mathbb{R}$-tree such that $\lim_{\rho\to\epsilon} L(\rho) = L(\sigma) = L(p)$ in $\mathbb{P}\mathrm{LF}(G)$*

PROOF. By (3.3) there is a subsequence ρ_{n_i} and an action σ of G on an $\mathbb{R}$-tree such that $\lim_{i\to\infty} L(\rho_{n_i}) = L(\sigma)$ in $\mathbb{P}\mathrm{LF}(G)$. From the above discussion $\lim_{i\to\infty} L(\rho_{n_i}) = \lim_{\rho\to\epsilon} L(\rho) = L(p)$ □

In particular this means that we may scale metric on the $\mathbb{R}$-tree $(\Gamma, \overline{d})$ obtained as a limit of the representations ρ_n so that the translation length function for it is integer valued.

LEMMA 4.3. *Suppose G is a finitely generated group and C is a connected component of a curve in $\mathrm{Hom}(G, \mathrm{SL}_2\,\mathbb{C})$ and Γ is an $\mathbb{R}$-tree and suppose*

- *ρ_n is a sequence in C.*
- *$\alpha \in G$ and $\mathrm{trace}(\rho_n\alpha) \to \infty$.*
- *$\exists\ \rho' \in C$ and $\rho'(G)$ is not virtually solvable.*
- *$\exists\ \sigma : G \longrightarrow \mathrm{Isom}(\Gamma)$ and $\sigma(\Gamma)$ has no global fixed point.*
- *$L(\rho_n)$ converge to $L(\sigma)$ in $\mathbb{P}\mathrm{LF}(G)$.*

Then σ is irreducible.

PROOF. We may suppose that a subsequence of ρ_n has been chosen which converges to some end ϵ of C. Since C contains a representation which is not virtually solvable then for ρ near ϵ we have that ρ is not virtually solvable. The Tits alternative says that a subgroup of a linear group is either virtually solvable or else contains a non-abelian free subgroup. (The proof of this for $\mathrm{SL}_2\,\mathbb{C}$ is elementary.) Thus there is a conjugate, β, of α such that the fixed points of $\rho\beta$ in S^2_∞ are both distinct from those of $\rho\alpha$. Conjugate ρ so that

$$A = \rho(\alpha) = \begin{pmatrix} a & 0 \\ 0 & 1/a \end{pmatrix} \qquad B = \rho(\beta) = \begin{pmatrix} b+c & b^2-c^2-1 \\ 1 & b-c \end{pmatrix}.$$

This may be done so that $a \to \infty$ as $\rho \to \epsilon$. Then

$$t_p \equiv \mathrm{trace}(A^pBA^{-p}B^{-1}) = (1-b^2+c^2)(a^{2p}+a^{-2p}) + 2(b^2-c^2)$$

Observe that $\mathrm{trace}(\rho\beta) = 2b$ thus b may be regarded as a polynomial function on C. Similarly $\mathrm{trace}(\rho\alpha\beta) = a(b+c) - (b-c)/a$ so that c may also be regarded as a rational function on C.

We claim that for all sufficiently large p that $t_p \to \infty$ near ϵ. To see this, note that if $(1-b^2+c^2) = 0$ then B is lower triangular hence fixes $0 \in S^2_\infty$. But A fixes 0 also, and this contradicts the choice of β. Let V_ϵ be the valuation assoicated to ϵ. Given two functions, one of which grows faster than the other, then the rate of growth of their sum equals the rate of growth of the larger one, thus

$$V_\epsilon(f) > V_\epsilon(g) \quad \Longrightarrow \quad V_\epsilon(f+g) = V_\epsilon(f)$$

For p sufficiently large we have

$$V_\epsilon((1-b^2+c^2)(a^{2p}+a^{-2p})) = V_\epsilon(1-b^2+c^2)+V_\epsilon(a^{2p}+a^{-2p}) > V_\epsilon(2(b^2-c^2))$$

This is because $a \to \infty$ and so $V_\epsilon(a^{2p}+a^{-2p}) \geq 2p$. Thus for p large, $V_\epsilon(t_p) > 0$ so $t_p \to \infty$ near ϵ as claimed.

It follows that $\sigma\alpha^p$ and $\sigma\beta$ are hyperbolics and that their commutator is also hyperbolic. Thus the axes of these two elements intersect in at most a compact interval. Then $\sigma\alpha^p$ and a conjugate of it by any sufficiently large power of $\sigma\beta$ have disjoint axes. Hence σ is irreducible □

In summary we have the main theorem about compactifying curves of representations:

THEOREM 4.4. *Let C be a connected component of an algebraic curve in the representation variety* $\mathrm{Hom}(G, \mathrm{SL}_2\,\mathbb{C})$ *such that*

- $\exists\ \rho \in C$ *such that* $\rho(G)$ *is not solvable.*
- $\exists\ \rho, \rho' \in C$ *that are not conjugate.*

If ϵ is an end of C then there is a metric simplicial tree T, and an irreducible action $\sigma : G \longrightarrow \mathrm{Isom}(T)$ with the following property. Suppose that ρ_n is any sequence in C converging to ϵ. Then $\lim L(\rho_n) = L(\sigma)$ in $\mathbb{P}\mathrm{LF}(G)$

We remark that if all the representations on C are virtually solvable then it is easy to show that there is an action σ of G by isometries on $\mathbb{R}$ such that $\lim L(\rho_n) = L(\sigma)$.

References

[1] M. Bestvina, *Degenerations of the hyperbolic space*, Duke Math. J. **56** (1988), no. 1, 143–161, DOI 10.1215/S0012-7094-88-05607-4. MR932860

[2] D. Cooper, M. Culler, H. Gillet, D. D. Long, and P. B. Shalen, *Plane curves associated to character varieties of 3-manifolds*, Invent. Math. **118** (1994), no. 1, 47–84, DOI 10.1007/BF01231526. MR1288467

[3] D. Cooper and D. D. Long, *Roots of unity and the character variety of a knot complement*, J. Austral. Math. Soc. Ser. A **55** (1993), no. 1, 90–99. MR1231696

[4] M. Culler, C. McA. Gordon, J. Luecke, and P. B. Shalen, *Dehn surgery on knots*, Ann. of Math. (2) **125** (1987), no. 2, 237–300, DOI 10.2307/1971311. MR881270

[5] M. Culler and J. W. Morgan, *Group actions on **R**-trees*, Proc. London Math. Soc. (3) **55** (1987), no. 3, 571–604, DOI 10.1112/plms/s3-55.3.571. MR907233

[6] M. Culler and P. B. Shalen, *Varieties of group representations and splittings of 3-manifolds*, Ann. of Math. (2) **117** (1983), no. 1, 109–146, DOI 10.2307/2006973. MR683804

[7] M. Culler and P. B. Shalen, *Bounded, separating, incompressible surfaces in knot manifolds*, Invent. Math. **75** (1984), no. 3, 537–545, DOI 10.1007/BF01388642. MR735339

[8] M. Gromov, *Hyperbolic manifolds, groups and actions*, Riemann surfaces and related topics: Proceedings of the 1978 Stony Brook Conference (State Univ. New York, Stony Brook, N.Y., 1978), Ann. of Math. Stud., vol. 97, Princeton Univ. Press, Princeton, N.J., 1981, pp. 183–213. MR624814

[9] J. W. Morgan and P. B. Shalen, *Valuations, trees, and degenerations of hyperbolic structures. I*, Ann. of Math. (2) **120** (1984), no. 3, 401–476, DOI 10.2307/1971082. MR769158

[10] D. Mumford, *Algebraic geometry. I: Complex projective varieties*, Springer-Verlag, Berlin-New York, 1976. Grundlehren der Mathematischen Wissenschaften, No. 221. MR0453732

[11] J.-P. Otal, *Compactification of spaces of representations after Culler, Morgan and Shalen*, Berkovich spaces and applications, Lecture Notes in Math., vol. 2119, Springer, Cham, 2015, pp. 367–413, DOI 10.1007/978-3-319-11029-5_7. MR3330768

[12] J.-P. Serre, *Arbres, amalgames,* SL_2 (French), Société Mathématique de France, Paris, 1977. Avec un sommaire anglais; Rédigé avec la collaboration de Hyman Bass; Astérisque, No. 46. MR0476875

[13] P. B. Shalen, *Dendrology of groups: an introduction*, Essays in group theory, Math. Sci. Res. Inst. Publ., vol. 8, Springer, New York, 1987, pp. 265–319, DOI 10.1007/978-1-4613-9586-7_4. MR919830

Math Dept, University of California , Santa Barbara, California 93106
Email address: cooper@math.ucsb.edu

Contemporary Mathematics
Volume **760**, 2020
https://doi.org/10.1090/conm/760/15288

Taut foliations from double-diamond replacements

Charles Delman and Rachel Roberts

ABSTRACT. Suppose M is an oriented 3-manifold with connected boundary a torus, and suppose M contains a properly embedded, compact, oriented, surface R with a single boundary component that is Thurston norm minimizing in $H_2(M, \partial M)$. We define a readily recognizable type of sutured manifold decomposition, which for notational reasons we call *double-diamond taut*, and show that if R admits a *double-diamond taut* sutured manifold decomposition, then for every boundary slope except one, there is a co-oriented taut foliation of M that intersects ∂M transversely in a foliation by curves of that slope. In the case that M is the complement of a knot κ in S^3, the exceptional filling is the meridional one; in particular, restricting attention to rational slopes, it follows that every manifold obtained by non-trivial Dehn surgery along κ admits a co-oriented taut foliation. As an application, we show that if R is a Murasugi sum of surfaces R_1 and R_2, where R_2 is an unknotted band with an even number $2m \geq 4$ of half-twists, then every manifold obtained by non-trivial surgery on $\kappa = \partial R$ admits a co-oriented taut foliation.

1. Introduction

Taut foliations have long played an important role in the study of 3-manifolds, informed by the relationship among various geometric and algebraic properties. We present here the relevant definitions, known results, and conjectures. For simplicity of exposition, we assume all 3-manifolds are compact, oriented and are not homeomorphic to $S^1 \times S^2$.

DEFINITION 1.1. Call a 3-manifold *foliar* if it supports a co-oriented taut foliation.

The presence of a taut foliation (co-oriented or not) in a closed 3-manifold imposes restrictions on its fundamental group: it is not a non-trivial free product (indeed, M is irreducible) [**33, 43, 45**], and it contains elements of infinite order [**18, 20, 33**]– indeed, the manifold is covered by $\mathbb{R}^3$[**42**].

Another well-studied property of groups is *orderability*: that is, an ordering that respects the group operation. Focusing on transformations by group elements acting on the left:

2010 *Mathematics Subject Classification*. Primary 57M50.
Key words and phrases. Taut foliation, sutured manifold decomposition, L-space Conjecture, Dehn filling, branched surface, disk decomposition, L-space, L-space knot, persistently foliar knot, foliar.
This work was partially supported by National Science Foundation Grant DMS-1612475.

DEFINITION 1.2. A nontrivial group G is *left-orderable* if its elements can be given a strict total ordering $<$ which is left invariant, meaning that $g < h$ implies $fg < fh$ for all $f, g, h \in G$.

Another invariant of 3-manifolds is Heegaard Floer Homology, introduced by Ozsváth and Szabó [**39, 40**]. In particular, the Heegaard Floer homology of a rational homology sphere Y satisfies

$$\mathrm{Rank}(\widehat{HF}(Y;\mathbb{Z}/2)) \geq |H_1(M,\mathbb{Z})|.$$

If equality holds, Y is called an *L-space*; elliptic manifolds, such as lens spaces, are examples [**41**]. In contrast, Ozsváth and Szabó proved foliar manifolds cannot be L-spaces [**2, 26, 27, 38**].

Boyer, Gordon, and Watson have conjectured that an irreducible rational homology 3-sphere is an L-space if and only if its fundamental group is not left-orderable and also asked if the presence of a *co-oriented* taut foliation implies that a 3-manifold's fundamental group is left-orderable [**4**], leading to the *L-space Conjecture*:

CONJECTURE 1.3 (L-space Conjecture [**4, 25, 40**]). *Let Y be an irreducible rational homology 3-sphere. Then the following three statements are equivalent:*

(1) *Y is not an L-space,*
(2) *$\pi_1(Y)$ is left-orderable, and*
(3) *Y is foliar.*

Namely, in the spirit of Haken, Waldhausen, and Thurston [**21, 48, 50**], these are three measures of what it means for a 3-manifold to be "large," and the conjecture is that they are all equivalent. (We note that Gabai includes co-orientability in his original definition of *taut*.) It is known only that (3) implies (1). This paper addresses the conjecture that (1) implies (3).

DEFINITION 1.4. Let M be a 3-manifold with boundary a torus. A *boundary slope* is an isotopy class of curves in ∂M.

Following Boyer and Clay we introduce:

DEFINITION 1.5 ([**3**]). A foliation $\mathcal{F}$ *strongly realizes* a slope if $\mathcal{F}$ intersects ∂M transversely in a foliation by curves of that slope.

Note that if a foliation strongly realizes a compact slope, it may be capped off by disks to obtain a co-oriented taut foliation in the manifold obtained by Dehn filling along that slope.

Suppose ∂M is a torus and M contains a properly embedded, compact, oriented, Thurston norm minimizing [**49**] surface R with a single boundary component. Call the isotopy class of ∂R the *longitude* and any slope having geometric intersection number one with this longitude a *meridian*. In this paper, we define an easily-recognized criterion that implies such a 3-manifold supports co-oriented taut foliations strongly realizing all boundary slopes except a single *distinguished meridian* ν. Thus, the manifolds obtained by Dehn surgery along all slopes other than ν are foliar.

Focusing attention on manifolds obtained by Dehn surgery on knots, we introduce the following notation, definition, and conjecture:

NOTATION 1.6. Recall that a knot in a 3-manifold P is a submanifold of the interior of P homeomorphic to S^1. Given a knot κ in a closed manifold P, let $N(\kappa)$ denote an open regular neighborhood of κ with closure contained in the interior of P, and let X_κ denote the knot complement $P \setminus N(\kappa)$. In the case $M = X_\kappa$, we show in the proof of Corollary 4.11 that the distinguished meridian indicated above is the meridian of κ, namely the unique slope μ whose elements bound disks in the closure of $N(\kappa)$.

DEFINITION 1.7 ([**7**]). A knot κ in a 3-manifold is *persistently foliar* if, for each non-meridional boundary slope of X_κ, there is a co-oriented taut foliation strongly realizing that slope.

Many knots in S^3 are known to be persistently foliar, including all alternating and Montesinos knots with no L-space surgeries ([**8**] and [**9, 10**], respectively), all fibered knots with fractional Dehn twist coefficient zero [**7, 44**], and all composite knots in which each of two summands is alternating, Montesinos, or fibered [**7**]. (Among the alternating and Montesinos knots, only the $(2, n)$-torus knots [**41**] and $(-2, 3, n)$-pretzel knots, where n is (odd and) positive [**1, 28**], have L-space surgeries.) As a direct derivative of the L-space conjecture, we have the following pair of conjectures about knots:

CONJECTURE 1.8 (Classical L-space knot conjecture). *A knot in S^3 is persistently foliar if and only if it has no non-trivial L-space or reducible surgeries.*

CONJECTURE 1.9 (L-space knot conjecture). *A knot in an L-space is persistently foliar if and only if it has no non-trivial L-space or reducible surgeries.*

In Section 2 we provide all essential background on spines and branched surfaces. In Section 3 we provide the necessary definitions and results from sutured manifold theory. Some familiarity with laminations and foliations is assumed, or would at least be helpful to the reader. In Section 4, we prove a technical theorem (Theorem 4.10) about a particular type of sutured manifold decomposition with associated distinguished meridian, which we call *double-diamond taut*. We defer the statement of this theorem and note the following direct consequence:

COROLLARY (4.11). *Suppose the sutured manifold decomposition $(M|_R, \partial M|_{\partial R}) \overset{S}{\rightsquigarrow} (M', \gamma')$ is double-diamond taut, with associated distinguished meridian ν. If the branched surface associated to this decomposition has no sink disk disjoint from ∂M, then there are co-oriented taut foliations that strongly realize all boundary slopes except ν. Moreover, if $M = X_\kappa$ for some knot $\kappa \subset S^3$, κ is persistently foliar.*

We end with an application to knots in S^3. By Ghiggini [**19**], Ni [**31, 32**] and Juhasz [**23, 24**], non-fibered knots do not admit L-space surgeries; hence, if the L-space conjecture holds, all non-fibered knots that do not admit a reducible surgery are persistently foliar. If in addition the cabling conjecture holds, then by a result of Scharlemann [**46**] all non-fibered hyperbolic knots are persistently foliar. Although knots with Seifert surface obtained by plumbing Hopf bands have long been known to be fibered [**47**], knots obtained by plumbing a Seifert surface with a $(2, 2n)$ torus link with $|n| \geq 2$ are never fibered [**15**]. Thus we expect such knots to be persistently foliar, and indeed this is the case:

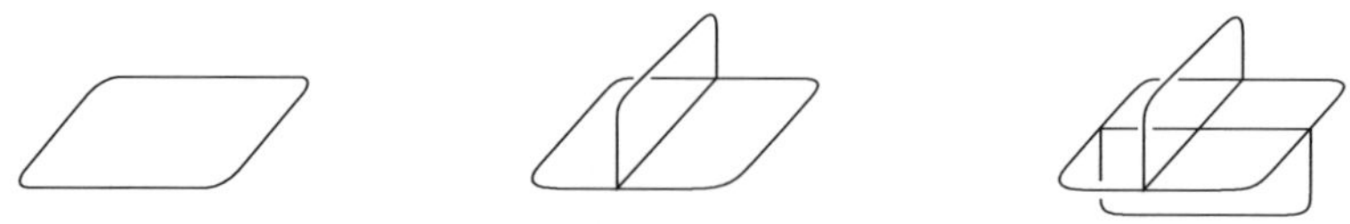

FIGURE 1. Local models of a standard spine at interior points.

COROLLARY (Corollary 5.3.). *Suppose κ is a knot in S^3 with minimal genus Seifert surface R. If R is a plumbing of surfaces R_1 and R_2, where R_2 is an unknotted band with an even number $2m \geq 4$ of half-twists, then the decomposing disk for R_2 can be chosen to be double-diamond taut. Thus, κ is persistently foliar.*

2. Spines and branched surfaces

For completeness we include the relevant standard definitions and results on spines and branched surfaces, as also found in [**7**].

NOTATION 2.1. Given any metric space X and any subset $A \subset X$, the *closed complement* of A in X, denoted $X|_A$, is the metric completion of $X \setminus A$.

REMARK 2.2. The space $X|_A$ is obtained by "cutting" X along A. Although other authors have used the notation X_A, we feel the inclusion of a vertical slash evokes the notion of "cutting."

REMARK 2.3. We introduce this notation to unify several related concepts and simplify the notation that arises in examples; however, the reader may prefer a different perspective. In the context of this paper, X will be a 3-manifold and A an embedded surface, or X will be a surface and A an embedded simple closed curve or arc. For these examples, X is retrieved as a quotient space of $X|_A$ by identifying corresponding points of the two copies of A. Also, in these examples, $X|_A$ is homeomorphic to $X \setminus \text{int}N(A)$, where $N(A)$ is a regular neighborhood; furthermore, if $N(A)$ is I-fibered (see Definition 2.6, which extends in the obvious way to simple curves in surfaces) collapsing fibers gives a quotient map from $X \setminus \text{int}N(A)$ to a manifold homeomorphic to X. However, the definition above extends to laminations, for which regular neighborhoods do not apply, and our notation avoids the cumbersome notation of the regular neighborhood description.

DEFINITION 2.4. A *standard spine* [**6**] is a space Σ locally modeled on one of the spaces of Figure 1. A standard spine with boundary has the additional local models shown in Figure 2. The *critical locus* Γ of Σ is the 1-complex of points of Σ where the spine is not locally a manifold. The critical locus is a stratified space (graph) consisting of triple points Γ^0 and arcs of double points $\Gamma^1 = \Gamma \setminus \Gamma^0$. The components of $\Sigma|_\Gamma$ are called the *sectors* of Σ.

DEFINITION 2.5. A *branched surface (with boundary)* ([**51**]; see also [**11, 34, 36**]) is a space B locally modeled on the spaces of Figure 3 (along with those in Figure 4); that is, B is homeomorphic to a spine, with the additional structure of a well-defined tangent plane at each point. The *branching locus* Γ of B is the 1-complex of points of B where B is not locally a manifold; such points are called *branching points*. The

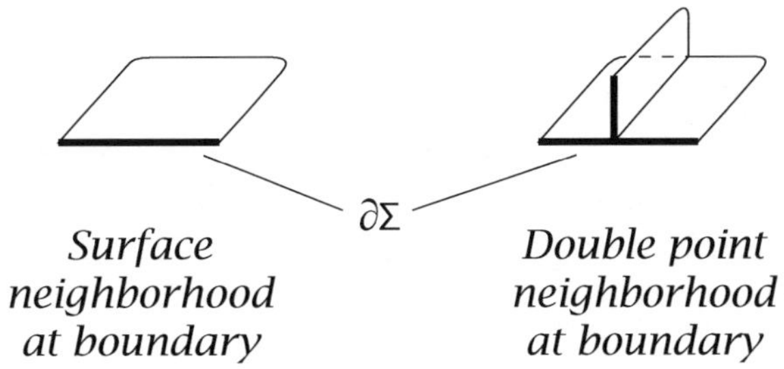

FIGURE 2. Local models of a standard spine at boundary points.

FIGURE 3. Local model of a branched surface at interior points.

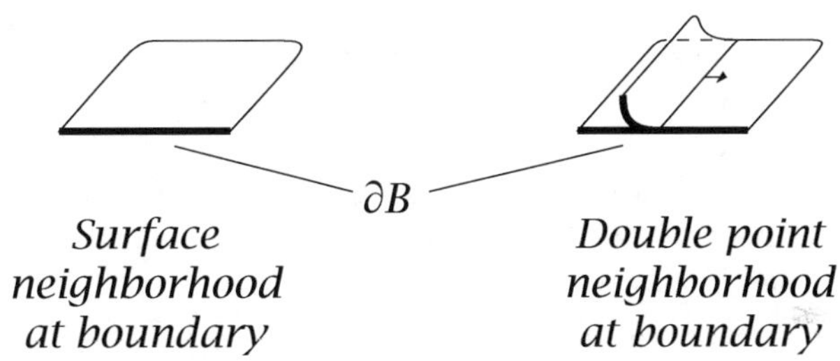

FIGURE 4. Local model of a branched surface at boundary points.

branching locus is a stratified space (graph) consisting of triple points Γ^0 and arcs of double points $\Gamma^1 = \Gamma \setminus \Gamma^0$. The components of $B|_\Gamma$ are called the *sectors* of B.

DEFINITION 2.6 ([**11**]). An *I-fibered neighborhood* of a branched surface B in a 3-manifold M is a regular neighborhood $N(B)$ foliated by interval fibers that intersect B transversely, as locally modeled by the spaces in Figure 5 at interior points; if the ambient manifold M has non-empty boundary, all spines and branched surfaces are assumed to be properly embedded, with $N(B) \cap \partial M$ a union (possibly empty) of I-fibers. The surface $\partial N(B) \setminus \partial M$ is a union of two subsurfaces, $\partial_v N(B)$ and $\partial_h N(B)$, where $\partial_v N(B)$, the *vertical boundary*, is a union of sub-arcs of I-fibers, and $\partial_h N(B)$, the *horizontal boundary*, is everywhere transverse to the I-fibers.

DEFINITION 2.7. The *complementary regions* of B are the components of $M \setminus \text{int}N(B)$; the *complement* of B is the union of the complementary regions.

DEFINITION 2.8. A surface is *carried by* B if it is contained in $N(B)$ and is everywhere transverse to the one-dimensional foliation of $N(B)$. A surface is *fully carried by* B if it carried by B and has nonempty intersection with every I-fiber of

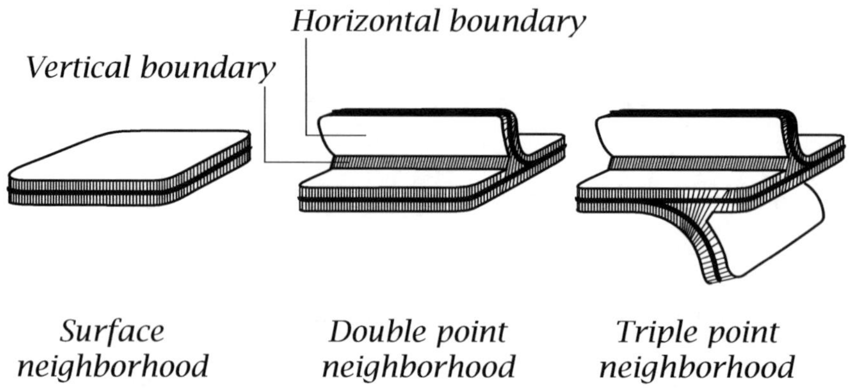

FIGURE 5. Local models for $N(B)$ (at interior points).

$N(B)$. A lamination $\mathcal{L}$ is *carried* by B if each leaf of $\mathcal{L}$ is carried by B, and *fully carried* if, in addition, each I-fiber of $N(B)$ has nonempty intersection with some leaf of $\mathcal{L}$. A foliation is *fully carried* by B if there is a Denjoy splitting of $\mathcal{F}$ that is fully carried by B.

Let π be the retraction of $N(B)$ onto the quotient space obtained by collapsing each fiber to a point. The branched surface B is obtained, modulo a small isotopy, as the image of $N(B)$ under this retraction. We will freely identify B with this image [**34**] and the core of each component of vertical boundary with its image in Γ. Double points of the branching locus are *cusps* with *cusp direction* pointing inward from the vertical boundary if B is viewed as the quotient of $N(B)$ obtained by collapsing the vertical fibers to points.

NOTATION 2.9. Cusp directions will be indicated by arrows, as in Figures 3 and 4. Call a sector S of B a *source* (*sink*) if all cusp directions along ∂S point out of (into) S.

DEFINITION 2.10 ([**18**]). A branched surface B in a closed 3-manifold M is called an *essential* branched surface if it satisfies the following conditions:

(1) $\partial_h N(B)$ is incompressible in $M \setminus \text{int}(N(B))$, no component of $\partial_h N(B)$ is a sphere, and $M \setminus \text{int}(N(B))$ is irreducible.
(2) There is no monogon in $M \setminus \text{int}(N(B))$; i.e., no disk $D \subset M \setminus \text{int}(N(B))$ with $\partial D = D \cap N(B) = \alpha \cup \beta$, where $\alpha \subset \partial_v N(B)$ is in an interval fiber of $\partial_v N(B)$ and $\beta \subset \partial_h N(B)$.
(3) There is no Reeb component; i.e., B does not carry a torus that bounds a solid torus in M.

In the spirit of earlier definitions of Gabai, Oertel, Sullivan and others, we introduce:

DEFINITION 2.11. A branched surface is *taut* if it is co-oriented, has taut sutured manifold complement (see Definition 3.3), and through every sector there is a closed oriented curve that is positively transverse to B.

Observe that a taut branched surface is, in particular, essential; furthermore, if a taut branched surface fully carries a lamination, this lamination is a sublamination of a taut foliation (see Corollary 3.9 below). However, in practice,

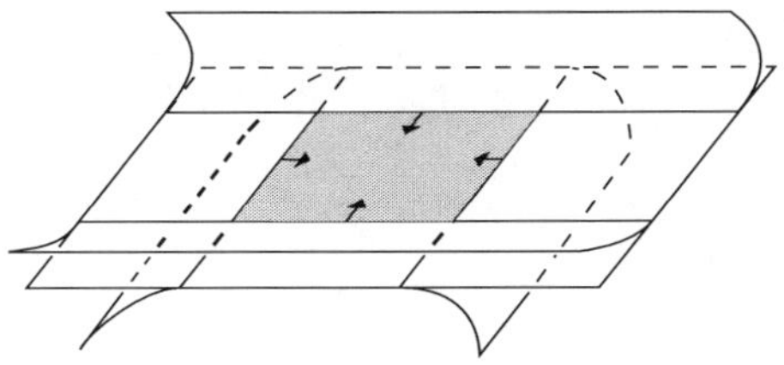

FIGURE 6. A sink disk.

it can be difficult to determine whether an essential branched surface fully carries a lamination. In [**29, 30**], Li defines the notion of *laminar*, a very useful criterion that is sufficient (although not necessary) to guarantee that an essential branched surface fully carries a lamination. We recall the necessary definitions here.

DEFINITION 2.12 ([**29, 30**]). Let B be a branched surface in a 3-manifold M. A *sink disk* is a disk sector of B that is a sink. A *half sink disk* is a sink disk which has nonempty intersection with ∂M.

Sink disks and half sink disks play a key role in Li's notion of laminar branched surface. A sink disk or half sink disk D can be eliminated by splitting D open along a disk in its interior; these trivial splittings must be ruled out:

DEFINITION 2.13 ([**29**], [**30**]). Let D_1 and D_2 be the two disk components of the horizontal boundary of a $D^2 \times I$ region in $M \setminus \mathrm{int}(N(B))$. If the projection $\pi : N(B) \to B$ restricted to the interior of $D_1 \cup D_2$ is injective, i.e, the intersection of any I-fiber of $N(B)$ with $\mathrm{int}(D_1) \cup \mathrm{int}(D_2)$ is either empty or a single point, then we say that $\pi(D_1 \cup D_2)$ forms a *trivial bubble* in B.

DEFINITION 2.14 ([**29**], [**30**]). An essential branched surface B in a compact 3-manifold M is called *laminar* if it satisfies the following conditions:

(1) B has no trivial bubbles.
(2) B has no sink disk or half sink disk.

THEOREM 2.15 ([**29**], [**30**]). *Suppose M is a compact and orientable 3-manifold.*

(a) *Every laminar branched surface in M fully carries an essential lamination.*
(b) *Any essential lamination in M that is not a lamination by planes is fully carried by a laminar branched surface.*

3. Sutured manifold decompositions

We assume the reader is familiar with the basics of co-oriented taut foliations. Precise definitions and terminology as used here can be found in [**7**]. The basics of Gabai's theory of sutured manifolds [**13, 16, 17**] play a key role in this paper, and in this section we give some necessary sutured manifold background. Where noted, we use the definitions of [**13**] as stated in [**5**].

DEFINITION 3.1 (Definition 2.6 of [**13**],[**5**]). A pair (M, γ) is a *sutured manifold* if M is a compact, oriented 3-manifold and $\gamma \subset \partial M$ is a compact subsurface such that

$$\gamma = A(\gamma) \cup T(\gamma),$$

where $A(\gamma) \cap T(\gamma) = \emptyset$, $A(\gamma)$ is a disjoint union of annuli and $T(\gamma)$ a disjoint union of tori. The components of $A(\gamma)$ are called *annular sutures*, the components of $T(\gamma)$

are called *toral sutures*, and the closure of $\partial M \setminus \gamma$ is denoted by $R(\gamma)$. The surface $R(\gamma)$ comes with a transverse orientation; let $R_+(\gamma)$ be the subsurface oriented outward from M and $R_-(\gamma)$ be the inwardly oriented subsurface. For each annular suture A, it is assumed that one component of ∂A is also a component of $\partial R_+(\gamma)$, the other a component of $\partial R_-(\gamma)$.

As noted in [**17**], a branched surface gives rise to a sutured manifold. If B is a co-oriented branched surface in M, and ∂M is a union of tori, then the regions $(\partial M \setminus \text{int} N(B), \partial_v N(B) \cap \partial M)$ are products. Setting $\gamma = \partial_v N(B) \cup (\partial M \setminus \text{int} N(B))$, the pair $(M \setminus \text{int} N(B), \gamma)$ is a sutured manifold, with $R_+(\gamma)$ (respectively, $R_-(\gamma)$) consisting of the components of $\partial_h N(B)$ with co-orientation pointing into (respectively, out of) N(B).

DEFINITION 3.2 ([**13**]). A sutured manifold (M, γ) is a *product sutured manifold* if (M, γ) is homeomorphic to $(F \times [0,1], \partial F \times [0,1])$, where F is a compact, not necessarily connected, surface with nonempty boundary.

DEFINITION 3.3 ([**5**,**13**]). A sutured manifold (M, γ) is *taut* if M is irreducible and $R(\gamma)$ is both incompressible and Thurston norm minimizing [**49**] in $H_2(M, \gamma)$.

In particular, product sutured manifolds are taut, as are sutured manifolds of the form $(M, \gamma) = (X_\kappa|_R, \partial X_\kappa|_{\partial R})$, for R a (co-oriented) minimal genus Seifert surface for κ, or more generally of the form $(M|_R, \partial M|_{\partial R})$, where ∂M is a torus and R is a properly embedded, co-oriented, compact, Thurston norm minimizing surface with a single boundary component. In the latter cases we will denote $R_+(\gamma)$ simply by R_+ and, similarly, $R_-(\gamma)$ simply by R_-.

DEFINITION 3.4 (Definition 3.1 of [**13**],[**5**]). Let (M, γ) be a sutured manifold and let $S \subset M$ be a compact, properly imbedded, transversely oriented surface. Suppose that one of the following holds for each component λ of $S \cap \gamma$:

(1) λ is a properly imbedded, nonseparating arc in $A(\gamma)$.
(2) λ is a simple closed curve in a component A of $A(\gamma)$ parallel to and coherently oriented with each component of ∂A.
(3) If T is a component of $T(\gamma)$ and $\lambda \subset T$, then λ is an essential circle in T and every component of $S \cap T$ is parallel to and coherently oriented with λ.

Further suppose that, if α is a circle component of $S \cap R(\gamma)$, α does not bound a disk in $R(\gamma)$ nor is α the boundary of a disk component D of S. Finally, if the component α of $S \cap R(\gamma)$ is a properly imbedded arc, suppose that it is not boundary compressible in $R(\gamma)$. Then S is called a *decomposing surface* for (M, γ).

A decomposing surface S defines a *sutured manifold decomposition*

$$(M, \gamma) \overset{S}{\rightsquigarrow} (M', \gamma')$$

where

$$M' = M|_S$$

and

$$\begin{aligned} \gamma' &= (\gamma \cap M') \cup N(S'_+ \cap R_-(\gamma)) \cup N(S'_- \cap R_+(\gamma)), \\ R'_+(\gamma') &= ((R_+(\gamma) \cap M') \cup S'_+)|_{A(\gamma)}, \text{ and} \\ R'_-(\gamma') &= ((R_-(\gamma) \cap M') \cup S'_-)|_{A(\gamma)}. \end{aligned}$$

where S'_+ (S'_-) is that component of $\partial N(S) \cap M'$ whose normal vector points out of (into) M'. This sutured manifold decomposition is called *taut* if both (M, γ) and (M', γ') are taut.

DEFINITION 3.5 (Definition 3.4 of [**13**]). A sutured manifold (M_0, γ_0) is *decomposable* if there is a sequence of taut sutured manifold decompositions

$$(M_0, \gamma_0) \overset{S_1}{\rightsquigarrow} (M_1, \gamma_1) \overset{S_2}{\rightsquigarrow} \cdots \overset{S_n}{\rightsquigarrow} (M_n, \gamma_n)$$

such that (M_n, γ_n) is a product sutured manifold. If each component of (M_n, γ_n) is homeomorphic to $(D^2 \times [0,1], \partial D_2 \times [0,1])$, then the sequence is called a *taut sutured manifold hierarchy*.

NOTATION 3.6 (Construction 4.16 of [**17**]). Suppose $n \geq 1$ and

$$(M_0, \gamma_0) \overset{S_1}{\rightsquigarrow} (M_1, \gamma_1) \overset{S_2}{\rightsquigarrow} \cdots \overset{S_n}{\rightsquigarrow} (M_n, \gamma_n)$$

is a sequence $\mathcal{S}$ of sutured manifold decompositions. Let $B^G(\mathcal{S}) = \langle S_1, \cdots, S_n \rangle$ denote the co-oriented branched surface with underlying co-oriented spine $S_1 \cup \cdots \cup S_n$.

These branched surfaces $B^G(\mathcal{S})$ will play a key role in the constructions found in this paper. The superscript G is used both to remind the reader that this is the branched surface described by Gabai and to distinguish this family of branched surfaces from the branched surfaces $B(\mathcal{S}')$ described in Section 4.

Note that, by definition of decomposing surface, $B^G(\mathcal{S})$ has no trivial bubbles. It follows immediately from Theorem 3.7 below that $B^G(\mathcal{S})$ is taut if $\mathcal{S}$ is a sequence of taut sutured manifold decompositions.

We will use extensively the following result of Gabai:

THEOREM 3.7 (Theorems 3.13, 4.2 and 5.1 of [**13**]). *Suppose (M, γ) is taut. Then (M, γ) is decomposable, with a choice $\mathcal{S}$ of taut sutured manifold hierarchy such that Construction 4.17 of* [**17**] *applied to $B^G(\mathcal{S})$ yields a co-oriented taut foliation fully carried by $B^G(\mathcal{S})$.*

We are primarily interested in the case that (M, γ) is the complement of an I-fibered neighbourhood of a taut co-oriented branched surface B that fully carries a lamination $\mathcal{L}$. In this setting, we are interested in applying Theorem 3.7 to extend $\mathcal{L}$ to a co-oriented taut foliation. Construction 4.17 of [**17**] guarantees that this is always possible. This follows from the fact that there is a branched surface B', obtained from B by splitting along at most finitely many compact surfaces of contact, such that B' fully carries $\mathcal{L}$ and $(B', \mathcal{L})$ satisfies the following *noncompact extension property*.

DEFINITION 3.8 (Noncompact extension property; see [**8**] for details). Suppose B fully carries a lamination $\mathcal{L}$. A component A of $\partial_v N(B)$ satisfies the *noncompact extension property* relative to the pair $(B, \mathcal{L})$ if there is a copy of $[0, \infty) \times [0,1]$ properly embedded in $N(B) \cap M|_{\mathcal{L}}$ with $\{0\} \times [0,1]$ contained in an I-fiber of $N(B)$ and containing an I-fiber of A, and $[0, \infty) \times \{0,1\}$ contained in leaves of $\mathcal{L}$.

$(B, \mathcal{L})$ satisfies the *noncompact extension property* if each component of $\partial_v N(B)$ does.

COROLLARY 3.9 ([**8**]). *Suppose B is a co-oriented branched surface that fully carries a lamination $\mathcal{L}$. If B is taut, then there is a co-oriented taut foliation that* *contains $\mathcal{L}$ as a sublamination. Indeed, if $\mathcal{S}$ is the taut sutured manifold hierarchy*

$$(M \setminus \text{int}N(B), \gamma) \overset{S_1}{\rightsquigarrow} (M_1, \gamma_1) \overset{S_2}{\rightsquigarrow} (M_2, \gamma_2) \overset{S_3}{\rightsquigarrow} \cdots \overset{S_n}{\rightsquigarrow} (M_n, \gamma_n),$$

where $\gamma = \partial_v N(B) \cup (\partial M \setminus intN(B))$, then the branched surface obtained by applying Gabai's Construction 4.16 to $\mathcal{S}$, starting with B, yields a taut branched surface that fully carries a taut foliation containing $\mathcal{L}$ as a sublamination.

The following proposition shows that if the complementary region containing a torus boundary component of M is of a particular type and also has vertical boundary whose components all satisfy the noncompact extension property, then $\mathcal{L}$ has extensions to taut, co-oriented foliations realizing every slope, except one, on that boundary component.

PROPOSITION 3.10 ([**8**]). *Let $\partial_0 M$ denote a torus boundary component of M. Suppose B is a taut co-oriented branched surface that fully carries a lamination $\mathcal{L}$ and is disjoint from $\partial_0 M$. If the complementary region $(Y, \partial_v Y)$ of $N(B)$ containing $\partial_0 M$ is homeomorphic to*

$$(\partial_0 M \times [0,1], V_1 \cup \cdots \cup V_{2n}),$$

where $\partial_0 M \times \{0\} = \partial M$ and $V_1, \cdots, V_{2n}$ are disjoint essential annuli in $\partial_0 M \times \{1\}$ that satisfy the noncompact extension property relative to $(B, \mathcal{L})$, then there are co-oriented taut foliations extending $\mathcal{L}$ that strongly realize all boundary slopes on $\partial_0 M$ except the one isotopic in Y to the core of any V_i.

4. Modifying sutured manifold decompositions

Some sutured manifold hierachies are better than others. In this section, we focus on the case that there is a sutured manifold hierarchy

$$(M_0, \partial M_0) \overset{R}{\rightsquigarrow} (M_1, \gamma_1) \overset{S}{\rightsquigarrow} (M_2, \gamma_2) \overset{S_3}{\rightsquigarrow} \cdots \overset{S_n}{\rightsquigarrow} (M_n, \gamma_n)$$

in which R is a minimal Seifert surface, and S is *double-diamond taut* (see Definition 4.8). In this case, it is possible to construct an associated branched surface that fully carries co-oriented taut foliations that strongly realize all boundary slopes except one. For simplicity of exposition, we assume in what follows that ∂M_0 is connected.

Roughly speaking, the branched surface $B^G(\mathcal{S})$ of Gabai is modified as follows. Before beginning the hierarchy, a boundary-parallel torus, which we denote by T, is added, separating M into two components. The component homeomorphic to M_0 then replaces M_0 in the sutured manifold hierarchy above. The union of surfaces $T \cup R \cup S \cdots \cup S_n$ is smoothed to a branched surface $B(\mathcal{S}')$ in two steps. First, the forthcoming special property of S (the *double-diamond* condition) is used to smooth $T \cup R \cup S$, in such a way that the resulting branched surface is laminar and has taut sutured manifold complement. This complement consists of two components: one is isomorphic to the complement of the branched surface $\langle R, S \rangle$ and the other is isomorphic to $(T^2 \times [0,1], (V_1 \cup V_2) \times \{0\})$, where $T^2 \times \{0\} = T$, $T^2 \times \{1\} = \partial M_0$, and V_1 and V_2 are disjoint annuli with slope a meridian ν, which in the case of a knot complement is necessarily the usual meridian. In the second step, the addition of the surfaces $S_1, \cdots, S_n$ and the smoothing to $B(\mathcal{S}')$ proceed as described in Gabai's Construction 4.16; see Notation 3.6.

The branched surface $B(\mathcal{S}')$ fully carries a lamination $\mathcal{L}$ and has complement the disjoint union of (M_n, γ_n) with $(T^2 \times [0,1], (V_1 \cup V_2) \times \{0\})$. It follows that

$\mathcal{L}$ admits extensions to a family of foliations strongly realizing all boundary slopes except ν.

NOTATION 4.1. Let $M = M_0$ be a compact oriented 3-manifold with ∂M a torus. Let T be a boundary parallel torus properly embedded in M, and let X and Y denote the components of $M|_T$, with X the component homeomorphic to M. Associated to a sequence $\mathcal{S}$ of sutured manifold decompositions

$$(M_0, \gamma_0) \overset{S_1}{\rightsquigarrow} (M_1, \gamma_1) \overset{S_2}{\rightsquigarrow} \cdots \overset{S_n}{\rightsquigarrow} (M_n, \gamma_n),$$

(where we assume that $S_i \cap Y = \partial S_i \times I$, for each i) is the isomorphic sequence $\mathcal{S}'$

$$(X, T) \overset{S_1'}{\rightsquigarrow} (X_1, \gamma_1') \overset{S_2'}{\rightsquigarrow} \cdots \overset{S_n'}{\rightsquigarrow} (X_n, \gamma_n'),$$

where $S_i' = S_i \cap X$. Let $\Sigma(\mathcal{S})$ denote the co-oriented spine $S_1 \cup \cdots \cup S_n$, and let $\Sigma(\mathcal{S}')$ denote the unoriented spine $T \cup S_1' \cup \cdots \cup S_n'$.

DEFINITION 4.2. Two 1-manifolds properly embedded in a common surface intersect *efficiently* if any intersections are transverse and no isotopy through properly embedded 1-manifolds reduces the number of points of intersection.

DEFINITION 4.3. Let M be an oriented 3-manifold with connected boundary ∂M a torus, and suppose M contains a properly embedded, compact, oriented, Thurston norm minimizing surface R with a single boundary component. Call a decomposing surface S for $(M|_R, \partial M|_{\partial R})$ *tight* if

(1) the images of the 1-manifolds $\alpha_+ = S \cap R_+$ and $\alpha_- = S \cap R_-$ under the quotient map $M|_R \to M$ intersect efficiently in R, and
(2) the sutured manifold decomposition $(M|_R, \partial M|_{\partial R}) \overset{S}{\rightsquigarrow} (M', \gamma')$ is taut.

DEFINITION 4.4 ([**15**]). A decomposing surface D is called a *product* disk if D is a disk and $|D \cap A(\gamma)| = 2$.

NOTATION 4.5. Let S be a tight decomposing surface for $(M|_R, \partial M|_{\partial R})$. Let $\partial_1 S, \cdots, \partial_s S$ be the components of ∂S that have nonempty intersection with $A(\gamma)$. Each such component can be written as a concatenation of arcs that lie alternately in R_+, ∂M, and R_-. Letting $*$ denote concatenation, and denoting the (even) number of arcs of $\partial_i S$ on ∂M by $2n_i$, we write:

$$\partial_i S = \alpha_{i,1} * \tau_{i,1} * \cdots \alpha_{i,2n_i} * \tau_{i,2n_i}, \quad 1 \leq i \leq s,$$

where $\alpha_{i,2j-1} \subset R_+$, $\alpha_{i,2j} \subset R_-$, and $\tau_{i,k} \subset \partial M$, for $j = 1, 2, \ldots n_i$ and $k = 1, 2, \ldots, 2n_i$.

DEFINITION 4.6. We call the arcs $\tau_{i,k}$, as well as the corresponding arcs in the spine $\Sigma(\mathcal{S}')$, *transition arcs.*

DEFINITION 4.7. Given a framing on ∂M with longitude given by the oriented curve ∂R and arbitrary fixed meridian μ_0, we may talk of the *sign* of the transition $\tau_{i,k}$ relative to μ_0. This sign is defined as follows. Orient $\tau_{i,k}$ so that it is a path from R_+ to R_-. Associated to the framing $\partial M = S^1 \times S^1$, with $S^1 \times \{1\}$ representing the preferred longitude and $\{1\} \times S^1$ representing the preferred meridian, is a retraction $r_{\mu_0} : S^1 \times S^1 \to S^1$ given by $r_{\mu_0}(z, w) = z$. Define a transition arc to be *positive (negative) rel* μ_0 if $\tau_{i,k}$ can be isotoped in ∂M rel endpoints so that the restriction of r_{μ_0} to $\tau_{i,k}$ is orientation preserving (reversing). If the meridian μ_0 is understood, refer to the associated retraction as r and to a transition arc as being positive or

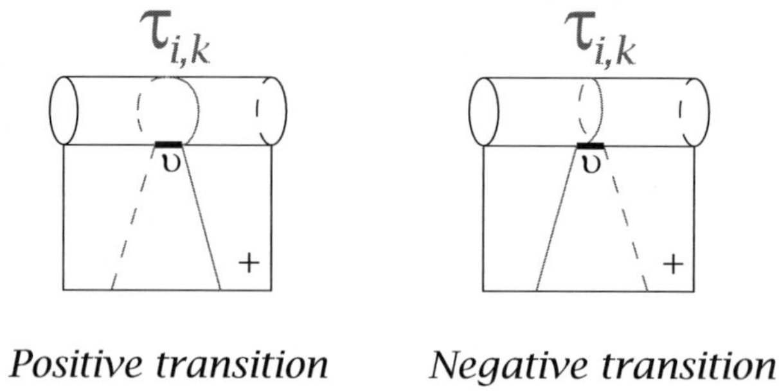

FIGURE 7. Positive and negative transition models.

negative. This designation of positive or negative is independent of the choice of orientation chosen on R.

For an alternative viewpoint that is often useful, observe that associated to a transition arc $\tau_{i,k}$ (oriented as above), there are exactly two choices of meridian with representatives disjoint from $\tau_{i,k}$. These are obtained from the two simple closed curves obtained by pasting $\tau_{i,k}$ with one of the two components of $\partial R|_{\partial \tau_{i,k}}$. For either such choice of meridian, $\upsilon = r(\tau_{i,k})$ is a proper arc of $S^1 \times \{1\}$, and $\tau_{i,k}$ is positive or negative, respectively, if υ inherits positive or negative orientation along ∂R from $\tau_{i,k}$, as illustrated in Figure 7. The transition arc is positive with respect to one choice and negative with respect to the other.

DEFINITION 4.8. Let S be a tight decomposing surface for $(M|_R, \partial M|_{\partial R})$ that is not a product disk. For simplicity of notation, identify arcs in $R_+ \cup R_-$ with their images in R under the quotient map $M|_R \to M$. Suppose that $s \geq 1$ and there exist i, j such that $\alpha_{i,j}$ and $\alpha_{i,j+1}$ are isotopic through proper embeddings in R, and let Δ be the component of $R|_{\alpha_{i,j} \cup \alpha_{i,j+1}}$ that contains $\tau_{i,j}$. If Δ is a source when considered as a sector of the branched surface $\langle R, S \rangle$, then call S and the sutured manifold decomposition $(M|_R, \partial M|_{\partial R}) \overset{S}{\rightsquigarrow} (M', \gamma')$ *double-diamond taut* (with respect to $\alpha_{i,j} * \tau_{i,j} * \alpha_{i,j+1}$).

In particular, a double-diamond taut decomposition is taut. In addition, a double-diamond taut decomposition determines a *distinguished meridian*, which we denote by ν, defined as follows.

DEFINITION 4.9. Suppose S is double-diamond taut with respect to $\alpha_{i,j} * \tau_{i,j} * \alpha_{i,j+1}$. Define the distinguished meridian ν on ∂M to be the isotopy class of

$$\nu_{i,j} = \tau_{i,j} \cup \upsilon_{i,j},$$

where $\upsilon_{i,j}$ in the unique arc in $\partial R \cap \Delta$ joining the endpoints of $\tau_{i,j}$.

Note that $\nu_{i,j}$ is a simple closed curve that has geometric intersection number one with ∂R and hence is a meridian. Since S is tight and $\alpha_{i,j}$ and $\alpha_{i,j+1}$ are isotopic, the transition arcs $\tau_{i,j-1}$, $\tau_{i,j}$, and $\tau_{i,j+1}$ have the same sign rel ν. For each transition arc $\tau_{i,k}$, let $\nu_{i,k}$ denote the representative $\tau_{i,k} \cup r(\tau_{i,k})$ of ν. It will follow from Corollary 4.11 that if $M = X_\kappa$ for a knot κ, then $\nu = \mu$, the meridian of κ

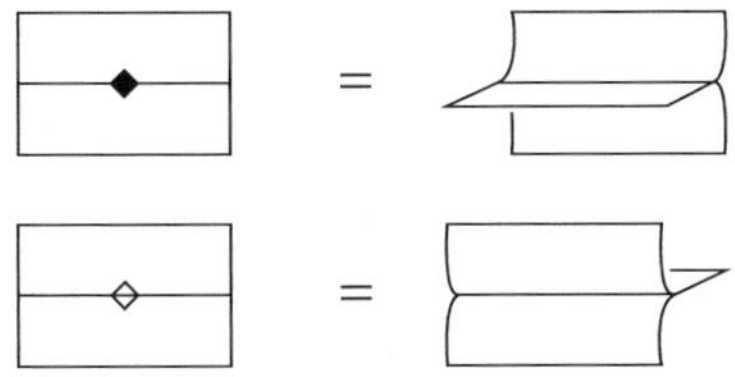

FIGURE 8. Diamond notation

In the following theorem statement and proof, we use Notation 4.1. In particular, T is a boundary parallel torus properly embedded in M. We also refer to the curve on T isotopic to ν as the distinguished meridian and use the same notation for it.

THEOREM 4.10 (Double-diamond replacement). *Let $\mathcal{S}_0$ be a sequence of taut sutured manifold decompositions*

$$(M, \partial M) \overset{R}{\rightsquigarrow} (M_1, \gamma_1) \overset{S}{\rightsquigarrow} (M_2, \gamma_2)$$

such that ∂M is a torus, ∂R is connected and nonempty, and S is double-diamond taut with respect to

$$\alpha_{i,j} * \tau_{i,j} * \alpha_{i,j+1} \subset \partial S.$$

Let Δ be the component of $R|_{\alpha_{i,j} \cup \alpha_{i,j+1}}$ that contains $\tau_{i,j}$, and suppose each sector of $\Sigma(\mathcal{S}_0') \cap \Sigma(\mathcal{S}_0)$ except Δ inherits its co-orientation from that of $\Sigma(\mathcal{S}_0)$. Give Δ the opposite orientation from R. There is a choice of orientations on the sectors of T in $\Sigma(\mathcal{S}_0')$ so that $\Sigma(\mathcal{S}_0')$ admits a smoothing to a (co-oriented) taut branched surface $B(\mathcal{S}_0')$ that has complement comprised of two sutured manifold components: one is $B^G(\mathcal{S}_0)$ and the other is $(\partial M \times [0,1], V_1 \cup V_2)$, where $\partial M \times \{0\} = \partial M$ and V_1 and V_2 are disjoint annuli of slope ν, the distinguished meridian determined by S.

PROOF. In order to simplify diagrams, we enhance arrow notation for cusp directions with diamond notation, as originally introduced by Wu [**52**] and depicted in Figure 8.

The desired co-orientations on the sectors of T, along with the resulting smoothing and the paths of the sutures that result, are indicated for negative transitions in Figure 9; the construction for positive transitions is given by the mirror image. A schematic treatment for negative transitions is provided in Figure 10, and for positive transitions in Figure 11. (The reason for the mysterious term "double-diamond" should now be clear!) The local models for co-orientations and associated smoothings in neighborhoods of $\nu_{i,j-1}, \nu_{i,j}$ and $\nu_{i,j+1}$ are highlighted in Figures 10 and 11. It is straightforward to check that the resulting co-orientation on $\Sigma(\mathcal{S}_0')$ admits a smoothing to a co-oriented branched surface, which we denote by $B(\mathcal{S}_0')$.

The local smoothings at $\nu_{i,j-1}$ and $\nu_{i,j}$ (marked C in Figures 10, 11 and 12) each introduce a cusp of slope ν on T with sink direction pointing into R and S, resulting in a suture on the boundary of the complementary region Y containing ∂M. As shown in Figures 9 and 12, the remaining complementary region is isotopic to the complement of $B^G(\mathcal{S}_0)$. The complementary regions of $B(\mathcal{S}_0')$ are therefore taut sutured manifolds.

It remains only to show that through every sector of $B(\mathcal{S}_0')$ there is a closed oriented curve that is positively transverse to $B(\mathcal{S}_0')$. For points in Δ, such a

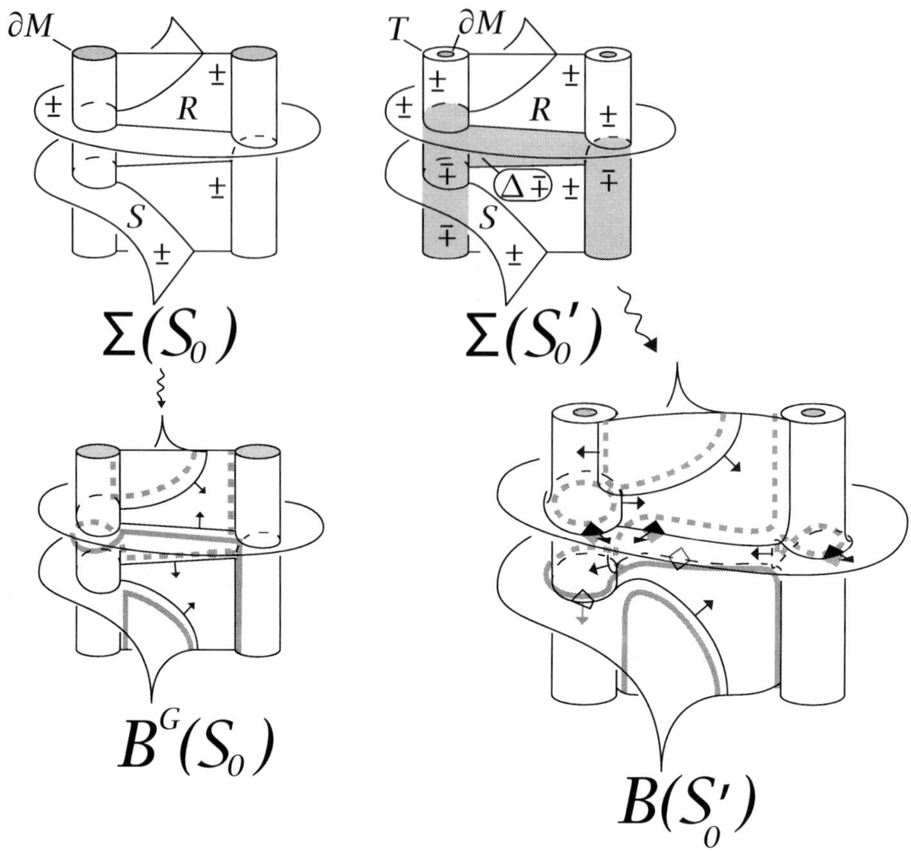

FIGURE 9. The transverse orientations on $\Sigma(\mathcal{S}_0)$ and $\Sigma(\mathcal{S}_0')$ and the resulting branched surfaces $B^G(\mathcal{S}_0)$ and $B(\mathcal{S}_0')$, with the paths of sutures in red. Sutures behind surface R, as well as the meridional sutures in component Y, are dashed; those in front of R are solid.

transversal is provided by a curve α parallel to ν, appropriately oriented. It follows from Theorem 3.7 and Definition 4.3 that $B^G(\mathcal{S}_0)$ is taut; therefore, through any point in a sector disjoint from T and Δ there is a closed positive transversal to $B^G(\mathcal{S}_0)$. It is possible that this transversal passes through Δ, in which case such intersections may be removed by inserting copies of α. Finally, let Γ be any sector of $B(\mathcal{S}_0')$ that is contained in T. Without loss of generality, we may assume that a positive transversal to Γ points into a complementary region disjoint from ∂M. Let β_1 be an arc properly embedded in the complement of $B(\mathcal{S}_0')$ that runs from Γ to S_- (viewed as pushed slightly off the negative side of S), let β_2 be a simple arc in S_- from the endpoint of β_1 to a point in a sector of T adjoining a meridional cusp, and let β_3 be a simple arc properly embedded in the complementary region containing ∂M that connects β_3 to β_1. The simple closed curve $\beta = \beta_1 * \beta_2 * \beta_3$ is the desired transversal. (See Figure 13.) $\square$

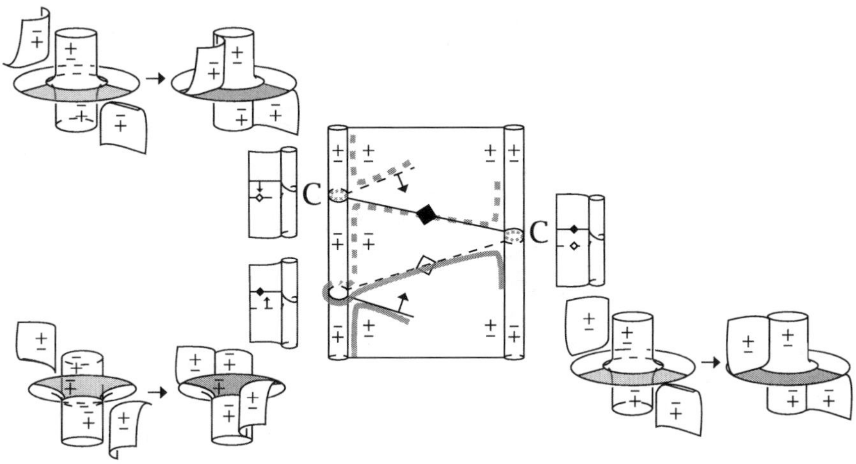

FIGURE 10. The local models in neighborhoods of $\nu_{i,j-1}, \nu_{i,j}$ and $\nu_{i,j+1}$.

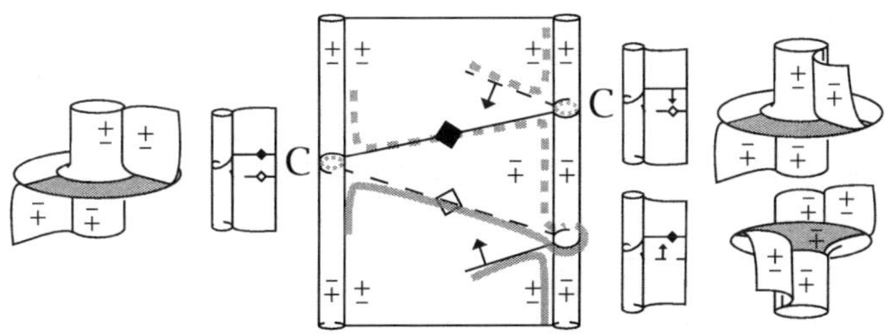

FIGURE 11. The local models in neighborhoods of $\nu_{i,j-1}, \nu_{i,j}$ and $\nu_{i,j+1}$.

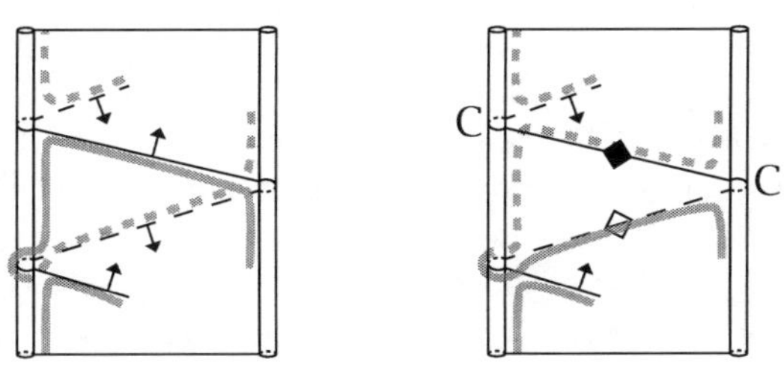

FIGURE 12. Comparison of the sutures in the complements of $B^G(\mathcal{S}_0)$ and $B(\mathcal{S}_0')$.

We have now established the desired taut branched surface, but in contrast to Gabai's branched surface $B^G(\mathcal{S}_0)$, it is not inherent in the construction of $B(\mathcal{S}_0')$ that it fully carries a lamination. However, an easily checked technical condition on $B^G(\mathcal{S}_0)$ suffices to ensure that it does. We then apply Corollary 3.9 and Proposition 3.10 to show this lamination extends to taut foliations that strongly realize all boundary slopes except the distinguished meridian ν.

COROLLARY 4.11. *Let $\mathcal{S}_0$ be a sequence of taut sutured manifold decompositions*

$$(M, \partial M) \overset{R}{\rightsquigarrow} (M_1, \gamma_1) \overset{S}{\rightsquigarrow} (M_2, \gamma_2)$$

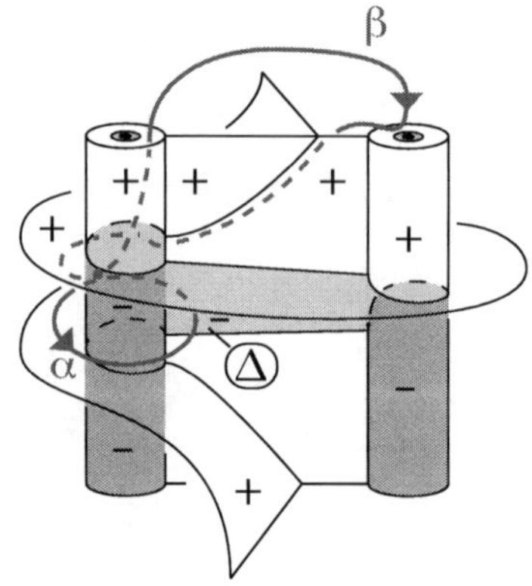

FIGURE 13. Positive closed transversals for Δ and the sectors of T in $B(\mathcal{S}')$.

such that ∂M is a torus, ∂R is connected and nonempty, and S is double-diamond taut with respect to

$$\alpha_{i,j} * \tau_{i,j} * \alpha_{i,j+1} \subset \partial S.$$

If $B^G(\mathcal{S}_0)$ has no sink disks disjoint from ∂M, then there are co-oriented taut foliations that strongly realize all boundary slopes except ν, the distinguished meridian determined by S. Moreover, if $M = X_\kappa$ for some knot κ in an L-space, κ is persistently foliar.

PROOF. Let $B(\mathcal{S}_0')$ be the taut, and hence essential, branched surface obtained from $B^G(\mathcal{S}_0)$ by double-diamond replacement, as described in Theorem 4.10. Since $B^G(\mathcal{S}_0)$ contains no trivial bubbles, neither does $B(\mathcal{S}_0')$. We show that $B(\mathcal{S}_0')$ contains no sink disks, and hence is laminar.

The sectors of $B(\mathcal{S}_0')$ that are contained in T correspond to digon complementary regions of $B^G(\mathcal{S}_0) \cap \partial M$ and hence are not sink disks. We therefore restrict attention to the sectors of $B(\mathcal{S}_0')$ that are not contained in T. These are in bijective correspondence with, and isomorphic to, those in $B^G(\mathcal{S}_0)$, with arrows replaced by diamonds at $\alpha_{i,j} \cup \alpha_{i,j+1}$. In addition, outward pointing arrows are introduced along boundary arcs on T except for those adjacent to one of the two meridian cusps, marked C in Figures 10 and 12. Any sector with boundary that has nonempty intersection with T therefore has outward pointing arrow along T; in particular, this is true for the sectors that have nonempty intersection with $\alpha_{i,j} \cup \alpha_{i,j+1}$. Moreover, by assumption, any sector that has boundary disjoint from T is not a sink disk. Thus, $B(\mathcal{S}_0')$ is laminar, and therefore fully carries a lamination $\mathcal{L}$.

Since any leaf of $\mathcal{L}$ that passes through the I-fibered neighbourhood of the branch S is asymptotic to T, it is noncompact. Hence the conditions of Proposition 3.10 are satisfied. It therefore follows from Corollary 3.9 and Proposition 3.10 that M supports co-oriented taut foliations that strongly realize all boundary slopes except ν. In the case $M = X_\kappa$ for a knot κ in an L-space, it must be the case that $\nu = \mu$, the meridian of κ, since by [**2, 26, 27, 38**] an L-space supports no co-oriented taut foliation; hence κ is persistently foliar. □

REMARK 4.12. If S is a tight product disk and $\alpha_1 = \alpha_{1,1}$ and $\alpha_2 = \alpha_{1,2}$ are isotopic through proper embeddings in R, then after isotoping D so that $\alpha_1 = \alpha_2$,

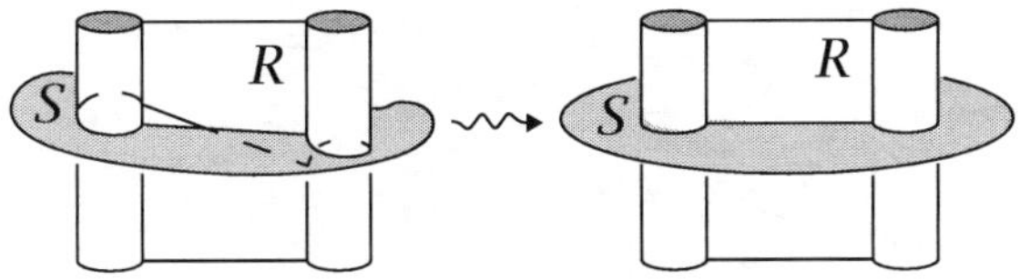

FIGURE 14. If S is a tight product disk and $\alpha_1 = \alpha_{1,1}$ and $\alpha_2 = \alpha_{1,2}$ are isotopic through proper embeddings in R, then κ is composite.

D determines a meridional annulus, and hence κ is composite, as illustrated in Figure 14. For some results on persistently foliar composite knots, see [**7**].

REMARK 4.13. Definition 4.3 admits a straightforward generalization to sutured manifold decompositions of greater length; namely, to appropriately constrained decompositions of the form

$$\Big((M,\partial M) \overset{S_1}{\rightsquigarrow} (M_1,\gamma_1) \overset{S_2}{\rightsquigarrow} \cdots \overset{S_n}{\rightsquigarrow} (M_n,\gamma_n)\Big) \overset{S}{\rightsquigarrow} (M',\gamma')\,.$$

This generalization then leads naturally to a generalization of Theorem 4.10. As these generalizations require significantly more cumbersome notation, and have yet to prove useful in explicit applications, we leave their statements as exercises for the interested reader.

5. Application to Murasugi sums

DEFINITION 5.1 ([**12**]). Let $n \geq 1$. The oriented surface R in S^3 is a *Murasugi sum* along P of oriented surfaces R_1 and R_2 if

(1) there is a sphere S in S^3 such that $P = S \cap R$ is a $2n$-gon embedded in R with boundary a concatenation $\partial P = \alpha_1 * \beta_1 * \cdots \alpha_n * \beta_n$ such that each α_i is a subarc of ∂R and each β_i is properly embedded in R, and
(2) if B_1 and B_2 are the two embedded 3-balls with common boundary S, then for each i, R_i is homeomorphic to the subsurface $B_i \cap R$.

When $n = 2$, a Murasugi sum is also known as a *plumbing* (see Figure 15). For a nice history of the definitions of plumbing and Murasugi sum, see [**37**].

THEOREM 5.2 ([**12, 14**]). *Suppose $R \subset S^3$ is a Murasugi sum along P of oriented surfaces R_1 and R_2. Then R is a minimal genus Seifert surface for ∂R if and only if each R_i, $i = 1, 2$, is a minimal genus Seifert surface for ∂R_i.*

COROLLARY 5.3. *Suppose κ is a knot in S^3 with minimal genus Seifert surface R. If R is a plumbing of surfaces R_1 and R_2, where R_2 is an unknotted band with an even number $2m \geq 4$ of half-twists, then the decomposing disk for R_2 can be chosen to be double-diamond taut. Thus, κ is persistently foliar.*

PROOF. The decomposing disk D dual to the band R_2 is double-diamond taut, as indicated in Figure 15. Moreover, $B^G(\mathcal{S}_0) = \langle R, D\rangle$ has no sink disks disjoint from ∂X_κ. The result therefore follows immediately from Corollary 4.11. The relevant parts of $B^G(\mathcal{S}_0)$ and $B(\mathcal{S}_0')$ are shown in Figure 16. □

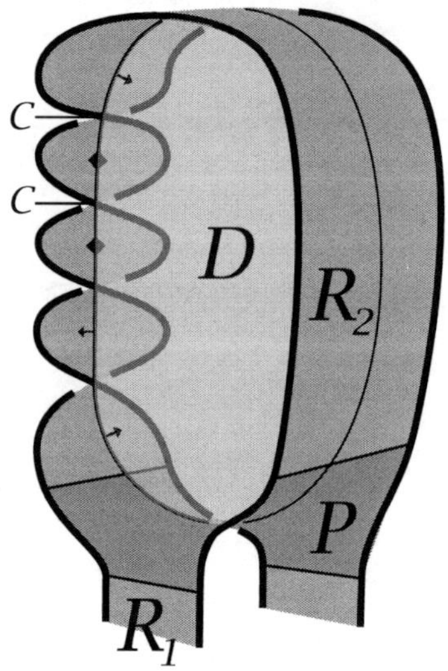

FIGURE 15. $B(\mathcal{S}_0')$ for a positive plumbed band.

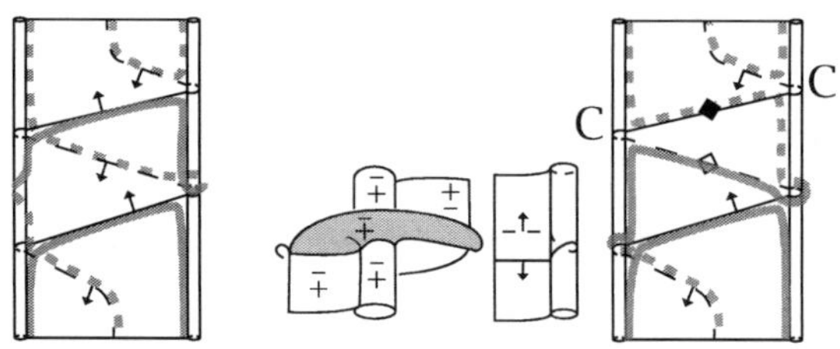

FIGURE 16. Comparison of the sutures in the complements of $B^G(\mathcal{S}_0)$ and $B(\mathcal{S}_0')$.

Acknowledgment

It is a pleasure and a privilege to honor Steven Boyer for his many contributions over the years. We thank the referee for providing a thorough report that helped us improve the exposition significantly.

References

[1] K. L. Baker and A. H. Moore, *Montesinos knots, Hopf plumbings, and L-space surgeries*, J. Math. Soc. Japan **70** (2018), no. 1, 95–110, DOI 10.2969/jmsj/07017484. MR3750269

[2] J. Bowden, *Approximating C^0-foliations by contact structures*, Geom. Funct. Anal. **26** (2016), no. 5, 1255–1296, DOI 10.1007/s00039-016-0387-2. MR3568032

[3] S. Boyer and A. Clay, *Foliations, orders, representations, L-spaces and graph manifolds*, Adv. Math. **310** (2017), 159–234, DOI 10.1016/j.aim.2017.01.026. MR3620687

[4] S. Boyer, C. McA. Gordon, and L. Watson, *On L-spaces and left-orderable fundamental groups*, Math. Ann. **356** (2013), no. 4, 1213–1245, DOI 10.1007/s00208-012-0852-7. MR3072799

[5] J. Cantwell and L. Conlon, *Taut Foliations of Sutured 3-manifolds and Regularity*, preprint.

[6] B. G. Casler, *An imbedding theorem for connected 3-manifolds with boundary*, Proc. Amer. Math. Soc. **16** (1965), 559–566, DOI 10.2307/2033878. MR178473

[7] C. Delman and R. Roberts, *Persistently foliar composite knots*, `arXiv:1905.04838`.

[8] C. Delman and R. Roberts, *Non-torus alternating knots are persistently foliar*, preprint.

[9] C. Delman and R. Roberts, *Montesinos knots satisfy the L-space knot conjecture*, preprint.

[10] C. Delman and R. Roberts, *Persistently foliar Montesinos knots*, preprint.

[11] W. Floyd and U. Oertel, *Incompressible surfaces via branched surfaces*, Topology **23** (1984), no. 1, 117–125, DOI 10.1016/0040-9383(84)90031-4. MR721458

[12] D. Gabai, *The Murasugi sum is a natural geometric operation*, Low-dimensional topology (San Francisco, Calif., 1981), Contemp. Math., vol. 20, Amer. Math. Soc., Providence, RI, 1983, pp. 131–143, DOI 10.1090/conm/020/718138. MR718138

[13] D. Gabai, *Foliations and the topology of* 3*-manifolds*, J. Differential Geom. **18** (1983), no. 3, 445–503. MR723813

[14] D. Gabai, *The Murasugi sum is a natural geometric operation. II*, Combinatorial methods in topology and algebraic geometry (Rochester, N.Y., 1982), Contemp. Math., vol. 44, Amer. Math. Soc., Providence, RI, 1985, pp. 93–100, DOI 10.1090/conm/044/813105. MR813105

[15] D. Gabai, *Detecting fibred links in* S^3, Comment. Math. Helv. **61** (1986), no. 4, 519–555, DOI 10.1007/BF02621931. MR870705

[16] D. Gabai, *Foliations and the topology of* 3*-manifolds. II*, J. Differential Geom. **26** (1987), no. 3, 461–478. MR910017

[17] D. Gabai, *Foliations and the topology of* 3*-manifolds. III*, J. Differential Geom. **26** (1987), no. 3, 479–536. MR910018

[18] D. Gabai and U. Oertel, *Essential laminations in* 3*-manifolds*, Ann. of Math. (2) **130** (1989), no. 1, 41–73, DOI 10.2307/1971476. MR1005607

[19] P. Ghiggini, *Knot Floer homology detects genus-one fibred knots*, Amer. J. Math. **130** (2008), no. 5, 1151–1169, DOI 10.1353/ajm.0.0016. MR2450204

[20] A. Haefliger, *Variétés feuilletées* (French), Ann. Scuola Norm. Sup. Pisa Cl. Sci. (3) **16** (1962), 367–397. MR189060

[21] W. Haken, *Theorie der Normalflächen. Ein Isotopiekriterium für den Kreisknoten*, Acta Math., **105 (3–4)** (1961), 245–375.

[22] K. Honda, W. H. Kazez, and G. Matić, *Right-veering diffeomorphisms of compact surfaces with boundary*, Invent. Math. **169** (2007), no. 2, 427–449, DOI 10.1007/s00222-007-0051-4. MR2318562

[23] A. Juhász, *Floer homology and surface decompositions*, Geom. Topol. **12** (2008), no. 1, 299–350, DOI 10.2140/gt.2008.12.299. MR2390347

[24] A. Juhász, *The sutured Floer homology polytope*, Geom. Topol. **14** (2010), no. 3, 1303–1354, DOI 10.2140/gt.2010.14.1303. MR2653728

[25] A. Juhász, *A survey of Heegaard Floer homology*, New ideas in low dimensional topology, Ser. Knots Everything, vol. 56, World Sci. Publ., Hackensack, NJ, 2015, pp. 237–296, DOI 10.1142/9789814630627_0007. MR3381327

[26] W. H. Kazez and R. Roberts, *Approximating* $C^{1,0}$*-foliations*, Interactions between low-dimensional topology and mapping class groups, Geom. Topol. Monogr., vol. 19, Geom. Topol. Publ., Coventry, 2015, pp. 21–72, DOI 10.2140/gt. MR3609903

[27] W. H. Kazez and R. Roberts, C^0 *approximations of foliations*, Geom. Topol. **21** (2017), no. 6, 3601–3657, DOI 10.2140/gt.2017.21.3601. MR3693573

[28] T. Lidman and A. H. Moore, *Pretzel knots with L-space surgeries*, Michigan Math. J. **65** (2016), no. 1, 105–130. MR3466818

[29] T. Li, *Laminar branched surfaces in 3-manifolds*, Geom. Topol. **6** (2002), 153–194, DOI 10.2140/gt.2002.6.153. MR1914567

[30] T. Li, *Boundary train tracks of laminar branched surfaces*, Topology and geometry of manifolds (Athens, GA, 2001), Proc. Sympos. Pure Math., vol. 71, Amer. Math. Soc., Providence, RI, 2003, pp. 269–285, DOI 10.1090/pspum/071/2024638. MR2024638

[31] Y. Ni, *Knot Floer homology detects fibred knots*, Invent. Math. **170** (2007), no. 3, 577–608, DOI 10.1007/s00222-007-0075-9. MR2357503

[32] Y. Ni, *Erratum: Knot Floer homology detects fibred knots [MR2357503]*, Invent. Math. **177** (2009), no. 1, 235–238, DOI 10.1007/s00222-009-0174-x. MR2507641

[33] S. P. Novikov, *The topology of foliations* (Russian), Trudy Moskov. Mat. Obšč. **14** (1965), 248–278. MR0200938

[34] U. Oertel, *Incompressible branched surfaces*, Invent. Math. **76** (1984), no. 3, 385–410, DOI 10.1007/BF01388466. MR746535

[35] U. Oertel, *Homology branched surfaces: Thurston's norm on $H_2(M^3)$*, Low-dimensional topology and Kleinian groups (Coventry/Durham, 1984), London Math. Soc. Lecture Note Ser., vol. 112, Cambridge Univ. Press, Cambridge, 1986, pp. 253–272. MR903869

[36] U. Oertel, *Measured laminations in 3-manifolds*, Trans. Amer. Math. Soc. **305** (1988), no. 2, 531–573, DOI 10.2307/2000878. MR924769

[37] B. Ozbagci and P. Popescu-Pampu, *Generalized plumbings and Murasugi sums*, Arnold Math. J. **2** (2016), no. 1, 69–119, DOI 10.1007/s40598-015-0033-3. MR3460043

[38] P. Ozsváth and Z. Szabó, *Holomorphic disks and genus bounds*, Geom. Topol. **8** (2004), 311–334, DOI 10.2140/gt.2004.8.311. MR2023281

[39] P. Ozsváth and Z. Szabó, *Holomorphic disks and topological invariants for closed three-manifolds*, Ann. of Math. (2) **159** (2004), no. 3, 1027–1158, DOI 10.4007/annals.2004.159.1027. MR2113019

[40] P. Ozsváth and Z. Szabó, *Holomorphic disks and three-manifold invariants: properties and applications*, Ann. of Math. (2) **159** (2004), no. 3, 1159–1245, DOI 10.4007/annals.2004.159.1159. MR2113020

[41] P. Ozsváth and Z. Szabó, *On knot Floer homology and lens space surgeries*, Topology **44** (2005), no. 6, 1281–1300, DOI 10.1016/j.top.2005.05.001. MR2168576

[42] C. F. B. Palmeira, *Open manifolds foliated by planes*, Ann. of Math. (2) **107** (1978), no. 1, 109–131, DOI 10.2307/1971256. MR501018

[43] G. Reeb, *Sur certaines propriétés topologiques des variétés feuilletées* (French), Actualités Sci. Ind., no. 1183, Hermann & Cie., Paris, 1952. Publ. Inst. Math. Univ. Strasbourg 11, pp. 5–89, 155–156. MR0055692

[44] R. Roberts, *Taut foliations in punctured surface bundles. II*, Proc. London Math. Soc. (3) **83** (2001), no. 2, 443–471, DOI 10.1112/plms/83.2.443. MR1839461

[45] H. Rosenberg, *Foliations by planes*, Topology **7** (1968), 131–138, DOI 10.1016/0040-9383(68)90021-9. MR228011

[46] M. Scharlemann, *Producing reducible 3-manifolds by surgery on a knot*, Topology **29** (1990), no. 4, 481–500, DOI 10.1016/0040-9383(90)90017-E. MR1071370

[47] J. R. Stallings, *Constructions of fibred knots and links*, Algebraic and geometric topology (Proc. Sympos. Pure Math., Stanford Univ., Stanford, Calif., 1976), Proc. Sympos. Pure Math., XXXII, Amer. Math. Soc., Providence, R.I., 1978, pp. 55–60. MR520522

[48] W. P. Thurston, *Three-dimensional manifolds, Kleinian groups and hyperbolic geometry*, Bull. Amer. Math. Soc. (N.S.) **6** (1982), no. 3, 357–381, DOI 10.1090/S0273-0979-1982-15003-0. MR648524

[49] W. P. Thurston, *A norm for the homology of 3-manifolds*, Mem. Amer. Math. Soc. **59** (1986), no. 339, i–vi and 99–130. MR823443

[50] F. Waldhausen, *On irreducible 3-manifolds which are sufficiently large*, Ann. of Math. (2) **87** (1968), 56–88, DOI 10.2307/1970594. MR224099

[51] R. F. Williams, *Expanding attractors*, Inst. Hautes Études Sci. Publ. Math. **43** (1974), 169–203. MR348794

[52] Y.-Q. Wu, *Persistently laminar branched surfaces*, Comm. Anal. Geom. **20** (2012), no. 2, 397–434, DOI 10.4310/CAG.2012.v20.n2.a7. MR2928718

Department of Mathematics and Computer Science, Eastern Illinois University, Charleston, IL 61920

Email address: `cidelman@eiu.edu`

Department of Mathematics, Washington University, St. Louis, MO 63105

Email address: `roberts@wustl.edu`

Contemporary Mathematics
Volume **760**, 2020
https://doi.org/10.1090/conm/760/15289

A census of exceptional Dehn fillings

Nathan M. Dunfield

ABSTRACT. This paper describes the complete list of all 205,822 exceptional Dehn fillings on the 1-cusped hyperbolic 3-manifolds that have ideal triangulations with at most 9 ideal tetrahedra. The data is consistent with the standard conjectures about Dehn filling and suggests some new ones.

1. Introduction

1.1. Dehn filling. Suppose M is a compact orientable 3-manifold with ∂M a torus. A *slope* on ∂M is an unoriented isotopy class of simple closed curve, or equivalently a primitive element of $H_1(\partial M; \mathbb{Z})$ modulo sign. The set of all slopes will be denoted $Sl(M)$, which can be viewed as the rational points in the projective line $P^1\big(H_1(\partial M; \mathbb{R})\big) \cong \mathbb{R}\mathrm{P}^1$. The Dehn fillings of M are parameterized by $\alpha \in Sl(M)$, with $M(\alpha)$ being the Dehn filling where α bounds a disk in the attached solid torus. When the interior of M admits a hyperbolic metric of finite volume, it is called a *1-cusped hyperbolic 3-manifold.* For such hyperbolic M, Thurston showed that all but finitely many $M(\alpha)$ are also hyperbolic [**Thu**]. The nonhyperbolic Dehn fillings are called *exceptional*, and the corresponding slopes the *exceptional slopes.* Understanding the possible exceptional fillings has been a major topic in the study of 3-manifolds over the past 40 years; see the surveys [**Gor1, Gor2, Gor3, Gor4**] for further background.

This paper gives a census of all exceptional Dehn fillings on a certain collection of 1-cusped hyperbolic 3-manifolds. Specifically, let $\mathcal{C}_t$ be the set of all orientable 1-cusped hyperbolic 3-manifolds that have ideal triangulations with at most t ideal tetrahedra. For $t \leq 9$, the set $\mathcal{C}_t$ has been enumerated by [**HW, CHW, Thi, Bur2**] and is included with SnapPy [**CDGW**], whose nomenclature for these manifolds (e.g. $m004$, $s011$, $v1002$, $t12345$, and $o9_{60000}$) I will use freely throughout. Each manifold M in $\mathcal{C}_t$ has a preferred basis for $H_1(\partial M; \mathbb{Z})$, and so I will denote slopes in $Sl(M)$ by elements in $\mathbb{Z}^2$. See Figure 1 for some basic statistics on the 59,107 manifolds in $\mathcal{C}_9$. The main result of this paper is:

1.2. THEOREM. *There are precisely 205,822 exceptional Dehn fillings on the manifolds in $\mathcal{C}_9$, that is, pairs (M, α) where $M(\alpha)$ is not hyperbolic, of the types listed in Table* 2 *and distributed as in Figure* 3.

2020 *Mathematics Subject Classification.* Primary 57K35; Secondary 57K32.
Key words and phrases. exceptional Dehn filling, Dehn surgery.

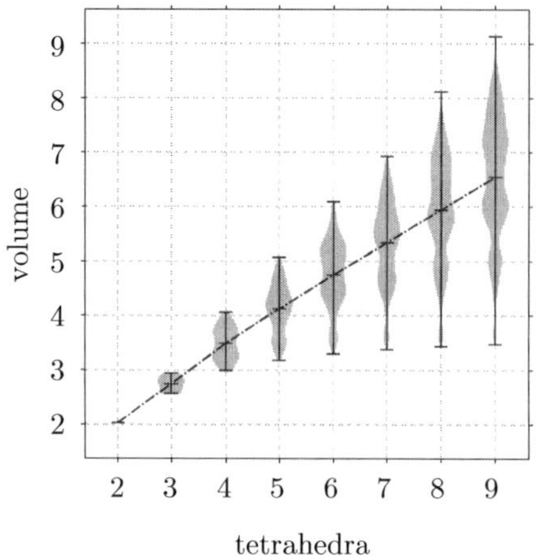

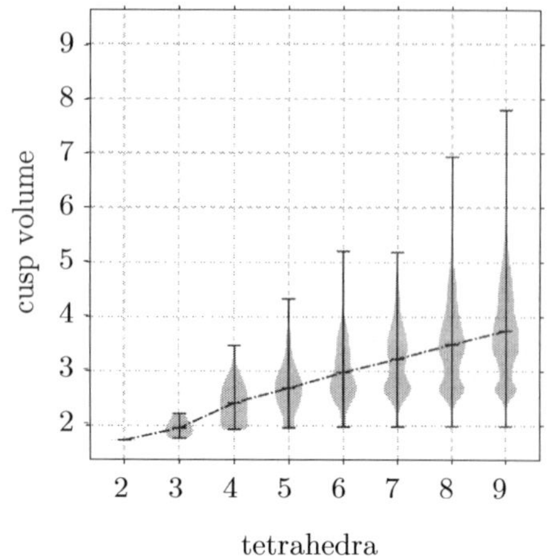

tets	#manifolds
2	2
3	9
4	52
5	223
6	913
7	3,388
8	12,241
9	42,279
total	59,107

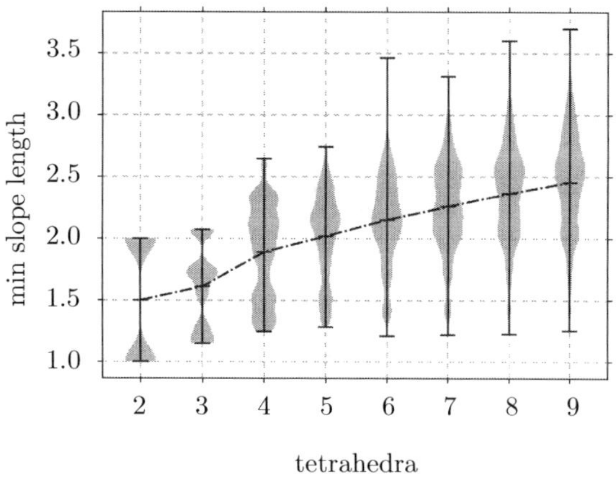

FIGURE 1. Some basic statistics on the manifolds in $\mathcal{C}_9$. The table in the lower left shows the number of manifolds in $\mathcal{C}_t \setminus \mathcal{C}_{t-1}$, that is, those whose minimal ideal triangulations have exactly t tetrahedra. In the top left, the hyperbolic volume is shown via a violin plot. Here, each "violin" shows the distribution of volume for $\mathcal{C}_t \setminus \mathcal{C}_{t-1}$, where the top and bottom horizontal bars are the min and max, the middle horizontal bar is the mean, and the violin "body" is a smoothed histogram of the volumes. The plot in the upper right shows the cusp volume, i.e. the volume of a maximal cusp neighborhood bounded by an embedded horotorus. The plot in the lower right shows the minimal slope length, that is, the shortest essential curve in the maximal horotorus. See Section 4.3 for the relevance of this cusp data.

	$b_1(M(\alpha)) = 0$		$b_1(M(\alpha)) > 0$
atoroidal			
S^3	1,267		
lens space	44,487		
finite π_1	13,446	$S^2 \times S^1$	242
Seifert fibered	71,111	Seifert fibered	118
connected sum	4,296	connected sum	169
toroidal			
Seifert fibered	1,730	Seifert fibered	159
graph manifold	63,325	graph manifold	3,043
hyperbolic piece	2,136	hyperbolic piece	74
		Sol torus bundle	219
totals	201,798		4,024

TABLE 2. Summary of the topological types of the 205,822 exceptional fillings from Theorem 1.2, broken down by betti number and whether the filling is toroidal. See Section 2 for precise definitions. Here, each filling is listed only once in the most restricted category possible from Figure 4, and there are no fillings that are both toroidal and connected sums.

$e(M)$	#manifolds
0	3
1	1,824
2	9,730
3	18,003
4	19,009
5	8,751
6	1,776
7	8
8	2
9	0
10	1

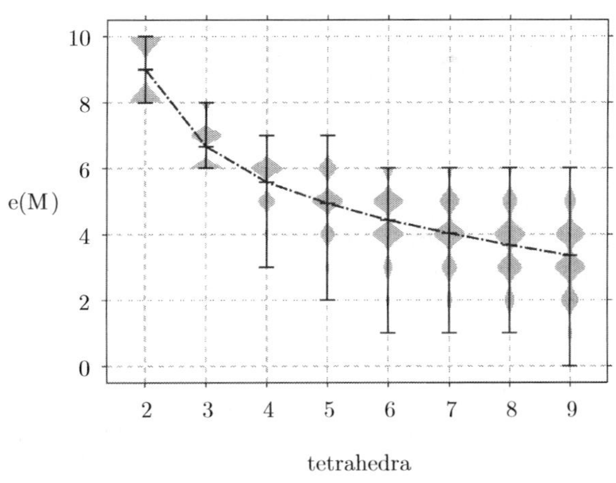

FIGURE 3. This figure describes how the exceptional fillings of Theorem 1.2 are distributed over the manifolds in $\mathcal{C}_9$. The number of exceptional fillings on M is denoted $e(M)$, and the table at left shows the number of manifolds in $\mathcal{C}_9$ having each possible value of $e(M)$, which is at most 10 by [**LM**]. At right is a violin plot of $e(M)$ as a function of the number of tetrahedra of M; see Figure 1 for more on violin plots.

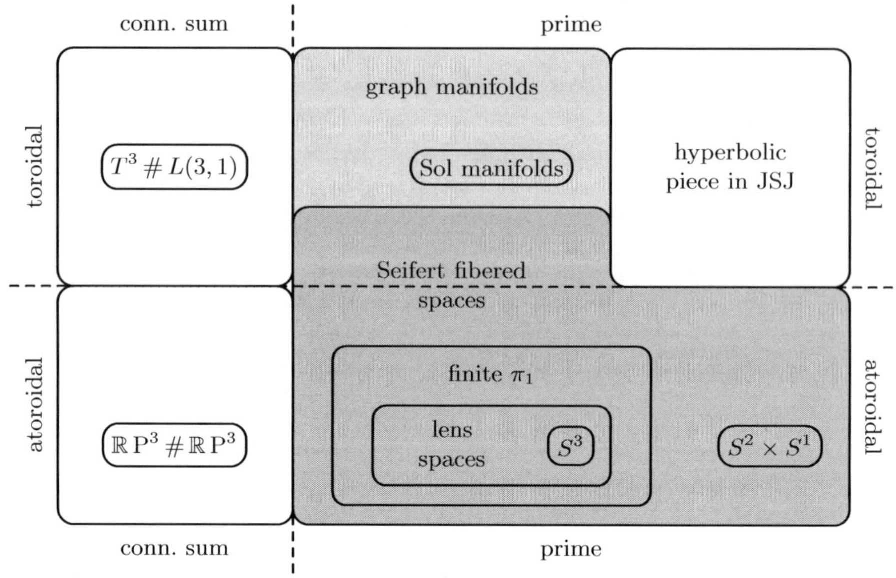

FIGURE 4. The different types of nonhyperbolic 3-manifolds, and hence different types of exceptional Dehn fillings. See Section 2 for definitions.

The list of these exceptional (M, α) together with the precise topology of each $M(\alpha)$ is available at [**Dun**]. Here, in addition to describing the proof of Theorem 1.2 in Section 5, I will give summaries of this data as it relates to known results and open questions about Dehn filling in Sections 3 and 4.

1.3. Prior work. In the 1990s, Hodgson and Weeks studied the exceptional Dehn fillings on the 286 manifolds in $\mathcal{C}_5$; this work was never published but is referred to extensively in [**Gor1**] and provided many key examples in the subject. The series of papers [**MP**, **MPR**, **Mar**] classified all exceptional fillings on an important series of chain links with as many as 7 components; as noted in [**Mar**, §3], this determines the exceptional fillings on more than 95% of the 4,587 manifolds in $\mathcal{C}_7$. John Berge (personal communication) independently did a search for exceptional fillings on $\mathcal{C}_9$ using a new version of his program Heegaard [**Ber**], and found more than 99.3% of the exceptional fillings included in Theorem 1.2.

1.4. Acknowledgements. I thank Ken Baker, Bruno Martelli, and Jake Rasmussen for helpful discussions on the topic of this paper and also thank John Berge for sharing his unpublished data on this topic with me. This work was done at the University of Illinois, the University of Melbourne, and IAS, and was funded in part by the Simons Foundation and the US National Science Foundation, the latter under grants DMS-1510204, DMS-1811156, and the GEAR Network (DMS-1107452). I also thank the referee for their helpful comments on this paper, especially their observation in Section 4.3.

2. Background and conventions

I first review the different types of nonhyperbolic 3-manifolds to establish my conventions on the kinds of exceptional Dehn fillings one can study. Sources vary slightly on the latter point, and here I use a relatively fine-grained division, which is illustrated in Figure 4. The summary for experts, who may safely skip this section, is that here atoroidal means geometrically atoroidal, the manifold $S^2 \times S^1$ is neither reducible nor an honorary lens space, and the term Seifert fibered will not include $\mathbb{R}P^3 \# \mathbb{R}P^3$. I will assume familiarity with basic 3-manifold topology, the Geometrization Theorem, and the resulting general structure of 3-manifolds, see e.g. [**Hat, Sco, Bon**] for details. Throughout, all 3-manifolds will be compact, orientable, and be either closed or have boundary that is a union of tori; the symbol M will always refer to such a manifold.

Our first two kinds of nonhyperbolic 3-manifolds are those containing certain spheres and tori. An embedded 2-sphere in M is *essential* if it does not bound a 3-ball. If there are no essential spheres then M is *irreducible*, and this includes all hyperbolic M. Those M containing *separating* essential spheres are called *connected sums*; here, I avoid the more common term reducible for this as for some authors reducible is the complement of irreducible and so includes $S^2 \times S^1$ whose only essential sphere is nonseparating. When M is not a connected sum it is *prime*. An M is *toroidal* when it contains an embedded essential torus T, that is, one where $\pi_1 T \to \pi_1 M$ is injective and T is not isotopic to a component of ∂M; this is sometimes called *geometrically toroidal*. When M is not toroidal it is *atoroidal*. All hyperbolic M are atoroidal.

When M has a foliation by circles it is *Seifert fibered* and called a *Seifert fibered space*; I will shortly revise this definition to exclude a particularly unusual such manifold. The Seifert fibered manifolds are exactly those admitting these six of the eight possible geometries: S^3, $\mathbb{E}^3$, $\mathbb{H}^2 \times \mathbb{R}$, $S^2 \times \mathbb{R}$, Nil, and $\widetilde{\mathrm{PSL}_2\mathbb{R}}$ and in particular are not hyperbolic. Those with spherical geometry are precisely the 3-manifolds with finite π_1, including S^3 itself and the lens spaces $L(p,q)$ which are all quotients of S^3 by a cyclic group. There are only two M with an $S^2 \times \mathbb{R}$ geometry, namely $S^2 \times S^1$ and $\mathbb{R}P^3 \# \mathbb{R}P^3$, which are both rather special. First, $S^2 \times S^1$ is the only closed 3-manifold with Heegaard genus one that is not a lens space, and also the only one whose fundamental group is infinite cyclic. Second, $\mathbb{R}P^3 \# \mathbb{R}P^3$ is the unique Seifert fibered manifold that is a connected sum. I henceforth adopt the nonstandard convention that $\mathbb{R}P^3 \# \mathbb{R}P^3$ is not Seifert fibered; this way, all Seifert fibered spaces are prime.

Any irreducible M has a collection of disjoint essential tori that cut it up into pieces that are either Seifert fibered or hyperbolic. The minimal such collection is unique up to isotopy and gives the *JSJ decomposition* of M. A *graph manifold* is one where all the pieces in the JSJ decomposition are Seifert fibered. So Seifert fibered manifolds are graph manifolds as are those that admit the Sol geometry; the latter are virtually torus bundles over the circle with Anosov monodromy. An irreducible M that is neither hyperbolic nor a graph manifold has a nontrivial JSJ decomposition where at least one piece is hyperbolic; such M have a *hyperbolic piece*.

Figure 4 summarizes all the different types of nonhyperbolic 3-manifolds. Of course, many manifolds satisfy several of these conditions, and in certain tables I will want each nonhyperbolic manifold to have a single type. In such cases, the type used will be the most restricted possible in Figure 4; for example, the type of $L(3,1)$ will be a lens space, even though it also has finite π_1, is Seifert fibered, and is a graph manifold. (This always makes sense because I set up the various definitions to minimize overlaps that are not containments.) This more restrictive convention is used in Tables 2 and Tables 6 only; Table 5 is correct with either convention.

3. Evidence for standard conjectures

A great deal has been proven about the possibilities for exceptional Dehn fillings; with regards to the gaps in our knowledge, the fillings of Theorem 1.2 are consistent with the standard conjectures as I now describe.

3.1. Knots in the 3-sphere. I start with the 1,267 manifolds in $\mathcal{C}_9$ that are exteriors of knots in S^3, which collectively have some 2,615 additional exceptional fillings.

(1) There are no Dehn fillings that are connected sums, consistent with the Cabling Conjecture [**Gor3**, §2.2].
(2) The Berge Conjecture [**Gor3**, §3.2] holds for the 178 nontrivial lens space fillings.
(3) All 1,143 nontrivial Seifert fibered fillings are along integral slopes and have the form $S^2(q_1, q_2, q_3)$ or $\mathbb{R}P^2(q_1, q_2)$; compare [**Gor3**, §3.3]. In particular, all fillings with finite fundamental group are integral.

We now turn to considering all of the manifolds in $\mathcal{C}_9$.

3.2. Distances between exceptional slopes. A key topological invariant of a pair of slopes α and β on a torus is their geometric intersection number $\Delta(\alpha, \beta)$. When α and β are exceptional slopes for a particular M, then $\Delta(\alpha, \beta) \leq 8$ by [**LM**]. Gordon conjectured there are only four possible M with exceptional slopes where $\Delta \geq 5$ [**Gor1**, Conjecture 3.4], and this holds for $\mathcal{C}_9$. Much of the work on exceptional fillings has focused on understanding the maximum possible $\Delta(\alpha, \beta)$ where $M(\alpha)$ and $M(\beta)$ are particular types of exceptional fillings. I summarize what is observed for $\mathcal{C}_9$ and how it relates to the known upper bounds on $\Delta(\alpha, \beta)$ in Table 5. In all cases, the maximum value of $\Delta(\alpha, \beta)$ for $\mathcal{C}_9$ is the same as that already found in the literature, compare with [**Gor1, Gor2, Gor3, Gor4**] and also page 971 and Section A.2 of [**MP**]. I think it very likely that all possible maximum values of $\Delta(\alpha, \beta)$ have been observed at this point.

3.3. Atoroidal Seifert fibered and finite π_1 fillings. There are two cases of M in $\mathcal{C}_9$ with slopes α and β with $\Delta(\alpha, \beta) = 4$ where $M(\alpha)$ is an atoroidal Seifert fibered space with $\pi_1\big(M(\alpha)\big)$ infinite and $\pi_1\big(M(\beta)\big)$ is finite and noncyclic. These are already contained in [**MP**], but this aspect is not highlighted there and so is worth describing here. The first example is $m007$ where $m007(-2,1)$ is the Seifert fibered space $S^2\big((2,1),(3,1),(9,-7)\big)$ and $m007(2,1)$ is $S^2\big((2,1),(3,2),(3,-1)\big)$ which has nonabelian fundamental group of order 120; this example is $M3_2$ with slopes -4 and 0 in Table A.3 of [**MP**]. The second is $m034$ with $m034(2,1) = S^2\big((2,1),(3,1),(11,-9)\big)$ and $m034(-2,1) = S^2\big((2,1),(3,2),(5,-3)\big)$ where the latter has nonabelian fundamental group of order 2,040; it is the example described in Table A.8 [**MP**], with $r/s = 2$ and slopes -4 and 0. Here, my conventions for describing Seifert fibered spaces follow Regina [**BBP+**].

3.4. Many exceptional fillings. For a 1-cusped manifold M, let $e(M)$ denote the number of exceptional fillings. The distribution of $e(M)$ is shown in Figure 3. There are only 11 manifolds in $\mathcal{C}_9$ where $e(M) \geq 7$, namely $m003$, $m004$, $m006$, $m007$, $m009$, $m016$, $m017$, $m023$, $m035$, $m038$, and $m039$. According to Gordon [**Gor1**, pages 136–7], these 11 were first noticed by Hodgson when he examined the 286 manifolds in $\mathcal{C}_5$. Gordon writes there that "In view of this data it is tempting to believe that these eleven manifolds are the only ones with $e(M) \geq 7$", and it was later shown [**LM**] that one always has $e(M) \leq 10$. These 11 are also the only manifolds with $e(M) \geq 7$ among all Dehn fillings on the magic manifold [**MP**]. In light of the additional data here, it is safe to promote this temptation to a conjecture.

3.5. Connected sums. The connected sums in this census are all built of quite simple pieces. Specifically, the summands all have finite π_1 or are $S^2 \times S^1$; there are only two summands in all but three cases: the filling $o9_{39343}(1,0)$ is $\mathbb{R}\mathrm{P}^3 \# \mathbb{R}\mathrm{P}^3 \# \mathbb{R}\mathrm{P}^3$ and both $o9_{41447}(1,0)$ and $o9_{43255}(1,0)$ are the manifold $L(3,1) \# \mathbb{R}\mathrm{P}^3 \# \mathbb{R}\mathrm{P}^3$. While there are infinite families with two connected sum fillings [**EMW**], it is an open question whether there is a manifold with three such fillings, see [**HM**, §4]. In $\mathcal{C}_9$ there are only 14 manifolds with two distinct Dehn fillings that are connected sums, and none with more than two. Another question from [**HM**, §4] is when there are two such fillings, must both have at least one summand that is $\mathbb{R}\mathrm{P}^3 = L(2,1)$, $L(3,1)$, or $L(4,1)$? The answer is yes for the 14 such manifolds in $\mathcal{C}_9$.

4. New observations

Here are some interesting patterns that I couldn't find in the existing literature. I encourage you to download the complete data at [**Dun**] and find others that I have missed.

4.1. Finite nonabelian fillings. The maximum number of fillings on M in $\mathcal{C}_9$ where the fundamental group is finite and nonabelian is three. There are only four such M, namely $m011$, $s757$, $v2702$, and $v2797$. I conjecture that these are the only four manifolds with this property.

4.2. Toroidal fillings. The maximum number of toroidal fillings on M in $\mathcal{C}_9$ is 4, and there only 27 such M, namely $s772$, $s778$, $s911$, $v2640$, $t08282$, $t11538$, $t12033$, $t12035$, $t12036$, $t12041$, $t12043$, $t12045$, $t12050$, $t12548$, $t12648$, $o9_{35259}$, $o9_{36732}$, $o9_{37030}$, $o9_{38039}$, $o9_{39094}$, $o9_{40054}$, $o9_{41000}$, $o9_{41004}$, $o9_{41006}$, $o9_{41007}$, $o9_{41008}$, $o9_{43799}$. Are there are infinitely many such examples? Perhaps we should expect there to be since the previous list includes manifolds with 6, 7, 8, and 9 ideal tetrahedra. None of the examples with 4 toroidal Dehn fillings is the exterior of a knot in S^3, consistent with a conjecture of [**EM**, Page 60].

4.3. Lengths of exceptional slopes. Hoffman and Purcell [**HP**] studied the length of exceptional slopes α in the horotorus cutting off a maximal cusp for M. By the 6-Theorem, the length $\ell(\alpha)$ of such α is at most 6. Table 6 details the longest exceptional slopes of each type observed in $\mathcal{C}_9$; compare with Table 1 of [**HP**]. The new feature is slopes of length more than 4 yielding manifolds with finite fundamental group; these are listed in Table 7. Can some of these be made

	S^3	lens	finite	#sum	$S^2 \times S^1$	SFS(A)	tor
S^3	$-\infty$	1	1	$-\infty$	$-\infty$	1	2
lens space	[**CGLS**]	1	2	1	1	2	3
finite π_1	§3(3)	[**BZ1**]	3	1	1	4	5
conn. sum	§3(1)	[**BZ2**]	[**BGZ1**]	1	$-\infty$	2	3
$S^2 \times S^1$	[**Gab2**]	[**CGLS**]	[**BGZ1**]	[**GL3**]	$-\infty$	1	2
Seifert (ator.)	§3(3)			[**BCSZ**]		6	7
toroidal	[**GL2**]	[**Lee2**]		[**Oh, Wu**]	[**Lee1**]	[**BGZ2**]	8

TABLE 5. This table lists the maximum value of $\Delta(\alpha, \beta)$ where $M \in \mathcal{C}_9$ and the fillings $M(\alpha)$ and $M(\beta)$ have the indicated types; here it is implicit that $\alpha \neq \beta$ and $-\infty$ is used when no such pair exists. As these values form a symmetric matrix, only the entries on or above the diagonal are given. Below the diagonal is a reference for the strongest proven upper bound on $\Delta(\alpha, \beta)$ for that pair of types, with references for the diagonal being, in order, [**GL1, CGLS, BZ3, GL3, Gab1, LM, Gor**]. Cases where the value for $\mathcal{C}_9$ is *strictly smaller* than the best proven upper bound are indicated by the shaded boxes. Thus in the unshaded cases the maximum possible value of $\Delta(\alpha, \beta)$ for *all* 3-manifolds has been established exactly by the indicated reference.

Type of $M(\alpha)$	$\min\big(\ell(\alpha)\big)$	$\max\big(\ell(\alpha)\big)$
S^3	1.000	3.323
$S^2 \times S^1$	1.288	3.328
connected sum	1.398	3.707
lens space	1.189	3.928
Sol torus bundle	2.288	4.185
finite π_1	1.520	4.443
Seifert (toroidal)	1.906	4.583
Seifert (atoroidal)	1.935	4.841
graph manifold	2.178	5.318
hyperbolic piece	3.520	6.000

TABLE 6. This table gives minimum and maximum lengths of each type of exceptional slope in Theorem 1.2 as measured in the torus bounding a maximal cusp. Lengths have been rounded to three decimal places, and here type refers to the most restricted category possible from Figure 4, which is why the $\min\big(\ell(\alpha)\big)$ for lens spaces is bigger than that for S^3. Compare with [**HP**, Table 1].

M	α	$M(\alpha)$	$\lvert\pi_1(M(\alpha))\rvert$	$\ell(\alpha)$
$t05002$	$(-1,1)$	$S^2\big((2,1),(3,2),(3,-1)\big)$	120	4.004139
$o9_{28194}$	$(-1,1)$	$S^2\big((2,1),(3,2),(4,-1)\big)$	528	4.017192
$o9_{35417}$	$(1,1)$	$S^2\big((2,1),(2,1),(8,3)\big)$	352	4.021047
$o9_{35418}$	$(1,1)$	$S^2\big((2,1),(2,1),(11,-3)\big)$	352	4.021047
$v3479$	$(1,1)$	$S^2\big((2,1),(3,2),(5,-1)\big)$	3,480	4.028619
$o9_{36221}$	$(1,1)$	$S^2\big((2,1),(2,1),(10,3)\big)$	520	4.033399
$o9_{36224}$	$(-1,1)$	$S^2\big((2,1),(2,1),(13,-3)\big)$	520	4.033399
$v2420$	$(-1,1)$	$S^2\big((2,1),(3,2),(4,-1)\big)$	528	4.060890
$m342$	$(1,1)$	$S^2\big((2,1),(3,2),(3,-1)\big)$	120	4.067597
$m011$	$(2,1)$	$S^2\big((2,1),(3,2),(5,-3)\big)$	2,040	4.085768
$o9_{41134}$	$(-1,1)$	$S^2\big((2,1),(3,2),(4,-1)\big)$	528	4.184451
$o9_{12592}$	$(-1,1)$	$S^2\big((2,1),(3,2),(5,-3)\big)$	2,040	4.195283
$s954$	$(1,1)$	$S^2\big((2,1),(3,2),(4,-1)\big)$	528	4.207000
$s546$	$(-1,1)$	$S^2\big((2,1),(3,2),(5,-3)\big)$	2,040	4.442966

TABLE 7. All pairs (M,α) with $M \in \mathcal{C}_9$ and $\pi_1\big(M(\alpha)\big)$ finite where $\ell(\alpha) > 4$.

into an infinite family of finite exceptional slopes with $\ell(\alpha) \to 5$ analogous to Proposition 4.2 of [**HP**]?

The referee kindly pointed out that one can use a covering trick to create even longer slopes for some types starting with the examples in Table 7. Specifically, the extreme example for lens spaces comes from $M = o9_{18855}$ and the slope $\alpha = (1,1)$ where $M(\alpha) = L(39,16)$ and $\ell(\alpha) \approx 3.92794$. The core curve of the Dehn filling turns out to generate $H_1(M(\alpha);\mathbb{Z}) \cong \mathbb{Z}/39\mathbb{Z}$, and consequently one can take a 39-fold cyclic cover of M to get the exterior of a knot in S^3 whose meridian μ also has $\ell(\mu) \approx 3.92794$. As discussed in [**HP**], it is conjectured that for an S^3 filling one always has $\ell(\mu) \leq 4$, and there are several families of such where $\ell(\mu) \to 4$ from below.

One can apply the same trick to $s546(-1,1)$ from Table 7 to produce a hyperbolic knot in the Poincaré homology sphere where the meridian has length about 4.442966; specifically, take the 17-fold cyclic cover of $s546$ corresponding to the kernel of the map $\pi_1\big(s546(-1,1)\big) \to H_1\big(s546(-1,1);\mathbb{Z}\big) \cong \mathbb{Z}/17\mathbb{Z}$.

4.4. A cabling conjecture for $S^2 \times S^1$. As per Table 5, there are no known hyperbolic knot exteriors in $S^2 \times S^1$ with a Dehn filling that is a connected sum. Thus, as in Section 3(1) for the case of S^3, I conjecture that none exist, i.e. that no hyperbolic knot in $S^2 \times S^1$ has a Dehn surgery yielding a connected sum.

5. Outline of the proof of Theorem 1.2

I turn now to the proof of Theorem 1.2. Initially, I found a candidate $\mathcal{E}$ for the list of all exceptional fillings as a byproduct of another project. However, the proof of the correctness of $\mathcal{E}$ follows the approach of [**MPR**]. The list $\mathcal{E}$, related data, and the code used in the proof can all be obtained from [**Dun**]; to run the code, which requires using several software packages together in consort, the Docker image [**ComTop**] may be helpful.

PROOF OF THEOREM 1.2. The set $\mathcal{E}$ consists of 205,822 pairs (M, α) where $M \in \mathcal{C}_9$ and $\alpha \in Sl(M)$. There are two things to show: that every $M(\alpha)$ is not hyperbolic of the type claimed in Table 2 and that all other fillings on $M \in \mathcal{C}_9$ are hyperbolic.

For the latter task, for each $M \in \mathcal{C}_9$ I found an embedded cusp neighborhood so that I could measure the lengths of slopes in its horotorus boundary; this was done rigorously in SnapPy [**CDGW**] running inside [**Sage**] using the approach of [**HIKMOT**] and [**DHL**, §3.6]. By the 6-Theorem of [**Ago, Lac**], it suffices to examine all slopes $\beta \in Sl(M)$ where $\ell(\beta) \leq 6$. For the cusp neighborhoods I used, overall there were some 355,128 such slopes. For the 149,306 pairs (M, β) that were not in $\mathcal{E}$, I checked that $M(\beta)$ was hyperbolic using the method of [**HIKMOT**] as reimplemented in SnapPy. As in the proof of Theorem 5.2 of [**HIKMOT**], it was sometimes necessarily to search around for a triangulation that could be used to certify the existence of a hyperbolic structure. This completes the proof that filling along any slope not in $\mathcal{E}$ yields a hyperbolic manifold.

In the other direction, to show that each $M(\alpha)$ in $\mathcal{E}$ is not hyperbolic, I primarily used Regina [**BBP+**], specifically its combinatorial recognition methods [**Bur1**, §4]. These work when the input triangulation has the very particular form associated to a standard triangulation of a Seifert fibered space or graph manifold. Of course, there are many triangulations of such manifolds which do not have this structure, so I generated many different 1-vertex triangulations of each $M(\alpha)$ and fed them into Regina until it succeeded in recognizing the topology. This worked for all but 2,890 of the $M(\alpha)$. Of those remaining, in 680 cases the Recognizer program of [**Mat, MT**] showed that they were graph manifolds. (Currently, Regina can only identify graph manifolds where the graph in question is either a segment with two or three vertices or a loop with one vertex. These 680 all have slightly more complicated graphs, for example a loop with either two or three vertices.) For each of the remaining 2,210 manifolds, Regina found at least one essential normal torus. Cutting along a suitable collection of such essential tori gave pieces that always included a cusped hyperbolic 3-manifold with an ideal triangulation with at most 6 ideal tetrahedra; in particular, each of these 2,210 manifolds is nonhyperbolic with a non-trivial JSJ decomposition with a hyperbolic piece. Thus every $M(\alpha)$ in $\mathcal{E}$ is not hyperbolic. This completes the proof that $\mathcal{E}$ is precisely the list of exceptional fillings on the manifolds in $\mathcal{C}_9$.

To prove the correctness of Table 2, the hard part is ensuring that nothing listed as a (proper) graph manifold is actually Seifert fibered. Everything else can be read off from the Seifert/graph descriptions found in the previous step, though I double-checked much of it in other ways. For example, I used Magma [**BCP**] to give an independent check that the 59,200 spherical manifolds had the claimed type of fundamental group (this could also be done with GAP [**GAP**]). I also had Regina compute directly which manifolds are toroidal using normal surface techniques and this matched what follows from the Seifert/graph descriptions. As mentioned, Regina identifies structure in the given triangulation, which might well be a graph manifold structure that can be simplified after the fact. For example, it will sometimes return graph manifolds where one of the nodes is a solid torus. In such instances, additional triangulations were examined until a more concise description was found. To certify a graph description as minimal, I just checked that all Seifert pieces have incompressible boundary (i.e. no solid tori) and that no two Seifert pieces are glued together so that the fibers match up; here, I took care to consider the possibility of switching the Seifert fibration for the exceptional piece which is both $D^2\big((2,1),(2,1)\big)$ and the twisted circle bundle over the Möbius band. $\square$

References

[Ago] I. Agol, *Bounds on exceptional Dehn filling*, Geom. Topol. **4** (2000), 431–449, DOI 10.2140/gt.2000.4.431. MR1799796

[Ber] J. Berge. (2018). *Heegaard3: a program for studying heegaard splittings.* `http://bitbucket.org/t3m/heegaard3`

[Bon] F. Bonahon, *Geometric structures on 3-manifolds*, Handbook of geometric topology, North-Holland, Amsterdam, 2002, pp. 93–164. MR1886669

[BCP] W. Bosma, J. Cannon, and C. Playoust, *The Magma algebra system. I. The user language*, J. Symbolic Comput. **24** (1997), no. 3-4, 235–265, DOI 10.1006/jsco.1996.0125. Computational algebra and number theory (London, 1993). MR1484478

[BCSZ] S. Boyer, M. Culler, P. B. Shalen, and X. Zhang, *Characteristic subsurfaces, character varieties and Dehn fillings*, Geom. Topol. **12** (2008), no. 1, 233–297, DOI 10.2140/gt.2008.12.233. MR2390346

[BGZ1] S. Boyer, C. McA. Gordon, and X. Zhang, *Reducible and finite Dehn fillings*, J. Lond. Math. Soc. (2) **79** (2009), no. 1, 72–84, DOI 10.1112/jlms/jdn063. MR2472134

[BGZ2] S. Boyer, C. McA. Gordon, and X. Zhang, *Characteristic submanifold theory and toroidal Dehn filling*, Adv. Math. **230** (2012), no. 4-6, 1673–1737, DOI 10.1016/j.aim.2012.03.029. MR2927352

[BZ1] S. Boyer and X. Zhang, *Finite Dehn surgery on knots*, J. Amer. Math. Soc. **9** (1996), no. 4, 1005–1050, DOI 10.1090/S0894-0347-96-00201-9. MR1333293

[BZ2] S. Boyer and X. Zhang, *On Culler-Shalen seminorms and Dehn filling*, Ann. of Math. (2) **148** (1998), no. 3, 737–801, DOI 10.2307/121031. MR1670053

[BZ3] S. Boyer and X. Zhang, *A proof of the finite filling conjecture*, J. Differential Geom. **59** (2001), no. 1, 87–176. MR1909249

[Bur1] B. A. Burton, *Computational topology with Regina: algorithms, heuristics and implementations*, Geometry and topology down under, Contemp. Math., vol. 597, Amer. Math. Soc., Providence, RI, 2013, pp. 195–224, DOI 10.1090/conm/597/11877. MR3186674

[Bur2] B. A. Burton. *The cusped hyperbolic census is complete.* Preprint 2014, 32 pages.

[BBP+] B. A. Burton, R. Budney, W. Pettersson, et al. (1999). *Regina: Software for low-dimensional topology.* `http://regina-normal.github.io/`

[CHW] P. J. Callahan, M. V. Hildebrand, and J. R. Weeks, *A census of cusped hyperbolic 3-manifolds*, Math. Comp. **68** (1999), no. 225, 321–332, DOI 10.1090/S0025-5718-99-01036-4. With microfiche supplement. MR1620219

[ComTop] *Computop sagemath docker image*, The Computop Group. `https://hub.docker.com/r/computop/sage`

[CDGW] M. Culler, N. M. Dunfield, M. Goerner, and J. R. Weeks. *SnapPy, a computer program for studying the topology and geometry of* 3*-manifolds, version 2.6.1.* Available at `http://snappy.computop.org`

[CGLS] M. Culler, C. McA. Gordon, J. Luecke, and P. B. Shalen, *Dehn surgery on knots*, Ann. of Math. (2) **125** (1987), no. 2, 237–300, DOI 10.2307/1971311. MR881270

[Dun] N. M. Dunfield. *Data and code to accompany this paper.* `https://doi.org/10.7910/DVN/6WNVGO`

[DHL] N. M. Dunfield, N. R. Hoffman, and J. E. Licata, *Asymmetric hyperbolic L-spaces, Heegaard genus, and Dehn filling*, Math. Res. Lett. **22** (2015), no. 6, 1679–1698, DOI 10.4310/MRL.2015.v22.n6.a7. MR3507256

[EM] M. Eudave-Muñoz, *Non-hyperbolic manifolds obtained by Dehn surgery on hyperbolic knots*, Geometric topology (Athens, GA, 1993), AMS/IP Stud. Adv. Math., vol. 2, Amer. Math. Soc., Providence, RI, 1997, pp. 35–61. MR1470720

[EMW] M. Eudave-Muñoz and Y.-Q. Wu, *Nonhyperbolic Dehn fillings on hyperbolic 3-manifolds*, Pacific J. Math. **190** (1999), no. 2, 261–275, DOI 10.2140/pjm.1999.190.261. MR1722893

[Gab1] D. Gabai, *Foliations and the topology of 3-manifolds. II*, J. Differential Geom. **26** (1987), no. 3, 461–478. MR910017

[Gab2] D. Gabai, *Foliations and the topology of 3-manifolds. III*, J. Differential Geom. **26** (1987), no. 3, 479–536. MR910018

[GAP] (2019). *GAP – Groups, Algorithms, and Programming, Version 4.10.2*, The GAP Group. `https://www.gap-system.org`

[Gor] C. McA. Gordon, *Boundary slopes of punctured tori in 3-manifolds*, Trans. Amer. Math. Soc. **350** (1998), no. 5, 1713–1790, DOI 10.1090/S0002-9947-98-01763-2. MR1390037

[GL1] C. McA. Gordon and J. Luecke, *Knots are determined by their complements*, J. Amer. Math. Soc. **2** (1989), no. 2, 371–415, DOI 10.2307/1990979. MR965210

[GL2] C. McA. Gordon and J. Luecke, *Dehn surgeries on knots creating essential tori. I*, Comm. Anal. Geom. **3** (1995), no. 3-4, 597–644, DOI 10.4310/CAG.1995.v3.n4.a3. MR1371211

[GL3] C. McA. Gordon and J. Luecke, *Reducible manifolds and Dehn surgery*, Topology **35** (1996), no. 2, 385–409, DOI 10.1016/0040-9383(95)00016-X. MR1380506

[Gor1] C. McA. Gordon, *Dehn filling: a survey*, Knot theory (Warsaw, 1995), Banach Center Publ., vol. 42, Polish Acad. Sci. Inst. Math., Warsaw, 1998, pp. 129–144. MR1634453

[Gor2] C. McA. Gordon, *Small surfaces and Dehn filling*, Proceedings of the Kirbyfest (Berkeley, CA, 1998), Geom. Topol. Monogr., vol. 2, Geom. Topol. Publ., Coventry, 1999, pp. 177–199, DOI 10.2140/gtm.1999.2.177. MR1734408

[Gor3] C. McA. Gordon, *Dehn filling*, Low dimensional topology, New Stud. Adv. Math., vol. 3, Int. Press, Somerville, MA, 2003, pp. 41–59. MR2052245

[Gor4] C. McA. Gordon, *Exceptional Dehn filling*, Introductory lectures on knot theory, Ser. Knots Everything, vol. 46, World Sci. Publ., Hackensack, NJ, 2012, pp. 124–134, DOI 10.1142/9789814313001_0006. MR2885233

[Hat] A. Hatcher. (2007). *Notes on basic 3-manifold topology.* `http://pi.math.cornell.edu/~hatcher/3M/3Mdownloads.html`

[HW] M. Hildebrand and J. Weeks, *A computer generated census of cusped hyperbolic 3-manifolds*, Computers and mathematics (Cambridge, MA, 1989), Springer, New York, 1989, pp. 53–59. MR1005959

[HM] J. A. Hoffman and D. Matignon, *Examples of bireducible Dehn fillings*, Pacific J. Math. **209** (2003), no. 1, 67–83, DOI 10.2140/pjm.2003.209.67. MR1973934

[HIKMOT] N. Hoffman, K. Ichihara, M. Kashiwagi, H. Masai, S. Oishi, and A. Takayasu, *Verified computations for hyperbolic 3-manifolds*, Exp. Math. **25** (2016), no. 1, 66–78, DOI 10.1080/10586458.2015.1029599. MR3424833

[HP] N. R. Hoffman and J. S. Purcell, *Geometry of planar surfaces and exceptional fillings*, Bull. Lond. Math. Soc. **49** (2017), no. 2, 185–201, DOI 10.1112/blms.12000. MR3656288

[Lac] M. Lackenby, *Word hyperbolic Dehn surgery*, Invent. Math. **140** (2000), no. 2, 243–282, DOI 10.1007/s002220000047. MR1756996

[LM] M. Lackenby and R. Meyerhoff, *The maximal number of exceptional Dehn surgeries*, Invent. Math. **191** (2013), no. 2, 341–382, DOI 10.1007/s00222-012-0395-2. MR3010379

[Lee1] S. Lee, *Toroidal Dehn surgeries on knots in $S^1 \times S^2$*, J. Knot Theory Ramifications **14** (2005), no. 5, 657–664, DOI 10.1142/S0218216505004007. MR2162119

[Lee2] S. Lee, *Lens spaces and toroidal Dehn fillings*, Math. Z. **267** (2011), no. 3-4, 781–802, DOI 10.1007/s00209-009-0646-0. MR2776058

[Mar] B. Martelli. *Dehn surgery on the minimally twisted seven-chain link.* Preprint 2018, 33 pages.

[MP] B. Martelli and C. Petronio, *Dehn filling of the "magic" 3-manifold*, Comm. Anal. Geom. **14** (2006), no. 5, 969–1026. MR2287152

[MPR] B. Martelli, C. Petronio, and F. Roukema, *Exceptional Dehn surgery on the minimally twisted five-chain link*, Comm. Anal. Geom. **22** (2014), no. 4, 689–735, DOI 10.4310/CAG.2014.v22.n4.a4. MR3263935

[MT] S. V. Matveev and V. V. Tarkaev. (2015). *The 3-manifold recognizer.* `http://www.matlas.math.csu.ru/`

[Mat] S. V. Matveev, *Computer recognition of three-manifolds*, Experiment. Math. **7** (1998), no. 2, 153–161. MR1677162

[Oh] S. Oh, *Reducible and toroidal 3-manifolds obtained by Dehn fillings*, Topology Appl. **75** (1997), no. 1, 93–104, DOI 10.1016/S0166-8641(96)00065-X. MR1425387

[Sco] P. Scott, *The geometries of 3-manifolds*, Bull. London Math. Soc. **15** (1983), no. 5, 401–487, DOI 10.1112/blms/15.5.401. MR705527

[Sage] W. Stein et al. (2018). *Sagemath, the Sage Mathematics Software System (Version 8.3).* `http://www.sagemath.org`

[Thi] M. Thistlethwaite. (2010). *Cusped hyperbolic 3-manifolds constructed from 8 ideal tetrahedra.* `https://www.math.utk.edu/~morwen/8tet`

[Thu] W. P. Thurston. (1980). *Geometry and topology of three-manifolds.* Unpublished lecture notes. `http://library.msri.org/books/gt3m/`

[Wu] Y.-Q. Wu, *Dehn fillings producing reducible manifolds and toroidal manifolds*, Topology **37** (1998), no. 1, 95–108, DOI 10.1016/S0040-9383(97)00016-5. MR1480879

Dept. of Math., MC-382, University of Illinois, 1409 W. Green St., Urbana, Illinois 61801

Email address: `nathan@dunfield.info`

URL: `http://dunfield.info`

Contemporary Mathematics
Volume **760**, 2020
https://doi.org/10.1090/conm/760/15290

Parallelism on Lie groups and Fox's free differential calculus

William M. Goldman

Dedicated to Steven Boyer on his sixtieth birthday

ABSTRACT. Fox's free differential calculus arose from calculating the Alexander polynomial of a knot. It arises naturally from differentiating word maps on Lie groups, and thus intimately relates to computing the tangent spaces to spaces of representations. This paper expounds the theory of derivatives of word maps on Lie groups and relates Fox's calculus to character varieties and Alexander invariants.

Introduction

In a series of papers [**1, 4–6**], R. H. Fox develops the theory of derivations in the group rings of free groups. He was apparently motivated by computing the Alexander polynomials of knots; see for example Crowell and Fox [**2, 3**]. Denote the free group based on a set $\{x_1, \dots, x_n\}$ by F_n. Fox defines for any word $w(x_1, \dots, x_n) \in \mathsf{F}_n$ elements

$$\frac{\partial w}{\partial x_i} \in \mathbb{Z}[\mathsf{F}_n]$$

where $\mathbb{Z}[\mathsf{F}_n]$ denotes the integral group ring of the rank n free group F_n.

The space $\mathsf{Hom}(\pi, G)$ of representations of a finitely generated group π in a Lie group G give rise to word maps as follows. Let $\gamma_1, \dots, \gamma_n$ be a generating set for π, and let $R_\alpha(\gamma_1, \dots, \gamma_n) = 1$ be defining relations for π, where α ranges over an index set $\mathscr{A}$. Then the map

$$G^n \xrightarrow{R_\alpha} G$$
$$(g_1, \dots, g_n) \longmapsto R_\alpha(g_1, \dots, g_n)$$

is an analytic map and $\mathsf{Hom}(\pi, G)$ embeds in G^n as the subset comprising $(g_1, \dots, g_n)$ such that $R_\alpha(g_1, \dots, g_n) = 1$. Compare Goldman [**8**], Labourie [**12**], and Sikora [**16**] for descriptions of the general theory and Raghunathan [**15**] for a careful description of the relation of $\mathsf{Hom}(\pi, G)$ with the space of cocycles of π in the adjoint representation of G. Fried [**7**] gives a dynamical study of word maps. Compare also Lubotzky-Magid [**13**] for a description of the affine variety of representations of finitely generated groups. The relation of the Fox calculus to word maps is somewhat implicit in Weil [**17**]. However the lack of explicit statements of results in Weil's paper has perhaps lessened its impact in this subject.

2010 *Mathematics Subject Classification.* 22E15, 22E40, 53C15, 57M50.

The Alexander polynomial of knots relates directly to representations of fundamental groups. Milnor [**14**] describes the Alexander invariants in terms of infinite cyclic coverings which can be specified by scalar characters $\pi \xrightarrow{\rho} \mathbb{C}^*$. The roots of the Alexander polynomial correspond to those characters for which the corresponding representation into diagonal matrices in $\mathsf{SL}(2,\mathbb{C})$ admits nontrivial deformations, which is equivalent to the twisted cohomology $H^1(\pi,\mathbb{C}_\rho)$ being nonzero. For a detailed discussion, with nontrivial applications, see Heusener and Porti [**9**, **10**]. A related application is Hironaka [**11**], where the Fox calculus is used to study the stratification of the character variety.

I first learned this theory in discussing the geometry of the space of surface group representations in [**8**], and was surprised to learn that the application of the Fox calculus to representations was not better known. The purpose of this note is to describe how the Fox derivatives arise naturally from differentiating word maps on Lie groups, and can be applied to explicit calculations with representation varieties and character varieties.

I am grateful to E. Hironaka, M. Kapovich, S. Lomanaco, S. Tillmann and the anonymous referee for helpful conversations and suggestions on the Fox calculus.

1. Smooth manifolds and Lie groups

Let M be a smooth manifold and denote its tangent bundle by $\mathsf{T}M \to M$. If $M \xrightarrow{f} N$ is a smooth map, denote its differential at $x \in M$ by:

$$\mathsf{T}_x M \xrightarrow{(\mathsf{D}f)_x} \mathsf{T}_{f(x)} N.$$

The differential of a composition

$$M \xrightarrow[f]{} N \xrightarrow[g]{} P, \qquad M \xrightarrow{g \circ f} P$$

satisfies the *chain rule:*

$$\mathsf{T}_x M \xrightarrow[(\mathsf{D}f)_x]{} \mathsf{T}_{f(x)} N \xrightarrow[(\mathsf{D}g)_{f(x)}]{} P, \qquad \mathsf{T}_x M \xrightarrow{\left(\mathsf{D}(g \circ f)\right)_x} P$$

that is, $\mathsf{D}(g \circ f)_x = (\mathsf{D}g)_{f(x)} \circ (\mathsf{D}f)_x$.

Denote the group of diffeomorphisms

$$M \longrightarrow M$$

by $\mathsf{Diff}(M)$ and the Lie algebra of vector fields on M by $\mathsf{Vec}(M)$. If $\xi \in \mathsf{Vec}(M)$, denote its *value* at $p \in M$ by $\xi_p \in \mathsf{T}_p M$. A diffeomorphism $f \in \mathsf{Diff}(M)$ acts on a vector field $\xi \in \mathsf{Vec}(M)$ by the *pushforward* $f_*(\xi) \in \mathsf{Vec}(M)$, whose value at $p \in M$ equals:

$$f_*(\xi)_p := (\mathsf{D}f)_{f^{-1}(p)}\big(\xi_{f^{-1}(p)}\big).$$

Let G be a Lie group with identity element $e \in G$. If $a \in G$, define operations $\mathfrak{L}^a, \mathfrak{R}^a$ of *left-* and *right- multiplication*, respectively, by:

$$ab = \mathfrak{L}^a(b) = \mathfrak{R}^b(a).$$

Associativity implies $\mathfrak{L}^a \circ \mathfrak{L}^b = \mathfrak{L}^{ab}$ and $\mathfrak{R}^a \circ \mathfrak{R}^b = \mathfrak{R}^{ba}$. A vector field $\xi \in \mathsf{Vec}(G)$ is *right-invariant* if and only if $(\mathfrak{R}^a)_*\xi = \xi$ for all $a \in G$. That is,

$$\xi(ba) = (\mathsf{D}\mathfrak{R}^a)_b(\xi_b)$$

for all $a, b \in G$. Taking $b = e$, a right-invariant vector field ξ is determined by its value $\xi_e \in \mathsf{T}_e G$ by:

$$\xi_a = (\mathsf{D}\mathfrak{R}^a)_e(\xi_e)$$

for any $a \in G$. Right-invariant vector fields comprise a Lie subalgebra $\mathsf{Vec}(G)^{\mathfrak{R}} \subset \mathsf{Vec}(G)$, which is isomorphic to $\mathsf{T}_e G$ as a vector space:

$$\begin{aligned} \mathfrak{g} := \mathsf{Vec}(G)^{\mathfrak{R}} &\xrightarrow{\cong} \mathsf{T}_e G \\ \xi &\longmapsto \xi_e \end{aligned}$$

For $a \in G$, define the *inner automorphism* $\mathsf{Inn}(a)$:

$$\begin{aligned} G &\xrightarrow{\mathsf{Inn}(a)} G \\ b &\longmapsto aba^{-1} \end{aligned}$$

The *adjoint representation* of $a \in G$ is defined as the differential of $\mathsf{Inn}(a)$ at $e = \mathsf{Inn}(a)e$:

$$\mathsf{Ad}(a) := (\mathsf{D}\,\mathsf{Inn}(a))_e : \mathsf{T}_e G \circlearrowleft$$

which we henceforth simply write as $\mathfrak{g} \xrightarrow{\mathsf{Ad}(a)} \mathfrak{g}$. The adjoint representation is a representation of Lie groups $G \xrightarrow{\mathsf{Ad}} \mathsf{Aut}(\mathfrak{g})$.

Equivalently, the adjoint representation is the action of G by left-multiplication on the Lie algebra of right-invariant vector fields: If $\xi \in \mathfrak{g}$ is a right-invariant vector field, then

$$(\mathfrak{L}^a)_*\xi = \mathsf{Ad}(a)\xi \tag{1.1}$$

$$(\mathfrak{R}^a)_*\xi = \xi \tag{1.2}$$

Thus $\mathfrak{g}$ becomes a *bimodule* over G, with trivial right action and left action given by Ad.

Each tangent space $\mathsf{T}_a G$ identifies with $\mathfrak{g}$ by the *right-parallelism*, where each $v \in \mathsf{T}_a G$ extends to a right-invariant vector field $\xi \in \mathfrak{g}$ by:

$$b \overset{\xi}{\longmapsto} (\mathsf{D}\,\mathfrak{R}^{a^{-1}b})_a(v) \ \in \mathsf{T}_b G$$

for $b \in G$. The graph of the corresponding map $TG \longrightarrow \mathfrak{g}$ is a *parallelism,* as it defines a trivialization

$$\mathsf{T}G \xrightarrow{\cong} G \times \mathfrak{g}$$

of the tangent bundle $\mathsf{T}G \longrightarrow G$. The map $\mathsf{T}G \longrightarrow \mathfrak{g}$ is characterized as being constant on sections arising as right-invariant vector fields.

Equivalently, this defines a (flat) affine connection on G where the right-invariant vector fields are parallel (and conversely).

1.1. The differential notation. Let $n > 0$ and consider the n-th Cartesian power G^n. If $X = (X_1, \dots, X_n) \in G^n$, then the right-parallelism identifies the tangent space $\mathsf{T}_X G^n$ with the direct sum $\oplus^n \mathfrak{g}$. We denote the i-th coordinate projection by dX_i:

$$\mathsf{T}_{(X_1,\dots,X_n)} G^n \cong \mathfrak{g} \oplus \cdots \oplus \mathfrak{g} \xrightarrow{dX_i} \mathfrak{g}. \tag{1.3}$$

If f is an $\mathsf{End}(\mathfrak{g})$-valued function on G^n, the composition $f dX_i$ maps the tangent vector $v \in T_X G^n$ to $f(X)\big(dX_i(v)\big)$. In this notation the differential of the multiplication map

$$\begin{aligned} G^2 &\longrightarrow G \\ (X_1, X_2) &\longmapsto X_1 X_2 \end{aligned}$$

is $dX_1 + \mathsf{Ad}(X_1) dX_2$ by applying (1.2) and (1.1) to $X_1 X_2 = \mathfrak{R}(X_2) X_1$ and $\mathfrak{L}_{X_1}(X_2)$ respectively.

2. Deformations and cohomology

Now let π be a group and $\pi \xrightarrow{\rho} G$ a representation. The composition

$$\pi \xrightarrow{\rho} G \xrightarrow{\mathsf{Ad}} \mathsf{Aut}(\mathfrak{g})$$

defines a left π-module, denoted $\mathfrak{g}_{\mathsf{Ad}\rho}$. (As above, we can extend this structure to a bimodule with trivial right action.)

Suppose that ρ_t is a path in $\mathsf{Hom}(\pi, G)$, with $\rho_0 = \rho$ and depending differentiably on t:

$$\rho_t(xy) = \rho_t(x)\rho_t(y) = \mathfrak{L}^{\rho_t(x)}\big(\rho_t(y)\big) = \mathfrak{R}^{\rho_t(y)}\big(\rho_t(x)\big) \tag{2.1}$$

for $x, y \in \pi$. Then the *velocity vector*

$$U(x) := \frac{d}{dt}\Big|_{t=0} \rho_t(x) \in \mathsf{T}_{\rho(x)} G$$

extends to a right-invariant vector field which we denote $u(x) \in \mathfrak{g}$. Furthermore differentiating (2.1) at $t = 0$ yields:

$$\begin{aligned} U(xy) = &\big(\mathsf{D}\mathfrak{R}^{\rho(y)}\big)_{\rho(x)} U(x) \\ &+ \big(\mathsf{D}\mathfrak{L}^{\rho(x)}\big)_{\rho(y)} U(y) \end{aligned}$$

Applying the right-parallelism implies the *cocycle identity:*

$$u(xy) = u(x) + \mathsf{Ad}\rho(x) u(y),$$

that is, $\pi \xrightarrow{u} \mathfrak{g}$ is a 1-cocycle in $Z^1(\pi, \mathfrak{g}_{\mathsf{Ad}\rho})$. When the context is clear, simply write xv for $\mathsf{Ad}\rho(x)(v)$ when $x \in \pi$ and $v \in \mathfrak{g}$. Furthermore extend the action ρ of π by linearity to an action of the group ring $\mathbb{Z}\pi$ on the vector space $\mathfrak{g}$. The above cocycle identity becomes:

$$u(xy) = u(x) + x u(y) \tag{2.2}$$

In particular the (Zariski) tangent space at ρ to the affine variety $\mathsf{Hom}(\pi, G)$ equals:

$$\mathsf{T}_\rho \mathsf{Hom}(\pi, G) = \mathsf{Z}^1(\pi, \mathfrak{g}_{\mathsf{Ad}\rho}).$$

A *trivial deformation* is a path ρ_t induced by a path $h_t \in G$:

$$\rho_t := \mathsf{Inn}(h_t) \circ \rho$$

Then the corresponding cocycle u is a 1-*coboundary* $\delta\eta$ in $\mathsf{B}^1(\pi, \mathfrak{g}_{\mathsf{Ad}\rho})$:

$$x \xmapsto{\delta\eta} \eta_t - \mathsf{Ad}\rho(x)(\eta)$$

where

$$\eta \in \mathfrak{g} \longleftrightarrow \frac{d}{dt}\Big|_{t=0} h_t \in \mathsf{T}_{h_t} G$$

by the right-parallelism, which we write:

$$u(x) = v - xv \tag{2.3}$$

for $v = \eta \in \mathfrak{g}$. In particular the tangent space to the orbit $\mathsf{Inn}(G)\cdot\rho$ of $\rho \in \mathsf{Hom}(\pi, G)$ equals:

$$\mathsf{T}_\rho\big(\mathsf{Inn}(G)\cdot\rho\big) = \mathsf{B}^1(\pi, \mathfrak{g}_{\mathsf{Ad}\rho}) \subset \mathsf{Z}^1(\pi, \mathfrak{g}_{\mathsf{Ad}\rho}) = \mathsf{T}_\rho \mathsf{Hom}(\pi, G).$$

In the nicest cases, we can say more. In general $\mathsf{Hom}(\pi, G)$ will be singular, although it is stratified by smooth manifolds, and the top-dimensional stratum is smooth. In general, $\mathsf{Inn}(G)$ will neither act freely nor properly, so that the quotient will generally neither be smooth nor Hausdorff (even when restricted to the top stratum). However, in the nicest cases — when restricted to an open invariant smooth subset upon which the action is proper and free, the quotient map

$$\mathsf{Hom}(\pi, G) \longrightarrow \mathsf{Hom}(\pi, G)/\mathsf{Inn}(G)$$

restricts to a smooth fiber bundle over a smooth (Hausdorff) manifold $\mathscr{X}$. For a point ρ in the a fiber over $\mathscr{X}$, the tangent space $\mathsf{T}_{[\rho]}\mathscr{X}$ identifies with the cohomology group

$$\mathsf{H}^1(\pi, \mathfrak{g}_{\mathsf{Ad}\rho}) := \mathsf{Z}^1(\pi, \mathfrak{g}_{\mathsf{Ad}\rho})/\mathsf{B}^1(\pi, \mathfrak{g}_{\mathsf{Ad}\rho}).$$

(Compare Weil [**17**] and Raghunathan [**15**].)

3. Cocycles, free groups and derivations

We relate Eilenberg-MacLane group cocycles to the Fox calculus. As this applies to π-modules — not necessarily arising as $\mathsf{V} = \mathfrak{g}_{\mathsf{Ad}\rho}$ for $\rho \in \mathsf{Hom}(\pi, G)$, in this section we consider the case V is an arbitrary π-module. We also temporarily restrict to the case when $\pi = \mathsf{F}_n$ is an n-generator free group. We start by exploring the consequences of the cocycle identity (2.2) for arbitrary words $w \in \mathsf{F}_n$.

Let $u \in \mathsf{Z}^1(\mathsf{F}_n, \mathsf{V})$ be a V-valued 1-cocycle, that is, a map $\pi \xrightarrow{u} \mathsf{V}$ satisfying the cocycle identity (2.2). This identity readily implies that $u(e) = 0$, and for every $n > 0$,

$$\begin{aligned} u(x^n) &= u(x) + xu(x) + \cdots + x^{n-1}u(x) \\ &= (e + x + \cdots + x^{n-1})\ u(x). \end{aligned} \tag{3.1}$$

Furthermore

$$u(x^{-1}) = -x^{-1}u(x),$$

and for $n > 0$,

$$\begin{aligned} u(x^{-n}) &= -(x^{-1} + \cdots + x^{-n})\ u(x) \\ &= -x^{-1}(e + \cdots + x^{1-n})u(x) \end{aligned} \tag{3.2}$$

Furthermore

$$\begin{aligned} u(xyx^{-1}) &= u(x) + xu(y) - xyx^{-1}u(x) \\ &= (e - xyx^{-1})u(x) + xu(y). \end{aligned} \tag{3.3}$$

and

$$u(xyx^{-1}y^{-1}) = \begin{aligned}&(e - xyx^{-1}) \quad u(x)\\ &+ x(e - yx^{-1}y^{-1})\, u(y).\end{aligned} \tag{3.4}$$

In general, the value $u(w)$ of a cocycle u on any word

$$w = w(x_1, \dots, x_n) \in \mathsf{F}_n$$

equals:

$$u(w) = \sum_{i=1}^{n} \frac{\partial w}{\partial x_i} u(x_i),$$

where

$$\frac{\partial w}{\partial x_i} \in \mathbb{Z}\mathsf{F}_n.$$

These expressions are the *Fox partial derivatives* of w in the free group ring $\mathbb{Z}\mathsf{F}_n$.

THEOREM 3.1 (Fox [**4**]). *Let* $\mathsf{Der}(\mathsf{F}_n)$ *denote the set of* derivations $\mathsf{F}_n \xrightarrow{D} \mathbb{Z}\mathsf{F}_n$, *that is, mappings D which satisfy:*

$$D(ab) = D(a) + aD(b). \tag{3.5}$$

Then

- $\mathsf{Der}(\mathsf{F}_n)$ *is a left* $\mathbb{Z}\mathsf{F}_n$*-module where* $y \in \mathbb{Z}\mathsf{F}_n$ *maps* $D \in \mathsf{Der}(\mathsf{F}_n)$ *to* $yD \in \mathsf{Der}(\mathsf{F}_n)$ *by:*

$$\begin{aligned}\mathsf{F}_n &\xrightarrow{yD} \mathbb{Z}\mathsf{F}_n\\ w &\longmapsto yD(w)\end{aligned}$$

- $\mathsf{Der}(\mathsf{F}_n)$ *is a free* $\mathbb{Z}\mathsf{F}_n$*-module, with basis*

$$\frac{\partial}{\partial x_1}, \dots, \frac{\partial}{\partial x_n}$$

where the basic derivations $\frac{\partial}{\partial x_i}$ *are characterized by:*

$$\frac{\partial x_i}{\partial x_j} = \delta_{i,j}. \tag{3.6}$$

The derivation properties (3.5) and (3.6) permit an easy calculation of the Fox derivatives. Suppose that $\frac{\partial w'}{\partial x_i}$ is known for all words w' of length $\ell(w') < m$. If $\ell(w) = m$, then $\exists w', \sigma$ with $\ell(w') = m - 1$ and $\sigma = \pm 1$ such that $w = x_j^{\sigma} w'$. Then

$$\frac{\partial w}{\partial x_i} = \begin{cases} \delta_{ij} e & + x_j \frac{\partial w'}{\partial x_i} \quad \text{if } \sigma = +1 \\ -\delta_{ij} x_j^{-1} & + x_j^{-1} \frac{\partial w'}{\partial x_i} \quad \text{if } \sigma = -1 \end{cases}$$

as desired.

In analogy with the differential notation in §1.1, write

$$dw = \sum_{i=1}^{n} \frac{\partial w}{\partial x_i} dx_i.$$

For brevity we henceforth denote $\frac{\partial}{\partial x_i}$ by ∂_i.

In general suppose that $w = w(x_1, \dots, x_n) \in \mathsf{F}_n$ is a word, and V is a π-module, for some group π. The values of cocycles $u \in \mathsf{Z}^1(\pi, \mathsf{V})$ on words $w \in \mathsf{F}_n$ can be computed by Fox derivatives as follows. Let $\gamma_1, \dots, \gamma_n \in \pi$. Since the π-action on

V extends linearly to an action of the group ring $\mathbb{Z}\pi$ on V, the Fox derivatives $\partial_i w$ define linear transformations $\partial_i w(\gamma_1,\dots,\gamma_n)$ of V.

PROPOSITION 3.2. *Let* $\pi \xrightarrow{u} \mathsf{V}$ *be a* 1*-cocycle and* $w(x_1,\dots,x_n) \in \mathsf{F}_n$. *Then*

$$u\big(w(\gamma_1,\dots,\gamma_n)\big) = \sum_{i=1}^{n} \Big(\partial_i w(\gamma_1,\dots,\gamma_n)\Big) u(\gamma_i) \tag{3.7}$$

PROOF. The derivation property (3.5) implies that

$$\begin{aligned} \pi &\longrightarrow \mathsf{V} \\ w(\gamma_1,\dots,\gamma_n) &\longmapsto \sum_{i=1}^{n} \Big(\partial_i w(\gamma_1,\dots,\gamma_n)\Big) u(\gamma_i) \end{aligned}$$

satisfies the cocycle identity (2.2). Furthermore this cocycle agrees with u on the generators, and therefore equals u as claimed. □

A 1-cochain $\pi \xrightarrow{u} \mathsf{V}$ uniquely extends to a linear map $\mathbb{Z}\pi \xrightarrow{u} \mathsf{V}$. The linear extension of the trivial homomorphism $\pi \twoheadrightarrow \{e\} \hookrightarrow \pi$ is the *augmentation* homomorphism $\mathbb{Z}\pi \xrightarrow{\varepsilon} \mathbb{Z}$. Then 1-cocycles identify with linear maps $\mathbb{Z}\pi \xrightarrow{u} \mathsf{V}$ such that

$$u(ab) = u(a)\varepsilon(b) + au(b)$$

where now $a, b \in \mathbb{Z}\pi$ are $\mathbb{Z}$-linear combinations of elements of π.

When $\pi = \mathsf{F}_n = \langle x_1,\dots,x_n\rangle$, a representation $\rho \in \mathsf{Hom}(\pi, G)$ is equivalent to an n-tuple

$$(X_1,\dots,X_n) \in G^n$$

where $X_i = \rho(x_i)$.

THEOREM 3.3. *Let* $w \in \mathsf{F}_n$ *be a word* $w = w(x_1,\dots,x_n)$. *Using right-parallelism, the differential of the word map*

$$G^n \xrightarrow{w} G$$

at $(X_1,\dots,X_n) \in G^n$

$$\mathsf{T}_{X_1}G \times \dots \mathsf{T}_{X_n}G \xrightarrow{\mathsf{D}w} \mathsf{T}_{w(X_1,\dots,X_n)}G$$

identifies with the map

$$\begin{aligned} \mathfrak{g} \oplus \dots \oplus \mathfrak{g} &\longrightarrow \mathfrak{g} \\ \xi_1 \oplus \dots \oplus \xi_n &\longmapsto \mathsf{Ad}\ \partial_1 w(X_1,\dots,X_n)\ (\xi_1) \\ &\qquad + \dots + \ \mathsf{Ad}\ \partial_n w(X_1,\dots,X_n)\ (\xi_n), \end{aligned}$$

that is, the differential

$$(\mathsf{D}w)_{(X_1,\dots,X_n)} = \mathsf{Ad}\big(dw(X_1,\dots,X_n)\big).$$

4. The tangent space to the representation variety

Now let π be a finitely generated group. An n-generator presentation of π consists of an isomorphism $\mathsf{F}_n/\mathscr{R} \cong \pi$ where $\mathscr{R} \triangleleft \mathsf{F}_n$ is a normal subgroup. Choose a set of normal generators $R_\alpha \in \mathscr{R}$, where $\alpha \in \mathscr{A}$.

Then $\mathsf{Hom}(\pi, G)$ identifies with subvariety of $G^n \cong \mathsf{Hom}(\mathsf{F}_n, G)$ comprising $(X_1, \dots, X_n)$ satisfying $R_\alpha(X_1, \dots X_n) = 1$, for $\alpha \in \mathscr{A}$.

The Zariski tangent space of $\mathsf{Hom}(\pi, G)$ at ρ is

$$\mathsf{T}_\rho\mathsf{Hom}(\pi, G) \;=\; \mathsf{Z}^1(\pi, \mathfrak{g}_{\mathsf{Ad}\rho}).$$

By Theorem 3.3, this equals the intersection of the kernels of the differentials $(\mathsf{D}R_\alpha)_{(X_1,\dots,X_n)}$ which are given as the solutions $(\xi_1, \dots, \xi_n) \in \mathfrak{g} \oplus \dots \oplus \mathfrak{g}$ of

$$\begin{aligned} &dw_{(X_1,\dots,X_n)}(\xi_1, \dots, \xi_n) \;= \\ &\left(\mathsf{Ad}\frac{\partial w}{\partial x_1}(X_1, \dots, X_n)\right)(\xi_1) + \dots + \left(\mathsf{Ad}\frac{\partial w}{\partial x_n}(X_1, \dots, X_n)\right)(\xi_n) \;=\; 0. \end{aligned}$$

The Fox calculus permits an explicit description of the Zariski tangent space $\mathsf{T}_\rho\mathsf{Hom}(\pi, G)$ as the space $\mathsf{Z}^1(\pi, \mathfrak{g}_{\mathsf{Ad}\rho})$ of 1-cocycles $\pi \longrightarrow \mathfrak{g}_{\mathsf{Ad}\rho}$, where $\mathfrak{g}_{\mathsf{Ad}\rho}$ denotes the π-module $\mathfrak{g}$ defined by the composition

$$\pi \xrightarrow{\rho} G \xrightarrow{\mathsf{Ad}} \mathsf{Aut}(\mathfrak{g})$$

in terms of a free presentation

$$\pi = \langle \gamma_1, \dots, \gamma_n \mid R_\alpha(\gamma_1, \dots, \gamma_n) = e, \alpha \in \mathscr{A} \rangle$$

of π. Thus $\pi = \mathsf{F}_n/\mathscr{R}$, where $\mathscr{R} \triangleleft \mathsf{F}_n$ is a normal subgroup of *relations* among a fixed set of generators $\gamma_1, \dots, \gamma_n$, that is, the normal closure in F_n of the subset

$$\{R_\alpha(x_1, \dots, x_n) \mid \alpha \in \mathscr{A}\} \subset \mathsf{F}_n.$$

In particular, $\mathsf{Hom}(\pi, G)$ identifies with the subset of $\mathsf{Hom}(\mathsf{F}_n, G) \approx G^n$ comprising $(g_1, \dots, g_n)$ satisfying

$$R_\alpha(X_1, \dots, X_n) = 1$$

for $\alpha \in \mathscr{A}$.

5. Examples

As an application, we consider the fundamental group of a closed surface Σ of genus $p \geq 1$. When Σ is orientable, then its fundamental group $\pi = \pi_1(\Sigma)$ admits a presentation

$$\pi \;=\; \langle A_1, B_1, \dots, A_p, B_p \mid R_p(A_1, B_1, \dots, A_p, B_p) = e \rangle$$

where

$$R_p(A_1, B_1, \dots, A_p, B_p) := [A_1, B_1] \dots [A_p, B_p]$$

is the defining relation and $[A, B] := ABA^{-1}B^{-1}$. Then the Fox differential is

$$\begin{aligned} dR_p &= (e - A_1B_1^{-1}A_1^{-1})\, dA_1 + \\ &\qquad A_1(e - B_1^{-1}A_1^{-1}B_1^{-1})\, dB_1 + \dots + \\ &\qquad\qquad [A_1, B_1] \dots [A_{p-1}, B_{p-1}](e - A_pB_p^{-1}A_p^{-1})\, dA_p + \\ &\qquad\qquad\quad [A_1, B_1] \dots [A_{p-1}, B_{p-1}]A_p(e - B_p^{-1}A_p^{-1}B_p^{-1})\, dB_p \end{aligned}$$

and the tangent spaces to $\mathsf{Hom}(\pi, G)$ identify with the kernel of this linear map composed with the adjoint representation of G.

When Σ is nonorientable, then its fundamental group admits a presentation

$$\pi = \langle C_1, \ldots, C_p \mid \widehat{R}_p(C_1, \ldots, C_p) = e \rangle$$

with defining relation

$$\widehat{R}_p(C_1, \ldots, C_p) := (C_1)^2 \ldots (C_p)^2.$$

Then the Fox differential is

$$d\widehat{R}_p = (e + C_1)\, dC_1 + \cdots + (C_1)^2 \ldots (C_{p-1})^2 (e + C_p)\, dC_p$$

and the tangent spaces to $\mathsf{Hom}(\pi, G)$ identify with the kernel of this linear map composed with the adjoint representation of G.

References

[1] K.-T. Chen, R. H. Fox, and R. C. Lyndon, *Free differential calculus. IV. The quotient groups of the lower central series*, Ann. of Math. (2) **68** (1958), 81–95, DOI 10.2307/1970044. MR102539

[2] R. H. Crowell and R. H. Fox, *Introduction to knot theory*, Based upon lectures given at Haverford College under the Philips Lecture Program, Ginn and Co., Boston, Mass., 1963. MR0146828

[3] ———, *Introduction to knot theory*, Springer-Verlag, New York-Heidelberg, 1977, Reprint of the 1963 original, Graduate Texts in Mathematics, No. 57. MR0445489

[4] R. H. Fox, *Free differential calculus. I. Derivation in the free group ring*, Ann. of Math. (2) **57** (1953), 547–560, DOI 10.2307/1969736. MR53938

[5] R. H. Fox, *Free differential calculus. II. The isomorphism problem of groups*, Ann. of Math. (2) **59** (1954), 196–210, DOI 10.2307/1969686. MR62125

[6] R. H. Fox, *Free differential calculus. III. Subgroups*, Ann. of Math. (2) **64** (1956), 407–419, DOI 10.2307/1969592. MR95876

[7] D. Fried, *Word maps, isotopy and entropy*, Trans. Amer. Math. Soc. **296** (1986), no. 2, 851–859, DOI 10.2307/2000391. MR846609

[8] W. M. Goldman, *The symplectic nature of fundamental groups of surfaces*, Adv. in Math. **54** (1984), no. 2, 200–225, DOI 10.1016/0001-8708(84)90040-9. MR762512

[9] M. Heusener and J. Porti, *Deformations of reducible representations of 3-manifold groups into* $\mathrm{PSL}_2(\mathbb{C})$, Algebr. Geom. Topol. **5** (2005), 965–997, DOI 10.2140/agt.2005.5.965. MR2171800

[10] M. Heusener and J. Porti, *Representations of knot groups into* $\mathrm{SL}_n(\mathbb{C})$ *and twisted Alexander polynomials*, Pacific J. Math. **277** (2015), no. 2, 313–354, DOI 10.2140/pjm.2015.277.313. MR3402353

[11] E. Hironaka, *Alexander stratifications of character varieties*, Ann. Inst. Fourier (Grenoble) **47** (1997), no. 2, 555–583. MR1450425

[12] F. Labourie, *Lectures on representations of surface groups*, Zurich Lectures in Advanced Mathematics, European Mathematical Society (EMS), Zürich, 2013. MR3155540

[13] A. Lubotzky and A. R. Magid, *Varieties of representations of finitely generated groups*, Mem. Amer. Math. Soc. **58** (1985), no. 336, xi+117, DOI 10.1090/memo/0336. MR818915

[14] J. W. Milnor, *Infinite cyclic coverings*, Conference on the Topology of Manifolds (Michigan State Univ., E. Lansing, Mich., 1967), Prindle, Weber & Schmidt, Boston, Mass., 1968, pp. 115–133. MR0242163

[15] M. S. Raghunathan, *Discrete subgroups of Lie groups*, Springer-Verlag, New York-Heidelberg, 1972. Ergebnisse der Mathematik und ihrer Grenzgebiete, Band 68. MR0507234

[16] A. S. Sikora, *Character varieties*, Trans. Amer. Math. Soc. **364** (2012), no. 10, 5173–5208, DOI 10.1090/S0002-9947-2012-05448-1. MR2931326

[17] A. Weil, *Remarks on the cohomology of groups*, Ann. of Math. (2) **80** (1964), 149–157, DOI 10.2307/1970495. MR169956

Department of Mathematics, University of Maryland, College Park, Maryland 20742

Email address: `wmg@math.umd.edu`

URL: `http://www.math.umd.edu/~wmg`

Contemporary Mathematics
Volume **760**, 2020
https://doi.org/10.1090/conm/760/15291

Culler-Shalen seminorms of fillings of the Whitehead link exterior

Gabriel Indurskis

ABSTRACT. We determine the total Culler-Shalen seminorms for the 3-manifolds $W_{p/q} := W(p/q, \cdot\,)$ obtained by Dehn filling with slope p/q on one boundary component of the Whitehead link exterior W when p is odd. As part of the proof, we use an explicit parametrization of the eigenvalue variety of W to find a one-variable polynomial whose roots characterize characters of p-reps of $\pi_1(W_{p/q})$, i.e. representations with values in $\mathrm{SL}_2(\mathbb{C})$ which are parabolic on the peripheral subgroup.

1. Introduction

Let W be the exterior of the right-handed Whitehead link in S^3 with the canonical framing (μ_0, λ_0) and (μ_1, λ_1) as indicated in figure 1. Let $W_{p/q} =: W(p/q, \cdot\,)$ be the manifold obtained from W by Dehn filling the first boundary torus along the slope p/q, i.e. by attaching a solid torus by identifying its meridian with the curve $\mu_0^p\lambda_0^q$. Here p and q are coprime integers, and unless otherwise noted we will fix the convention $q > 0$. (Note that there is an ambient isotopy taking one component of the link to the other, so in fact $W(p/q, \cdot\,) \cong W(\,\cdot\,, p/q)$. To avoid confusion, we will nevertheless always consider the boundary with curves μ_0, λ_0 to be filled, and the curves μ_1 and λ_1 to lie on the remaining boundary torus of $W_{p/q}$ after filling.)

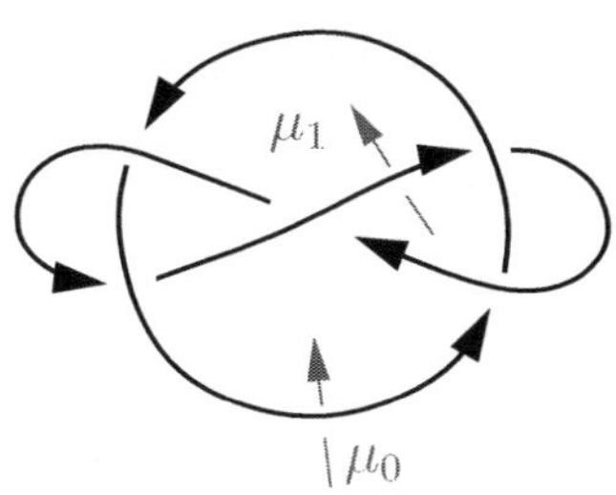

FIGURE 1. The right-handed Whitehead link

2010 *Mathematics Subject Classification.* Primary 57M27; Secondary 20C15, 57N10, 57R65.
Key words and phrases. 3-manifold, Dehn surgery, character variety, Culler-Shalen seminorm.
This work was supported in part by a scholarship of the *Institut des Sciences Mathématiques*, Montréal.

The manifolds $W_{p/q}$ constitute a rich class of 3-manifolds whose boundary consists of one torus, and contain important infinite subfamilies of well-known manifolds such as the twist knot exteriors (for $p/q = 1/n$, $n \in \mathbb{Z}$) and an important class of once-punctured torus bundles (for $p/q = m \in \mathbb{Z}$, cf. [**HMW92**]). Note that for example the figure-8 knot complement ($p/q = -1$) and its so-called sister manifold ($p/q = 5$) are part of this family. Since the Whitehead link exterior admits a complete hyperbolic structure of finite volume, the same is true for all but a finite number of the manifolds $W_{p/q}$ (cf. [**Thu79**]). Furthermore, unless $p/q \in \{0, 4\}$, all manifolds $W_{p/q}$ are *small* manifolds, i.e. they do not contain any closed essential surfaces (cf. [**BGZ01**, Appendix]).

The goal of this paper is to determine the total Culler-Shalen seminorm of the filled manifold $W_{p/q}$ when p is odd, in terms of the filling slope p/q. This work can be seen as a generalization of similar work previously done for twist knots by Boyer, Mattman and Zhang ([**BMZ97**]). The results given in this paper completely include these previous results, and extend them to a much larger class of fillings.

THEOREM 1. *Let p and q be coprime integers with $q > 0$ and p odd, $p/q \neq 3$. The total Culler-Shalen seminorm of the manifold $W_{p/q} = W(p/q, \cdot)$ is given by*

$$\|\gamma\| = \sum_{j=1}^{3} a_j \Delta(\gamma, \beta_j),$$

where $\Delta(\cdot, \cdot)$ denotes the geometric intersection number of two slopes, the slopes β_j ($j = 1, 2, 3$) are the possible boundary slopes for $W_{p/q}$ given in table 1 below, and the coefficients a_1, a_2, and a_3 are given in table 2. Furthermore, the minimal non-zero total norm s is as given in the last column of table 2 on page 169.

We remark that the condition that p be odd (which is required in the proofs of section 5 to avoid a possible obstruction to lifting $\mathrm{PSL}_2(\mathbb{C})$ representations) automatically excludes the special cases W_0 and W_4 (both *large*, toroidal manifolds which do not admit a complete hyperbolic structure), as well as W_6: Filling this manifold a second time with slope 1 produces the closed manifold $L_2 \# L_3$, a connected sum of lens spaces, causing the arguments in section 5 to fail. The same argument applies to W_3, which therefore has to be excluded separately.

$r = p/q \in$	β_1	β_2		β_3
$[-\infty, 0]$	4	$4r^{-1} = \dfrac{4q}{p}$	$\in [-\infty, 0]$	0
$[0, 2]$	4	$4r^{-1} + 2 = \dfrac{2p + 4q}{p}$	$\in [4, \infty]$	0
$[2, 4]$	4	$6 - r = \dfrac{-p + 6q}{q}$	$\in [2, 4]$	0
$[4, \infty]$	4	$4(r-2)^{-1} = \dfrac{4q}{p - 2q}$	$\in [0, 2]$	0

TABLE 1. Possible boundary slopes of $W_{p/q}$

$p/q \in$	a_1	a_2	a_3	s
$(-\infty, 0)$	$-p+2q-1$	2	$2q-2$	$-3p+4q-3$
$(0,2)$	$-p+2q-1$	2	$2q-2$	$p+4q-3$
$(2,3)\cup(3,4)$	$p-2q-1$	4	$2q-2$	$p+4q-3$
$(4,6)\cup(6,\infty)$	$p-2q-1$	2	$2q-2$	$3p-4q-3$

TABLE 2. The parameters for the total Culler-Shalen seminorm of $W_{p/q}$ for p odd, $q > 0$

The knowledge of the coefficients a_j given in Theorem 1 has immediate consequences: If $a_j > 0$, the fundamental polygon (the boundary of the norm-ball of radius s) of the Culler-Shalen seminorm has a vertex corresponding to β_j, and Lemma 6.1 of [**BZ96**] then shows that in this case β_j is a *strongly detected*, strict boundary slope, i.e. the slope of an essential surface associated to some ideal point of the character variety of $W_{p/q}$, which is neither a fibre nor semi-fibre. Inspection of table 2 therefore implies:

COROLLARY 2. *Let p and q be coprime integers with $q > 0$, p odd, and $p/q \neq 3$.*

(1) *$\beta_1 = 4$ is a strongly detected slope of $W_{p/q}$* unless $p = 2q \pm 1$.
(2) *β_2 is* always *a strongly detected slope of $W_{p/q}$.*
(3) *$\beta_3 = 0$ is a strongly detected slope of $W_{p/q}$ iff $q > 1$, otherwise it is the slope of a fibre in the once-punctured torus bundle $W(p)$.*

An interesting by-product of the proof of Theorem 1 is the complete determination of the so-called p-reps of the fundamental group of $W_{p/q}$, which are of interest in their own right. These are representations into $SL_2(\mathbb{C})$ which are parabolic on the peripheral subgroup, and were first studied for knot groups by Riley (cf. [**Ril72**]). (To be precise, the generator corresponding to the meridian of the boundary torus is sent to a non-trivial element of trace ± 2 in $SL_2(\mathbb{C})$. As a consequence, the generator corresponding to the longitude will be either parabolic as well, or $\pm I$.) In his classical work, he showed how under certain conditions on the group presentation (e.g. for the fundamental group of a 2-bridge knot) all such p-reps can be characterized by the complex roots of a single one-variable polynomial. This is in contrast to the general situation, where only a characterization by a system of *several* polynomial equations can be expected.

Passing through the so-called *eigenvalue variety* (as defined by Tillmann in [**Til05**]) of the unfilled Whitehead link exterior W, we are able to find a similar approach which seems naturally adapted to the situation of Dehn filling to determine a *single* one-variable polynomial whose roots are in two-to-one correspondence with the set of eigenvalues of p-reps. This in turn is shown to be in bijective correspondence with the set of p-reps itself, and we obtain a complete characterization of all p-reps of $W_{p/q}$:

THEOREM 3. *The conjugacy classes of irreducible p-reps of the manifold $W_{p/q}$ (p can be odd or even) are in two-to-one correspondence with the non-trivial roots $s \notin \{0, \pm 1\}$ of the Laurent polynomial*

$$\mathrm{res}_{p,q}(s) = s^{p-2q} + (-1)^{q+1} 2T_q(y) + s^{-p+2q},$$

where $y = \frac{1}{2}(-s^2 + 4 - s^{-2})$, and $T_q(y)$ is the q-th Chebyshev polynomial of first type. The conjugacy classes of non-abelian reducible p-reps are in two-to-one correspondence with the non-trivial roots of unity $s \neq \pm 1$, $s^p = 1$.

We remark that whereas it seems a hard analytic problem to determine if the non-trivial roots of these polynomials are always simple, this is in fact a consequence of Theorem 1 when p is odd:

COROLLARY 4. *When p is odd (including $p/q = 3$), all non-trivial roots $s \notin \{0, \pm 1\}$ of $\mathrm{res}_{p,q}$ are simple.*

PROOF. Note that even though Theorem 1 excludes $p/q = 3$, we can include it here simply by manually inspecting $\mathrm{res}_{3,1}(s) = -s^3(s+1)^2(s^2 - 3s + 1)$.

For all other cases with p odd, Theorem 1 shows that the upper bounds for the count of distinct non-trivial roots of $\mathrm{res}_{p,q}$ obtained in section 3.5 are actually attained, which can only happen if these roots are all simple. □

We note that experimental evidence seems to *suggest* that this also holds when p is even. Furthermore, we would like to point out that all our results would immediately extend to the case when p is even if a work-around for the lifting obstruction in section 5 could be found.

The organization of this paper is as follows:

Section 2: After discussing some preliminaries about character varieties and the Culler-Shalen seminorm, we give the general outline for the proof of Theorem 1. We also determine the set of possible boundary slopes of $W_{p/q}$.

Section 3: In this section, we identify the p-reps of the manifold $W_{p/q}$ with points in the eigenvalue variety of the unfilled manifold W characterized by the roots of a single one-variable Laurent polynomial $\mathrm{res}_{p,q}$, proving Theorem 3. A detailed analysis of the properties and symmetries of these recursively defined polynomials is given, including an upper bound for the number of their roots in terms of p and q, which in turn gives an upper bound for the number of conjugacy classes of p-reps.

Section 4: We show that the *minimal* nonzero value of the Culler-Shalen seminorm of $W_{p/q}$ is precisely equal to the number of conjugacy classes of p-reps, and therefore satisfies the same bounds as shown in the previous section: Using group cohomology calculations and the characterization of p-reps from section 3, it is shown that each character of a p-rep is a smooth point on a non-trivial curve in the character variety of $W_{p/q}$, and that it furthermore is a simple root of the trace function f_{μ_1}.

Section 5: We explicitly determine the Culler-Shalen seminorm of slopes σ resulting in Seifert-fibered manifolds $W_{p/q}(\sigma)$ when p is odd. This is done by a detailed study of $\mathrm{PSL}_2(\mathbb{C})$ representations of the orbifold fundamental groups of their base orbifolds, which can always be lifted to $\mathrm{SL}_2(\mathbb{C})$ representations when p is odd, but not necessarily if p was even.

Section 6: The total Culler-Shalen seminorm is determined for p odd: Combining the results from the previous sections, we solve the linear systems relating the seminorms obtained for the Seifert fillings with the coefficients determining the seminorm.

2. Preliminaries and outline

We briefly recall the construction of the Culler-Shalen norm developed in [**CGLS87**], and its generalization to a seminorm by Boyer and Zhang in [**BZ98**].

Let M be a connected, orientable, compact, irreducible and boundary-irreducible 3-manifold with boundary ∂M consisting of one torus. Unless noted otherwise, we will generally suppress the base point of fundamental groups.

Denote by $\mathfrak{R}(M) := \mathrm{Hom}(\pi_1(M); \mathrm{SL}_2(\mathbb{C}))$ the *representation variety* of M, and by $\mathfrak{X}(M)$ its *character variety*, i.e. the set of characters of representations in $\mathfrak{R}(M)$ (cf. [**CS83**]). These two sets are in fact affine algebraic varieties and as such can be identified with subsets of $\mathbb{C}^N$ (for appropriate $N \in \mathbb{N}$), defined by a set of polynomial equations. We will occasionally need to consider representations into $\mathrm{PSL}_2(\mathbb{C})$, which we will denote by $\overline{\rho} \in \overline{\mathfrak{R}}(M) := \mathrm{Hom}(\pi_1(M); \mathrm{PSL}_2(\mathbb{C}))$.

Since M is boundary-irreducible, we can identify $\pi_1(\partial M)$ with its image under the natural inclusion $\pi_1(\partial M) \to \pi_1(M)$ (defined up to conjugation). Denote the first homology of the boundary with integer coefficients $H_1(\partial M; \mathbb{Z})$ by L and regard it as a lattice in the 2-dimensional real vector space $V = H_1(\partial M; \mathbb{R})$. Using the Hurewicz homomorphism, we can identify L with $\pi_1(\partial M)$.

A *slope* r is the isotopy class of an essential, unoriented, simple closed curve on ∂M. It can be identified with a pair of elements $\pm\delta = \pm(p, q) \in L$, or with the rational number $p/q \in \mathbb{Q} \cup \{\infty\}$, where $\infty = 1/0$ and $-\infty = -1/0$ are identified. Let $\alpha(r)$ be either one of $\pm\delta$. We will often slightly abuse notation and interchange freely between these notations; for example writing $M(\delta)$ or $M(p/q)$ instead of $M(r)$ to denote the manifold obtained by Dehn filling M along the slope r. A slope is called a *boundary slope* if it represents the isotopy class of the boundary components (on one component) of ∂M of an essential surface $S \subset M$. A boundary slope is called *strict* if it is the boundary slope of an essential surface S which is neither a fibre or a semi-fibre in a fibre or semi-fibre bundle structure on M. For two elements $\delta_1, \delta_2 \in L$, we denote their geometric intersection number, called their *distance*, by $\Delta(\delta_1, \delta_2)$. If $\delta_i = (p_i, q_i)$, we have $\Delta(\delta_1, \delta_2) = |p_1q_2 - p_2q_1|$.

Let $\mathfrak{X}_0 \subset \mathfrak{X}$ be a curve in the character variety $\mathfrak{X} = \mathfrak{X}(M)$, i.e. an irreducible algebraic component of complex dimension 1. Call $\mathfrak{X}_0$ *non-trivial* if it contains the character of an irreducible representation.

For $\gamma \in \pi_1(M)$, define the regular function $I_\gamma \colon \mathfrak{X}_0 \to \mathbb{C}$ by $I_\gamma(\chi_\rho) = \chi_\rho(\gamma) = \operatorname{trace}\rho(\gamma)$. Use this to define the map $f_\gamma := I_\gamma^2 - 4$, which is also regular and can hence be pulled back to the smooth projective model (i.e. the unique non-singular projective variety birationally equivalent to the curve $\mathfrak{X}_0$, cf. [**Sha94**, Ch. II, § 4.5 and 5.3]) $\widetilde{\mathfrak{X}}_0$ of $\mathfrak{X}_0$ (we will denote this map also by f_γ).

For $\gamma \in L$, define $\|\gamma\|_{\mathfrak{X}_0}$ as the degree of the map $f_\gamma \colon \widetilde{\mathfrak{X}}_0 \to \mathbb{CP}^1$. This can be extended naturally to define a function on the real vector space $V := H_1(\partial M; \mathbb{R})$:

$$\|\cdot\|_{\mathfrak{X}_0} : H_1(\partial M; \mathbb{R}) \to [0; \infty) \tag{1}$$

This function is a seminorm on $H_1(\partial M; \mathbb{R})$, called the *Culler-Shalen seminorm* associated to the curve $\mathfrak{X}_0$ (cf. [**CGLS87**, Proposition 1.1.2] for the case of a norm, and [**BZ98**, Section 5] for the general case). Seen as a seminorm on a 2-dimensional real vector space, $\|\cdot\|_{\mathfrak{X}_0}$ is either a norm, a nonzero indefinite seminorm, or identically zero. When M is small, the last possibility cannot occur (cf. [**BZ98**, Proposition 5.5]). Furthermore $\dim \mathfrak{X}(M) \leq 1$ in this case (cf. [**CCG⁺94**]), and it is therefore natural to consider the sum of the Culler-Shalen seminorms over all *non-trivial* curves $\mathfrak{X}_i \subset \mathfrak{X}(M)$, which we will call the *total* Culler-Shalen seminorm:

$$\|\gamma\| := \sum_i \|\gamma\|_i \tag{2}$$

We denote the union of all non-trivial curves in $\mathfrak{X}(M)$ by $\mathfrak{X}^{\mathrm{irr}}(M)$.

If f_γ is non-constant on $\mathfrak{X}_0$ for some $\gamma \in L$, define

$$s_{\mathfrak{X}_0} := \min\{\|\gamma\|_{\mathfrak{X}_0} \mid \gamma \in L,\ \gamma \neq 0,\ f_\gamma \text{ not constant on } \mathfrak{X}_0\}, \tag{3}$$

otherwise set $s_{\mathfrak{X}_0} := 0$. We call $s_{\mathfrak{X}_0}$ the *minimal* seminorm on $\mathfrak{X}_0$, the corresponding value for the total seminorm is the *total minimal seminorm*, denoted by s.

The following lemma enables us to determine the Culler-Shalen seminorm from only a few known values:

LEMMA 5 ([**BZ96**, Lemma 6.2]). *If β_j are the boundary slopes of M and $\mathfrak{X}_i \subset \mathfrak{X}(M)$ is a non-trivial curve (i.e. a curve which contains the character of an irreducible representation), then*

$$\|\gamma\|_{\mathfrak{X}_i} = \sum_j a_{ij} \Delta(\gamma, \beta_j),$$

where the a_{ij} are non-negative, even integers independent of $\gamma \in \pi_1(M)$. □

For the total seminorm, we therefore get

$$\|\gamma\| = \sum_j a_j \Delta(\gamma, \beta_j), \tag{4}$$

with non-negative even integers a_j.

To determine the total seminorm for the manifolds $W_{p/q}$, it therefore suffices to determine the boundary slopes β_j and the corresponding coefficients a_j. Following a technique due to Mattman (cf. [**Mat02**]), we will use explicit calculations of the seminorm of certain slopes (cf. section 5) together with equation (4) to set up a system of linear equations with unknowns a_j and s. In our case, this system turns out to be underdetermined, requiring one more piece of information to successfully solve it. We therefore first study the set of p-reps of $W_{p/q}$ (cf. section 3) and use this knowledge to explicitly determine the minimal total seminorm s (cf. section 4). With this information, we are then able to solve the linear systems in section 6, completing the proof of Theorem 1.

We finish this section by determining the set of possible boundary slopes of $W_{p/q}$:

2.1. Boundary slopes of the unfilled Whitehead link exterior. In [**HS07**], Hoste and Shanahan give an explicit algorithm, building on work of Floyd and Hatcher (cf. [**FH88**]) and on the unpublished dissertation of Lash (cf. [**Las93**]), to determine the boundary slopes of 2-bridge links. In particular, they show that up to symmetry between the two boundary tori, the *left-handed* Whitehead link ($L_{3/8}$ in 2-bridge notation) has boundary slope pairs $(0,0)$, $(0,\varnothing)$, $(-4,-2)$, $(-4,\varnothing)$ (here $\varnothing$ signifies that the intersection with one of the boundary components is empty) and the continuous spectrum of slope pairs $(2t^{-1}, 2t)$, $(-2t^{-1}, -2-2t)$, $(-3+s, -3-s)$ parametrized by $t \in (\mathbb{Q} \cup \{\infty\}) \cap [0,\infty]$ and $s \in \mathbb{Q} \cap [-1,1]$ (cf. table 4 in [**HS07**], using $k=1$).

After applying an orientation-reversing homeomorphism to switch to the *right-handed* Whitehead link (which simultaneously changes the signs of all slopes), we use the symmetry between the two boundary components and reparametrize the slope pairs such that the first is always equal to $r = p/q$. The results are shown in table 3.

$r = p/q$ on T_0	p'/q' on T_1	
0	0 or $\varnothing$	
4	2 or $\varnothing$	
$\varnothing$	0	
2 or $\varnothing$	4	
$r \in [-\infty, 0]$	$4r^{-1} = \dfrac{4q}{p}$	$\in [-\infty, 0]$
$r \in [0, 2]$	$4r^{-1} + 2 = \dfrac{2p + 4q}{p}$	$\in [4, \infty]$
$r \in [2, 4]$	$6 - r = \dfrac{-p + 6q}{q}$	$\in [2, 4]$
$r \in [4, \infty]$	$4(r-2)^{-1} = \dfrac{4q}{p - 2q}$	$\in [0, 2]$

TABLE 3. Boundary slope pairs of the right-handed Whitehead link W

2.2. Boundary slopes of the once-filled Whitehead link exterior. Denote the two boundary components of W by T_0 and T_1. We now proceed to determine the boundary slopes of the once-filled Whitehead link exterior $W_{p/q} = W(p/q, \,\cdot\,) = W \cup_{\mu_0^p \lambda_0^q} V$, where V is the attached solid torus. Note that we can view W as lying inside $W_{p/q}$.

LEMMA 6. *Let S be an essential surface in $W_{p/q}$. Then either S is in fact an essential surface in $W \subset W_{p/q}$, or it is the cap-off of an essential surface S' in $W \subset W_{p/q}$. In particular, S has non-empty intersection with at least one of T_0 and T_1.*

PROOF. Let S be an essential surface in $W_{p/q}$. We can isotope S so it transversely intersects $T_0 \subset W \subset W_{p/q}$ and $T_1 \subset W_{p/q}$, as well as the core of the filling torus V. If $S \cap T_0 = S \cap T_1 = \varnothing$, S can be viewed as a closed essential surface in W, which is a contradiction, since W is small. We therefore see that S has to intersect at least one of T_0 or T_1. □

We now have two cases to consider:

$S \cap \operatorname{int} V = \varnothing$: In this case, S is in fact an essential surface in $W \subset W_{p/q}$ with empty intersection with T_0. According to table 3, it therefore must have boundary slope on T_1 equal to either 0 or 4.

$S \cap \operatorname{int} V \neq \varnothing$: In this case, the intersection $S \cap T_0$ must consist of simple closed curves of slope p/q. This implies that $S' = S \cap W$ is an essential surface in W with boundary slope p/q on T_0. Unless $p/q \in \{0, 4\}$, this implies that $S' \cap T_1 \neq \varnothing$, and we get the corresponding boundary slope on T_1 from table 3, dependent on p/q. We remark that if $p/q \in \{0, 4\}$, S could in fact be a closed essential torus in $W_{p/q}$.

We therefore have at most three *candidates* for boundary slopes in the manifold $W_{p/q}$, as summarized in table 1 on page 168: We have two fixed slopes $\beta_1 = 4$ and $\beta_3 = 0$, and one varying slope β_2 which depends on p and q. The parametrization of the latter depends on the range of the filling slope p/q.

Note that we have at this point not determined if these slopes are in fact boundary slopes of $W_{p/q}$. For the following, we will rather only assume that the set of boundary slopes of $W_{p/q}$ is *contained* in the set $\{\beta_1,\ \beta_2,\ \beta_3\}$. The proof of Theorem 1 will then in turn imply Corollary 2, showing that unless $p = 2q \pm 1$ or $q = 1$, all three slopes are indeed strongly detected, strict boundary slopes.

3. The set of p-reps of a once-filled Whitehead link exterior

The goal of this section is to prove Theorem 3 (in sections 3.1–3.4), as well as to provide upper bounds for the number of non-trivial roots of the Laurent polynomials $\operatorname{res}_{p,q}$ (in section 3.5), which in turn then gives upper bounds for the number of conjugacy classes of p-reps (in section 3.6).

The fundamental group of the Whitehead link exterior admits the presentation

$$\pi_1(W) = \langle\ \mu_0,\ \mu_1 \mid \mu_0\mu_1\mu_0^{-1}\mu_1^{-1}\mu_0^{-1}\mu_1\mu_0\mu_1 = \mu_1\mu_0\mu_1\mu_0^{-1}\mu_1^{-1}\mu_0^{-1}\mu_1\mu_0\ \rangle \tag{5}$$

which then gives the following presentation for the fundamental group of the filled manifold:

$$\begin{aligned}\pi_1(W_{p/q}) = \langle\ \mu_0,\ \mu_1 \mid \mu_0\mu_1\mu_0^{-1}\mu_1^{-1}\mu_0^{-1}\mu_1\mu_0\mu_1 &= \mu_1\mu_0\mu_1\mu_0^{-1}\mu_1^{-1}\mu_0^{-1}\mu_1\mu_0,\\ \mu_0^p\lambda_0^q &= 1\ \rangle,\end{aligned} \tag{6}$$

where

$$\lambda_0 = \mu_0\mu_1\mu_0\mu_1^{-1}\mu_0^{-1}\mu_1^{-1}\mu_0\mu_1\mu_0^{-2} = \mu_1\mu_0\mu_1^{-1}\mu_0^{-1}\mu_1^{-1}\mu_0\mu_1\mu_0^{-1} \tag{7}$$

(since it commutes with μ_0) and

$$\lambda_1 = \mu_0\mu_1\mu_0^{-1}\mu_1^{-1}\mu_0^{-1}\mu_1\mu_0\mu_1^{-1}. \tag{8}$$

Following [**Til02**, section 5.1.9] (see also [**Til20**, section 5.6] in these proceedings), we first describe a convenient subset of the representation variety $\mathfrak{R}(W)$: Let $\rho \in \mathfrak{R}(W)$ be a non-abelian representation. Then ρ can be conjugated to be of the following normal form:

$$\rho(\mu_0) = \begin{pmatrix} s & c \\ 0 & s^{-1} \end{pmatrix} \qquad \text{and} \qquad \rho(\mu_1) = \begin{pmatrix} u & 0 \\ 1 & u^{-1} \end{pmatrix} \tag{9}$$

for $s, u \in \mathbb{C}^*$ and $c \in \mathbb{C}$ (note that $c = 0$ when ρ is non-abelian reducible). We will denote the subset of representations in $\mathfrak{R}(W_{p/q})$ of this form by $\mathfrak{C}(W_{p/q})$.

Applying (9) to the left-hand side w_1 and right-hand side w_2 of the defining relation in (5) yields

$$0 = \rho(w_1) - \rho(w_2) = \begin{pmatrix} -cf & c(u-u^{-1})f \\ (s-s^{-1})f & cf \end{pmatrix}, \tag{10}$$

where

$$\begin{aligned} f(s,u,c) = (s-s^{-1})(u-u^{-1}) &+ c(s^{-2}u^{-2} - u^{-2} - s^{-2} + 4 - s^2 - u^2 + s^2u^2)\\ &+ c^2(2s^{-1}u^{-1} - su^{-1} - s^{-1}u + 2su) + c^3. \end{aligned} \tag{11}$$

This shows that $\mathfrak{C}(W) \subset \mathbb{C}^3$ is defined by the equation $f(s,u,c) = 0$.

Let now $\rho \in \mathfrak{R}(W_{p/q})$ be a representation which sends the meridian μ_1 to a parabolic element (a matrix $A \neq \pm I$ with $|\operatorname{trace} A| = 2$), and which has non-abelian image in $\mathrm{SL}_2(\mathbb{C})$. Since μ_1 and λ_1 commute, this implies that $\rho(\lambda_1)$ is either $\pm I$ or also parabolic. Following Riley (cf. [**Ril72**] and [**Ril75**]) we call such a representation an $\mathrm{SL}_2(\mathbb{C})$ *p-representation*, or, for short, *p-rep*. (We remark that Riley actually used the name p-rep for $\mathrm{PSL}_2(\mathbb{C})$ representations only, whereas a $\mathrm{SL}_2(\mathbb{C})$ representation which projects to a $\mathrm{PSL}_2(\mathbb{C})$ p-rep was called *sl-rep*, c.f. [**Ril75**].)

Denote by $\mathfrak{P} \subset \mathfrak{C}(W_{p/q})$ the subset of p-reps of $\pi_1(W_{p/q})$ of the form (9). Note that by construction, $\mathfrak{C}(W_{p/q})$ projects generically 4-to-1 to $\mathfrak{X}(W_{p/q})$, corresponding to the action of the Kleinian four-group (generated here by the inversions $(s,u,c) \mapsto (s^{-1}, u^{-1}, c)$ and $(s,u,c) \mapsto (s, u^{-1}, c + (s - s^{-1})(u - u^{-1})))$ on the set of possible bases for the normal form (9). However, for a p-rep $\rho \in \mathfrak{C}(W_{p/q})$, we have $u^2 = 1$ (and $s^2 \neq 1$) and hence the projection is 2-to-1. Each conjugacy class of p-reps has therefore precisely two representatives in $\mathfrak{P}$.

Since $\mathfrak{R}(W_{p/q}) \subset \mathfrak{R}(W)$, each p-rep $\rho \in \mathfrak{R}(W_{p/q})$ induces a non-abelian representation $\rho' \in \mathfrak{R}(W)$ satisfying $\rho'(\mu_0^p \lambda_0^q) = I$ which still sends μ_1 to a parabolic element. We will call such a representation ρ' a *partial* p-rep. Conversely, each partial p-rep ρ' factors through $\pi_1(W_{p/q})$, and we therefore have a bijection between the set $\mathfrak{P} \subset \mathfrak{C}(W_{p/q})$ of p-reps in normal form (9) and the set $\mathfrak{P}' \subset \mathfrak{C}(W)$ of partial p-reps in normal form.

Instead of trying to translate the filling relation $\mu_0^p \lambda_0^q = 1$ into defining equations for the parameters s, c and u in $\mathfrak{C}(W_{p/q})$ (which is difficult for general p and q), we will now restrict our attention to the eigenvalues of p-reps.

3.1. Detour through the eigenvalue variety. The eigenvalue variety of a manifold whose boundary consists of a finite number of tori was introduced by Tillmann in [**Til05**] as a generalization of the A-polynomial for knots, and an explicit parametrization of the eigenvalue variety of the Whitehead link exterior can be found in his thesis (cf. [**Til02**, section 5.4.3]) and in [**Til20**, Appendix B.1] in these proceedings.

Let $\rho \in \mathfrak{C}(W)$. For $\gamma \in \pi_1(W)$, let

$$\begin{aligned} \theta_\gamma \colon \mathfrak{C}(W) &\to \mathbb{C} \\ \rho &\mapsto \rho(\gamma)_{1,1} \end{aligned} \tag{12}$$

be the map taking ρ to the upper left entry of the image $\rho(\gamma)$. Now define the eigenvalue map to be

$$\begin{aligned} e \colon \mathfrak{C}(W) &\to \mathbb{C}^4 \\ \rho &\mapsto (\theta_{\mu_0}(\rho), \theta_{\lambda_0}(\rho), \theta_{\mu_1}(\rho), \theta_{\lambda_1}(\rho)) \,. \end{aligned} \tag{13}$$

Note that this is well-defined, as each $\rho \in \mathfrak{C}(W)$ is triangular on the peripheral subgroup.

The image $e(\mathfrak{C}(W)) \subset \mathbb{C}^4$ is a subset of the eigenvalue variety which we will denote by $\mathfrak{E}_0(W)$. Tillmann showed in [**Til02**] that the map e has degree one, and that furthermore:

LEMMA 7 (Lemma 5.6 in [**Til02**] or Lemma 17 in [**Til20**]). *The varieties $\mathfrak{C}(W)$ and $\mathfrak{E}_0(W)$ are birationally equivalent.* □

In particular, the map e has the following inverse map for $s \neq \pm 1$ and $u^2 \neq s^2$:

$$e^{-1} : \mathfrak{E}_0(W) \to \mathfrak{C}(W), \qquad (s,t,u,v) \mapsto \left(s, u, \frac{s^2(t-1) + u^2(1-v)}{s^{-1}u^{-1}(s^2 - u^2)}\right) \tag{14}$$

As shown in [**Til02**], one can in fact find a pair of inverse maps which together are defined on all but eight points of $\mathfrak{E}_0(W)$:

$$e_1^{-1} : \mathfrak{E}_0(W) \to \mathfrak{C}(W), \qquad (s,t,u,v) \mapsto \left(s, u, \frac{(s^2 - v)(1-u^2)}{su(1+v)}\right) \tag{15}$$

$$e_2^{-1} : \mathfrak{E}_0(W) \to \mathfrak{C}(W), \qquad (s,t,u,v) \mapsto \left(s, u, \frac{(u^2 - t)(1-s^2)}{su(1+t)}\right) \tag{16}$$

Four of the points which are not defined by these two maps correspond to the discrete faithful representations of $\pi_1(W)$ associated to the complete hyperbolic structure of W, parametrized by $t = v = -1$ and $s^2 = u^2 = 1$. The remaining four points are given by $t = v = -1$ and $s^2 = u^2 = -1$.

Since a partial p-rep $\rho' \in \mathfrak{C}(W)$ satisfies $u^2 = 1$ and $s^2 \neq 1$, we see that the eigenvalue map e sends the set $\mathfrak{P}'$ bijectively onto a set $\mathfrak{P}'' \subset \mathfrak{E}_0$, as shown in the following diagram:

$$\begin{array}{ccccc} \mathfrak{C}(W_{p/q}) & \xrightarrow{\subset} & \mathfrak{C}(W) & \xrightarrow{\simeq} & \mathfrak{E}_0(W) \\ \uparrow & & \uparrow & & \uparrow \\ \mathfrak{P} & \xrightarrow{\cong} & \mathfrak{P}' & \xrightarrow{\cong} & \mathfrak{P}'' \end{array} \tag{17}$$

We therefore have a bijective correspondence between the set $\mathfrak{P}$ of p-reps in normal form and a set $\mathfrak{P}''$ of eigenvalue tuples in $\mathfrak{E}_0$. Since each p-rep $\rho \in \mathfrak{C}(W_{p/q})$ and each partial p-rep $\rho' \in \mathfrak{C}(W)$ satisfies $\rho(\mu_0^p \lambda_0^q) = I$, we obtain that the set $\mathfrak{P}''$ lies in the intersection $\mathfrak{F}$ of the following subvarieties of $\mathfrak{E}_0(W)$:

$$\mathfrak{P}'' \subsetneq \mathfrak{F} := \mathfrak{E}_0(W) \cap \{s^p t^q = 1\} \cap \{u^2 = 1\}. \tag{18}$$

It is important to note that $\mathfrak{P}''$ is a proper subset of $\mathfrak{F}$, as there are indeed points in $\mathfrak{E}_0(W) \cap \{s^p t^q = 1\}$ corresponding to representations which do not factor through $\pi_1(W_{p/q})$ (we will see that these are precisely points which satisfy $s^2 = 1$).

In summary, we have reduced the generally hard problem of determining the set $\mathfrak{P}$ to the problem of determining the set $\mathfrak{P}''$. This task is made feasible by the fact that we can first intersect $\mathfrak{E}_0(W)$ with the set $\{u^2 = 1\}$ (which will significantly simplify the defining polynomials involved), and *then* intersect the resulting set with the subset $\{s^p t^q = 1\}$ corresponding to the Dehn filling.

3.2. The subset of the eigenvalue variety corresponding to p-reps. Using the defining polynomial (11) of $\mathfrak{C}(W)$ and using elementary elimination theory, one finds that $e(\mathfrak{C}(W)) = \mathfrak{E}_0(W) \subset \mathbb{C}^4$ can be defined as an algebraic set by the following three polynomials (cf. [**Til02**, section 5.4.3] or [**Til20**, Appendix

B.1]):

(19a) $$h_1 = t - s^2t + s^2t^2 - s^4t^2 - u^2 - 2s^2tu^2 + s^4tu^2 - t^2u^2 + 2s^2t^2u^2 + s^4t^3u^2 + tu^4 - s^2tu^4 + s^2t^2u^4 - s^4t^2u^4,$$

(19b) $$h_2 = s^2 - v - s^4v + u^2v + 2s^2u^2v + s^4u^2v - s^2u^4v + s^2v^2 - u^2v^2 - 2s^2u^2v^2 - s^4u^2v^2 + u^4v^2 + s^4u^4v^2 - s^2u^4v^3,$$

(19c) $$h_3 = s^4t - s^6t - s^2tu^2 + s^4tu^2 + s^6t^2u^2 - s^2u^2v + u^4v + s^2u^4v - 2s^4tu^4v - u^6v + s^2u^6v^2.$$

Intersecting with the subvariety given by $\{u^2 = 1\}$ then simplifies the defining polynomials to

(20a) $$g_1 = (t-1)\left[t(t-1)s^4 + 4ts^2 + (1-t)\right],$$

(20b) $$g_2 = s^2(1-v)(v+1)^2,$$

(20c) $$g_3 = s^2\left[t(t-1)s^4 + 2t(1-v)s^2 + (v^2 - t)\right].$$

Since $s \neq 0$, it can be easily verified that the subvariety defined by these polynomials splits into the following two irreducible components:

(21) $$\mathfrak{E}_0(W) \cap \{u^2 = 1\} = \{u^2 = 1,\ t = v = 1\} \cup \{u^2 = 1,\ v = -1,\ s^4t^2 + (-s^4 + 4s^2 - 1)t + 1 = 0\}$$

Note that points in the first component correspond to reducible representations.

Intersecting this set further, we obtain the set $\mathfrak{F} = \mathfrak{E}_0(W) \cap \{u^2 = 1\} \cap \{s^pt^q = 1\}$ as

(22) $$\mathfrak{F} = \mathfrak{F}^r \cup \mathfrak{F}^i = \{u^2 = t = v = 1,\ s^p = 1\} \cup \{u^2 = 1, v = -1,\ s^4t^2 + (-s^4 + 4s^2 - 1)t + 1 = 0,\ s^pt^q = 1\}.$$

Let us denote the subsets of $\mathfrak{P}''$ corresponding to reducible and irreducible p-reps by $(\mathfrak{P}'')^r \subsetneq \mathfrak{F}^r$ and $(\mathfrak{P}'')^i \subsetneq \mathfrak{F}^i$, respectively, so that $\mathfrak{P}'' = (\mathfrak{P}'')^r \cup (\mathfrak{P}'')^i$.

3.3. Eigenvalues of reducible p-reps. From equation (22), we can immediately read off the points corresponding to reducible representations as

(23) $$\mathfrak{F}^r = \{(s, 1, \pm 1, 1) \mid s^p = 1\},$$

i.e. the eigenvalue s is necessarily a $|p|$-th root of unity.

We now need to determine which of these points also correspond to p-reps $\rho \in \mathfrak{R}(W_{p/q})$, or equivalently to partial p-reps $\rho' \in \mathfrak{R}(W)$.

The reader can easily verify (for example by using the inverse map defined in (15)) that a reducible representation $\rho' \in \mathfrak{C}(W)$ corresponding to the eigenvalues $(\pm 1, 1, \pm 1, 1)$ is necessarily abelian and hence *not* a partial p-rep. On the other hand, all points $(s, 1, \pm 1, 1) \in \mathfrak{F}^r$ with $s \neq \pm 1$ give indeed rise to partial p-reps: Since $u^2 = t = v = 1$ implies $\rho'(\lambda_0) = I$, we have $\rho'(\mu_0^p\lambda_0^q) = \rho'(\mu_0^p) = I$ and hence ρ' factors through $\pi_1(W_{p/q})$. Furthermore, $s^2 \neq 1$ implies that ρ' is indeed non-abelian.

This means that the subset $(\mathfrak{P}'')^r \subsetneq \mathfrak{F}^r$ is given by

(24) $$(\mathfrak{P}'')^r = \{(s, 1, \pm 1, 1) \mid s^p = 1,\ s \neq \pm 1\}.$$

To obtain the cardinality of this set, first note that -1 is a $|p|$-th root of unity iff p is even. Since we always have two choices for $u = \pm 1$, we therefore obtain

$$\#(\mathfrak{P}'')^r = \begin{cases} 2(|p|-1) & \text{if } p \text{ is odd,} \\ 2(|p|-2) & \text{if } p \text{ is even.} \end{cases} \tag{25}$$

3.4. Eigenvalues of irreducible p-reps. Let us denote the defining polynomials of $\mathfrak{F}^i$ by

$$k_1 = s^p t^q - 1, \tag{26}$$

$$k_2 = s^4 t^2 + (-s^4 + 4s^2 - 1)t + 1, \tag{27}$$

so that $\mathfrak{F}^i = \{u^2 = 1,\ v = -1,\ k_1 = k_2 = 0\}$.

To determine the solutions of $k_1 = k_2 = 0$ we will use the resultant of k_1 and k_2 with respect to t, which we determine by induction on $q \geq 1$ as the determinant of the corresponding Sylvester matrix:

(28)

$$\begin{aligned} \operatorname{res}_{p,q}(s) &:= \operatorname{Resultant}(k_1, k_2, t) \\ &= \det \begin{pmatrix} s^p & 0 & s^4 & & & \\ 0 & s^p & -s^4+4s^2-1 & s^4 & & \\ & 0 & 1 & -s^4+4s^2-1 & \ddots & \\ & & & 1 & \ddots & \\ \vdots & \vdots & & & \ddots & \\ & & & & \ddots & s^4 \\ -1 & 0 & & & & -s^4+4s^2-1 \\ 0 & -1 & & & & 1 \end{pmatrix}_{(q+2)\times(q+2)} \\ &= s^{p+2q}\left(s^{p-2q} + (-1)^{q+1} 2T_q(y) + s^{-p+2q}\right) \\ &\doteq s^{p-2q} + (-1)^{q+1} 2T_q(y) + s^{-p+2q}, \end{aligned}$$

where $y = y(s) = \frac{1}{2}(-s^2 + 4 - s^{-2})$, and $T_q(y)$ is the q-th Chebyshev polynomial of first type, satisfying $T_0(y) = 1$, $T_1(y) = y$ and $T_{k+1}(y) = 2y\, T_k(y) - T_{k-1}(y)$ for $k \geq 1$ (see standard literature on orthogonal polynomials, e.g. [**Sze75**]).

Note that $\operatorname{res}_{p,q}$ is in fact a Laurent polynomial in $\mathbb{Z}[s^{\pm 1}]$ when $p < 0$, but we will often simply use the word polynomial to also include Laurent polynomials. Since we are only interested in solutions with $s \neq 0$, we will always regard this polynomial as being defined up to multiplication by a unit s^k ($k \in \mathbb{Z}$), which we indicate with the symbol $\doteq$. Since $T_q(y)$ is a polynomial of degree q in y, the middle terms of $\operatorname{res}_{p,q}$ are formed by a symmetrical Laurent polynomial of span $4q$ in the variable s, containing only even powers. Note that we can consider equation (28) as a formal definition for the case $p/q = 1/0$, as then $\operatorname{res}_{1,0} \doteq (s-1)^2 = 0$ is equivalent to $k_1 = 0$, implying $s = 1$, $t = -1$.

For us, the main importance of this resultant is that its roots characterize the solutions of the system $\{k_1 = k_2 = 0\}$: The extension theorem of elimination theory (cf. [**CLO97**, §1, Theorem 3]) shows that the system $\{k_1 = 0,\ k_2 = 0\}$ has a solution (s, t) if and only if $s \neq 0$ is a root of this resultant. We will call s a *partial* solution in this case.

Furthermore note that given a partial solution s, the value for t is *uniquely* determined by $k_1 = 0$ and $k_2 = 0$: To see this, note that $k_2 = 0$ implies $t^2 =$

$-s^{-4}\left((-s^4+4s^2-1)t+1\right)$, and hence higher powers of t can recursively be replaced by linear terms in t. The equation $k_1 = s^p t^q - 1 = 0$ can therefore always be expanded to an equation which is linear in t.

Denote the unique solution for t for a given s by t_s — each root $s \neq 0$ of the resultant therefore corresponds to two eigenvalue points $(s, t_s, \pm 1, 1)$ in $\mathfrak{F}^i \subset \mathfrak{E}_0(W)$, so that we have

$$\mathfrak{F}^i = \{(s, t_s, \pm 1, 1) \mid \operatorname{res}_{p,q}(s) = 0\}. \tag{29}$$

As in the reducible case, not all of these points correspond to representations which will factor through $\pi_1(W_{p/q})$ to give p-reps in $\mathfrak{R}(W_{p/q})$. This is due to the fact that the eigenvalue condition $s^p t^q = 1$ does not necessarily imply that $\rho(\mu_0)^p \rho(\lambda_0)^q = I$. In fact, we have:

PROPOSITION 8. *The roots $s \notin \{0, 1, -1\}$ of $\operatorname{res}_{p,q}$ correspond to irreducible p-reps of $\pi_1(W_{p/q})$; the roots $s = \pm 1$ (when they arise) correspond to discrete faithful representations of $\pi_1(W)$ which therefore do not factor through $\pi_1(W_{p/q})$.*

PROOF. For $s = \pm 1$, a direct calculation shows that an irreducible representation with $s^2 = u^2 = 1$ is conjugate to the form

$$\rho(\mu_0) = \begin{pmatrix} s & -su \pm i \\ 0 & s \end{pmatrix}, \ \rho(\mu_1) = \begin{pmatrix} u & 0 \\ 1 & u \end{pmatrix}, \quad \text{and } \rho(\lambda_0) = \begin{pmatrix} -1 & -4u \\ 0 & -1 \end{pmatrix}, \tag{30}$$

where $s, u \in \{\pm 1\}$. (These are in fact, up to conjugation, all discrete faithful representations into $\mathrm{SL}_2(\mathbb{C})$ corresponding to the unique complete hyperbolic structure of the unfilled Whitehead link complement.) For each choice of $s = \pm 1$, there are four such representations. As the subgroup generated by the parabolic elements $\rho(\mu_0), \rho(\lambda_0) \in \mathrm{SL}_2(\mathbb{C})$ is therefore isomorphic to $\mathbb{Z}^2$, the equation $\rho(\mu_0)^p \rho(\lambda_0)^q = I$ can only be satisfied by $p = q = 0$ (which is not a valid filling slope).

For $s \neq \pm 1$, first note that the inverse map (14) of the eigenvalue map is well defined (since $u^2 = 1$). The two points $(s, t_s, \pm 1, -1)$ associated to a root s therefore correspond to two partial p-reps in $\mathfrak{C}(W)$. To see that these indeed factor through $\pi_1(W_{p/q})$, note that for $s \neq \pm 1$, both $\rho(\mu_0)$ and $\rho(\lambda_0)$ are simultaneously diagonizable, since they are not parabolic and commute. The condition $s^p t^q = 1$ then immediately implies $\rho(\mu_0)^p \rho(\lambda_0)^q = I$. □

This concludes the proof of Theorem 3. For the following application to the minimal total Culler-Shalen seminorm in section 4, we now need to study the number of conjugacy classes of p-reps.

3.5. Roots of the resultant. We summarize some key facts about the roots of the resultant $\operatorname{res}_{p,q}$:

LEMMA 9 (Trivial roots). *If q is even (and hence p odd), $s = 1$ is a root of order 2 of $\operatorname{res}_{p,q}$. If both q and p are odd, $s = -1$ is a root of order 2 of $\operatorname{res}_{p,q}$. In the remaining case (q odd and p even), neither of ± 1 is a root of $\operatorname{res}_{p,q}$.*

PROOF. This can be directly verified from the equations $k_1 = k_2 = 0$ (cf. (26) and (27)): If $s = \pm 1$, equation $k_2 = 0$ necessarily implies $t_s = -1$. Using $k_1 = 0$, we then immediately see that $(1, -1)$ is a solution iff q is even, and that $(-1, -1)$ is a solution iff $p + q$ is even (which — under the assumption that p and q are coprime

— is equivalent to both p and q being odd). To show that the order is equal to 2, note that

$$\mathrm{res}'_{p,q} := \frac{d}{ds}\,\mathrm{res}_{p,q} = (p-2q)\left(s^{p-2q-1} - s^{-p+2q-1}\right) + (-1)^{q+1} 2q U_{q-1}(y) y'$$

and

$$\mathrm{res}''_{p,q} = (p-2q)\left((p-2q-1)s^{p-2q-2} + (-p+2q-1)s^{-p+2q-2}\right) \\ +(-1)^{q+1}2q\left(y'' U_{q-1}(y) + (y')^2 \tfrac{\mathrm{d}}{\mathrm{d}y} U_{q-1}(y)\right),$$

where $U(y)$ is the Chebyshev polynomial of second type, satisfying $U_0(y) = 1$, $U_1(y) = 2y$ and $U_{k+1}(y) = 2yU_k(y) - U_{k-1}(y)$ for $k \geq 1$ (cf. [**Sze75**]). Since $y'(\pm 1) = 0$, we immediately get $\mathrm{res}'_{p,q}(\pm 1) = 0$. Furthermore, $\mathrm{res}''_{p,q}(\pm 1) \neq 0$ unless $p/q = 0/1$ or $p/q = 4/1$, but in those cases $\mathrm{res}_{p,q}$ degenerates and has no non-zero roots. □

LEMMA 10 (Symmetries).

(1) $\mathrm{res}_{p,q}(s) = 0 \iff \mathrm{res}_{p,q}(s^{-1}) = 0.$
(2) $\mathrm{res}_{p,q}(s) = 0 \iff \mathrm{res}_{p,q}(\bar{s}) = 0.$
(3) $(\mathrm{res}_{p,q}(s) = 0 \iff \mathrm{res}_{p,q}(-s) = 0)$ *iff p is even.*
(4) $\mathrm{res}_{p,q}(s) \doteq \mathrm{res}_{-p+4q,q}(s).$

PROOF. For (1), observe that $y(s^{-1}) = y(s)$. For (2), note that $\mathrm{res}_{p,q}$ is a Laurent polynomial with real (in fact integer) coefficients. For (3), first note that $y(-s) = y(s)$ and hence

$$\mathrm{res}_{p,q}(-s) - \mathrm{res}_{p,q}(s) = ((-1)^p - 1)\left(s^{p-2q} + s^{-p+2q}\right).$$

This is clearly 0 when p is even. Conversely, assume this difference to be 0 but p to be odd. Then $s^{2(p-2q)} = -1$, which implies $s = e^{i\phi}$ for some $0 \leq \phi < 2\pi$. But then $\mathrm{res}_{p,q}(s) = (-1)^{q+1} 2T_q(y(s)) = (-1)^{q+1}2T_q(2 - \cos(2\phi)) \neq 0$, since the roots of $T_q(y)$ lie in the interval $(0, 1)$ (cf. [**Sze75**]). The last statement is obvious from (28). □

LEMMA 11 (Real roots). *When $p > 4q > 0$ or $p < 0$, the only possible real-valued roots of* $\mathrm{res}_{p,q}$ *are* ± 1. *When $0 < p < 4q$ and p is odd, there are two additional real-valued, positive roots. When $0 < p < 4q$ and p is even (and hence q odd), there are precisely four real-valued roots in total, none of which is* ± 1.

PROOF. Assume $\mathrm{res}_{p,q}(s) = 0$ with real-valued $s > 0$. Then

$$s^{p-2q} + s^{-(p-2q)} = (-1)^q 2T_q(y(s)).$$

Denote the left-hand side of this equation by $\phi(s)$, the right-hand side by $\psi(s)$. We will study intersections of these two real-valued functions. Note that $\phi(s)$ is strictly monotonously decreasing for $0 < s < 1$, has a global minimum of value 2 at $s = 1$, and is strictly monotonously increasing for $s > 1$. Furthermore, note that the range of $y(s)$ is $(-\infty, 1]$, with global maximum at $s = 1$. This implies that the range of $\psi(s) = (-1)^q 2T_q(y(s))$ is $(-\infty, 2)$, using the fact that $T_q(y)$ is a polynomial of degree q in y, whose zeros and relative extrema of value ± 1 are all contained in the interval $[-1, 1]$. More specifically, ψ oscillates with values between -2 and 2 for $s \in [1 - \sqrt{2}, 1 + \sqrt{2}]$ (with $\psi(1) = (-1)^q 2$) and tends strictly monotonously to $+\infty$ for $s \to 0$ and $s \to \infty$.

Apart from the possibility of an intersection at $s = 1$ (in the case that q is even), other intersections between ϕ and ψ can therefore only occur outside the interval $(1-\sqrt{2}, 1+\sqrt{2})$ when the maximal degree in s of ψ is larger than that of ϕ. But $\operatorname{maxdeg}_s \phi = |p-2q| = -\operatorname{mindeg}_s \phi$ and $\operatorname{maxdeg}_s \psi = 2q = -\operatorname{mindeg}_s \psi$. It is then easy to see that for $p > 4q > 0$ or $p < 0$, $\operatorname{maxdeg}_s \psi < \operatorname{maxdeg}_s \phi$ and there are no such intersections. The same argument then shows that for $0 < p < 4q$, there are precisely two positive real-valued roots different from 1 (which are inverses of each other, by Lemma 10 (1)).

To analyze the remaining case $s < 0$, note that $\phi(-s) = (-1)^p\phi(s)$ while $\psi(-s) = \psi(s)$. If p is odd, this means that there cannot be any intersection apart from possibly $s = -1$ (which arises if q is odd), as the two functions diverge in opposite directions. If p is even (and hence q odd), we get as roots the negatives of the roots found for $s > 0$. In total, there are either 0 or 4 real roots in this case, as $\operatorname{res}_{p,q}(\pm 1) \neq 0$. □

Note that this includes the case $p/q = 2/1$, when there are only four non-zero roots in total, all of which are real and distinct from ± 1.

LEMMA 12 (Imaginary roots). *When $p > 4q > 0$ or $p < 0$, $\operatorname{res}_{p,q}$ has imaginary roots iff $p \in 4\mathbb{Z}$. In that case, there are four such roots. When $0 < p < 4q$, there are no imaginary roots.*

PROOF. Similar to the proof of Lemma 11. Assume $\operatorname{res}_{p,q}(ix) = 0$ with $x > 0$. Then

$$i^{-p+2q}\left((-1)^p x^{p-2q} + x^{-p+2q}\right) = (-1)^q 2T_q\left(\tfrac{x^2+4+x^{-2}}{2}\right).$$

Now note that the right-hand side is always real-valued, while the same for the left-hand side is only true when p is even. In that case, the equation is equivalent to

$$\phi(x) := (-1)^{p/2}\left(x^{p-2q} + x^{-p+2q}\right) = 2T_q\left(\tfrac{x^2+4+x^{-2}}{2}\right) =: \psi(x).$$

As $\psi(x) > 0$ for all $x \in \mathbb{R}$, this equation can only be satisfied if $p/2$ is even, i.e. if $p \in 4\mathbb{Z}$. Under this assumption, comparison of the maximal degrees then shows that there is no intersection of ϕ and ψ when $0 < p < 4q$, and precisely two intersections (which are inverses of each other) when $p > 4q > 0$ or $p < 0$. The symmetry with respect to complex conjugation then shows that there are in total four pure imaginary roots in this case, and none when $p \notin 4\mathbb{Z}$. □

LEMMA 13. *None of the roots $s \notin \{\pm 1\}$ of $\operatorname{res}_{p,q}$ lies on the unit circle.*

PROOF. Simply note that for $\phi \in [0, 2\pi)$,

$$\operatorname{res}_{p,q}(e^{i\phi}) = 2(\underbrace{\cos(p-2q)\phi}_{\in[-1,1]} + (-1)^{q+1}\underbrace{T_q(2-\cos 2\phi)}_{\geq 1}) = 0$$

implies $\phi = 0$ or $\phi = \pi$. □

Examples of the typical distribution of the roots are shown in figures 2(a), 2(b) and 2(c) on page 182.

We will now determine the number of non-zero roots of the resultant. The number of non-zero roots (with multiplicities) of a Laurent polynomial is given by its span, i.e. the difference of the maximal and minimal degrees. From (28), we see that unless $p/q \in \{0, 4\}$, the span of $\operatorname{res}_{p,q}$ is $2\max(|p-2q|, 2q)$ (recall that we always assume $q > 0$). When $p/q \in \{0, 4\}$, $\operatorname{res}_{p,q} \doteq 1$, and there are therefore no

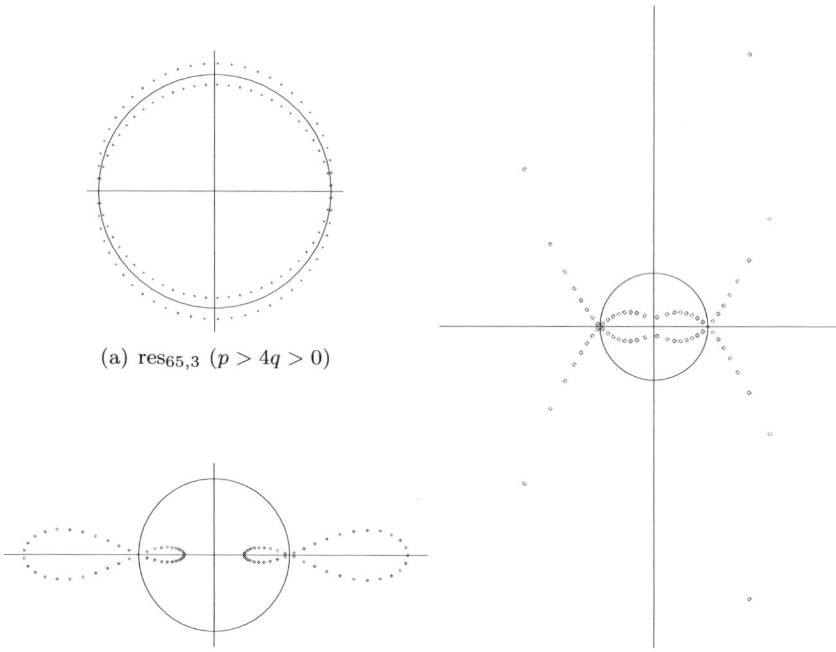

(a) $\mathrm{res}_{65,3}$ $(p > 4q > 0)$

(b) $\mathrm{res}_{65,23}$ $(0 < p < 4q)$

(c) $\mathrm{res}_{65,16}$ $(p - 4q > 0$, but small)

FIGURE 2. Examples of the distribution of the non-zero roots of $\mathrm{res}_{p,q}$

irreducible p-preps — matching the fact that W_0 and W_4 are toroidal manifolds which do not admit a complete hyperbolic structure.

Discarding 0, 1 and -1 from the roots of $\mathrm{res}_{p,q}$, we therefore obtain an upper bound for the number of *distinct* non-trivial roots as given in table 4. We remark that it is a hard problem to analytically show that all non-trivial roots are in fact simple — but the proof of Theorem 1 will in fact imply this for odd p (cf. Corollary 4).

$p/q \in$	$(-\infty, 0)$	$(0, 4)$	$(4, \infty)$
p odd	$2\lvert p\rvert + 4\lvert q\rvert - 2$	$4\lvert q\rvert - 2$	$2\lvert p\rvert - 4\lvert q\rvert - 2$
p even	$2\lvert p\rvert + 4\lvert q\rvert$	$4\lvert q\rvert$	$2\lvert p\rvert - 4\lvert q\rvert$

TABLE 4. Upper bound for the number of distinct roots $s \notin \{0, \pm 1\}$ of $\mathrm{res}_{p,q}$

3.6. The number of conjugacy classes of p-reps. We will now put together the results of the previous sections to obtain the count of conjugacy classes of p-reps for the manifold $W_{p/q}$.

We have determined the two components of the set $\mathfrak{P}''$ of eigenvalue tuples corresponding to p-reps as

$$(\mathfrak{P}'')^r = \{(s, \pm 1, 1, 1) \mid s^p = 1,\ s \neq \pm 1\}, \tag{31}$$

$$(\mathfrak{P}'')^i = \{(s, \pm 1, t_s, -1) \mid \mathrm{res}_{p,q}(s) = 0,\ s \neq \pm 1\}, \tag{32}$$

with their cardinalities given in equation (25) and table 4, respectively.

Now recall that by construction, the cardinality of $\mathfrak{P}''$ is equal to the cardinality of $\mathfrak{P}'$ and that of $\mathfrak{P}$. Furthermore, we saw that each conjugacy class of p-reps in $\mathfrak{C}(W_{p/q})$ is represented by *two* elements in $\mathfrak{C}(W_{p/q})$ (corresponding to the pair of eigenvalue points (s,t,u,v) and (s^{-1},t^{-1},u,v)). To obtain the number of conjugacy classes of p-reps, we therefore have to divide the cardinality of $\mathfrak{P}$ by 2. Since the two representatives for a conjugacy class in $\mathfrak{C}(W_{p/q})$ clearly have the same character independent of whether they are reducible or irreducible, we see that this number also coincides with the *number of characters* of p-reps.

This then shows:

PROPOSITION 14. *The number of distinct conjugacy classes of* $\mathrm{SL}_2(\mathbb{C})$ *p-reps of* $\pi_1(W_{p/q})$ *is bounded above by the numbers shown in table* 5*:*

$p/q \in$	$(-\infty,0)$	$(0,4)$	$(4,\infty)$
p odd	$3\lvert p\rvert+4\lvert q\rvert-3$	$\lvert p\rvert+4\lvert q\rvert-3$	$3\lvert p\rvert-4\lvert q\rvert-3$
p even	$3\lvert p\rvert+4\lvert q\rvert-2$	$\lvert p\rvert+4\lvert q\rvert-2$	$3\lvert p\rvert-4\lvert q\rvert-2$

TABLE 5. Upper bound for the number of conjugacy classes of $\mathrm{SL}_2(\mathbb{C})$ p-reps of $\pi_1(W_{p/q})$

4. The minimal total Culler-Shalen seminorm and p-reps

The goal of this section is to show:

THEOREM 15. *The minimal total Culler-Shalen seminorm of the manifold* $W_{p/q}$ *is equal to the number of characters of p-reps of* $\pi_1(W_{p/q})$*. In particular, it is less than or equal to the bounds given in Table* 5*.*

Since Dehn filling along the meridian μ_1 of $W_{p/q}$ gives a closed manifold with cyclic fundamental group, while μ_1 $(=\frac{1}{0}=\infty)$ is not a boundary slope, Corollary 1.1.4 of [**CGLS87**] implies that its Culler-Shalen seminorm is in fact equal to the minimal total Culler-Shalen seminorm s:

$$s = \|\mu_1\| = \|\infty\|\,. \tag{33}$$

We can therefore find s by explicitly determining $\|\mu_1\|$.

Recall that $\|\mu_1\| = \deg(f_{\mu_1} : \widetilde{\mathfrak{X}}^{\mathrm{irr}}(W_{p/q}) \to \mathbb{CP}^1)$. We will determine this degree as a count of the roots of f_{μ_1}. First note that we can in fact restrict our attention to $\mathfrak{X}(W_{p/q})$, since f_{μ_1} has poles at each ideal point in $\widetilde{\mathfrak{X}}(W_{p/q})$ (see e.g. the remarks after Corollary 9.2.7 in [**Sha02**]).

Since $f_{\mu_1}=0$ is equivalent to $\chi_\rho(\mu_1) = \operatorname{trace}\rho(\mu_1) = \pm 2$, we immediately see that zeroes of this function are characters of representations $\rho \in \mathfrak{R}(W_{p/q})$ which send the meridian $\mu_1 \in \pi_1(W_{p/q})$ to $\pm I$ or a parabolic element of $\mathrm{SL}_2(\mathbb{C})$. Since we only regard non-trivial curves in $\mathfrak{X}(W_{p/q})$, the representations in question have to be non-abelian, and are therefore *p-reps.*

Theorem 15 will therefore be a direct consequence of the following two propositions, which are the subject of the bulk of this section:

PROPOSITION 16. *Each character of a p-rep of* $\pi_1(W_{p/q})$ *is a smooth point of a unique non-trivial curve in* $\mathfrak{X}^{irr}(W_{p/q})$.

PROPOSITION 17. *Each character of an irreducible p-rep of $\pi_1(W_{p/q})$ is a simple zero of f_{μ_1}. If $p/q \neq 0$, also each character of a non-abelian reducible p-rep of $\pi_q(W_{p/q})$ is a simple zero of f_{μ_1}.*

We will begin in section 4.1 by using work of Heusener and Porti (cf. [**HP05**]) to show Proposition 16 for a *general non-abelian reducible* representation (therefore including the case of reducible p-reps). We remark that this is a subtle question, as it asks when the character of a non-abelian reducible representation can be deformed into the character of an irreducible representation. This result will also be needed in section 5.

To show Proposition 16 for an *irreducible* p-rep, we then exhibit a useful partially-diagonal normal form for p-reps in section 4.2, and use this in section 4.3 to determine an explicit presentation of the cohomology of $\pi_1(W_{p/q})$ with coefficients in $\mathfrak{sl}_2^\rho$ (the Lie algebra $\mathfrak{sl}_2(\mathbb{C})$ viewed as a $\pi_1(W_{p/q})$ module via the map $\mathrm{Ad} \circ \rho$ for a given p-rep ρ). The knowledge of the dimension of this cohomology group then implies our claim using Theorem A of [**Boy02**].

Finally, Proposition 17 will be shown in section 4.4 by studying the degree of the trace function f_{μ_1} in relation to deformations of representations.

4.1. Smoothness of the characters of general non-abelian reducible representations. We will show:

PROPOSITION 18. *Let $\overline{\rho} \in \overline{\mathfrak{R}}(W_{p/q})$ be a non-abelian reducible* $\mathrm{PSL}_2(\mathbb{C})$ *representation. For $p/q \neq 4$, the character $\chi_{\overline{\rho}}$ of $\overline{\rho}$ is contained in precisely two irreducible components of $\overline{\mathfrak{X}}(W_{p/q})$, both of which are curves, one trivial, the other non-trivial. In addition, $\chi_{\overline{\rho}}$ is a smooth point of both curves and the intersection at $\chi_{\overline{\rho}}$ is transverse. When $p/q = 4$, the above is true for all non-abelian reducible representations $\overline{\rho}$ with* $\mathrm{trace}\,\overline{\rho}(\mu_0) \neq \pm 2$.

Since every non-abelian reducible $\mathrm{SL}_2(\mathbb{C})$ representation $\rho \in \mathfrak{R}(W_{p/q})$ induces a non-abelian reducible $\mathrm{PSL}_2(\mathbb{C})$ representation $\overline{\rho} \in \overline{\mathfrak{R}}(W_{p/q})$, this then implies the following stronger version of Proposition 16 in the case of non-abelian reducible representations:

COROLLARY 19. *Every non-abelian reducible* $\mathrm{SL}_2(\mathbb{C})$ *character $\chi_\rho \in \mathfrak{X}(W_{p/q})$ is a smooth point on a unique non-trivial curve $\mathfrak{X}_0 \subset \mathfrak{X}(W_{p/q})$.*

PROOF. Let $\rho \in \mathfrak{R}(W_{p/q})$ be a non-abelian reducible representation, and let $\overline{\rho} \in \overline{\mathfrak{R}}(W_{p/q})$ be its image under the natural projection $\pi \colon \mathfrak{R}(W_{p/q}) \to \overline{\mathfrak{R}}(W_{p/q})$. Proposition 18 now shows that $\overline{\rho}$ lies as a smooth point on a unique non-trivial curve $\overline{\mathfrak{X}}_0 \subset \overline{\mathfrak{X}}(W_{p/q})$, and we only need to show that this curve lifts to a non-trivial curve $\mathfrak{X}_0 \subset \mathfrak{X}(W_{p/q})$. But this is immediately clear, since $\overline{\rho}$ lifts to ρ by construction, and π is a regular covering map onto its image (cf. [**BZ98**, Section 3]). □

We start by determining all possible non-abelian reducible representations of $\pi_1(W_{p/q})$ up to conjugation:

LEMMA 20. *Let $\overline{\rho}$ be a non-abelian reducible representation in $\overline{\mathfrak{R}}(W_{p/q})$. Then up to conjugation, $\overline{\rho}$ has the form*

$$\overline{\rho}(\mu_0) = \pm \begin{pmatrix} s & 0 \\ c & s^{-1} \end{pmatrix} \qquad \text{and} \qquad \overline{\rho}(\mu_1) = \pm \begin{pmatrix} u & 0 \\ d & u^{-1} \end{pmatrix},$$

where either

(1) $s^2 = 1$, $c \neq 0$ *and* $u^2 = \frac{1}{2q}(-p + 2q \pm \sqrt{p(p-4q)})$ *(where* $p/q = 0$ *cannot arise), or*
(2) $u^2 = 1$, $d \neq 0$ *and* $s^p = \pm 1$, *but* $s^2 \neq 1$.

PROOF. Let $\overline{\rho} \in \overline{\mathfrak{R}}(W_{p/q})$ be non-abelian reducible. Since it is reducible, we can conjugate it into lower triangular form as in the claim. The first group relation of $\pi_1(W_{p/q})$ (cf. equation (6)) now implies that either $du(1-s^2) + cs(u^2-1) = 0$ (in which case $\overline{\rho}$ would be abelian, a contradiction), or one of $s^2 = 1$ or $u^2 = 1$ — but not both (otherwise $\overline{\rho}$ would again be abelian).

If $s^2 = 1$, we hence have $u^2 \neq 1$ and $c \neq 0$. We can now conjugate by diagonal matrices to let $c = 1$. Then $\overline{\rho}(\lambda_0) = \pm \begin{pmatrix} 1 & 0 \\ s^{-1}u^{-2}(u^2-1)^2 & 1 \end{pmatrix}$, and we get

$$\pm I = \overline{\rho}(\mu_0)^p \overline{\rho}(\lambda_0)^q = \pm \begin{pmatrix} s^p & 0 \\ s^{p-1}p & s^p \end{pmatrix} \begin{pmatrix} 1 & 0 \\ qs^{-1}u^{-2}(u^2-1)^2 & 1 \end{pmatrix}$$
$$= \pm \begin{pmatrix} s^p & 0 \\ ps^{p-1} + s^p qs^{-1}u^{-2}(u^2-1)^2 & s^p \end{pmatrix}.$$

Since $s^2 = 1$, the lower left entry implies $qu^4 + u^2(p-2q) + q = 0$, and therefore $u^2 = \frac{1}{2q}(-p+2q \pm \sqrt{p(p-4q)})$ as claimed. Note that these roots are distinct unless $p/q = 0$ or $p/q = 4$. When $p/q = 0$, this implies that $u^2 = 1$, a contradiction since $\overline{\rho}$ is non-abelian. When $p/q = 4$, we get $u^2 = -1$, which yields a valid non-abelian reducible representation.

In the remaining case, when $s^2 \neq 1$, we have $u^2 = 1$ and can assume $d = 1$. Then $\overline{\rho}(\lambda_0) = \pm I$ and we immediately get $\pm I = \overline{\rho}(\mu_0)^p \overline{\rho}(\lambda_0)^q = \overline{\rho}(\mu_0)^p$ which implies $s^p = \pm 1$. □

Following [**HP05**], we now determine the twisted Alexander polynomial associated to the diagonal representation $\overline{\rho}_\alpha$ obtained from a non-abelian reducible representation $\overline{\rho} \in \overline{\mathfrak{R}}(W_{p/q})$ as in Lemma 20 by setting all non-diagonal matrix entries to be zero. In particular, this representation has the same character as $\overline{\rho}$, i.e. $\chi_{\overline{\rho}_\alpha} = \chi_{\overline{\rho}}$.

LEMMA 21. *Let* $\overline{\rho}$ *be a non-abelian reducible representation in* $\overline{\mathfrak{R}}(W_{p/q})$, *and let* $\overline{\rho}_\alpha$ *be the diagonal representation with the same character as* $\overline{\rho}$ *as described above. The twisted Alexander polynomial associated to* ρ_α *is given by:*

$$\Delta^{\Phi_\sigma}(t) \doteq \begin{cases} t-1 & \text{if } s^2 \neq 1, \\ qt^2 + (p-2q)t + q & \text{if } s^2 = 1. \end{cases}$$

PROOF. We follow the notation of [**HP05**] to facilitate comparison for the reader. First note that $H_1(W_{p/q};\mathbb{Z}) \cong \mathbb{Z}/_p \oplus \mathbb{Z} = \langle \mu_0 \rangle \oplus \langle \mu_1 \rangle$, where we use the same symbols for the generators of $\pi_1(W_{p/q})$ and their images under abelianization. We now define the following homomorphisms:

$$\alpha\colon \pi_1(W_{p/q}) \to H_1(W_{p/q};\mathbb{Z}) \to \mathbb{C}^*, \qquad \mu_0 \mapsto s^2, \quad \mu_1 \mapsto u^2,$$
$$\phi\colon \pi_1(W_{p/q}) \to \mathbb{Z} = \langle t \rangle, \qquad \mu_0 \mapsto 1, \quad \mu_1 \mapsto t,$$
$$\sigma\colon \pi_1(W_{p/q}) \to \operatorname{tors} H_1(W_{p/q};\mathbb{Z}) \xrightarrow{\alpha} \mathbb{C}^*, \qquad \mu_0 \mapsto s^2, \quad \mu_1 \mapsto 1,$$

where s and u are the upper left entries of $\overline{\rho}(\mu_0)$ and $\overline{\rho}(\mu_1)$, respectively. Using these maps, we can now define a map $\Phi_\sigma\colon \pi_1(W_{p/q}) \to \mathbb{C}[t^{\pm}]^*$ to the units of the ring of Laurent polynomials $\mathbb{C}[t^{\pm 1}]$ by $\Phi_\sigma(\gamma) := \sigma(\gamma)t^{\phi(\gamma)}$ for $\gamma \in \pi_1(W_{p/q})$. This map can be linearly extended to the group ring $\mathbb{Z}\pi_1(W_{p/q})$ which allows us to define the twisted Alexander polynomial $\Delta^{\Phi_\sigma}(t) \in \mathbb{C}[t^{\pm 1}]$ (cf. [**HP05**, Section 2] or e.g. [**Wad94**] for more background). Since we have a presentation with two generators μ_0 and μ_1 and two relators

$$r_1 = \mu_0\mu_1\mu_0^{-1}\mu_1^{-1}\mu_0^{-1}\mu_1\mu_0\mu_1\mu_0^{-1}\mu_1^{-1}\mu_0\mu_1\mu_0\mu_1^{-1}\mu_0^{-1}\mu_1^{-1}$$

and $r_2 = \mu_0^p\lambda_0^q$, the twisted Alexander polynomial can be obtained as the greatest common divisor of the entries of the twisted Fox Jacobian J^{Φ_σ}:

$$\begin{aligned} J^{\Phi_\sigma} &= \begin{pmatrix} \dfrac{\partial r_1}{\partial \mu_0} & \dfrac{\partial r_1}{\partial \mu_1} \\ \dfrac{\partial r_2}{\partial \mu_0} & \dfrac{\partial r_2}{\partial \mu_1} \end{pmatrix}^{\Phi_\sigma} \\ &= \begin{pmatrix} s^{-2}t^{-1}(t-1)^2(s^2-1) & s^{-2}t^{-1}(1-t)(s^2-1)^2 \\ \sum_{k=0}^{p-1} s^{2k} + s^{2p}qt^{-1}(t-1)^2 & s^{2p}qt^{-1}(1-t)(s^2-1) \end{pmatrix}. \end{aligned}$$

This matrix can be obtained by hand by first determining the Fox derivatives of the group relators (cf. [**Fox53**, Ch. 9B] or [**BZ03**] for background on Fox free differential calculus), and then applying the map Φ_σ to the obtained elements of the group ring $\mathbb{Z}\pi_1(W_{p/q})$.

We now have two cases: If $s^2 = 1$, we get

$$J^{\Phi_\sigma} = \begin{pmatrix} 0 & 0 \\ p + qt^{-1}(t-1)^2 & 0 \end{pmatrix},$$

and thus that $\Delta^{\Phi_\sigma}(t) = p + qt^{-1}(t-1)^2 \doteq qt^2 + (p-2q)t + q$ (where $\doteq$ denotes equivalence up to multiplication by units of $\mathbb{C}[t^{\pm 1}]$). Note that this is in fact the *untwisted* Alexander polynomial of $W_{p/q}$, as $s^2 = 1$ implies that σ is the trivial 1-dimensional representation (cf. also e.g. [**HMW92**], taking into account that the authors work with the *left-handed* Whitehead link).

If on the other hand $s^2 \neq 1$, we have $s^{2p} = 1$ and $\sum_{k=0}^{p-1} s^{2k} = 0$ which yields

$$J^{\Phi_\sigma} = \begin{pmatrix} s^{-2}t^{-1}(t-1)^2(s^2-1) & s^{-2}t^{-1}(1-t)(s^2-1)^2 \\ qt^{-1}(t-1)^2 & qt^{-1}(1-t)(s^2-1) \end{pmatrix}.$$

As a result, $\Delta^{\Phi_\sigma}(t) = t^{-1}(t-1) \doteq t-1$ in this case. □

We can now proceed to prove the main result:

Proof of Proposition 18. Let $\overline{\rho} \in \overline{\mathfrak{R}}(W_{p/q})$ be non-abelian reducible. Lemma 20 shows that $s^2 \neq 1$ implies $u^2 = 1$ (i.e. u^2 is the root of the polynomial $t-1$), and $s^2 = 1$ implies u^2 is a root of $qt^2 + (p-2q)t + q$. But these are precisely the twisted Alexander polynomials found in Lemma 21 for these cases, and so we have that u^2 is a simple root of the twisted Alexander polynomial, unless $s^2 = 1$ and $p/q = 4$ (in which case $u^2 = -1$ is a double root). Apart from this exception, Theorems 1.2 and 1.3 of [**HP05**] (with $\delta = \mu_1$) then show that the character of the diagonal representation $\overline{\rho}_\alpha$ (which is equal to the character of $\overline{\rho}$) satisfies the claim of the Proposition. □

4.2. A partially diagonal parametrization of p-reps. We determine a parametrization of p-reps similar to the normal form (9) from section 3.1, but which is better suited to the following cohomology calculations.

Let $\rho \in \mathfrak{R}(W_{p/q}) \subset \mathfrak{R}(W)$ be a p-rep. We therefore can assume that $\rho(\mu_0)$ is not parabolic, and that furthermore $\operatorname{trace}\rho(\mu_1) = \pm 2$.

For clarity of exposition, we will in the following restrict ourselves to the case $\operatorname{trace}\rho(\mu_1) = +2$*, and we attach the symbol "+" to all related sets and expressions to indicate this choice of sign. The negative trace case proceeds completely analogously and will be omitted here.*

As a consequence, ρ can be conjugated into the form

$$\rho(\mu_0) = \begin{pmatrix} s & 0 \\ 0 & s^{-1} \end{pmatrix} \quad \text{and} \quad \rho(\mu_1) = \begin{pmatrix} a & -(a-1)^2 \\ 1 & 2-a \end{pmatrix}, \tag{34}$$

where $s \in \mathbb{C}^*$, $a \in \mathbb{C}$, subject to the condition (obtained from applying (34) to the first group relation)

$$r_1^+ = (a-1)\left[-(s^2-1)^2a^2 + (s^2-1)(s^2-3)a - 2\right] = 0. \tag{35}$$

Note that this implies $a \neq 0$ and $a \neq 2$ (otherwise $r_1^+ = 2s^4 \neq 0$).

We will in the following use this condition to simplify all arising polynomials modulo r_1^+, i.e. we will replace each polynomial f by its canonical form f_0 modulo the ideal generated by r_1^+, in notation $f \equiv_{r_1^+} f_0$ (see e.g. [**CLO97**]). We also extend this to rational functions by setting $f/g \equiv_{r_1^+} f_0/g$ if $f \equiv_{r_1^+} f_0$.

Using this parametrization, we then obtain for the longitude that $\rho(\lambda_0) = \left(\begin{smallmatrix} t & 0 \\ 0 & t^{-1} \end{smallmatrix}\right)$ with $t \equiv_{r_1} a/(2-a)$, which implies that ρ is diagonal on the whole peripheral subgroup of the first boundary torus. This fact will be of great importance in section 4.3.

Applying (34) to the filling relation, $\rho(\mu_0^p\lambda_0^q) = I$, now yields

$$r_2^+ \equiv_{r_1^+} s^p\left(\tfrac{a}{2-a}\right)^q - 1 = 0. \tag{36}$$

Note that $t \neq 0$ implies $a \neq 0$ and $a \neq 2$.

Denote by $\mathfrak{D}^+(W_{p/q}) \subset \mathfrak{R}(W_{p/q})$ the subset of representations in partially-diagonal normal form (34) satisfying $r_1^+ = r_2^+ = 0$.

Now note that $\rho \in \mathfrak{D}^+(W_{p/q})$ is reducible iff $a = 1$. In this case the polynomial r_1^+ is identically zero, and poses therefore no further condition on s and a. Then $\rho \in \mathfrak{C}(W_{p/q}) \cap \mathfrak{D}^+(W_{p/q})$, and $a = u = 1$ is equal to the eigenvalue of $\rho(\mu_1)$. The condition $r_2^+ = 0$ then implies $s^p = 1$, just as we found in section 3.

If $\rho \in \mathfrak{D}^+(W_{p/q})$ is irreducible, we have $a \neq 1$, and we can further simplify r_1^+ to:

$$r_1^+ = -(s^2-1)^2a^2 + (s^2-1)(s^2-3)a - 2. \tag{37}$$

For simplicity, we will denote this polynomial again by r_1^+, since there will be no danger of confusion. Now note that the system $r_1^+ = r_2^+ = 0$ (for the coefficients s and a) is completely equivalent to the system $k_1 = k_2 = 0$ (for eigenvalues s and t) for p-reps (with positive trace on $\rho(\mu_1)$, i.e. $u = +1$) in section 3, where we showed that its solutions are characterized by the roots of the polynomial $\operatorname{res}_{p,q}$. Instead of explicitly solving this system in the new parametrization, we will therefore use the properties of $\operatorname{res}_{p,q}$ where appropriate (in particular in section 4.4).

4.3. The twisted cohomology and smoothness at p-reps. This section is dedicated to proving Proposition 16 for irreducible p-reps by determining an explicit presentation for the cohomology group $H^1 := H^1(W_{p/q}; \mathfrak{sl}_2^\rho) = H^1(\pi_1(W_{p/q}); \mathfrak{sl}_2^\rho)$. Good background references are [**Bro82**] and [**Por97**], we also recommend the articles [**HPSP01**] and [**HP05**] for a good summary in our context.

4.3.1. *Preliminaries.* Consider the Lie group $\mathrm{SL}_2(\mathbb{C})$ and its Lie algebra $\mathfrak{sl}_2(\mathbb{C})$, and denote the adjoint homomorphism by $\mathrm{Ad}\colon \mathrm{SL}_2(\mathbb{C}) \to \mathrm{Aut}(\mathfrak{sl}_2(\mathbb{C}))$ acting by conjugation, i.e. for $A \in \mathrm{SL}_2(\mathbb{C})$ and $U \in \mathfrak{sl}_2(\mathbb{C})$ we have $\mathrm{Ad}(A)(U) = AUA^{-1}$.

Let Γ be a finitely presented group, and let $\rho \in \mathfrak{R}(\Gamma) = \mathrm{Hom}(\Gamma; \mathrm{SL}_2(\mathbb{C}))$. The Lie algebra $\mathfrak{sl}_2(\mathbb{C})$ can then be viewed as a (left) Γ module via the map $\mathrm{Ad} \circ \rho$, i.e. for $\gamma \in \Gamma$ and $A \in \mathrm{SL}_2(\mathbb{C})$ we have the left action $\gamma \cdot A := \mathrm{Ad}(\rho(\gamma))(A)$. Denote $\mathfrak{sl}_2(\mathbb{C})$ with this left Γ module structure by $\mathfrak{sl}_2^\rho$.

Denote by $C^*(\Gamma; \mathfrak{sl}_2^\rho)$ the space of cochains with coefficients in $\mathfrak{sl}_2^\rho$. Similarly, let $B^*(\Gamma; \mathfrak{sl}_2^\rho)$ and $Z^*(\Gamma; \mathfrak{sl}_2^\rho)$ denote the coboundaries and cocycles, respectively, and $H^*(\Gamma; \mathfrak{sl}_2^\rho)$ the cohomology group with coefficients in $\mathfrak{sl}_2^\rho$.

Note that a cocycle $d \in Z^1(\Gamma; \mathfrak{sl}_2^\rho)$ is a map $d\colon \Gamma \to \mathfrak{sl}_2^\rho$ satisfying

$$d(\gamma_1\gamma_2) = d(\gamma_1) + \gamma_1 \cdot d(\gamma_2) \qquad \text{for all } \gamma_1, \gamma_2 \in \Gamma, \tag{38}$$

using the short notation for the $\mathrm{Ad} \circ \rho$ action of Γ on $\mathfrak{sl}_2(\mathbb{C})$. A cochain $b\colon \Gamma \to \mathfrak{sl}_2^\rho$ is a coboundary if there exists an $U \in \mathfrak{sl}_2(\mathbb{C})$ such that $b(\gamma) = \gamma \cdot U - U$ for all $\gamma \in \Gamma$.

It was an observation of Weil ([**Wei64**], cf. also [**Por97**]) that the Zariski tangent space $T_\rho^{\mathrm{Zar}}(\mathfrak{R}(\Gamma))$ at a point $\rho \in \mathfrak{R}(\Gamma)$ can be identified with a subspace of the space of 1-cocycles $Z^1(G; \mathfrak{sl}_2^\rho)$, and thus $\dim T_\rho^{\mathrm{Zar}}(\mathfrak{R}(\Gamma)) \leq \dim Z^1(\Gamma; \mathfrak{sl}_2^\rho)$. The knowledge of the dimension of the space of cocycles can therefore be used to determine whether a given representation or its character is a smooth point in the representation or character variety, respectively.

4.3.2. *Coboundaries at p-reps.* First we determine the coboundaries $B^1 := B^1(W_{p/q}; \mathfrak{sl}_2^\rho)$ for a p-rep in partially-diagonal normal form $\rho \in \mathfrak{D}^\pm(W_{p/q})$. As in section 4.2, we will restrict ourselves to the case $\rho \in \mathfrak{D}^+$, as the negative trace case proceeds analogously.

Let

$$e_1 = \begin{pmatrix} 0 & 1 \\ 0 & 0 \end{pmatrix}, \quad e_2 = \begin{pmatrix} 1 & 0 \\ 0 & -1 \end{pmatrix} \quad \text{and} \quad e_3 = \begin{pmatrix} 0 & 0 \\ 1 & 0 \end{pmatrix} \tag{39}$$

be a basis for the Lie algebra $\mathfrak{sl}_2 := \mathfrak{sl}_2(\mathbb{C})$. For a matrix $A \in \mathfrak{sl}_2$, let $u_A \in B^1$ be the 1-coboundary defined by $u_A(\gamma) = A - \gamma \cdot A$ for $\gamma \in \pi_1(W_{p/q})$. Given a representation $\rho \in \mathfrak{D}^+(W_{p/q})$, the coboundaries are then generated by

$$\text{(40)}\quad u_{e_1}: \quad \mu_0 \mapsto \begin{pmatrix} 0 & 1-s^2 \\ 0 & 0 \end{pmatrix}, \qquad \mu_1 \mapsto \begin{pmatrix} a & 1-a^2 \\ 1 & -a \end{pmatrix},$$

$$\text{(41)}\quad u_{e_2}: \quad \mu_0 \mapsto \begin{pmatrix} 0 & 0 \\ 0 & 0 \end{pmatrix}, \qquad \mu_1 \mapsto \begin{pmatrix} 2(a-1)^2 & -2a(a-1)^2 \\ 2(a-2) & -2(a-1)^2 \end{pmatrix},$$

$$\text{(42)}\quad u_{e_3}: \quad \mu_0 \mapsto \begin{pmatrix} 0 & 0 \\ 1-s^{-2} & 0 \end{pmatrix}, \quad \mu_1 \mapsto \begin{pmatrix} (2-a)(a-1)^2 & (a-1)^4 \\ -(a-1)(a-3) & -(2-a)(a-1)^2 \end{pmatrix}.$$

Since cochains are determined by their images on the generators, we will implicitly use the isomorphism $\tau : \mathfrak{sl}_2^2 \to \mathbb{C}^6$ given by

$$\left(\begin{pmatrix} x_1 & x_2 \\ x_3 & -x_1 \end{pmatrix}, \begin{pmatrix} y_1 & y_2 \\ y_3 & -y_1 \end{pmatrix}\right) \mapsto (x_1, x_2, x_3, y_1, y_2, y_3) \tag{43}$$

to identify a cochain u with an element of $\mathbb{C}^6$. We can therefore think of $Z^1 := Z^1(W_{p/q}; \mathfrak{sl}_2^\rho)$ and B^1 as subspaces of $\mathbb{C}^6$.

All coboundaries are now of the form $\alpha_1 u_{e_1} + \alpha_2 u_{e_2} + \alpha_3 u_{e_3}$ with $\alpha_i \in \mathbb{C}$. Seen as elements of $\mathbb{C}^6$, they hence lie in the span of the row vectors of the matrix

$$\mathcal{B} := \begin{pmatrix} 0 & 1-s^2 & 0 & a & 1-a^2 & 1 \\ 0 & 0 & 0 & 2(a-1)^2 & -2a(a-1)^2 & 2(a-2) \\ 0 & 0 & 1-s^{-2} & (2-a)(a-1)^2 & (a-1)^4 & -(a-1)(a-3) \end{pmatrix}. \tag{44}$$

Since $s \neq \pm 1$, one sees that the rank of this matrix, and hence the dimension of the subspace of coboundaries, is 3.

Note that this remains true when ρ is reducible, as then $a = 1$ and $\mathcal{B}$ becomes

$$\mathcal{B}^{\mathrm{red}} = \begin{pmatrix} 0 & 1-s^2 & 0 & 1 & 0 & 1 \\ 0 & 0 & 0 & 0 & 0 & -2 \\ 0 & 0 & 1-s^{-2} & 0 & 0 & 0 \end{pmatrix}. \tag{45}$$

4.3.3. *Cocycle condition from the first relation.* Let $(B^1)^\perp$ be the orthogonal complement of the subspace of coboundaries B^1 in $\mathbb{C}^6$ with respect to the standard inner product. For each cohomology element $\bar{u} \in H^1$ there then is a unique representative $u \in (B^1)^\perp$ such that $\bar{u} = u + B^1$. The orthogonal complement is given by the kernel of the matrix $\mathcal{B}$, and we can therefore identify H^1 with a subspace of $(B^1)^\perp$ in $\mathbb{C}^6$. In particular, we can obtain it as the kernel of the linear map given by the matrix $\mathcal{B}$ adjoined with additional rows corresponding to the cocycle versions of the group relations.

We therefore examine the effect of the first group relation on cocycles: Using $w_1 := \mu_1\mu_0\mu_1\mu_0^{-1}\mu_1^{-1}\mu_0^{-1}\mu_1\mu_0 = \mu_0\mu_1\mu_0^{-1}\mu_1^{-1}\mu_0^{-1}\mu_1\mu_0\mu_1 =: w_2$, we see that each cocycle $u \in Z^1$ has to satisfy the matrix equation $u(w_1) - u(w_2) = 0$. Assuming that $u(\mu_0) = \left(\begin{smallmatrix} x_1 & x_2 \\ x_3 & -x_1 \end{smallmatrix}\right)$ and $u(\mu_1) = \left(\begin{smallmatrix} y_1 & y_2 \\ y_3 & -y_1 \end{smallmatrix}\right)$ as in (43), we can iteratively use the cocycle condition (38) to expand this expression, yielding a matrix with entries which are polynomials in x_1, x_2, x_3, y_1, y_2 and y_3 with coefficients in $\mathbb{C}[a^{\pm 1}, s^{\pm 1}]$.

When ρ is reducible (but non-abelian), we find that only the lower left entry of $u(w_1) = u(w_2)$ gives a non-vanishing equation, which we write in vector form as

$$\begin{pmatrix} 0 \\ -2(s^2-1) \\ 0 \\ -2(s^2-1)^2 \\ s^{-2}(s^2-1)^2(s^4-s^2-1) \\ 0 \end{pmatrix}^T \begin{pmatrix} x_1 \\ x_2 \\ x_3 \\ y_1 \\ y_2 \\ y_3 \end{pmatrix} = 0. \tag{46}$$

This vector can now easily be seen to be linearly independent from the row vectors in the matrix $\mathcal{B}^{\mathrm{red}}$ in (45).

On the other hand, when ρ is irreducible (and hence $a \neq 1$), the matrix obtained is far more complicated. After dividing the entries by a common nonzero factor of $s^{-8}(s^2-1)(a-1)$ and reducing all coefficient polynomials in $\mathbb{C}[a^{\pm 1}, s^{\pm 1}]$ modulo

r_1^+, one finds that the upper left entry yields the equation

$$v^T \begin{pmatrix} x_1 & x_2 & x_3 & y_1 & y_2 & y_3 \end{pmatrix}^T = 0, \tag{47}$$

where the vector v is given by

$$\begin{pmatrix} -4(a^3s^2 - a^3 - 7a^2s^2 + 7a^2 + 2as^4 + 8as^2 - 14a + 4s^2 + 8) \\ -2(a^2s^2 - a^2 - 2as^4 - as^2 + 3a + s^4 - 2s^2 - 2) \\ 2(a^4s^2 - a^4 - 9a^3s^2 + 9a^3 + 24a^2s^2 - 29a^2 - as^6 - 16as^2 + 39a + s^6 - 2s^4 - 8s^2 - 18) \\ 2s^2(-6as^2 + 3a^2s^2 - as^4 - a^2 + 3a - 2) \\ as^6 + 2as^4 - 2as^2 + a - 2s^4 + 4s^2 - 2 \\ -3a^3s^2 + a^3 + 16a^2s^2 - 8a^2 - 4as^4 - 19as^2 + 17a - 2s^2 - 10 \end{pmatrix}, \tag{48}$$

and v^T denotes the transposed vector. Denote the first coefficient of v (corresponding to x_1) by ψ_1. One way to see that v is linearly independent from the row vectors in the matrix $\mathcal{B}$ in (44) is to verify that ψ_1 does not vanish at a representation $\rho \in \mathfrak{D}^+(W_{p/q})$: Using the computer algebra system Maple™ ([**Map**]) to calculate a reduced Groebner basis with respect to the pure lexicographic order $a > s$ for the ideal generated by ψ_1 and r_1^+, we find that its first generator is a power of s, implying that the only common zeros with r_1^+ lie in the subspace defined by $s = 0$, a contradiction to our assumptions.

We remark that the other entries of $u(w_1) - u(w_2)$ yield equations which can be shown to be consequences of the coboundary conditions and equation (48).

4.3.4. *Cocycle condition from the filling relation.* We now study the condition forced on the cocycles after applying p/q-filling on the first cusp. Letting $u(\mu_0) = \left(\begin{smallmatrix} x_1 & x_2 \\ x_3 & -x_1 \end{smallmatrix}\right)$ and $u(\mu_1) = \left(\begin{smallmatrix} y_1 & y_2 \\ y_3 & -y_1 \end{smallmatrix}\right)$ as before, we first determine the image $u(\lambda_0) = u(\mu_1\mu_0\mu_1^{-1}\mu_0^{-1}\mu_1^{-1}\mu_0\mu_1\mu_0^{-1})$. A direct calculation shows that

$$u(\lambda_0) =: \begin{pmatrix} l_1 & l_2 \\ l_3 & -l_1 \end{pmatrix}, \tag{49}$$

where the normalized entries l_1, l_2 and l_3 are given by

$$\begin{aligned} l_1 = &- 4s^{-6}(a^3s^2 - a^3 - 2a^2s^2 + 4a^2 - as^6 - 2as^4 - as^2 - 5a + s^6 + 2s^4 + 2s^2 + 2) \\ &- 2s^{-8}(a^2s^2 - a^2 - as^2 + 3a - 2s^6 - 2s^4 - 2s^2 - 2) \\ &+2s^{-4}(a^4s^2 - a^4 - 3a^3s^2 + 5a^3 - 3a^2s^2 - 7a^2 + 2as^4 \\ &\qquad +11as^2 + a - 2s^4 - 2s^2 + 2) \\ &- 4s^{-2}(a^3s^2 - a^3 - 3a^2s^2 + 4a^2 + 2as^2 - 5a + 2) \\ &- s^{-4}(2a^2s^2 - 2a^2 - as^6 - 3as^2 + 6a - 4s^4 - 2s^2 - 4) \\ &+2s^{-6}(a^4s^2 - a^4 - 6a^3s^2 + 7a^3 + 12a^2s^2 - 19a^2 \\ &\qquad -5as^2 + 23a - 2s^4 - 6s^2 - 10), \end{aligned} \tag{50}$$

$$\begin{aligned} l_2 = &4s^{-6}(a^4s^2 - a^4 - a^3s^2 + 3a^3 - 11a^2s^2 + 3a^2 + 17as^2 - 15a + 2s^2 + 10) \\ &+ s^{-8}(2a^3s^2 - 2a^3 + 4a^2 - 15as^4 + a - s^4 - 8s^2 - 3) \\ &-s^{-4}(2a^5s^2 - 2a^5 - 4a^4s^2 + 8a^4 - 12a^3s^2 - 6a^3 + 32a^2s^2 - 12a^2 \\ &\qquad -as^4 - 20as^2 + 21a + s^4 - 9) \\ &+ 2s^{-2}(2a^3s^2 - 2a^3 - 5a^2s^2 + 7a^2 + 2as^2 - 8a + s^2 + 3)a \\ &+ s^{-4}(2a^3s^2 - 2a^3 - 5a^2s^2 + 7a^2 - 5as^4 + 2as^2 - 7a + s^4 - s^2 + 2) \\ &-s^{-6}(2a^5s^2 - 2a^5 - 11a^4s^2 + 13a^4 + 18a^3s^2 - 32a^3 + 4a^2s^2 \\ &\qquad +30a^2 - 28as^2 + 2a - s^2 - 11) \end{aligned} \tag{51}$$

and

$$
\begin{aligned}
l_3 = &- 4s^{-6}(a^2s^2 - a^2 - 2as^8 + 2as^6 - as^2 + 3a + 2s^8 - 6s^6 - 2s^4 - 2s^2 - 2) \\
&+ s^{-2}(as^6 - 6as^4 + 7as^2 - 2a - s^6 + 8s^4 - 15s^2 + 4) \\
&+s^{-4}(2a^3s^2 - 2a^3 - 8a^2s^2 + 12a^2 + as^{12} - 6as^{10} + 11as^8 - 4as^6 - 4as^4 \\
&\qquad +2as^2 - 22a - s^{12} + 8s^{10} - 19s^8 + 12s^6 + 12s^4 + 12s^2 + 12) \\
&- 2s^{-2}(2a^2s^2 - 2a^2 - 2as^4 - 3as^2 + 7a + 2s^4 - 2s^2 - 6) \\
&+s^{-4}(as^{10} - 7as^8 + 13as^6 - 12as^4 + 6as^2 - a - s^{10} \\
&\qquad +9s^8 - 22s^6 + 22s^4 - 12s^2 + 2) \\
&+s^{-6}(2a^3s^2 - 2a^3 - 13a^2s^2 + 15a^2 + as^8 + 5as^6 + 11as^2 \\
&\qquad -31a - s^8 - 3s^6 + 10s^4 + 14s^2 + 18).
\end{aligned} \tag{52}
$$

The relation $\mu_0^p\lambda_0^q = 1$ then translates to $u(\mu_0^p\lambda_0^q) = \left(\begin{smallmatrix} 0 & 0 \\ 0 & 0 \end{smallmatrix}\right)$, where

$$
\begin{aligned}
u(\mu_0^p\lambda_0^q) &= u(\mu_0^p) + \mu_0^p \cdot u(\lambda_0^q) = \\
&= \left(1 + \mu_0 + \cdots + \mu_0^{p-1}\right) \cdot u(\mu_0) + \mu_0^p \left(1 + \lambda_0 + \cdots + \lambda_0^{q-1}\right) \cdot u(\lambda_0) \\
&= \begin{pmatrix} px_1 + ql_1 & x_2 \sum_{k=0}^{p-1} s^{2k} + s^{2p} l_2 \sum_{j=0}^{q-1} t^{2j} \\ x_3 \sum_{k=0}^{p-1} s^{-2k} + s^{-2p} l_3 \sum_{j=0}^{q-1} t^{-2j} & -px_1 - ql_1 \end{pmatrix}.
\end{aligned} \tag{53}
$$

We remark that this computation is the reason for our choice of the parametrization in section 4.2, as we could otherwise not have arrived at a general form in terms of p and q.

Now $s \neq \pm 1$ implies that we can use

$$\sum_{k=0}^{p-1} s^{2k} = \frac{s^{2p} - 1}{s^2 - 1} \tag{54}$$

and

$$\sum_{k=0}^{p-1} s^{-2k} = s^{-2(p-1)} \sum_{k=0}^{p-1} s^{2k} = s^{-2(p-1)} \frac{s^{2p} - 1}{s^2 - 1}. \tag{55}$$

We now need to distinguish between the reducible and irreducible case: When ρ is reducible, $a = 1$ implies $t = 1$ and hence $\sum_{j=0}^{q-1} t^{2j} = q$. But then $s^p t^q = 1$ also implies $s^p = 1$, and equation (53) becomes

$$\begin{pmatrix} 0 & 0 \\ 0 & 0 \end{pmatrix} = \begin{pmatrix} px_1 + ql_1 & ql_2 \\ ql_3 & -px_1 - ql_1 \end{pmatrix}. \tag{56}$$

But note that $a = 1$ implies that $l_1 = (s^2 - s^{-2})y_2$, $l_2 = 0$ and $l_3 = -2s^{-2}x_2 + 2(s^{-2} - 1)y_1 + (s^{-4} - 2 + s^2)y_2$. It is then easy to see that $ql_3 = 0$ is already a consequence of the coboundary conditions, and so the only new condition remaining from (56) is $(p/q)x_1 + l_1 = 0$. Adjoining this condition in row vector form together with the vector corresponding to equation (46) to the matrix from (45), we then obtain a presentation matrix for the cohomology group $H^1(W_{p/q}; \mathfrak{sl}_2^\rho)$, viewed as a subspace of $\mathbb{C}^6$:

$$\begin{pmatrix} 0 & 1 - s^2 & 0 & 1 & 0 & 1 \\ 0 & 0 & 0 & 0 & 0 & -2 \\ 0 & 0 & 1 - s^2 & 0 & 0 & 0 \\ 0 & 2(1 - s^2) & 0 & -2(1 - s^2)^2 & s^{-2}(s^4 - s^2 - 1)(1 - s^2)^2 & 0 \\ p/q & 0 & 0 & 0 & s^{-2}(s^4 - 1) & 0 \end{pmatrix}. \tag{57}$$

The rank of this matrix can easily be seen to be 5 for all $s \neq \pm 1$, and we therefore obtain

$$\dim_{\mathbb{C}} H^1(W_{p/q}; \mathfrak{sl}_2^\rho) = 1. \tag{58}$$

On the other hand, when ρ is irreducible, we can use $t \neq \pm 1$ to rewrite (53) as

$$\begin{aligned} u(\mu_0^p \lambda^q) &= \begin{pmatrix} px_1 + ql_1 & \frac{s^{2p}-1}{s^2-1}x_2 + s^{2p}\frac{t^{2q}-1}{t^2-1}l_2 \\ s^{-2(p-1)}\frac{s^{2p}-1}{s^2-1}x_3 + s^{-2p}t^{-2(q-1)}\frac{t^{2q}-1}{t^2-1}l_3 & -px_1 - ql_1 \end{pmatrix} \\ &= \begin{pmatrix} px_1 + ql_1 & \frac{s^{2p}-1}{s^2-1}x_2 + \frac{1-s^{2p}}{t^2-1}l_2 \\ s^2\frac{1-t^{2q}}{s^2-1}x_3 - t^2\frac{1-t^{2q}}{t^2-1}l_3 & -px_1 - ql_1 \end{pmatrix} \\ &= \begin{pmatrix} px_1 + ql_1 & (s^{2p}-1)\left(\frac{1}{s^2-1}x_2 - \frac{1}{t^2-1}l_2\right) \\ s^{-2p}(s^{2p}-1)\left(\frac{s^2}{s^2-1}x_3 - \frac{t^2}{t^2-1}l_3\right) & -px_1 - ql_1 \end{pmatrix}. \end{aligned} \tag{59}$$

From the upper right and lower left entries of this matrix we now see that either $s^{2p} = 1$ (which we know cannot be true by Lemma 13, since s is also a root of $\mathrm{res}_{p,q}$), or $(t^2-1)x_2 = (s^2-1)l_2$ and $s^2(t^2-1)x_3 = t^2(s^2-1)l_3$. After some careful algebraic manipulations, it can be verified that these last two conditions are in fact consequences of the coboundary conditions (44) and equation (48). The only new information comes therefore from the upper left matrix entry. Since $q > 0$, we can rewrite this in the form $l_1 + (p/q)x_1 = 0$, which gives us (after normalizing with respect to r_1^+) the condition

$$w^T \begin{pmatrix} x_1 & x_2 & x_3 & y_1 & y_2 & y_3 \end{pmatrix}^T = 0, \tag{60}$$

where the vector w is given by

$$\begin{pmatrix} p/q - 4s^{-6}(a^3s^2 - a^3 - 2a^2s^2 + 4a^2 - as^6 - 2as^4 - as^2 - 5a + s^6 + 2s^4 + 2s^2 + 2) \\ -2s^{-8}(a^2s^2 - a^2 - as^2 + 3a - 2s^6 - 2s^4 - 2s^2 - 2) \\ 2s^{-4}(a^4s^2 - a^4 - 3a^3s^2 + 5a^3 - 3a^2s^2 - 7a^2 + 2as^4 + 11as^2 + a - 2s^4 - 2s^2 + 2) \\ -4s^{-2}(a^3s^2 - a^3 - 3a^2s^2 + 4a^2 + 2as^2 - 5a + 2) \\ -s^{-4}(2a^2s^2 - 2a^2 - as^6 - 3as^2 + 6a - 4s^4 - 2s^2 - 4) \\ 2s^{-6}(a^4s^2 - a^4 - 6a^3s^2 + 7a^3 + 12a^2s^2 - 19a^2 - 5as^2 + 23a - 2s^4 - 6s^2 - 10) \end{pmatrix}. \tag{61}$$

Adjoining the two row vectors corresponding to equations (48) and (61) to the matrix B, we now obtain a presentation matrix for $H^1(W_{p/q}; \mathfrak{sl}_2^\rho)$. With some care and making sure no polynomial used in division can be zero on $\mathfrak{D}^+$, we can row-reduce this matrix to the form

$$\begin{pmatrix} \psi_1 & 0 & 0 & 0 & * & * \\ 0 & s^2-1 & 0 & 0 & -1 & -(a-1)^{-2} \\ 0 & 0 & s^{-2}-1 & 0 & s^2(a-1)^2 & s^2 \\ 0 & 0 & 0 & (a-1)^2 & -a(a-1)^2 & a-2 \\ 0 & 0 & 0 & 0 & * & * \end{pmatrix}, \tag{62}$$

where each $*$ denotes a polynomial in a and s (normalized modulo r_1^+) which does not vanish on $\mathfrak{D}^+$, and ψ_1 is the first coefficient of the vector in equation (48). This shows that the rank of this matrix is 5, and we therefore find that

$$\dim_{\mathbb{C}} H^1(W_{p/q}; \mathfrak{sl}_2^\rho) = 1 \tag{63}$$

or equivalently,

$$\dim_{\mathbb{C}} Z^1(W_{p/q}; \mathfrak{sl}_2^\rho) = 4. \tag{64}$$

Since Corollary 19 showed that a non-abelian reducible p-rep lies on a non-trivial curve in $\mathfrak{R}(W_{p/q})$, and an irreducible p-rep does so automatically, Theorem A of [**Boy02**] now shows:

PROPOSITION 22. *The character χ_ρ of a (reducible or irreducible) p-prep $\rho \in \mathfrak{R}(W_{p/q})$ is a smooth point of $\mathfrak{X}^{irr}(W)$ and is contained in a unique non-trivial curve. Furthermore, we have $T^{Zar}_{\chi_\rho}(\mathfrak{X}(W_{p/q})) \cong H^1(W_{p/q}; \mathfrak{sl}_2^\rho)$, presented as a subset of $\mathbb{C}^6$ as the kernel of the matrix* (57) *in the reducible, and the matrix* (62) *in the irreducible case.* □

This then concludes the proof of Proposition 16, and furthermore provides an explicit parametrization for non-trivial cocycles in $Z^1(M; \mathfrak{sl}_2^\rho)$ at a given representation $\rho \in \mathfrak{D}^+$. This will be needed in the following section, when we will study deformations of these representations.

4.4. Deformations of representations and degree of the trace function. We now turn our attention to the proof of Proposition 17.

Since Proposition 22 showed that $\mathfrak{X}^{\text{irr}}(W_{p/q})$ is smooth and positive-dimensional at the character of a p-rep ρ, we can use the existence of non-trivial cocycles to find a smooth deformation ρ_ν of $\rho_0 = \rho$, which admits an expansion of the form

$$\rho_\nu = \exp\left(\nu u_1 + \nu^2 u_2 + \nu^3 u_3 + \dots\right)\rho \tag{65}$$

where $u_j : \pi_1(G) \to \mathfrak{sl}_2(\mathbb{C})$, and $u_1 \in Z^1(G; \mathfrak{sl}_2^\rho)$ (see e.g. [**HPSP01**, § 3] or [**Gol84**]).

Developing $f_{\chi_{\rho_\nu}} = (\operatorname{trace} \rho_\nu(\gamma))^2 - 4$ using the expansion

$$\exp(A) = \sum_{j=1}^{\infty} A^j/j! \tag{66}$$

for $A \in \mathfrak{sl}_2(\mathbb{C})$, we find that the trace function $f_{\mu_1}(\chi_{\rho_\nu})$ has the following Taylor expansion at ρ_0 (see e.g. Lemma 1.3 in [**BAB01**]):

$$f_{\mu_1}(\chi_{\rho_\nu}) = f_{\mu_1}(\chi_{\rho_0}) + 2I_{\mu_1}(\chi_{\rho_0}) \operatorname{trace}\bigl(u_1(\mu_1)\rho_0(\mu_1)\bigr)\nu + O(\nu^2). \tag{67}$$

Note that in our case $G := \pi_1(W_{p/q})$ and $I_{\mu_1}(\chi_{\rho_0}) = (\operatorname{trace} \rho_0(\mu_1))^2 = 4$ for $\rho_0 \in \mathfrak{D}^\pm$. Since $f_{\mu_1}(\chi_{\rho_0}) = 0$ at a p-rep ρ_0, the order of this zero is 1 iff there is a non-trivial cocycle $u_1 \in Z^1(W_{p/q}; \mathfrak{sl}_2^\rho)$ satisfying

$$\operatorname{trace}\bigl(u_1(\mu_1)\rho_0(\mu_1)\bigr) \neq 0. \tag{68}$$

To prove this, note that for $\rho_0 \in \mathfrak{D}^+$ and an arbitrary cochain u_1 we have

$$\begin{aligned} \operatorname{trace}\bigl(u_1(\mu_1)\rho_0(\mu_1)\bigr) &= \operatorname{trace}\begin{pmatrix} y_1 & y_2 \\ y_3 & -y_1 \end{pmatrix}\begin{pmatrix} a & -(a-1)^2 \\ 1 & 2-a \end{pmatrix} \\ &= 2(a-1)y_1 + y_2 - (a-1)^2 y_3, \end{aligned} \tag{69}$$

and an analogous expression for $\rho_0 \in \mathfrak{D}^-$.

Let P be the matrix obtained by adjoining the row vector corresponding to this expression to the presentation matrix for $H^1(W_{p/q}; \mathfrak{sl}_2^\rho)$ from (57) or (62), respectively. The existence of a non-trivial cocycle u_1 such that (69) is equal to a non-zero value $c \in \mathbb{C}^*$ is now equivalent to finding a solution $u_1 := (x_1, x_2, x_3, y_1, y_2, y_3)$ to the linear equation system

$$P\,(x_1, x_2, x_3, y_1, y_2, y_3)^T = (0,0,0,0,0,c)^T \tag{70}$$

and hence a non-trivial element in $H^1(W_{p/q})$ represented by a cocycle u_1 which satisfies (68). In particular, if we can show that $\det P \neq 0$, the system (70) will in fact have a *unique* solution u_1 for any given $c \in \mathbb{C}^*$.

Again restricting attention to $\rho_0 \in \mathfrak{D}^+$, we will first discuss the case when ρ_0 is reducible: In this case, the matrix P is given by

$$P = \begin{pmatrix} 0 & 1-s^2 & 0 & 1 & 0 & 1 \\ 0 & 0 & 0 & 0 & 0 & -2 \\ 0 & 0 & 1-s^2 & 0 & 0 & 0 \\ 0 & 2(1-s^2) & 0 & -2(1-s^2)^2 & s^{-2}(s^4-s^2-1)(1-s^2)^2 & 0 \\ p/q & 0 & 0 & 0 & s^{-2}(s^4-1) & 0 \\ 0 & 0 & 0 & 0 & 1 & 0 \end{pmatrix}, \tag{71}$$

and we find $\det P = -(4p/q)(1-s^2)^2(s^4-2s^2+2)$. Now recall from section 3.3 that the eigenvalue s of a reducible p-rep must satisfy $s^p = 1$, $s \neq \pm 1$. None of the roots of $s^4-2s^2+2=0$ lie on the unit circle, and so $\det P \neq 0$ unless $p=0$: In that case, the trace in (69) will indeed be zero for all cocycles u_1, and hence the order of a zero of f_{μ_1} at the character of a reducible non-abelian p-rep is indeed greater than 1. This also reflects the fact that there is a whole curve of non-abelian reducible p-reps for $s \in \mathbb{C}^* - \{\pm 1\}$ in this case (since $p=0$, $q=1$ implies $s^p t^q = t = 1$, and there are no further conditions on s).

This proves Proposition 17 in the case of a non-abelian reducible p-rep, $p/q \neq 0$.

We will now discuss the remaining case, when $\rho_0 \in \mathfrak{D}^+$ is irreducible. Appending the row vector corresponding to the expression (69) to the matrix (62) and applying Gauss-Jordan to reduce the matrix, we obtain the 6×6-matrix

$$P = \begin{pmatrix} \psi_1 & & & & * & * \\ & * & & & * & * \\ & & * & & * & * \\ & & & * & * & * \\ & & & & * & * \\ & & & & 2a^2-2a+1 & (1-a)^{-1}(a^3-3a^2+5a-5) \end{pmatrix} \tag{72}$$

Denote by d the determinant of the lower right 2×2-block of this matrix. We wish to show that this determinant is not zero for $\rho_0 \in \mathfrak{D}^+$. To show this, we again used the computer algebra system Maple™ to calculate a Groebner basis of the ideal generated by r_1^+ and the numerator of d with respect to a lexicographic ordering with $a > s$ (we remark that this computation took more than 24 hours on a modern PC). We find that the first basis element is given by $s^{30}d_1d_2$ where

$$d_1 = p(p-4q)s^4 + (-6p^2+24pq-32q^2)s^2 + p(p-4q) \tag{73}$$

and

$$\begin{aligned} d_2 = {} & 22s^{40} - 438s^{38} + 4185s^{36} - 24868s^{34} + 101875s^{32} - 304088s^{30} \\ & + 683740s^{28} - 1182928s^{26} + 1598312s^{24} - 1708564s^{22} + 1466502s^{20} \\ & - 1027864s^{18} + 595850s^{16} - 286072s^{14} + 115452s^{12} - 41808s^{10} \\ & + 14586s^8 - 4582s^6 + 1097s^4 - 164s^2 + 11. \end{aligned} \tag{74}$$

If d were to vanish at an irreducible representation $\rho_0 \in \mathfrak{D}^+$, each of the polynomials in the Groebner basis, and in particular the first element, would have to vanish. We will therefore show that neither d_1 nor d_2 vanish at a representation $\rho_0 \in \mathfrak{D}^+$, which shows that d cannot vanish there. For this, we will make use of the fact that the parametrization of any irreducible representation $\rho_0 \in \mathfrak{D}^\pm$ also satisfies $\mathrm{res}_{p,q} = 0$ (as determined in section 3.4). It suffices therefore to show that $\mathrm{res}_{p,q}$ has no roots in common with d_1 and d_2.

As for d_1, we note that its four roots are

$$\pm\frac{\left(\left(3(p-2q)^2+4q^2\right) \pm \left(2|p-2q|\sqrt{2(p-2q)^2+8q^2}\right)\right)^{1/2}}{\sqrt{p(p-4q)}} \tag{75}$$

It is easy to see that in fact these roots are all real-valued when $p > 4q > 0$ or when $p < 0 < 4q$, and all imaginary when $0 < p < 4q$. To see this, simply note that

$$\left(3(p-2q)^2+4q^2\right)^2 - \left(2|p-2q|\sqrt{2(p-2q)^2+8q^2}\right)^2 = p^2(p-4q)^2 \geq 0 \tag{76}$$

which shows that the numerator in (75) is always real-valued.

But note that for $p > 4q > 0$ or $p < 0 < 4q$, apart from ± 1, none of the non-zero roots of $\mathrm{res}_{p,q}$ are real-valued by Lemma 11. Similarly for $0 < p < 4q$, none of its non-zero roots are pure imaginary by Lemma 12. So d_1 in fact does not share any roots with $\mathrm{res}_{p,q}$.

It now remains to be shown that the polynomial d_2 has no common roots with $\mathrm{res}_{p,q}$. To see this, first observe that d_2 is an irreducible polynomial, while its leading coefficient is $22 \neq 1$. On the other hand, while $\mathrm{res}_{p,q}$ is in general *not* irreducible (consider for example $q = 1$ and p even), it is a *monic* polynomial with integer coefficients — and therefore so are its irreducible factors: The Chebyshev polynomial $T_q(y)$ has leading term $2^{q-1}y^q$, and substituting $y = -s^2/2 + 2 - s^{-2}/2$ gives the leading term $1/2\, s^{2q}$. Now note that the "outer" terms of $\mathrm{res}_{p,q}$ are s^{p-2q} and s^{-p+2q}, and the middle terms are given by $2T_q(y)$.

We conclude that all roots of $\mathrm{res}_{p,q}$ are *algebraic integers*, while this is not true for any of the roots of d_2. This shows that d_2 and $\mathrm{res}_{p,q}$ have no roots in common. We show the 40 roots of d_2 in figure 3.

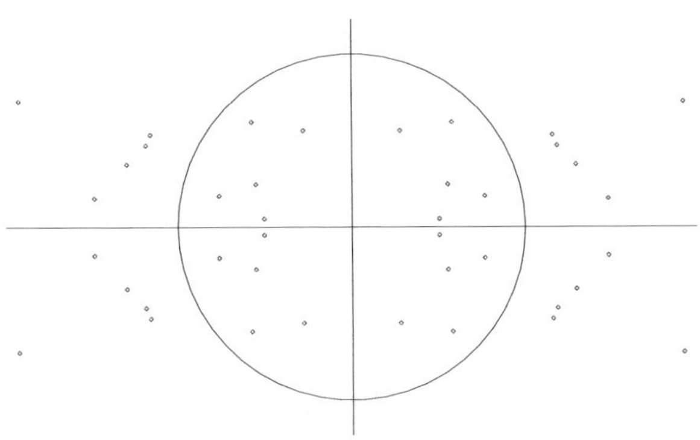

FIGURE 3. The roots of the polynomial d_2

We can therefore conclude that the 2×2 minor d is indeed non-zero at characters of p-reps, and that hence $\operatorname{rank} P = 6$ and $\det P \neq 0$, proving Proposition 17 for irreducible p-reps.

This then concludes the proof of Theorem 15, showing that the minimal seminorm s is given by the number of characters of p-reps.

5. Seifert fillings

The goal of this section is to prove:

PROPOSITION 23. *For p odd, the values of the total Culler-Shalen seminorm of the slopes* 1, 2 *and* 3 *of $W_{p/q}$ are given by*

$$\begin{aligned}\|1\| &= s + 2|p-6q| - 2 \quad \textit{(if } p/q \neq 6\textit{)},\\ \|2\| &= s + 3|p-4q| - 3 \quad \textit{(if } p/q \neq 4\textit{)},\\ \|3\| &= s + 4|p-3q| - 4 \quad \textit{(if } p/q \neq 3\textit{)},\end{aligned}$$

where s is the minimal total Culler-Shalen seminorm.

Note that while $p/q \neq 6$ and $p/q \neq 4$ are already satisfied when p is odd, we mention these conditions for completeness.

The proof is based on the well-known fact (see e.g. [**MP06**, § 9] or [**DT03**]) that for $\sigma \in \{1,2,3\}$, the closed manifold $W_{p/q}(\sigma) = W(p/q,\sigma)$ is Seifert fibered with base orbifold a sphere with three cone points (such manifolds are often called *small Seifert*):

- $W(p/q,1)$ is Seifert fibered over $S^2(2,3,|p-6q|)$ if $p/q \neq 6$
- $W(p/q,2)$ is Seifert fibered over $S^2(2,4,|p-4q|)$ if $p/q \neq 4$
- $W(p/q,3)$ is Seifert fibered over $S^2(3,3,|p-3q|)$ if $p/q \neq 3$

This can be seen from the fact that $W(1)$, $W(2)$ and $W(3)$ are all Seifert fibred with base orbifold a disc with two cone points: $D^2(2,3)$, $D^2(2,4)$ and $D^2(3,3)$, respectively. Furthermore, the corresponding fibre slopes are 6, 4 and 3, respectively. Therefore, applying a second Dehn filling generically produces a Seifert fibred manifold with base orbifold a sphere with three cone points, with only one exception each: When filling along the fibre slope, the resulting closed manifold is a connected sum of lens spaces: $W(1,6) \cong L_2\#L_3$, $W(2,4) \cong L_2\#L_4$ and $W(3,3) \cong L_3\#L_3$.

The methods from [**BAB01**] now enable us to explictly determine the Culler-Shalen seminorm of these small Seifert manifolds:

Let $\sigma \in \{1,2,3\}$ be one of the Seifert filling slopes for $W_{p/q}$. Theorem C of [**BAB01**] shows that in this situation $\|\sigma\| = s + 2A$, where A is the number of characters of non-abelian representations $\rho \in \mathfrak{R}(W_{p/q})$ satisfying $\rho(\sigma) = \pm I$ whose characters lie on non-trivial curves in $\mathfrak{X}(W_{p/q})$.

Note that such a representation necessarily induces a (non-abelian) $\mathrm{PSL}_2(\mathbb{C})$ representation $\overline{\rho} \in \overline{\mathfrak{R}}(W_{p/q})$ with $\overline{\rho}(\sigma) = \pm I$, which in turn factors through to give a representation $\overline{\rho}' \in \overline{\mathfrak{R}}(W_{p/q}(\sigma))$. (An abelian irreducible representation $\overline{\rho} \in \overline{\mathfrak{R}}$ necessarily has image isomorphic to $\mathbb{Z}/_2 \oplus \mathbb{Z}/_2$, generated by $\pm\left(\begin{smallmatrix} i & 0\\ 0 & -i\end{smallmatrix}\right)$ and $\pm\begin{pmatrix} 0 & -b\\ b^{-1} & 0\end{pmatrix}$ for some $b \in \mathbb{C}^*$. This possibility is excluded in our situation by the fact that p is odd.) Since $W_{p/q}(\sigma)$ has base orbifold $S^2(a,b,c)$, a sphere with three cone points, there is a presentation of $\pi_1(W_{p/q}(\sigma))$ of the form

$$\pi_1(W_{p/q}(\sigma)) = \left\langle x, y, h \mid h \text{ central},\ x^a = h^{a'},\ y^b = h^{b'},\ (xy)^c = h^{c'} \right\rangle \tag{77}$$

where $\gcd(a, a') = \gcd(b, b') = \gcd(c, c') = 1$ (cf. [**Jac80**, § VI]).

Lemma 3.1 from [**BAB01**] now shows that the representation $\overline{\rho}' \in \overline{\mathfrak{R}}(W_{p/q}(\sigma))$ again factors through to give a representation $\overline{\rho}'' \in \overline{\mathfrak{R}}(\Delta(a,b,c))$, where $\Delta(a,b,c) \cong \pi_1^{\mathrm{orb}}(S^2(a,b,c))$ is the (a,b,c) triangle group with presentation

$$\Delta(a,b,c) = \left\langle\, x, y \mid x^a = y^b = (xy)^c = 1 \,\right\rangle. \tag{78}$$

Conversely, a given representation $\overline{\rho}'' \in \overline{\mathfrak{R}}(\Delta(a,b,c))$ induces a unique representation $\overline{\rho}' \in \overline{\mathfrak{R}}(W_{p/q}(\sigma))$, which in turn induces a unique representation $\overline{\rho} \in \overline{\mathfrak{R}}(W_{p/q})$ satisfying $\overline{\rho}(\sigma) = \pm I$. We summarize the situation in the following diagram:

$$\begin{array}{ccc} \pi_1(W_{p/q}) & \xrightarrow{\ \rho\ } & \mathrm{SL}_2(\mathbb{C}) \\ \downarrow & \searrow^{\overline{\rho}} & \downarrow \\ \pi_1(W_{p/q}(\sigma)) & \xrightarrow{\ \overline{\rho}'\ } & \mathrm{PSL}_2(\mathbb{C}) \\ \downarrow & \nearrow_{\overline{\rho}''} & \\ \Delta(a,b,c) & & \end{array} \tag{79}$$

Using the following theorem from [**BAB01**], which we restate here in slightly adapted form, together with the fact that $H_1(W_{p/q}(r/s)) = \mathbb{Z}/_{|p|} \oplus \mathbb{Z}/_{|r|}$, one can now explicitly determine the number of $\mathrm{PSL}_2(\mathbb{C})$ characters of $\Delta(a,b,c)$, and hence of $\pi_1(W_{p/q})$, for each of the Seifert filling slopes:

THEOREM ([**BAB01**], Theorem D).

Let M be a non-Haken small Seifert manifold, with base orbifold of the form $S^2(a,b,c)$ where $a,b,c \geq 2$.

(1) *The number of reducible $\mathrm{PSL}_2(\mathbb{C})$-characters of $\pi_1(M)$ is*

$$\left\lfloor \frac{|H_1(M)|}{2} \right\rfloor + 1 + \begin{cases} 0 & \text{if } \gcd(a,b,c) \text{ is odd,} \\ 1 & \text{if } \gcd(a,b,c) \text{ is even.} \end{cases}$$

(2) *The number of $\mathrm{PSL}_2(\mathbb{C})$-characters of $\pi_1(M)$ representations with image a dihedral group of order at least 4 is*

$$\sigma(b,c)\left\lfloor \frac{a}{2} \right\rfloor + \sigma(a,c)\left\lfloor \frac{b}{2} \right\rfloor + \sigma(a,b)\left\lfloor \frac{c}{2} \right\rfloor - 2\sigma(a,b)\sigma(a,c)\sigma(b,c)$$

where $\sigma(m,n) = \left\lfloor \frac{\gcd(2,m,n)}{2} \right\rfloor$ is 1 if m and n are both even, and 0 otherwise.

(3) *The total number of $\mathrm{PSL}_2(\mathbb{C})$-characters of $\pi_1(M)$ is*

$$\left\lfloor \frac{|H_1(M)|}{2} \right\rfloor + 1 + \left\lfloor \frac{a}{2} \right\rfloor \left\lfloor \frac{b}{2} \right\rfloor \left\lfloor \frac{c}{2} \right\rfloor + \left\lfloor \frac{a-1}{2} \right\rfloor \left\lfloor \frac{b-1}{2} \right\rfloor \left\lfloor \frac{c-1}{2} \right\rfloor$$
$$- \left\lfloor \frac{\gcd(ab,ac,bc)}{2} \right\rfloor + \left\lfloor \frac{\gcd(a,b)}{2} \right\rfloor + \left\lfloor \frac{\gcd(a,c)}{2} \right\rfloor + \left\lfloor \frac{\gcd(b,c)}{2} \right\rfloor$$

□

The results are summarized in tables 6, 7 and 8 on page 198. Also included are the numbers of non-abelian reducible characters, which can be found by direct calculation.

$\gcd(6,p)$	total	irreducible	dihedral	reducible	non-ab. red.
1	$\frac{1}{2}(\lvert p\rvert + \lvert p-6q\rvert)$	$\frac{1}{2}(\lvert p-6q\rvert - 1)$	0	$\frac{1}{2}(\lvert p\rvert + 1)$	0
3	$\frac{1}{2}(\lvert p\rvert + \lvert p-6q\rvert)$	$\frac{1}{2}(\lvert p-6q\rvert - 1)$	0	$\frac{1}{2}(\lvert p\rvert + 1)$	0
2	$\frac{1}{2}(\lvert p\rvert + \lvert p-6q\rvert) + 1$	$\frac{1}{2}\lvert p-6q\rvert$	1	$\frac{1}{2}\lvert p\rvert + 1$	0
6	$\frac{1}{2}(\lvert p\rvert + \lvert p-6q\rvert)$	$\frac{1}{2}\lvert p-6q\rvert - 1$	1	$\frac{1}{2}\lvert p\rvert + 1$	1

TABLE 6. Number of $\mathrm{PSL}_2(\mathbb{C})$-characters of $\pi_1 W_{p/q}(1)$

$\gcd(4,p)$	total	irreducible	dihedral	reducible	non-ab. red.
1	$\lvert p\rvert + \lvert p-4q\rvert$	$\lvert p-4q\rvert - 1$	$\frac{1}{2}(\lvert p-4q\rvert - 1)$	$\lvert p\rvert + 1$	0
2	$\lvert p\rvert + \lvert p-4q\rvert + 2$	$\lvert p-4q\rvert$	$\frac{1}{2}\lvert p-4q\rvert + 1$	$\lvert p\rvert + 2$	0
4	$\lvert p\rvert + \lvert p-4q\rvert + 1$	$\lvert p-4q\rvert - 1$	$\frac{1}{2}\lvert p-4q\rvert + 1$	$\lvert p\rvert + 2$	1

TABLE 7. Number of $\mathrm{PSL}_2(\mathbb{C})$-characters of $\pi_1 W_{p/q}(2)$

$\gcd(6,p)$	total	irreducible	dihedral	reducible	non-ab. red.
1	$\frac{3}{2}(\lvert p\rvert - 1) + \lvert p-3q\rvert$	$\lvert p-3q\rvert - 1$	0	$\frac{3}{2}(\lvert p\rvert - 1) + 1$	0
3	$\frac{3}{2}(\lvert p\rvert - 1) + \lvert p-3q\rvert - 1$	$\lvert p-3q\rvert - 2$	0	$\frac{3}{2}(\lvert p\rvert - 1) + 1$	1
2	$\frac{3}{2}\lvert p\rvert + \lvert p-3q\rvert$	$\lvert p-3q\rvert - 1$	0	$\frac{3}{2}\lvert p\rvert + 1$	0
6	$\frac{3}{2}\lvert p\rvert + \lvert p-3q\rvert - 1$	$\lvert p-3q\rvert - 2$	0	$\frac{3}{2}\lvert p\rvert + 1$	1

TABLE 8. Number of $\mathrm{PSL}_2(\mathbb{C})$-characters of $\pi_1 W_{p/q}(3)$

It now remains to determine the number of non-abelian $\mathrm{SL}_2(\mathbb{C})$ characters, i.e. to determine which of the $\mathrm{PSL}_2(\mathbb{C})$ representations $\overline{\rho} \in \overline{\mathfrak{R}}(W_{p/q})$ lift to representations $\rho \in \mathfrak{R}(W_{p/q})$. There is in general an obstruction to this lifting, which is contained in the cohomology group

$$H^2(W_{p/q}; \mathbb{Z}/2) \cong \begin{cases} 0 & \text{if } p \text{ is odd,} \\ \mathbb{Z}/2 & \text{if } p \text{ is even.} \end{cases} \tag{80}$$

When p is even, we therefore possibly do have an obstruction to lifting, and it is to be expected that some representations do lift, but not others, making it impossible to directly make a statement about the number of $\mathrm{SL}_2(\mathbb{C})$ representations.

On the other hand, there is *no obstruction to lifting* when p is odd, *and we shall therefore restrict our attention to this case.*

The number of lifts of a given $\mathrm{PSL}_2(\mathbb{C})$ representation is generically given by the cardinality of

$$H^1(W_{p/q}; \mathbb{Z}/2) \cong \begin{cases} \mathbb{Z}/2 & \text{if } p \text{ is odd,} \\ \mathbb{Z}/2 \oplus \mathbb{Z}/2 & \text{if } p \text{ is even,} \end{cases} \tag{81}$$

but dihedral representations only lift half as many times as given by this number (see e.g. Lemma 5.5 of [**BZ96**] and the remarks following it for more details).

When p is odd, dihedral $\mathrm{PSL}_2(\mathbb{C})$ characters are therefore covered by only one $\mathrm{SL}_2(\mathbb{C})$ character, whereas the remaining non-abelian characters are covered by two, giving us the results listed in table 9, where the total number of non-abelian $\mathrm{SL}_2(\mathbb{C})$ characters shown in the last column gives the constant A needed to conclude the proof of Proposition 23.

	irred. non-dihedral	dihedral	non-ab. red.	total
$W_{p/q}(1)$	$\lvert p-6q\rvert-1$	0	0	$\lvert p-6q\rvert-1$
$W_{p/q}(2)$	$\lvert p-4q\rvert-1$	$\frac{1}{2}(\lvert p-4q\rvert-1)$	0	$\frac{3}{2}(\lvert p-4q\rvert-1)$
$W_{p/q}(3)$	$\begin{cases} 2\lvert p-3q\rvert-2 & \text{if } 3\nmid p \\ 2\lvert p-3q\rvert-4 & \text{if } 3\mid p \end{cases}$	0	$\begin{cases} 0 & \text{if } 3\nmid p \\ 2 & \text{if } 3\mid p \end{cases}$	$2(\lvert p-3q\rvert-1)$

TABLE 9. Number of non-abelian $\mathrm{SL}_2(\mathbb{C})$ characters for Seifert fillings of $W_{p/q}$ when p is odd

6. The total Culler-Shalen seminorm

Putting together the results of the previous sections, we now completely determine the total Culler-Shalen seminorm of the manifolds $W_{p/q}$ for p odd and $p/q \neq 3$, proving Theorem 1. In particular, we will show that the upper bounds established in Theorem 15 (cf. section 4) for the *minimal* total Culler-Shalen seminorm s are in fact attained, which in turn shows Corollary 4.

Using the possible boundary slopes β_j from section 2.2, we apply equation (4) with unknown coefficients a_j and s to the Seifert filling slopes 1, 2, 3 and compare with the explicit values (relative to s) obtained in Proposition 23.

From these equations, we then obtain a linear system which we can try to solve for the unknown coefficients a_j and s. Note that depending on the range of the filling slope p/q, the parametrization of the boundary slope β_2 changes (cf. table 3 on page 173), and hence we get different distances and accordingly different linear systems for each range. We will therefore discuss each range in sequence.

While there are some cases (when $p/q \in (3,4)$ or $p/q \in (4,6)$) when the linear system already determines a unique solution, the system is in general underdetermined and we need more information. This is where we use the upper bound for the total minimal seminorm s established in Theorem 15. In two sub-cases ($p/q \in (2,3)$ and $p/q > 6$), it turns out that even this is not yet sufficient, but using a symmetry argument about the resultant $\mathrm{res}_{p,q}$, we can reduce the case to one which was already proven before.

As before, we will always assume $q > 0$ and $p \in \mathbb{Z}$ with $(p,q) = 1$ and (due to the restrictions from section 5) p odd as well as $p/q \neq 3$.

6.1. Assuming $p/q \in (-\infty, 0)$, p odd. In this case, we obtain the following distances between the boundary, Seifert and meridional slopes:

Using the expressions of the seminorms for the Seifert slopes from Proposition 23 in terms of the minimal seminorm s on one hand and the corresponding linear combinations with the slope distances on the other, we get the following

Δ	$\beta_1 = 4$	$\beta_2 = \frac{4q}{p}$	$\beta_3 = 0$	1	2	3	$\mu = \infty$
β_1	0	$4(-p+q)$	4	3	2	1	1
β_2		0	$4q$	$-p+4q$	$-2p+4q$	$-3p+4q$	$-p$
β_3			0	1	2	3	1

TABLE 10. Distances between slopes in $W_{p/q}$ for $p/q \in (-\infty, 0)$, p odd, $q > 0$

linear system to solve for s, a_1, a_2 and a_3:

$$s + 2(-p + 6q - 1) = \|1\| = 3a_1 + (-p + 4q)a_2 + a_3 \tag{82a}$$

$$s + 3(-p + 4q - 1) = \|2\| = 2a_1 + (-2p + 4q)a_2 + 2a_3 \tag{82b}$$

$$s + 4(-p + 3q - 1) = \|3\| = a_1 + (-3p + 4q)a_2 + 3a_3 \tag{82c}$$

$$s = \|\infty\| = a_1 - pa_2 + a_3 \tag{82d}$$

which is equivalent in matrix form to

$$\underbrace{\begin{pmatrix} 3 & -p+4q & 1 & -1 \\ 2 & -2p+4q & 2 & -1 \\ 1 & -3p+4q & 3 & -1 \\ 1 & -p & 1 & -1 \end{pmatrix}}_{A} \begin{pmatrix} a_1 \\ a_2 \\ a_3 \\ s \end{pmatrix} = \begin{pmatrix} 2(-p-1)+12q \\ 3(-p-1)+12q \\ 4(-p-1)+12q \\ 0 \end{pmatrix} \tag{83}$$

Since the second row of A is equal to one half of the sum of the first and third rows, the rank of A is seen to be three, and the system is therefore underdetermined. We therefore need more information to determine the solution.

Using the bound from table 5, we have $s \leq -3p + 4q - 3$. In other words, we have

$$s = -3p + 4q - 3 - z, \tag{84}$$

where $z = z(p, q) \geq 0$ is some *positive*, integer-valued function in p and q.

Using this expression for s, we can formally solve the system (83) to obtain the following solution in terms of z:

$$\begin{pmatrix} a_1 \\ a_2 \\ a_3 \\ s \end{pmatrix} = \begin{pmatrix} -p + 2q - 1 - \frac{z}{2} \\ 2 + \frac{z}{4q} \\ 2q - 2 + z\frac{p-2q}{4q} \\ -3p + 4q - 3 - z \end{pmatrix}. \tag{85}$$

We can now use the fact that a_1, a_2, a_3 and s all are *even*, *non-negative* integers: The expression for a_2 then implies that z is divisible by $8q$, i.e. $z = 8qz'$ for some non-negative integer-valued $z' = z'(p, q)$. Using this in the expression for a_3 then shows that $2q - 2 + 2z'(p - 2q) \geq 0$, which implies that

$$z'(p - 2q) \geq 1 - q. \tag{86}$$

Since $p < 0$ and $q > 0$ implies $p - 2q < 0$, we then get

$$z' \leq \frac{1-q}{p-2q} = \frac{q-1}{2(q+|p|/2)} < 1. \tag{87}$$

But since $z' \in \mathbb{Z}_+$, we must conclude that in fact $z' = z = 0$.

This proves Theorem 1 and Corollary 4 for $p/q < 0$.

6.2. Assuming $p/q \in (0,2)$, p odd. In this case, we obtain the following distances between the relevant slopes:

Δ	$\beta_1 = 4$	$\beta_2 = \frac{2p+4q}{p}$	$\beta_3 = 0$	1	2	3	$\mu = \infty$
β_1	0	$-2p+4q$	4	3	2	1	1
β_2		0	$2p+4q$	$p+4q$	$4q$	$-p+4q$	p
β_3			0	1	2	3	1

TABLE 11. Distances between slopes in $W_{p/q}$ for $p/q \in (0,2)$, p odd

The corresponding linear system is in this case given by the equations:

$$s + 2(-p+6q-1) = \|1\| = 3a_1 + (p+4q)a_2 + a_3 \tag{88a}$$
$$s + 3(-p+4q-1) = \|2\| = 2a_1 + 4qa_2 + 2a_3 \tag{88b}$$
$$s + 4(-p+3q-1) = \|3\| = a_1 + (-p+4q)a_2 + 3a_3 \tag{88c}$$
$$s = \|\infty\| = a_1 + pa_2 + a_3 \tag{88d}$$

Using the upper bound $s \leq p+4q-3$ from table 5, we can assume

$$s = p + 4q - 3 - z, \tag{89}$$

where $z = z(p,q) \geq 0$ is some *positive*, integer-valued function in p and q.

Using this expression for s, we can formally solve the system (88) to obtain the following solution in terms of z:

$$\begin{pmatrix} a_1 \\ a_2 \\ a_3 \\ s \end{pmatrix} = \begin{pmatrix} -p+2q-1-\frac{qz}{2q-p} \\ 2+\frac{z}{2(2q-p)} \\ 2q-2-\frac{z}{2} \\ p+4q-3-z \end{pmatrix}. \tag{90}$$

Now recall that a_1, a_2, a_3 and s all are *even, non-negative* integers. The expression for a_2 then implies that z is divisible by $4(2q-p)$, i.e. $z = 4(2q-p)z'$ for some non-negative integer-valued $z' = z'(p,q)$. Using this in the expression for a_1 then shows that $-p+2q-1-4qz' \geq 0$, which implies that

$$z' \leq \frac{-p+2q-1}{4q} = 1 - \frac{p+1+2q}{4q} < 1, \tag{91}$$

so in fact $z' = z = 0$, proving Theorem 1 and Corollary 4 in the case $p/q \in (0,2)$.

6.3. Assuming $p/q \in (2,3) \cup (3,4)$, p odd. In this case, we obtain the following distances between the relevant slopes:

Δ	$\beta_1 = 4$	$\beta_2 = \frac{-p+6q}{q}$	$\beta_3 = 0$	1	2	3	$\mu = \infty$
β_1	0	$p-2q$	4	3	2	1	1
β_2		0	$-p+6q$	$-p+5q$	$-p+4q$	$\|p-3q\|$	q
β_3			0	1	2	3	1

TABLE 12. Distances between slopes in $W_{p/q}$ for $p/q \in (2,4)$, p odd

In this case, the corresponding linear system is:

$$s + 2(-p + 6q - 1) = \|1\| = 3a_1 + (-p + 5q)a_2 + a_3 \tag{92a}$$
$$s + 3(-p + 4q - 1) = \|2\| = 2a_1 + (-p + 4q)a_2 + 2a_3 \tag{92b}$$
$$s + 4(|p - 3q| - 1) = \|3\| = a_1 + |p - 3q|a_2 + 3a_3 \tag{92c}$$
$$s = \|\infty\| = a_1 + qa_2 + a_3 \tag{92d}$$

If $p/q \in (3, 4)$, this system already has rank 4, and we can determine the unique solution without further assumptions:

$$\begin{pmatrix} a_1 \\ a_2 \\ a_3 \\ s \end{pmatrix} = \begin{pmatrix} p - 2q - 1 \\ 4 \\ 2q - 2 \\ p + 4q - 3 \end{pmatrix}. \tag{93}$$

This proves Theorem 1 and Corollary 4 for $p/q \in (3, 4)$.

If on the other hand $p/q \in (2, 3)$, the system has only rank 3. Furthermore, using the upper bound for the minimal seminorm does *not* yet determine the system in this case.

But now recall from Lemma 10 (4) that $\mathrm{res}_{p,q} \doteq \mathrm{res}_{-p+4q,q}$. Furthermore, $2 < p/q < 3$ if and only if $1 < -p/q + 4 < 2$, and so the simplicity of the non-trivial roots of $\mathrm{res}_{p,q}$ when $p/q \in (2, 3)$ follows from the simplicity of the non-trivial roots of $\mathrm{res}_{-p+4q,q}$ when $\frac{-p+4q}{q} \in (1, 2)$, which we have already proven in section 6.2. We can therefore already assume the validity of Corollary 4 for $p/q \in (2, 3)$ and use that s actually *attains* the upper bound from table 5, i.e. $s = p + 4q - 3$. Solving the linear system with this information, we then obtain the same solution as given in equation (93), completing the proof of Theorem 1 in this case.

6.4. Assuming $p/q \in (4, 6) \cup (6, \infty)$, p odd. In this case, we obtain the following distances between the relevant slopes:

Δ	$\beta_1 = 4$	$\beta_2 = \frac{4q}{p-2q}$	$\beta_3 = 0$	1	2	3	$\mu = \infty$
β_1	0	$4p - 12q$	4	3	2	1	1
β_2		0	$4q$	$\lvert p - 6q\rvert$	$2p - 8q$	$3p - 10q$	$p - 2q$
β_3			0	1	2	3	1

TABLE 13. Distances between slopes in $W_{p/q}$ for $p/q \in (4, \infty)$, p odd

Similar to the previous cases, we obtain the corresponding linear system as:

$$s + 2(|p - 6q| - 1) = \|1\| = 3a_1 + |p - 6q|a_2 + a_3 \tag{94a}$$
$$s + 3(p - 4q - 1) = \|2\| = 2a_1 + (2p - 8q)a_2 + 2a_3 \tag{94b}$$
$$s + 4(p - 3q - 1) = \|3\| = a_1 + (3p - 10q)a_2 + 3a_3 \tag{94c}$$
$$s = \|\infty\| = a_1 + (p - 2q)a_2 + a_3 \tag{94d}$$

If $p/q \in (4,6)$, this system has rank 4, and we can determine the unique solution without further assumptions:

$$\begin{pmatrix} a_1 \\ a_2 \\ a_3 \\ s \end{pmatrix} = \begin{pmatrix} p-2q-1 \\ 2 \\ 2q-2 \\ 3p-4q-3 \end{pmatrix}. \tag{95}$$

This then proves Theorem 1 and Corollary 4 for $p/q \in (4,6)$.

If on the other hand $p/q \in (6,\infty)$, the system has only rank 3. Furthermore, using the upper bound for the minimal seminorm does *not* yet determine the system in this case.

But the same argument as used in the previous subsection holds here: By Lemma 10 (4), we have that $\mathrm{res}_{p,q} \doteq \mathrm{res}_{-p+4q,q}$. Furthermore, $p/q > 6$ if and only if $-p/q+4 < -2$, and so the simplicity of the non-trivial roots of $\mathrm{res}_{p,q}$ when $p/q > 6$ follows from the simplicity of the non-trivial roots of $\mathrm{res}_{-p+4q,q}$ when $\frac{-p+4q}{q} < -2$, which we have already proven in section 6.1. We can therefore already assume Corollary 4 for $p/q \in (6,\infty)$ and use that s actually *attains* the upper bound from table 5, i.e. $s = 3p-4q-3$. Solving the linear system with this information, we then obtain the same solution as given in equation (95), completing the proof of Theorem 1.

Acknowledgments

The work contained in this paper forms part of the author's dissertation and could not have been achieved without the never-waning support of his supervisor, Steven Boyer, for which the author would therefore like to express his sincerest thanks. Further thanks go to Stephan Tillmann for suggesting the "detour" through the eigenvalue variety and for many interesting discussions. Finally, the author would like to thank the eagle-eyed anonymous reviewer for catching a potentially show-stopping sign mistake, and whose comments and recommendations have greatly improved the readability of this paper.

References

[BAB01] L. Ben Abdelghani and S. Boyer, *A calculation of the Culler-Shalen seminorms associated to small Seifert Dehn fillings*, Proc. London Math. Soc. (3) **83** (2001), no. 1, 235–256, DOI 10.1112/plms/83.1.235. MR1829566

[BGZ01] S. Boyer, C. McA. Gordon, and X. Zhang, *Dehn fillings of large hyperbolic 3-manifolds*, J. Differential Geom. **58** (2001), no. 2, 263–308. MR1913944

[BMZ97] S. Boyer, T. Mattman, and X. Zhang, *The fundamental polygons of twist knots and the $(-2,3,7)$ pretzel knot*, KNOTS '96 (Tokyo), World Sci. Publ., River Edge, NJ, 1997, pp. 159–172. MR1664959

[Boy02] S. Boyer, *On the local structure of* $\mathrm{SL}(2,\mathbb{C})$*-character varieties at reducible characters*, Topology Appl. **121** (2002), no. 3, 383–413, DOI 10.1016/S0166-8641(01)00278-4. MR1909000

[Bro82] K. S. Brown, *Cohomology of groups*, Graduate Texts in Mathematics, vol. 87, Springer-Verlag, New York-Berlin, 1982. MR672956

[BZ96] S. Boyer and X. Zhang, *Finite Dehn surgery on knots*, J. Amer. Math. Soc. **9** (1996), no. 4, 1005–1050, DOI 10.1090/S0894-0347-96-00201-9. MR1333293

[BZ98] S. Boyer and X. Zhang, *On Culler-Shalen seminorms and Dehn filling*, Ann. of Math. (2) **148** (1998), no. 3, 737–801, DOI 10.2307/121031. MR1670053

[BZ03] G. Burde and H. Zieschang, *Knots*, 2nd ed., De Gruyter Studies in Mathematics, vol. 5, Walter de Gruyter & Co., Berlin, 2003. MR1959408

[CCG+94] D. Cooper, M. Culler, H. Gillet, D. D. Long, and P. B. Shalen, *Plane curves associated to character varieties of 3-manifolds*, Invent. Math. **118** (1994), no. 1, 47–84, DOI 10.1007/BF01231526. MR1288467

[CGLS87] M. Culler, C. McA. Gordon, J. Luecke, and P. B. Shalen, *Dehn surgery on knots*, Ann. of Math. (2) **125** (1987), no. 2, 237–300, DOI 10.2307/1971311. MR881270

[CLO97] D. Cox, J. Little, and D. O'Shea, *Ideals, varieties, and algorithms*, 2nd ed., Undergraduate Texts in Mathematics, Springer-Verlag, New York, 1997. An introduction to computational algebraic geometry and commutative algebra. MR1417938

[CS83] M. Culler and P. B. Shalen, *Varieties of group representations and splittings of 3-manifolds*, Ann. of Math. (2) **117** (1983), no. 1, 109–146, DOI 10.2307/2006973. MR683804

[DT03] N. M. Dunfield and W. P. Thurston, *The virtual Haken conjecture: experiments and examples*, Geom. Topol. **7** (2003), 399–441, DOI 10.2140/gt.2003.7.399. MR1988291

[FH88] W. Floyd and A. Hatcher, *The space of incompressible surfaces in a 2-bridge link complement*, Trans. Amer. Math. Soc. **305** (1988), no. 2, 575–599, DOI 10.2307/2000879. MR924770

[Fox53] R. H. Fox, *Free differential calculus. I. Derivation in the free group ring*, Ann. of Math. (2) **57** (1953), 547–560, DOI 10.2307/1969736. MR53938

[Gol84] W. M. Goldman, *The symplectic nature of fundamental groups of surfaces*, Adv. in Math. **54** (1984), no. 2, 200–225, DOI 10.1016/0001-8708(84)90040-9. MR762512

[HMW92] C. D. Hodgson, G. R. Meyerhoff, and J. R. Weeks, *Surgeries on the Whitehead link yield geometrically similar manifolds*, Topology '90 (Columbus, OH, 1990), Ohio State Univ. Math. Res. Inst. Publ., vol. 1, de Gruyter, Berlin, 1992, pp. 195–206. MR1184411

[HP05] M. Heusener and J. Porti, *Deformations of reducible representations of 3-manifold groups into* $\mathrm{PSL}_2(\mathbb{C})$, Algebr. Geom. Topol. **5** (2005), 965–997, DOI 10.2140/agt.2005.5.965. MR2171800

[HPSP01] M. Heusener, J. Porti, and E. Suárez Peiró, *Deformations of reducible representations of 3-manifold groups into* $\mathrm{SL}_2(\mathbf{C})$, J. Reine Angew. Math. **530** (2001), 191–227, DOI 10.1515/crll.2001.003. MR1807271

[HS07] J. Hoste and P. D. Shanahan, *Computing boundary slopes of 2-bridge links*, Math. Comp. **76** (2007), no. 259, 1521–1545, DOI 10.1090/S0025-5718-07-01936-9. MR2299787

[Jac80] W. Jaco, *Lectures on three-manifold topology*, CBMS Regional Conference Series in Mathematics, vol. 43, American Mathematical Society, Providence, R.I., 1980. MR565450

[Las93] A. E. Lash, *Boundary curve space of the Whitehead link complement*, ProQuest LLC, Ann Arbor, MI, 1993. Thesis (Ph.D.)–University of California, Santa Barbara. MR2689967

[Map] Maplesoft, a division of Waterloo Maple Inc., Waterloo, Ontario., *Maple*.

[Mat02] T. W. Mattman, *The Culler-Shalen seminorms of the* $(-2,3,n)$ *pretzel knot*, J. Knot Theory Ramifications **11** (2002), no. 8, 1251–1289, DOI 10.1142/S0218216502002232. MR1949779

[MP06] B. Martelli and C. Petronio, *Dehn filling of the "magic" 3-manifold*, Comm. Anal. Geom. **14** (2006), no. 5, 969–1026. MR2287152

[Por97] J. Porti, *Torsion de Reidemeister pour les variétés hyperboliques* (French, with English and French summaries), Mem. Amer. Math. Soc. **128** (1997), no. 612, x+139, DOI 10.1090/memo/0612. MR1396960

[Ril72] R. Riley, *Parabolic representations of knot groups. I*, Proc. London Math. Soc. (3) **24** (1972), 217–242, DOI 10.1112/plms/s3-24.2.217. MR300267

[Ril75] R. Riley, *Parabolic representations of knot groups. II*, Proc. London Math. Soc. (3) **31** (1975), no. 4, 495–512, DOI 10.1112/plms/s3-31.4.495. MR413078

[Sha94] I. R. Shafarevich, *Basic algebraic geometry. 1*, 2nd ed., Springer-Verlag, Berlin, 1994. Varieties in projective space; Translated from the 1988 Russian edition and with notes by Miles Reid. MR1328833

[Sha02] P. B. Shalen, *Representations of 3-manifold groups*, Handbook of geometric topology, North-Holland, Amsterdam, 2002, pp. 955–1044. MR1886685

[Sze75] G. Szegő, *Orthogonal polynomials*, 4th ed., American Mathematical Society, Providence, R.I., 1975. American Mathematical Society, Colloquium Publications, Vol. XXIII. MR0372517

[Thu79] W. P. Thurston, *Geometry and Topology of Three-Manifolds. Princeton lecture notes*, 1979.

[Til02] S. Tillmann, *On character varieties: Surfaces associated to mutation & deformation of hyperbolic 3-manifolds*, Ph.D. thesis, University of Melbourne, August 2002.

[Til05] S. Tillmann, *Boundary slopes and the logarithmic limit set*, Topology **44** (2005), no. 1, 203–216, DOI 10.1016/j.top.2004.07.002. MR2104008

[Til20] S. Tillmann, Tropical varieties associated to ideal triangulations: The Whitehead link complement, *Characters in Low-Dimensional Topology* (O. Collin, S. Friedl, C. Gordon, and S. Tillmann, eds.), Contemporary Mathematics, vol. 760, Amer. Math. Soc., Providence, RI, 2020, pp. 293–339.

[Wad94] M. Wada, *Twisted Alexander polynomial for finitely presentable groups*, Topology **33** (1994), no. 2, 241–256, DOI 10.1016/0040-9383(94)90013-2. MR1273784

[Wei64] A. Weil, *Remarks on the cohomology of groups*, Ann. of Math. (2) **80** (1964), 149–157, DOI 10.2307/1970495. MR169956

Département de mathématiques, Université du Québec à Montréal, 201, Avenue du Président-Kennedy, Montréal (Québec), Canada H2X 3Y7

Current address: Department of Mathematics, Champlain College St-Lambert, 900 Riverside Drive, St-Lambert (Québec), Canada J4P 3P2

Email address: gindurskis@crcmail.net

Contemporary Mathematics
Volume **760**, 2020
https://doi.org/10.1090/conm/760/15292

Alternating knots with large boundary slope diameter

Masaharu Ishikawa, Thomas W. Mattman, Kazuya Namiki, and Koya Shimokawa

Dedicated to Steve Boyer on his 65th birthday.

ABSTRACT. We show that, for an alternating knot, the ratio of the diameter of the set of boundary slopes to the crossing number can be arbitrarily large.

1. Introduction

Let K be a knot in S^3 and $c(K)$ the crossing number of K. The *diameter* of the set of boundary slopes of K, denoted by $\text{diam}(K)$, is the difference, as rational numbers, between the maximal boundary slope and the minimal boundary slope of K. We show that, for alternating knots, the ratio of the diameter of the boundary slopes to the crossing number can be arbitrarily large.

Curtis and Taylor [**CT**] showed the diameter of the boundary slopes of essential spanning surfaces of an alternating knot K is at least $2c(K)$. See also [**H**]. Ichihara and Mizushima [**IM**] argued that, for a Montesinos knot K, the ratio $\frac{\text{diam}(K)}{c(K)}$ is at most two, with equality if K is alternating. In fact, they also showed $\frac{\text{diam}(K)}{c(K)} \geq 2$ in the case of alternating knots (whether or not they are Montesinos). According to Culler's calculation of the A-polynomial [**C**], the 8–crossing non-Montesinos alternating knots 8_{17} and 8_{18} have boundary slopes -14 and 14 so that the diameter is at least 28, which is larger than twice the crossing number. Those calculations are confirmed by Kabaya, who, by applying his method for the deformation variety [**K2**], also demonstrated diameters in excess of twice the crossing number for the alternating knots 9_{32}, 9_{33}, 9_{34}, 9_{39}, 10_{82}, and 10_{94}. In the case of 9_{39}, he showed the diameter is at least 30, which is more than three times the crossing number [**K1**]. Dunfield and Garoufalidis [**CG**] also constructed interesting examples of alternating knots with rational boundary slopes using spun-normal surfaces.

Here we prove that, for an alternating knot, the ratio of the diameter of the boundary slopes to the crossing number can be arbitrarily large.

2010 *Mathematics Subject Classification.* 57M25.

Key words and phrases. Boundary slope, alternating knots, crossing number.

The first author is supported by JSPS KAKENHI Grant Numbers JP19K03499, JP17H06128 and Keio University Academic Development Funds for Individual Research. The second author thanks the Math Department of Saitama University for their hospitality during multiple visits related to this project. The last author is supported by JSPS KAKENHI Grant Numbers JP16K13751, JP16H03928, JP17H06460, JP17H06463 and JST CREST JPMJCR17J4.

THEOREM 1.1. *For any positive number r, there exists an alternating knot K such that the ratio $\frac{\mathrm{diam}(K)}{c(K)} > r$. In particular, there is a sequence of alternating knots $\{K_n\}_{n=1}^{\infty}$ such that $\lim_{c(K_n)\to\infty} \frac{\mathrm{diam}(K_n)}{c(K_n)} = \infty$.*

We adapt Hatcher and Oertel's [**HO**] method for Montesinos knots to arborescent knots formed as the product of two Montesinos tangles. In the next section we give a framework for such knots and in Section 3 we describe how to construct candidate surfaces from edgepaths in this setting. Section 4 shows how to calculate the corresponding boundary slopes. Finally, in Section 5 we prove our main theorem.

2. Arborescent knots

In this article, a *tangle* (B, T) is a pair of a 3-ball B and its properly embedded 1-submanifold T in B with 4 boundary points NE, NW, SE and SW. Sometimes a tangle is simply denoted by T. We assume that B is embedded in $\mathbb{R}^3$ and the four points NE, NW, SE and SW are on a plane P in $\mathbb{R}^3$. A *rational tangle* is a tangle that is homeomorphic to the pair $(D \times [0,1], \{p_1, p_2\} \times [0,1])$, where D is a disk and p_1 and p_2 are points in the interior of D.

DEFINITION 2.1. Let (B_1, T_1) and (B_2, T_2) be two tangles. We identify the right hemisphere of ∂B_1 and the left hemisphere of ∂B_2 to form a tangle $(B_1 \cup B_2, T_1 \cup T_2)$. The resulting tangle is denoted by $T_1 + T_2$ and called the *tangle sum* of T_1 and T_2. For rational tangles $(B_1, R_1), \cdots, (B_n, R_n)$, the tangle sum $R_1 + \cdots + R_n$ is called a *Montesinos tangle*. Let Δ_i be the disk $B_i \cap B_{i+1}$. Let A be $\partial\Delta_i$. The loop A is called the axis of the Montesinos tangle.

DEFINITION 2.2. Let R and T be tangles. Let R' denote the tangle that results from $\frac{\pi}{2}$-rotation of the reflection of R. Here the reflection means the tangle obtained by exchanging "over" and "under" at all crossings in the tangle. Then the *tangle product* of R and T, denoted $R \circ T$, is the sum $R' + T$. See Figure 1. We also define the axis of $R \circ T$ as the union of the boundaries of the left and right hemispheres of the attaching tangles.

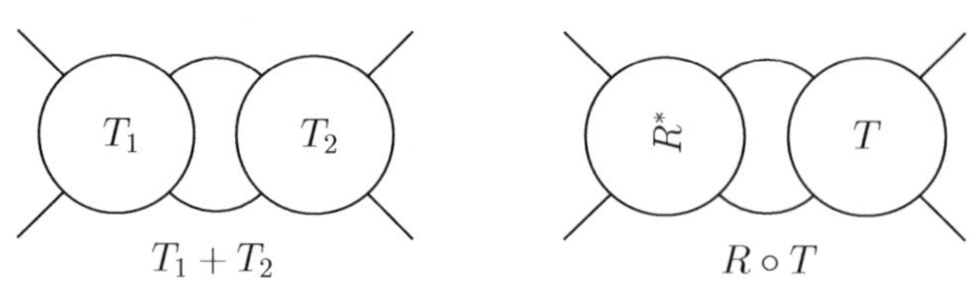

FIGURE 1. Tangle sum and product. R^* denotes the reflection of R. R' is the tangle that results from $\frac{\pi}{2}$-rotation of R^* .

DEFINITION 2.3. Let (B, T) be a tangle. We can construct a knot or a link by connecting NE and NW with an arc in $P \cap \partial B$ and SE and SW with an arc in $P \cap \partial B$. This knot or link is denoted by $N(T)$ and called the *numerator* of T.

DEFINITION 2.4. A knot K is an *SN knot* if there are Montesinos tangles T_1 and T_2 such that $K = N(T_1 \circ T_2)$. We call T_2 the *outer tangle*. The tangle obtained from T_1 by $\frac{\pi}{2}$-rotation of the reflection will be denoted T_1' and called the *inner tangle*.

Note that for an alternating Montesinos tangle T, the SN knot $N(T \circ T)$ is also alternating.

DEFINITION 2.5. A tangle is called an *algebraic tangle* if it is obtained by a finite sequence of tangle sums and tangle products of rational tangles. The numerator $N(T)$ of an algebraic tangle T is called an *arborescent knot or link*.

Note that the numerator of (B, T) can be obtained by identifying the left hemisphere of B with the right one. If an algebraic tangle T is made of rational tangles $(B_1, R_1), \cdots, (B_n, R_n)$, then the arborescent knot $N(T)$ is in the S^3 that is the union of the 3-balls $B_1 \cup \cdots \cup B_n$. The union of all the axes is an embedded graph in S^3.

3. Candidate surfaces and edgepaths

In this section we describe how to build candidate surfaces from edgepaths in the context of an arborescent knot. We closely follow the discussion of Hatcher and Oertel [**HO**] and begin with a review of some terminology from that paper. In Figure 2, we show the train track with weight (a, b, c). The triple (a, b, c) also labels a point of the Diagram $\mathcal{D}$ in [**HO**] with vertical coordinate "slope" $v = \frac{c}{a+b}$ and horizontal coordinate $u = \frac{b}{a+b}$. Let $\langle \frac{p}{q} \rangle$ denote the point $(1, q-1, p)$.

Given such a train track, Hatcher and Oertel describe how to construct an edgepath γ_i with the following four properties:

(E1) The starting point of γ_i lies on the edge $\langle \frac{p_i}{q_i}, \frac{p_i}{q_i} \rangle$, and if the starting point is not the vertex $\langle \frac{p_i}{q_i} \rangle$, then the edgepath γ_i is constant.

(E2) γ_i is minimal, i.e., it never stops and retraces itself, nor does it go along two sides of the same triangle of $\mathcal{D}$ in succession.

(E3) The ending points of the γ_i's are all rational points of $\mathcal{D}$ which all lie in one vertical line and whose vertical coordinates add up to zero.

(E4) γ_i proceeds monotonically from right to left, "monotonically" in a weak sense that motion along vertical edges is permitted.

As in [**HO**], from an edgepath γ_i we can construct a surface S_i in B_i with $\partial S_i \subset \partial B_i \cup K$, which we will call the *surface S_i given by the edgepath γ_i*. So far, except for property E3, we have described Hatcher and Oertel's construction for an individual rational tangle (B_i, R_i).

Now, let K be an arborescent knot made from rational tangles $(B_1, R_1), \cdots,$ (B_n, R_n). We will construct a properly embedded surface in the exterior of K using the S_i's given by the edgepaths γ_i. As an arborescent knot can be constructed by a sequence of tangle sums and tangle products, we take these operations in turn, starting with the sum.

LEMMA 3.1 ([**HO**]). *Let T_1 and T_2 be tangles with surfaces $\hat{S}_1$ and $\hat{S}_2$ having train tracks with weights (a_1, b_1, c_1) and (a_2, b_2, c_2) on the boundary, respectively. Suppose $a_1 = a_2$ and $b_1 = b_2$. Then we can construct a surface S in $T_1 + T_2$ for the triple $(a_1, b_1, c_1 + c_2)$ by gluing $\hat{S}_1$ and $\hat{S}_2$ canonically.*

Next we consider the tangle product. The tangle product can be obtained by a sequence of operations consisting of $\frac{\pi}{2}$-rotation, reflection, and tangle sum. In

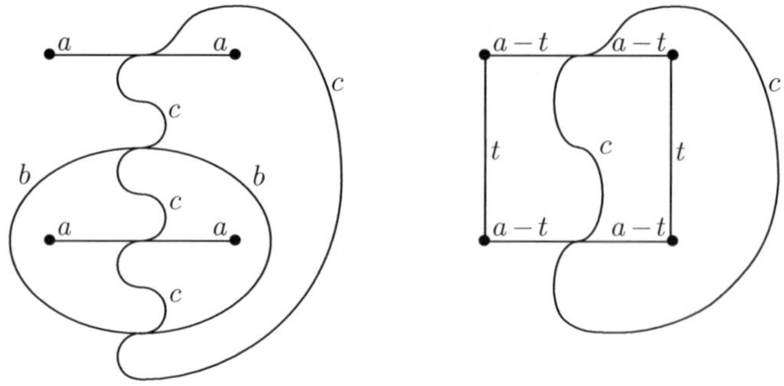

FIGURE 2. A train track with weight (a, b, c). The right figure has t ∞-edges.

the next lemma, we will consider how the weight of the train track is changed by $\frac{\pi}{2}$-rotation and reflection.

LEMMA 3.2. *Let T be a tangle with a surface having a train track with weight (a, b, c) on the boundary. Then the weight (x, y, z) of the train track of the tangle T' obtained from T by $\frac{\pi}{2}$-rotation and reflection satisfies the following in the case $c > 0$ (respectively, $c < 0$).*

1. *If T has neither slope 0 edges nor slope ∞ edges, then $(x, y, z) = (a, c - a, a + b)$ (resp. $(a, |c| - a, -(a + b))$).*
2. *If T has slope 0 edges and does not have slope ∞ edges, then, T' has $a - |c|$ slope ∞ edges and triple $(x, y, z) = (c, 0, b + c)$ (resp., $(|c|, 0, -(b + |c|))$).*
3. *If T does not have slope 0 edges and has t slope ∞ edges where $t < a$, then $(x, y, z) = (a, c - a + t, a - t)$ (resp. $(a, |c| - a + t, t - a)$).*
4. *If T has both slope 0 edges and t slope ∞ edges, where $t < a - |c|$, then T' has $a - t - |c|$ slope ∞ edges and triple $(x, y, z) = (c + t, 0, c)$ (resp. $(|c| + t, 0, -c)$).*

PROOF. We show the assertion in Case (1). In this case, the train track for T is the one shown on the left in Figure 2. If $c > 0$ then the $\frac{\pi}{2}$-rotation of its reflection becomes the train track on the top left in Figure 3. Then the assertion follows from the isotopy shown in Figure 3. If $c < 0$, the isotopy is given by the mirror image of the isotopy in Figure 3.

The assertions for Cases (2), (3) and (4) follow from the isotopies shown in Figures 4, 5 and 6, respectively. □

To construct a candidate surface of an arborescent knot K, we replace property E3 of Hatcher and Oertel with the following.

(E3') The ending points of the γ_i's are all rational points of $\mathcal{D}$ chosen so that the surfaces S_i given by the edgepaths γ_i constitute a surface bounded by the knot K. We can easily check if the S_i constitute a surface using Lemmas 3.1 and 3.2.

We call the surface obtained from the edgepaths as above a *candidate surface of the arborescent knot.*

Note that, in [**W**], Wu studied exceptional surgeries of large arborescent knots. The class of SN knots in our paper includes knots of type II in [**W**]. In that paper,

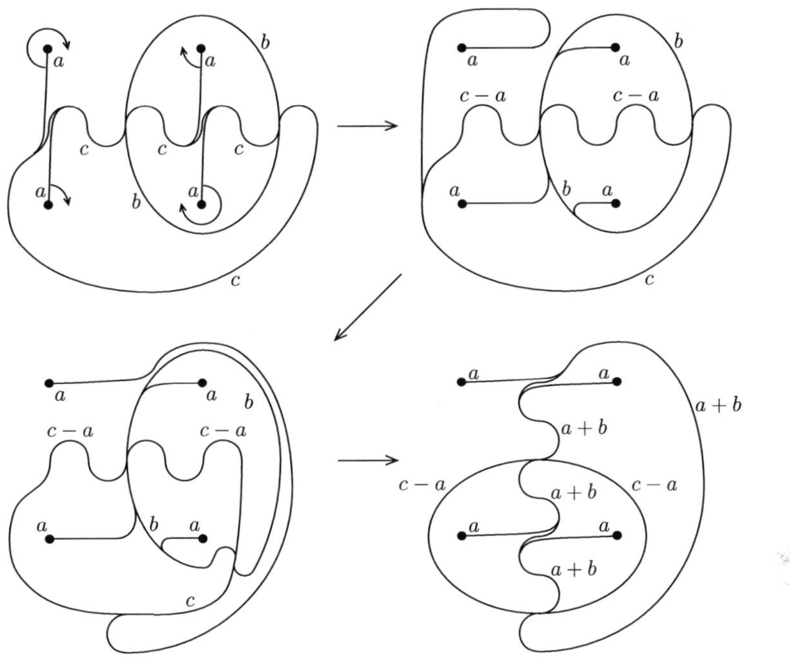

FIGURE 3. Case (1) : T has neither slope 0 edges nor slope ∞ edges.

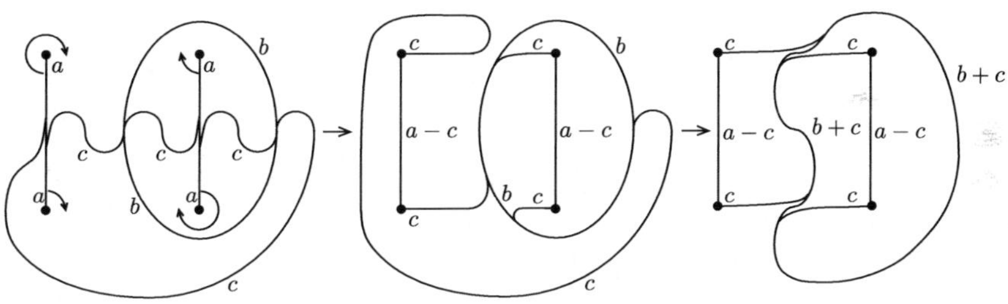

FIGURE 4. Case (2) : T has slope 0-edges. Then T' has $a - c$ slope ∞ edges.

he used a different train track that works especially well for knots of type II. In our paper, we follow the notation of Hatcher and Oertel to emphasize that the observations in Lemmas 3.1 and 3.2 work for any arborescent knot. In fact, these lemmas work for any tangle with a train track.

4. Calculation of boundary slopes

In this section we will calculate the boundary slope of a candidate surface S of an arborescent knot. Let $(B_1, R_1), \cdots, (B_n, R_n)$ be rational tangles constituting an arborescent knot. For each surface S_i in B_i given by an edgepath, we define $\tau(S_i)$ to be $\tau(S_i) = 2(e_- - e_+)$, where e_+ (resp. e_-) is the number of edges of the edgepath which increase (resp. decrease) slope, allowing fractional values for $e_\pm$ corresponding to a final edge traversing only a fraction of an edge of the diagram

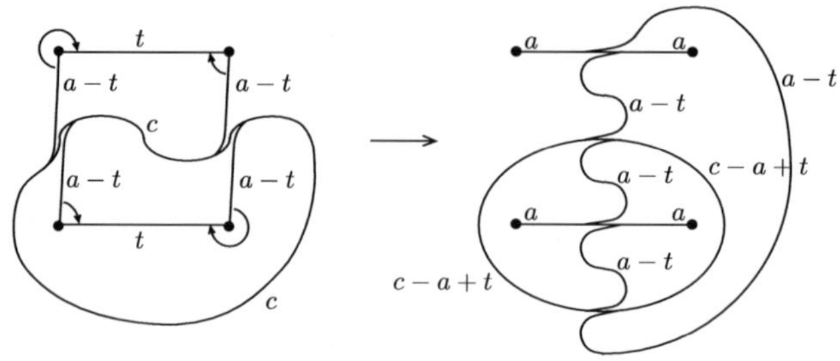

FIGURE 5. Case (3) : T has t slope ∞ edges with $t < a$.

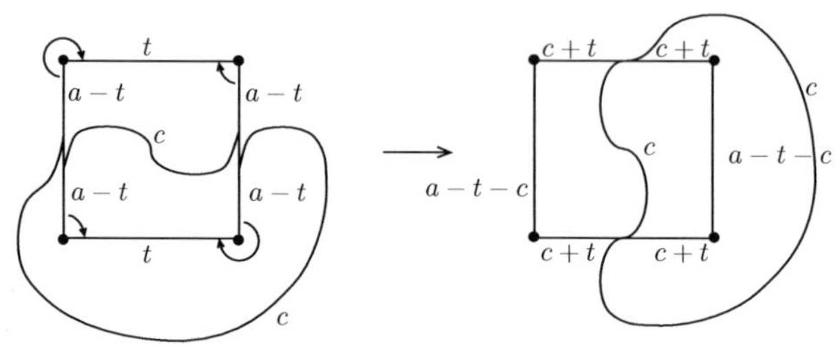

FIGURE 6. Case (4) : T has t slope ∞ edges with $t < a - c$. Then T' has $a - t - c$ slope ∞ edges.

$\mathcal{D}$. See [**HO**, page 460] for a precise explanation. This number $\tau(S_i)$ counts how many times the surface S_i rotates around the knot in B_i. In particular, if S_i is given by a constant edgepath then we have $\tau(S_i) = 0$. In [**HO**], $\tau(S)$ is defined to be the sum of these numbers for rational tangles constituting a Montesinos knot. We use the same idea for arborescent knots.

Let $\hat{S}$ be a surface in an algebraic tangle T obtained by gluing surfaces given by edgepaths in the rational tangles constituting T. We call $\hat{S}$ a *surface given by edgepaths*. We define $\tau(\hat{S})$ by applying Definitions 4.1 and 4.2 below inductively.

The first definition is for a tangle sum.

DEFINITION 4.1. Let T_1 and T_2 be algebraic tangles with surfaces $\hat{S}_1$ and $\hat{S}_2$, given by edgepaths, with the triples (a, b, c_1) and (a, b, c_2), respectively. Let $\hat{S}$ be the surface in $T_1 + T_2$ with the triple $(a_1, b_1, c_1 + c_2)$ obtained by gluing $\hat{S}_1$ and $\hat{S}_2$, whose existence is stated in Lemma 3.1. Then $\tau(\hat{S})$ is defined by $\tau(\hat{S}) = \tau(\hat{S}_1) + \tau(\hat{S}_2)$.

Before giving the second definition, we define $\tau'(\hat{S})$, which measures how much the surface rotates along the knot during the isotopies in Figures 3 through 6. Let T be a tangle with a candidate surface $\hat{S}$ having triple (a, b, c) and $\hat{S}'$ be the surface obtained from $\hat{S}$ by one of the isotopies shown in Figures 3 through 6. Set m to be the number of sheets that rotate around the knot during the isotopy. Note that $m = a$ in Case (1), $m = c$ in Cases (2) and (4) and $m = a - t$ in Case (3), where t is the number of slope ∞ edges in T. We define $\tau'(\hat{S})$ as follows.

$$\tau'(\hat{S}) = \begin{cases} -\frac{2m}{a} & \text{if} \quad c > 0 \\ \frac{2m}{a} & \text{if} \quad c < 0. \end{cases}$$

DEFINITION 4.2. Let T_1 and T_2 be algebraic tangles with surfaces $\hat{S}_1$ and $\hat{S}_2$ given by edgepaths, respectively. Suppose that $\hat{S}_1$ and $\hat{S}_2$ are glued in $T_1 \circ T_2$ canonically as in Lemma 3.1 after the $\frac{\pi}{2}$-rotation and reflection of T_1 for the tangle product. Let $\hat{S}$ denote this surface in $T_1 \circ T_2$. Then $\tau(\hat{S})$ is defined by $\tau(\hat{S}) = -\tau(\hat{S}_1) + \tau'(\hat{S}_1) + \tau(\hat{S}_2)$.

The value τ for a candidate surface of an arborescent knot is then defined as follows.

DEFINITION 4.3. Suppose that an arborescent knot K is the numerator of an algebraic tangle T, *i.e.*, $N(T) = K$, and let S be a candidate surface of K obtained from a surface $\hat{S}$ in T given by edgepaths. Then $\tau(S)$ is defined by $\tau(S) = \tau(\hat{S})$.

To get the boundary slope of the candidate surface S from $\tau(S)$, we need to determine $\tau(S_0)$ for a Seifert surface S_0. Remark that we use the same tangle decomposition of K when we calculate $\tau(S)$ and $\tau(S_0)$.

LEMMA 4.4. *Let K be an arborescent knot consisting of Montesinos tangles including a rational tangle with even denominator and let S_0 be a Seifert surface of K. Then $\tau(S_0)$ is the sum of $(-1)^{k_i}\tau(S_i)$ for all Montesinos tangles T_i constituting K, where k_i is the number of reflections applied to the tangle T_i.*

PROOF. As in [**HO**], we find a piece of the Seifert surface for each Montesinos tangle (B_i, T_i). In particular, the slopes of the surfaces in these tangles are ∞. When we make a tangle product, we take the reflection of B_i and rotate it by $\frac{\pi}{2}$. Therefore, its slope becomes 0 after the rotation. To glue this with other pieces, we add a saddle near ∂B_i that changes the slope from 0 to ∞. We can find suitable edgepaths whose surface can be glued with keeping the orientability when each Montesinos tangle contains a rational tangle with even denominator, see the explanation in [**HO**, page 461]. The surface so constructed is single sheeted and orientable, hence a Seifert surface. Since adding a saddle does not change the number of twists around the knot, the contribution to $\tau(S_0)$ of this adjustment is 0. Hence the assertion follows. □

LEMMA 4.5. *Let S be a candidate surface of an arborescent knot. Then $\tau(S) - \tau(S_0)$ is the boundary slope of S.*

PROOF. As in [**HO**], the value of τ can be determined by checking how the surface rotates around the knot. This rotation number is additive under tangle sum. Hence the rotation number of the surface in $T_1 + T_2$ obtained from surfaces $\hat{S}_1$ and $\hat{S}_2$ in T_1 and T_2, respectively, given by edgepaths, as mentioned in Lemma 3.1, is $\tau(\hat{S}_1) + \tau(\hat{S}_2)$. This is formulated in Definition 4.1. Next we observe the tangle product $T_1 \circ T_2$ with surfaces $\hat{S}_1$ and $\hat{S}_2$ given by edgepaths. Remember that the tangle product can be obtained by a sequence of operations consisting of $\frac{\pi}{2}$-rotation, reflection of T_1, and tangle sum. Since $\hat{S}'_1$ is the reflection of $\hat{S}_1$, the contribution of the twists of $\hat{S}_1$ to $\hat{S}'_1$ is $-\tau(\hat{S}_1)$. The contribution during the isotopy in Figures 3 through 6 is the value $\tau'(\hat{S}_1)$, which can be verified directly from those figures. Therefore, the rotation number becomes $-\tau(\hat{S}_1)+\tau'(\hat{S}_1)+\tau(\hat{S}_2)$ as in Definition 4.2. Thus $\tau(S)$ counts the rotation of S around the knot correctly. The boundary slope of S is given by the difference $\tau(S) - \tau(S_0)$ as in [**HO**]. □

5. Proof of Theorem 1.1

For $n \geq 2$, let K_n denote the SN knot $N((-\frac{1}{n}+\frac{1}{n+1})\circ(-\frac{1}{n}+\frac{1}{n+1}))$, see Figure 7. Note that K_n is achiral. Here $K_n = N(T_1 \circ T_2)$ and $(B_1, T_1) = (B_{1,1}, T_{1,1}) + (B_{1,2}, T_{1,2})$ and $(B_2, T_2) = (B_{2,1}, T_{2,1}) + (B_{2,2}, T_{2,2})$. Let A_i be the *axis* defined by $\partial(B_{i,1} \cap B_{i,2})$ $(i = 1, 2)$.

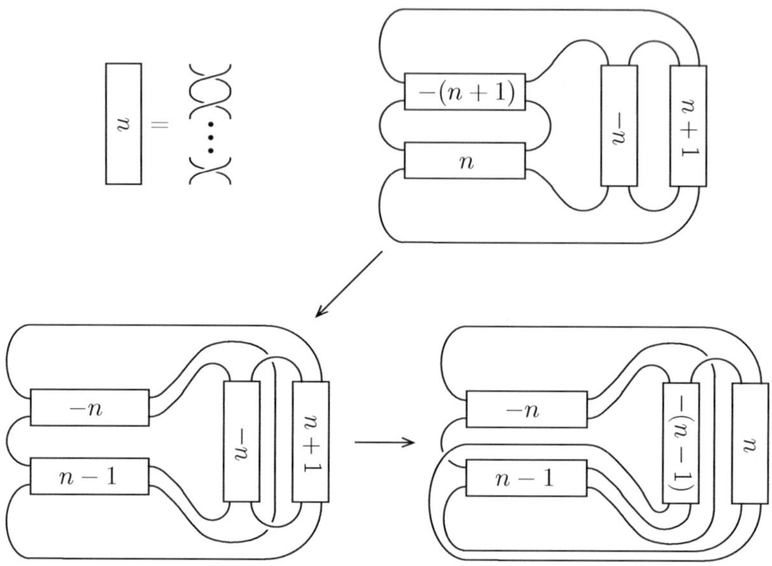

FIGURE 7. K_n is an alternating knot with crossing number $4n$. The number n in a box means the right-handed n half twists and $-n$ in a box means the left-handed n half twists.

PROPOSITION 5.1. *$-2(n+1)^2 + 4$ and $2(n+1)^2 - 4$ are boundary slopes of essential spanning surfaces of K_n.*

PROOF. Since K_n is achiral, it is enough to show that $-2(n+1)^2 + 4$ is the boundary slope of an essential spanning surface. For this, we describe the surface in terms of edgepaths in $T_1 = T_2 = -\frac{1}{n} + \frac{1}{n+1}$. In T_1, we use constant edgepaths in both the $-\frac{1}{n}$ and $\frac{1}{n+1}$ rational tangles. Let the triples for the two rational tangles be $(1, n^2+n-1, -n-1)$ and $(1, n^2+n-1, n)$ so that $(1, n^2+n-1, -1)$ is the triple for T_1. As there are neither 0-edges nor ∞-edges, by Lemma 3.2, the triple of T_1 after reflection, $\frac{\pi}{2}$-rotation and isotopy is $(1, 0, -n^2-n)$. Therefore the triple of T_2 must be $(1, 0, n^2+n)$. We choose the edgepaths of the $-\frac{1}{n}$ and $\frac{1}{n+1}$ tangles in T_2 as $\langle -\frac{1}{n}\rangle \to \langle 0\rangle$ and $\langle \frac{1}{n+1}\rangle \to \langle \frac{1}{n}\rangle \to \cdots \to \langle \frac{1}{2}\rangle \to \langle 1\rangle \to \cdots \to \langle n^2+n\rangle$, respectively.

Let S be the candidate surface obtained from these edgepaths. We must argue that the candidate surface S is incompressible. We adapt the arguments of Hatcher and Oertel [**HO**] to this case. Let D be a compressing disk of S in $E(K)$. We may assume that D meets ∂B_i and A_i transversely and misses the intersection points $A_i \cap A_j$. Set $G = D \cap (\partial B_1 \cup \partial B_2)$, which is a graph in D. We assume that the number of components of G is minimal among all compressing disks for S.

If G contains a loop that does not meet vertices, then by a surgery along an innermost disk of D, we will have another compressing disk whose graph has fewer components, a contradiction. Hence there is no loop without vertices in G.

We next argue that $D \cap B_1 \neq \emptyset$. For a contradiction, assume $D \cap B_1$ is empty. Let $(B_{2,1}, T_{2,1})$ and $(B_{2,2}, T_{2,2})$ denote the $-\frac{1}{n}$ and $\frac{1}{n+1}$ tangles in B_2, respectively.

Suppose that edgemost disks are contained in $B_{2,1}$. Let D' be an edgemost disk that is bounded by the union of an arc α on $B_{2,1} \cap B_{2,2}$ and an arc β on ∂D, see Figure 8 (left). Since the edgepath of $T_{2,1}$ is $\langle -\frac{1}{n} \rangle \to \langle 0 \rangle$, the piece $S \cap B_{2,1}$ of the candidate surface S in $B_{2,1}$ is a band as shown on the right in Figure 8. Then the disk D' either lies in the position shown in the figure or is bounded by $\alpha \cup \beta$ such that the endpoints of α lie on the same connected arc of $S \cap \partial B_{2,1}$. The latter case can be removed by isotopy of D'. Consider the former case. The assumption $D \cap B_1 = \emptyset$ implies that the arc α does not intersect the hemisphere $B_1 \cap B_{2,1}$. Since the band $S \cap B_{2,1}$ is twisted n times, there exists such a disk D' if and only if $|n| \leq 1$. In particular, if $n \geq 2$ then D' and B_1 should intersect. This contradicts the assumption that $D \cap B_1$ is an empty set.

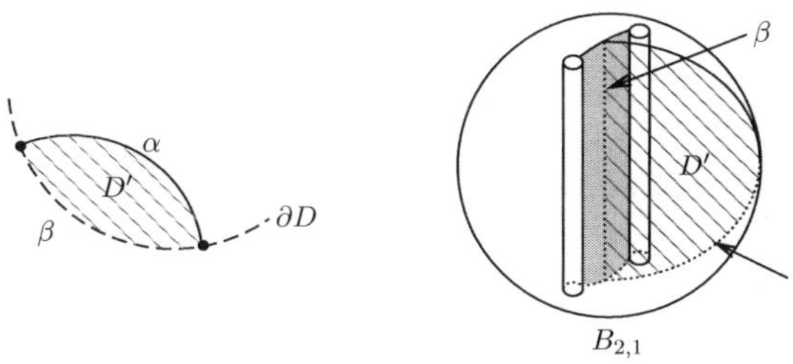

FIGURE 8. The edgemost disk D' in $B_{2,1}$.

Suppose that edgemost disks are in $B_{2,2}$. Then there is a disk Δ in D contained in $B_{2,1}$ whose boundary is $\alpha_1, \beta_1, \ldots, \alpha_k, \beta_k$, where the α_i are on $B_{2,1} \cap B_{2,2}$ and the β_i are contained in ∂D, see Figure 9. Since $D \cap B_1$ is empty, $\partial\Delta$ does not intersect the axis $\partial(B_{2,1} \cap B_{2,2})$. However, there is no such disk Δ since $S \cap B_{2,1}$ is a band as mentioned before.

Therefore, in either case, we have a contradiction.

Now, applying the argument of [**HO**, Proposition 2.1] to the tangles $B_{1,1}$, $B_{1,2}$ and B_2, where $T_{1,1}$ and $T_{1,2}$ are the $-\frac{1}{n}$ and $\frac{1}{n+1}$ tangles in B_1, respectively, $D \cap B_1$ includes at most one innermost disk D', which means there is at most one constant edgepath in T_1. However, by assumption, there are two. The contradiction shows there can be no such compressing disk and S is incompressible.

Finally, we calculate the slope associated to the incompressible surface S. Since both of the tangles in T_1 have constant edgepaths, $\tau(S_1) = 0$. Since there are neither 0 nor ∞ edges in the train track for T_1 and $c < 0$, $\tau'(S_1) = 2$. For $\tau(S_2)$, there is one upward edge on $\mathcal{D}$ for the $-\frac{1}{n}$ tangle and there are a further n^2+2n-1 upward edges for the $\frac{1}{n+1}$ tangle, so $\tau(S_2) = 2(-1-(n^2+2n-1)) = -2(n^2+2n)$. The twist of the surface is therefore $\tau(S) = -\tau(S_1) + \tau'(S_1) + \tau(S_2) = 0 + 2 - 2(n^2+2n) = -2(n+1)^2 + 4$ by Lemma 4.5.

Let $S_{0,1}$ and $S_{0,2}$ be pieces of a Seifert surface S_0 for T_1 and T_2, respectively, obtained as in [**HO**]. Since $S_{0,1}$ and $S_{0,2}$ are given by the same edgepaths, we have

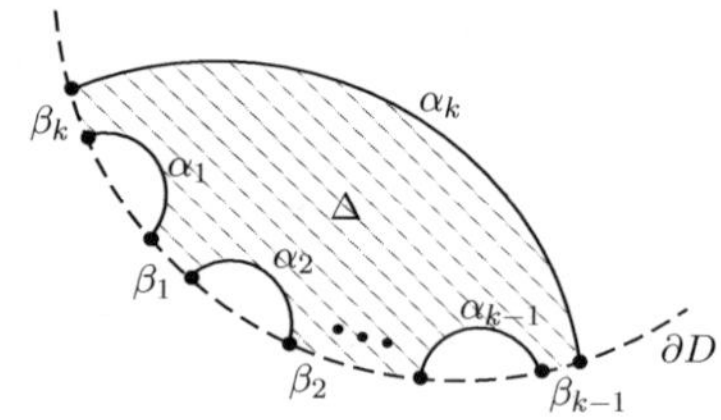

FIGURE 9. The disk Δ.

$\tau(S_0) = -\tau(S_{0,1}) + \tau(S_{0,2}) = 0$ by Lemma 4.4. Therefore the boundary slope of the incompressible surface S is $\tau(S) - \tau(S_0) = -2(n+1)^2 + 4$ by Lemma 4.5. □

PROOF OF THEOREM 1.1. As in Figure 7, K_n has an irreducible alternating diagram with $4n$ crossings, so that $c(K_n) = 4n$. By Proposition 5.1, $\mathrm{diam}(K_n) \geq 2(n+1)^2 - 4 - (-2(n+1)^2 + 4) = 4(n+1)^2 - 8$. Therefore, $\frac{\mathrm{diam}(K_n)}{c(K_n)} \geq \frac{(n+1)^2-2}{n}$ tends to infinity, as required. □

References

[C] M. Culler, A table of A-polynomials computed via numerical methods. http://www.math.uic.edu/~culler/Apolynomials/

[CT] C. L. Curtis and S. J. Taylor, *The Jones polynomial and boundary slopes of alternating knots*, J. Knot Theory Ramifications **20** (2011), no. 10, 1345–1354, DOI 10.1142/S0218216511009194. MR2851712

[CG] N. M. Dunfield and S. Garoufalidis, *Incompressibility criteria for spun-normal surfaces*, Trans. Amer. Math. Soc. **364** (2012), no. 11, 6109–6137, DOI 10.1090/S0002-9947-2012-05663-7. MR2946944

[HT] A. Hatcher and W. Thurston, *Incompressible surfaces in 2-bridge knot complements*, Invent. Math. **79** (1985), no. 2, 225–246, DOI 10.1007/BF01388971. MR778125

[HO] A. Hatcher and U. Oertel, *Boundary slopes for Montesinos knots*, Topology **28** (1989), no. 4, 453–480, DOI 10.1016/0040-9383(89)90005-0. MR1030987

[H] J. Howie, Boundary slopes of some non-Montesinos knots. arXiv:1401.2726

[IM] K. Ichihara and S. Mizushima, *Crossing number and diameter of boundary slope set of Montesinos knot*, Comm. Anal. Geom. **16** (2008), no. 3, 565–589. MR2429969

[K1] Y. Kabaya, Private communication, April, 2008.

[K2] Y. Kabaya, *A method to find ideal points from ideal triangulations*, J. Knot Theory Ramifications **19** (2010), no. 4, 509–524, DOI 10.1142/S0218216510007929. MR2646644

[W] Y.-Q. Wu, *Exceptional Dehn surgery on large arborescent knots*, Pacific J. Math. **252** (2011), no. 1, 219–243, DOI 10.2140/pjm.2011.252.219. MR2862149

DEPARTMENT OF MATHEMATICS, HIYOSHI CAMPUS, KEIO UNIVERSITY, 4-1-1 HIYOSHI, KOHOKU-KU, YOKOHAMA-SHI, KANAGAWA, 223-8521, JAPAN

Email address: ishikawa@keio.jp

DEPARTMENT OF MATHEMATICS AND STATISTICS, CALIFORNIA STATE UNIVERSITY, CHICO, CHICO, CA 95929-0525

Email address: TMattman@CSUChico.edu

DEPARTMENT OF MATHEMATICS, SAITAMA UNIVERSITY, 255 SHIMO-OKUBO, SAKURA-KU, SAITAMA-SHI, SAITAMA, 338-8570, JAPAN

DEPARTMENT OF MATHEMATICS, SAITAMA UNIVERSITY, 255 SHIMO-OKUBO, SAKURA-KU, SAITAMA-SHI, SAITAMA, 338-8570, JAPAN

Email address: kshimoka@rimath.saitama-u.ac.jp

Contemporary Mathematics
Volume **760**, 2020
https://doi.org/10.1090/conm/760/15293

Degeneration of strictly convex real projective structures on surfaces

Inkang Kim

ABSTRACT. In this paper we study the degeneration of convex real projective structures on bordered surfaces.

1. Introduction

Recently there has been intensive study of Anosov representations of Gromov hyperbolic groups into higher rank semisimple Lie groups extending the notion of convex cocompact representations in rank one semisimple Lie groups [**14,17**]. On the other hand, for free groups, there is a notion of primitive stable representations proposed by Minsky [**19**]. This notion works well for representations with parabolic elements. It is known that the restriction of a primitive stable representation to a non-trivial free factor is Anosov [**16**]. Hence a primitive stable representation is a very good generalization of Anosov representations, yet not all representations with parabolics are primitive stable. For example, a finite volume strictly convex real projective surface with more than two cusps has a holonomy representation which is not primitive stable. This can be easily seen by choosing a free generating set containing one of the elements representing a hole. Yet a finite volume strictly convex real projective surface with one cusp has a primitive stable holonomy representation [**15**].

Despite the intensive study of Anosov representations, its boundary of the space of Anosov representations is not well-understood. See [**20,21**] for degenerations of convex real projective structures on surface. For example, for the quasifuchsian space, it is known that the cusps are dense on the boundary of quasifuchsian space. This is no longer true for Anosov representations in general. Take the space $\mathcal{A}_2$ of Anosov representations from free group F_2 on 2 generators into $SL(3,\mathbb{R})$. Then $\mathcal{A}_2$ contains the space $\mathcal{P}$ of marked convex real projective structures on a pair of pants P with hyperbolic geodesic boundaries. It also contains the space $\mathcal{T}$ of marked convex real projective structures on a punctured torus T with a hyperbolic geodesic boundary. The real dimension of $\mathcal{A}_2$ is $-8\chi(T) = -8\chi(P) = 8$. Since both $\mathcal{T}$ and $\mathcal{P}$ have real dimension 8, they are open subsets of $\mathcal{A}_2$. The space

2010 *Mathematics Subject Classification.* Primary 51M10, 57S25.

Key words and phrases. Real projective surface, Goldman parameter, Bonahon-Dreyer parameters.

The author gratefully acknowledges the partial support of grant (NRF-2017R1A2A2A05001002) and KIAS Individual Grant (MG031408).

$\mathcal{T}_c$ consisting of convex real projective structures on T whose boundary curve is parabolic has real dimension 6. By cutting T along one simple closed curve C, we can see that any structure in $\mathcal{T}_c$ has two boundary parameters on C, two gluing parameters on C and two internal parameters on $T \setminus C$. This shows that $\mathcal{T}_c$ is included as a boundary of $\mathcal{A}_2$ of codimension 2. Similarly if we pinch one of the boundary of P to a parabolic, then the real dimension drops by 2. This suggests that the cusps appear in the codimension 2 strata. Hence cusps are not dense in $\partial\mathcal{A}_2$.

Such a phenomenon seems generic in higher rank case and it manifests a sharp contrast to the rank one case. In this paper, we are interested in such degeneration phenomena in terms of Fock-Goncharov or Bonahon-Dreyer coordinates.

In this note, the following theorem is proved in Sections 3 and 5. An explicit conventions and notations can be found in Section 5.

THEOREM 1.1. *The space of Anosov representations $\mathcal{A}_2$ contains $\mathcal{P}$ and $\mathcal{T}$ as components, and the boundaries of $\mathcal{P}$ and $\mathcal{T}$ in the $SL(3,\mathbb{R})$-character variety χ_2 of F_2 can be described explicitly using Bonahon-Dreyer parameters when the holonomy of the boundary loops degenerate to parabolic or quasi-hyperbolic elements.*
(i) For the boundary of $\mathcal{P}$

(1) *The space of convex projective structures on P with one boundary, say A_1, being parabolic is parametrized by*

$$\sigma_1(B_1), \sigma_2(B_1), \sigma_1(B_2), \sigma_1(B_3), \tau_{111}(T_+), \tau_{111}(T_-).$$

(2) *The space of convex projective structures on P with all boundaries being parabolic is parametrized by $\sigma_1(B_1), \tau_{111}(T_+)$.*
(3) *The space of convex projective structures on P with a boundary A_1 being quasi-hyperbolic is parametrized by*

$$\sigma_1(B_1), \sigma_2(B_1), \sigma_1(B_2), \sigma_2(B_2), \sigma_1(B_3), \tau_{111}(T_+), \tau_{111}(T_-).$$

(ii) For the boundary of $\mathcal{T}$

(1) *The space of convex projective structures on T with boundary being parabolic is parametrized by*

$$\sigma_1(B_1), \sigma_1(B_2), \sigma_1(B_3), \tau_{111}(T_+), \sigma_1(C), \sigma_2(C)$$

where C is a meridan curve so that $T \setminus C$ is a pair of pants.
(2) *The space of convex projective structures on T with boundary being quasi-hyperbolic is parametrized by*

$$\sigma_1(B_1), \sigma_1(B_2), \sigma_1(B_3), \tau_{111}(T_+), \tau_{111}(T_-), \sigma_1(C), \sigma_2(C).$$

We also address the relationship between area and divergence;

THEOREM 1.2. *If a triangle invariant or a difference of two shear coordinates along some geodesic diverges, then the Hilbert area of the convex projective surface tends to infinity.*

This theorem is proved in Section 4.

2. Preliminaries

2.1. Isometries in Hilbert metric. Suppose that Ω is a (not necessarily strictly) convex domain in $\mathbb{RP}^2$. Choose an affine set A containing Ω with a Euclidean norm $|\cdot|$. For $x \neq y \in \Omega$, let p, q be the intersection points of the line $\overline{xy}$ with $\partial\Omega$ such that p, x, y, q are in this order. The **Hilbert distance** is defined by

$$d_\Omega(x,y) = \frac{1}{2}\log\frac{|p-y||q-x|}{|p-x||q-y|}.$$

This metric coincides with the hyperbolic metric if $\partial\Omega$ is a conic. We introduce the notion of hyperbolic, quasi-hyperbolic and parabolic isometries as they will appear in the sequel. An isometry is **hyperbolic** if it can be represented by a diagonal matrix with diagonal entries $\lambda_1 > \lambda_2 > \lambda_3 > 0$ in $SL(3,\mathbb{R})$. It is **quasi-hyperbolic** if it is conjugate to $\begin{bmatrix} \mu & 1 & 0 \\ 0 & \mu & 0 \\ 0 & 0 & \nu \end{bmatrix}$ where $\mu > \nu > 0, \mu^2\nu = 1$. Finally it is parabolic if it is conjugate to $\begin{bmatrix} 1 & 1 & 0 \\ 0 & 1 & 1 \\ 0 & 0 & 1 \end{bmatrix}$.

2.2. Bonahon-Dreyer parameters. A flag in $\mathbb{R}^3$ is a family F of subspaces $F^0 \subset F^1 \subset F^2$ where F^i has dimension i. A pair of flags (E, F) is generic if every E^i is transverse to every F^j. Similarly a triple of flags (E, F, G) is generic if each triple E^i, F^j, G^k meets transversely.

For such a generic triple, one defines a triangle invariant

$$T(E,F,G) = \frac{e^2\wedge f^1}{f^1\wedge g^2}\frac{e^1\wedge g^2}{e^1\wedge f^2}\frac{f^2\wedge g^1}{e^2\wedge g^1},$$

where $0 \neq e^i \in \wedge^i(E^i)$ and similar for f^j, g^k. Note that for $i+j=3$, $e^i \wedge f^j$ can be identified with a real number by taking a determinant of corresponding 3×3 matrix (e^i, f^j). Easy calculation shows that

$$T(E,F,G) = T(F,G,E) = T(F,E,G)^{-1}.$$

Let (E, F, G, L) be a generic flags in the sense that every quadruple E^a, F^b, G^c, L^d meets transversely. In this case we have two double ratios

$$D_1(E,F,G,L) = -\frac{e^1\wedge f^1\wedge g^1}{e^1\wedge f^1\wedge l^1}\frac{f^2\wedge l^1}{f^2\wedge g^1}$$

$$D_2(E,F,G,L) = -\frac{e^2\wedge g^1}{e^2\wedge l^1}\frac{e^1\wedge f^1\wedge l^1}{e^1\wedge f^1\wedge g^1}.$$

Let $\rho : \pi_1(\Sigma)\to SL(3,\mathbb{R})$ be a holonomy representation which gives rise to a convex real projective structure on Σ. Give an ideal triangulation to Σ coming from the ideal triangulation of each pair of pants in a fixed pants decomposition of Σ. Let T_j be an ideal triangle and $\tilde{T}_j$ a lift to $\tilde{\Sigma}$. Let $\tilde{v}_j, \tilde{v}'_j, \tilde{v}''_j \in \partial_\infty\tilde{\Sigma}$ be vertices of $\tilde{T}_j$ in clockwise order. Define the triangle invariant of T_j to be

$$\tau^\rho_{111}(T_j, v_j) = \log T(\mathcal{F}_\rho(\tilde{v}_j), \mathcal{F}_\rho(\tilde{v}'_j), \mathcal{F}_\rho(\tilde{v}''_j)).$$

Here the flag $\mathcal{F}_\rho(\tilde{v}_j)$ is determined by the tangent line at $\tilde{v}_j$ to $\partial\Omega_\rho$ where Ω_ρ is a convex domain in $\mathbb{RP}^2$ determined by ρ and $\partial_\infty\tilde{\Sigma}$ is identified with $\partial\Omega_\rho$. Note that $\partial\Omega_\rho$ is C^1 since it is strictly convex.

Let $\tilde{\lambda}$ be an oriented leaf in $\tilde{\Sigma}$ such that two ideal triangles $\tilde{T}$ and $\tilde{T}'$ share $\tilde{\lambda}$ such that $\tilde{T}$ lies on the left of $\tilde{\lambda}$. Let x and y be the positive and negative end points of $\tilde{\lambda}$ and z, z' be third vertices of $\tilde{T}$ and $\tilde{T}'$ respectively. Then i-th shear invariant along the oriented leaf λ is

$$\sigma_i^\rho(\lambda) = \log D_i(\mathcal{F}_\rho x, \mathcal{F}_\rho y, \mathcal{F}_\rho z, \mathcal{F}_\rho z').$$

2.3. Goldman parameters for a pair of pants. Let A, B, C denote the boundary of a pair of pants satisfying $CBA = I$. By abuse of notations, A, B, C denote isometries corresponding to the curves. We adopt the notation that the eigenvalues of the matrix satisfy that $\lambda_1 > \lambda_2 > \lambda_3$ whose product equal to 1 and

$$\ell_1 = \log \lambda_1 - \log \lambda_2,\ \ \ell_2 = \log \lambda_2 - \log \lambda_3.$$

Then the Hilbert length of a closed geodesic represented by A is equal to $\ell_1(A) + \ell_2(A)$. In Goldman's notation [**12**], $\lambda(A) = \lambda_3(A)$ and $\tau(A) = \lambda_1(A) + \lambda_2(A)$.

Goldman showed that the convex structure on a pair of pants is determined by $(\lambda(A), \tau(A)), (\lambda(B), \tau(B)), (\lambda(C), \tau(C))$ plus two internal parameters s and t. For the precise definition of s and t, it is advised to consult [**4**] because some notations and conventions between [**4**] and [**12**] are different.

2.4. Bulging deformation. Even if all the triangle invariants are bounded, if the difference of shear coordinates along some infinite edge of an ideal triangle goes to infinity, then the Hilbert area tends to infinity as we will see in Proposition 4.2. We recall a **bulging deformation** along a geodesic [**13**].

Let Ω be a convex domain and l an oriented infinite geodesic in Ω. Draw two tangent lines to Ω at $l(\pm\infty)$ and denote $l^\perp$ the intersection of the two tangent lines. Then the bulging deformation along l is stretching the domain toward $l^\perp$. See [**10**] for details. More precisely, let $l(-\infty)$ be (1,0,0), $l(\infty)$ be $(0,0,1)$ and $l^\perp$ be $(0,1,0)$. Use the base $\{(1,0,0),(0,1,0),(0,0,1)\}$ to write a projective transformation in a matrix form. Then the bulging along l corresponds to a matrix

$$\begin{pmatrix} e^{-v} & 0 & 0 \\ 0 & e^{2v} & 0 \\ 0 & 0 & e^{-v} \end{pmatrix}.$$

Let $(1, y, 1)$ be the vertex of an ideal triangle whose one edge is l and on the right side of l. Under the bulging deformation along l, this vertex moves to $(1, e^{3v}y, 1)$ and two shear parameters along l under the bulging deformation changes:

$$\sigma_1 \rightarrow \sigma_1 - 3v, \sigma_2 \rightarrow \sigma_2 + 3v.$$

Hence the difference $\sigma_2 - \sigma_1$ measures the bulging parameter $6v$.

3. The moduli space of Anosov representations and the space of convex projective structures

3.1. Component of the moduli space of Anosov representations. As we have already seen in the introduction, the space $\mathcal{A}_2$ of Anosov representations from the free group F_2 to $SL(3,\mathbb{R})$ contains two disjoint spaces, $\mathcal{P}$ the space of convex real projective structures with geodesic boundary on the pair of pants P, and $\mathcal{T}$ the space of convex real projective structures with a geodesic boundary on the punctured torus T. Here by Anosov we means Borel-Anosov where Borel subgroup is the set of upper triangular matrices. As we have seen in a previous section, $\mathcal{P}$ has a strata of boundary of codimension 2 by pinching each boundary to cusps. On the other hands, $\mathcal{T}$ has a boundary of codimension 2, $\mathcal{T}_c$, where the boundary of T is a cusp.

It is obvious from the construction that $\overline{\mathcal{P}}$ and $\overline{\mathcal{T}}$ are disjoint where the closures are taken in the space of Anosov representations. Since both $\mathcal{P}$ and $\mathcal{T}$ are cells of dimension 8, they form two disjoint components of $\mathcal{A}_2$. The openness of $\mathcal{P}$ and $\mathcal{T}$ follow from the explicit coordinates.

In this section let Σ be a bordered surface of negative Euler characteristic with boundary. Then the set $\mathcal{C}$ of marked convex projective structures on Σ with hyperbolic geodesic boundaries form an open set in the moduli space of Anosov representations $\mathcal{A}$. To show the closedness of $\mathcal{C}$, suppose $\rho_i : \pi_1(\Sigma) \to SL(3,\mathbb{R})$ corresponding to convex projective structures in $\mathcal{C}$, converges to ρ in $\mathcal{A}$.

Let $X = SL(3,\mathbb{R})/SO(3)$ and fix a generating set S of $\pi_1(\Sigma)$. For a representation $\rho : \pi_1(\Sigma) \to SL(3,\mathbb{R})$, set $d_\rho(x) = \sup_{s\in S} d(x, \rho(s)x)$ and $\mu(\rho) = \inf_{x\in X} d_\rho(x)$ where d is a distance on X. Let $\mathrm{Min}_\rho = \{x \in X | d_\rho(x) = \mu(\rho)\}$. This is a closed convex set of X.

It is not difficult to see that Min_ρ is non-empty and bounded if and only if ρ is not parabolic, i.e., $\rho(\pi_1(\Sigma))$ does not fix a point in ∂X. Indeed, if Min_ρ is either empty or unbounded, one can find a sequence $x_i \to \eta \in \partial X$ so that

$$d(x_i, \rho(s)x_i) \le \mu(\rho) + \epsilon$$

for all large i and $s \in S$. This implies that η is fixed by ρ and hence ρ is parabolic.

Let's fix a base point $x_0 \in X$. By conjugating ρ_i if necessary, we may assume that $x_0 \in \mathrm{Min}_{\rho_i}$. Note that Min_{ρ_i} is not empty since ρ_i is not parabolic. Since $\limsup\{\mu(\rho_i) = d_{\rho_i}(x_0)\} \le \mu(\rho)$, one can extract a subsequence converging to τ. Then τ is again discrete and faithful. See [**11**] for example.

Suppose $\tau(\pi_1(\Sigma))$ is contained in a parabolic subgroup G_η of $SL(3,\mathbb{R})$. Here G_η means a parabolic subgroup stabilizing η in the visual boundary of X. In this case, Min_τ is either $\emptyset$ or unbounded. Since d_{ρ_i} converges uniformly to d_τ, $x_0 \in \mathrm{Min}_\tau$, hence Min_τ is unbounded. In this case one can show that τ fixes two end points of some geodesic l as follows. Let l be a geodesic emanating from η so that $\mathrm{Min}_\tau \cap l = L$ is noncompact. This is possible since Min_τ accumulates to η. We want to show that $L = l$. Choose $y \in L$ and take a generalized Iwasawa decomposition of the parabolic subgroup G_η as $G_\eta = N_\eta A_\eta K_\eta$ where K_η is an isotropy subgroup of K, the maximal compact subgroup stabilizing y (indeed K_η fixes l pointwise), $A_\eta y$ is the union of parallels to l, and N_η is the horospherical subgroup which is determined uniquely by η. See Proposition 2.17.5 (4) in [**9**]. Note for any $id \neq n \in N_\eta$, nl and l are not parallel, but they are asymptotic at η. Indeed $d(l(t), nl(t)) \to 0$ as $\lim_{t\to-\infty} l(t) = \eta$. This fact implies the following. For any $g \in \tau(S)$, where S is the fixed generating set of $\pi_1(\Sigma)$, if $g = nak$, we claim that $n = id$. If not, $d(nak(l(t)), l(t)) = d(na(l(t)), l(t))$ is strictly decreasing as $t \to -\infty$ since $al(t)$ is a geodesic emanating from η. Since $l(t) \in L$ for large negative t, this contradicts the definition of Min_τ. Hence any element in $\tau(S)$ sends l to a parallel geodesic. Consequently, $\tau(\pi_1(\Sigma))$ fixes two end points of l and $\tau(\pi_1(\Sigma)) \subset A_\eta K_\eta$.

Let W be the union of parallels to l, which is isometric to $l \times Y$ where Y is a closed convex complete subset of X. See Lemma 2.4 of [**2**].

If l is non-singular, W is a maximal flat containing l. If we take a Iwasawa decomposition $SL(3,\mathbb{R}) = KAN$ where K is the stabilizer of x_0, Ax_0 is the maximal flat containing l, N is the Nilpotent subgroup fixing $l(-\infty)$, then $\tau(\pi_1(\Sigma)) \subset MA$

since $\tau(\pi_1(\Sigma))$ stabiizes W, which is abelian. Here MAN is the Langlands decomposition of a parabolic subgroup. Since $\pi_1(\Sigma)$ is free and τ is faithful, this is impossible.

If l is singular, $W = l \times H^2$. Since τ preserves this splitting, it can be conjugated so that its image is contained in

$$\begin{pmatrix} M_{2\times 2} & 0 \\ 0 & \lambda \end{pmatrix}.$$

Then the projection

$$\pi : \tau(\pi_1(\Sigma)) \to Iso(H^2) = SL(2,\mathbb{R}), \begin{pmatrix} M & 0 \\ 0 & \lambda \end{pmatrix} \to \frac{1}{(detM)^{1/2}} M$$

is either discrete or solvable by Proposition 7.2.2 of [**9**]. The kernel of π will be central in $\tau(\pi_1(\Sigma))$ since it is of the form $\begin{pmatrix} \frac{1}{\sqrt{\lambda}} I & 0 \\ 0 & \lambda \end{pmatrix}$, and since τ is faithful and $\pi_1(\Sigma)$ has no center, π is injective. Hence $\pi\tau(\pi_1(\Sigma))$ is a discrete and faithul representation into $SL(2,\mathbb{R})$. Note that $\tau([\pi_1(\Sigma),\pi_1(\Sigma)])$ has image in $\begin{pmatrix} SL(2,\mathbb{R}) & 0 \\ 0 & 1 \end{pmatrix}$. Hence $\tau|_{[\pi_1(\Sigma),\pi_1(\Sigma)]}$ is a discrete and faithful representation into $SL(2,\mathbb{R})$. Then by Lemma 2 of [**6**], there exists an element $\gamma \in [\pi_1(\Sigma),\pi_1(\Sigma)]$ such that $tr(\tau(\gamma)) < 0$, i.e., its eigenvalues are $(-\lambda_1, -\lambda_2, 1)$ with $\lambda_i > 0$. Since $\rho_i \to \tau$, $\rho_i(\gamma)$ and $\tau(\gamma)$ will have the same sign for their eigenvalues for large i. But since ρ_i is a positive representation in the sense of Fock-Goncharov [**8**], the eigenvalues of $\rho_i(\gamma)$ must be all positive, a contradiction. This shows that $\tau(\pi_1(\Sigma))$ cannot be contained in a parabolic subgroup.

Suppose $\tau(\pi_1(\Sigma))$ is not contained in a parabolic subgroup. Then it is an irreducible representation in $\mathbb{R}^3$. Let Ω_i be a convex set in $\mathbb{RP}^2$ whose quotient by $\rho_i(\pi_1(\Sigma))$ is the corresponding real projective surface. Take a Hausdorff limit Ω of Ω_i, which is invariant under τ. Then Ω should be properly convex, otherwise τ will be reducible. Hence $\Omega/\tau(\pi_1(\Sigma))$ is a properly convex projective surface.

Since $\rho_i \to \rho$ before conjugating ρ_i, $[\rho_i] \to [\rho] = [\tau]$. If τ belongs to $\partial\mathcal{C}$, some boundary of Σ must be either parabolic or quasi-hyperbolic. For parabolic case, the orbit in X under such a parabolic element is not quasi-geodesic, which is not possible for Anosov representation. For quasi-hyperbolic case, if L is an axis of the quasi-hyperbolic isometry on the boundary of Ω, which connects the repelling point p_- and the attracting point p_+, then the tangent line at p_- contains the axis, while the tangent line at p_+ is different from the axis (see [**18**]). Hence the limit curve determined by Ω does not satisfy the antipodality that the Anosov representation should have. In either case, ρ is not Anosov. Since $[\rho]$ belongs to $\mathcal{A}$ by assumption, $[\tau] \in \mathcal{C}$.

This argument shows that:

THEOREM 3.1. *Let Σ be a punctured surface and $\mathcal{A}$ a moduli space of Anosov representations from $\pi_1(\Sigma)$ to $SL(3,\mathbb{R})$. Then $\mathcal{C}$, the set of marked convex projective structures on Σ with hyperbolic geodesic boundaries, is closed in $\mathcal{A}$. Hence $\mathcal{C}$ is a component of $\mathcal{A}$.*

3.2. Boundary of $\mathcal{C}$ in character variety. We saw in the previous section that $\mathcal{A}$ contains a component $\mathcal{C}$ consisting of convex real projective structures on Σ with hyperbolic boundaries. But $\mathcal{A}$ is just a subset of the character variety χ_n

of free group with n generators in $SL(3,\mathbb{R})$. In this section, we want to study $\partial\mathcal{A}$ in χ_n. By a theorem of Bonahon-Dreyer [**3**], $\mathcal{C}$ is the interior of a convex polytope in Euclidean space.

THEOREM 3.2. *The boundary of $\mathcal{C}$ consists of holonomies of real convex structures on Σ with either quasi-hyperbolic boundary or parabolic boundary. Each quasi-geodesic boundary determines a codimension 1 boundary in the polytope, and each parabolic boundary determines a codimension 2 boundary in the polytope.*

PROOF. Suppose $[\rho_i] \in \mathcal{C}$ converge to a finte point x in the boundary of the polytope. Then corresponding Fock-Goncharov coordinates are all finite for x. Then it is easy to see that all the traces of boundary curves of pairs of pants in pants decomposition of Σ are bounded. Since the trace of any other curve can be generated by the traces of these pants curve, x should correspond to a finite point in the character variety, i.e. x is represented by a representation $[\rho]$.

The proof goes in the exactly same way as in the previous section. One can show that, after conjugating ρ_i, $\rho_i \to \tau$ and $\tau(\pi_1(\Sigma))$ is not contained in a parabolic subgroup. Furthermore $\Omega_i \to \Omega$ in Hausdorff topology and $\Omega/\tau(\pi_1(\Sigma))$ is properly convex. This shows that $[\rho] = [\tau]$ is represented by a holonomy of a real convex structure. Hence the boundary curve should be either quasi-hyperbolic or parabolic. □

As will be described in Theorem 5.1, the space of convex real projective structures with all three boundary components of P being parabolic is parametrized by $(\sigma_1(B_1), \tau_{111}(T_+))$. Hence this codimension 6 space is homeomorphic to $\mathbb{R}^2$.

The space of convex real projective structures on T with the boundary curve being parabolic is parametrized by six parameters

$$\sigma_1(B_i), i = 1, 2, 3, \tau_{111}(T_+), \sigma_1(C), \sigma_2(C)$$

where C is a meridian curve so that $T \setminus C$ is a pair of pants. This space forms a codimension 2 boundary of $\mathcal{T}$ in χ_2. Hence this codimension 2 boundary set is homeomorphic to $\mathbb{R}^6$.

4. Area and degeneration

In this section, we are interested in the relationship between the Hilbert area and the degeneration of convex structures. It is expected if the convex structure diverges away from Teichmüller space, then the Hilbert area tends to infinity. Here are some supporting evidences.

THEOREM 4.1. *Suppose the Hilbert area of a strictly convex real projective structure is bounded. Then all the triangle invariants are bounded in terms of the area.*

PROOF. Let $\triangle$ be a lift of an ideal triangle in Ω. Then the tangent lines at the vertices of $\triangle$ to $\partial\Omega$ form a triangle T. Then

$$Area_T(\triangle) \leq Area_\Omega(\triangle) < R.$$

Up to affine transformation we can put $(T, \triangle)$ into the standard positions such that the vertices of T are $(0,1,1), (0,0,1), (1,0,1)$ and the vertices of $\triangle$ are

$$(0, 1/2, 1), (1/2, 1/2, 1), (\alpha, 0, 1)$$

where $0 < \alpha \leq 1/2$. Then the triangle invariant $T(\mathcal{E}, \mathcal{F}, \mathcal{G})$ of three flags

$$\mathcal{E} : \langle(0, 1/2, 1)\rangle \subset \langle(0, 0, 1), (0, 1, 1)\rangle$$

$$\mathcal{F} : \langle(1/2, 1/2, 1)\rangle \subset \langle(0, 1, 1), (1, 0, 1)\rangle$$

$$\mathcal{G} : \langle(\alpha, 0, 1)\rangle \subset \langle(0, 0, 1), (1, 0, 1)\rangle$$

is

$$\frac{1-\alpha}{\alpha}.$$

As $\alpha \to 0$, $Area_T(\triangle(\alpha)) \to \infty$ ([**7**]), hence $\alpha \geq \epsilon = \epsilon(R)$ for some ϵ. This implies that

$$1 \leq T(\mathcal{E}, \mathcal{F}, \mathcal{G}) \leq \frac{1-\epsilon}{\epsilon}.$$

□

The above theorem also can be deduced from [**1**].

PROPOSITION 4.2. *We fix a topological data of ideal triangulation of Σ coming from pants decomposition to define Bonahon-Dreyer coordinates. Suppose all triangle invariants, and all $|\sigma_1(l) + \sigma_2(l)|$ are bounded for any l, and some shear invariant on an oriented infinite leaf l satisfies that $|\sigma_1(l) - \sigma_2(l)| \to \infty$. Then the Hilbert area of the projective surface Σ defined by the Hilbert metric corresponding to Bonahon-Dreyer coordinates tends to infinity.*

PROOF. First note that the difference of two shear parameters along a geodesic gives a bulging parameter. Hence if the difference of two shear parameters along l tends to infinity, the part D of the domain Ω delimited by l, and the ideal triangle contained in D whose one edege is l converge to the triangle whose vertices are the end points $l(\pm\infty)$ of l and the intersection $l^{\perp}$ of two tangent lines at $l(\pm\infty)$ if the bulging parameter $v \to \infty$. The area of such triangles tends to infinity. If $v \to -\infty$, then D degenerates to l, then the area of the left ideal triangle tends to infinity. See [**7**] to see that the area is infinite for such a triangle. Hence the Hilbert area of one of two triangles sharing the infinite edge l tends to infinity in the limit domain. □

On the other hand, even if $\sigma_1 + \sigma_2 \to \infty$ and the triangle invariants are bounded then the area remains bounded since the sum represents the twisting parameter.

5. Bonahon-Dreyer coordinates and degeneration

In this section, we give explicit Bonahon-Dreyer coordinates for the degenerated convex structures on a pair of pants P and a punctured torus T.

Notations: For a pair of pants P, three boundary components are A_1, A_2, A_3. Three positive eigenvalues of the monodromy of A_i are $0 < \lambda_i < \mu_i < \nu_i$. In Goldman parametrization, $\tau_i = \mu_i + \nu_i$. P is decomposed into two triangles $T_{\pm}$ with edges B_1, B_2, B_3 which appear in the clock-wise order along T_+. See the Figure 1 in [**4**]. These edges are oriented counter-clockwise. The boundary component A_i is opposite to B_i. The indices i are modulo 3. Goldman's internal parameter for a convex projective structure on P are s, t and an explicit formula can be found in [**4**].

5.1. Explicit formula for P. Let P be a pair of pants. The space of marked convex projective structures on P with geodesic boundary has 8 dimensional parameter, 3×2 boundary parameters, and 2 more internal parameters s and t in Goldman parameters. Now suppose we pinch all the boundary curves to cusps. Then all the boundary parameters vanish and only 2 internal parameters survive in Goldman parameters. This can be justified by Figure 2 in [**4**] where verticies of Δ_+ are parabolic fixed points. This can be seen also by Proposition 4.1 of [**4**]. If all the boundary curves are parabolic, then $\lambda_i = 1, \tau_i = 2$ where λ_i is the smallest eigenvalue and τ_i is the sum of the rest eigenvalues of boundary curves. Then by the formula of Proposition 4.1 of [**4**],

$$\sigma_1(B_i) = \log(s\mu_{i-1}\sqrt{\frac{\lambda_{i-1}\lambda_{i+1}}{\lambda_i}}) = \log s$$

$$\sigma_2(B_i) = \log(\frac{\mu_{i+1}}{s}\sqrt{\frac{\lambda_{i-1}\lambda_{i+1}}{\lambda_i}}) = -\log s$$

where $\mu_i = \frac{\tau_i - \sqrt{\tau_i^2 - \frac{4}{\lambda_i}}}{2}$. And

$$\tau_{111}(T_+) = \log \frac{(e^{-\sigma_2(B_2)}+1)(e^{-\sigma_2(B_3)}+1)}{t(e^{\sigma_1(B_3)}+1)} = \log \frac{(s+1)}{t}.$$

PROPOSITION 5.1. *Suppose P is equipped with a convex projective structure with all three boundary components being parabolic. Then two numbers $(\sigma_1(B_1), \tau_{111}(T_+))$ parametrize the space of such convex projective structures on P with all three boundary components being parabolic.*

Similarly if one of the 3 boundary curves is pinched, then corresponding Bonahon-Dreyer coordinates can be calculated as follows.

PROPOSITION 5.2. *Suppose P is equipped with a convex projective structure with one boundary component A_1 being parabolic. Then Bonahon-Dreyer coordinates $\sigma_1(B_1), \sigma_2(B_1), \sigma_1(B_2), \sigma_1(B_3), \tau_{111}(T_+), \tau_{111}(T_-)$ are independent parameters to describe the space of convex projective structures on P with A_1 being parabolic.*

PROOF. The holonomy of A_1 is parabolic, hence $\lambda_1 = 1, \tau_1 = 2$ and $\mu_1 = 1$. Then

$$\sigma_1(B_2) = \log(s\sqrt{\frac{\lambda_3}{\lambda_2}}) = -\sigma_2(B_3) = -\log(\frac{1}{s}\sqrt{\frac{\lambda_2}{\lambda_3}}).$$

It can be checked that

$$\sigma_1(B_1) - \sigma_2(B_1) = \log(s^2\frac{\mu_3}{\mu_2}),$$

$$\sigma_1(B_3) - \sigma_2(B_2) = \log(s^2\frac{\mu_2}{\mu_3}).$$

This gives

$$\sigma_1(B_1) - \sigma_2(B_1) + \sigma_1(B_3) - \sigma_2(B_2) = \log(s^4).$$

Hence there are 6 independent parameters. □

The codimension 1 boundary of $\mathcal{P}$ is the set of properly convex projective structures on P with one boundary being quasi-hyperbolic. Recall that a quasi-hyperbolic isometry is a matrix in $SL(3,\mathbb{R})$ which is conjugate to $\begin{pmatrix} \mu & 1 & 0 \\ 0 & \mu & 0 \\ 0 & 0 & \nu \end{pmatrix}$ where $\mu > \nu > 0, \mu^2\nu = 1$.

Since quasi-hyperbolic elements are parametrized by one parameter only, if one of the boundary of P is quasi-hyperbolic, such a projective structures is parametrized by 7 parameters in total.

PROPOSITION 5.3. *Suppose P is equipped with a convex projective structure with one boundary component A_1 being quasi-hyperbolic. Then Bonahon-Dreyer coordinates*

$$\sigma_1(B_1), \sigma_2(B_1), \sigma_1(B_2), \sigma_2(B_2), \sigma_1(B_3), \tau_{111}(T_\pm)$$

are independent parameters to describe the space of convex projective structures on P with A_1 being quasi-hyperbolic.

PROOF. The holonomy of A_1 is quasi-hyperbolic, hence $\lambda_1 = \nu, \tau_1 = 2\mu$ and $\mu_1 = \mu$. Then

$$\sigma_1(B_2) = \log(s\mu\sqrt{\frac{\nu\lambda_3}{\lambda_2}})$$

$$\sigma_2(B_3) = \log(\frac{\mu}{s}\sqrt{\frac{\nu\lambda_2}{\lambda_3}}).$$

Hence

$$\sigma_1(B_2) + \sigma_2(B_3) = \log(\mu^2\nu) = 0$$

since $\mu^2\nu = 1$. Then $\sigma_1(B_1), \sigma_2(B_1), \sigma_1(B_2), \sigma_2(B_2), \sigma_1(B_3), \tau_{111}(T_\pm)$ are parameters. □

5.2. Explicit formula for T. In this section, we describe explicit parameters to describe the convex real projective structures on a punctured torus T whose boundary curve is B. Choose a meridian curve C so that $T\setminus C$ is a pair of pants. In this case, Goldman parameters are $\lambda(C), \tau(C), \lambda(B), \tau(B)$, two gluing parameters (u, v) along C together with two internal parameters (s, t). By Proposition 5.2 in [**4**], Bonahon-Dreyer shear coordinates $\sigma_1(C)$ and $\sigma_2(C)$ are given by

$$\sigma_1(C) = (u - u^0) - 3(v - v^0), \sigma_2(C) = (u - u^0) + 3(v - v^0),$$

once a starting point (u^0, v^0) is specified in Goldman's parameter. We normalize the coordinates that $(u^0, v^0) = (0, 0)$.

Set $B = A_1, C = A_2 = A_3$. When B becomes parabolic, $\lambda(B) = 1, \tau(B) = 2, \mu(B) = 1$. Then by the proof of Proposition 5.2, we get

$$\sigma_1(B_2) = -\sigma_2(B_3)$$

and

$$\sigma_1(B_1) - \sigma_2(B_1) = \log(s^2\frac{\mu_3}{\mu_2}),$$

$$\sigma_1(B_3) - \sigma_2(B_2) = \log(s^2\frac{\mu_2}{\mu_3}).$$

Furthermore

$$\sigma_1(B_1) - \sigma_2(B_1) + \sigma_1(B_3) - \sigma_2(B_2) = \log(s^4).$$

Hence $\sigma_1(B_1), \sigma_2(B_1), \sigma_1(B_2), \sigma_1(B_3), \tau_{111}(T_+), \tau_{111}(T_-)$ are independent parameters to describe projective structures on a pair of pants when B is parabolic. But in our case, since $B = A_1, C = A_2 = A_3$, $\lambda_2 = \lambda_3, \mu_2 = \mu_3$. Hence

$$\sigma_1(B_2) = \log s = -\sigma_2(B_3), \sigma_1(B_1) - \sigma_2(B_1) = 2\log s = \sigma_1(B_3) - \sigma_2(B_2).$$

Therefore if $\sigma_1(B_1), \sigma_1(B_2), \sigma_1(B_3)$ are given, then the others

$$\sigma_2(B_3), \sigma_2(B_1) = \log(\frac{\mu_2\lambda_2}{s}), \sigma_2(B_2) = \log(\frac{\mu_2}{s})$$

are known. Hence $s, \lambda_2 = \lambda_3, \mu_2 = \mu_3$ can be deduced from

$$\sigma_1(B_1), \sigma_1(B_2), \sigma_1(B_3).$$

By the formula in Proposition 4.1 of [**4**],

$$\tau_{111}(T_+) = \log\frac{(e^{-\sigma_2(B_2)}+1)(e^{-\sigma_2(B_3)}+1)}{t(e^{\sigma_1(B_3)}+1)}$$

$$\tau_{111}(T_-) = \log\frac{t\mu_1\mu_2\mu_3(e^{\sigma_1(B_3)}+1)}{(e^{-\sigma_2(B_2)}+1)(e^{-\sigma_2(B_3)}+1)}.$$

Hence if $\sigma_1(B_1), \sigma_1(B_2), \sigma_1(B_3), \tau_{111}(T_+)$ are given, then $\tau_{111}(T_-)$ is known. Therefore, $\sigma_1(B_1), \sigma_1(B_2), \sigma_1(B_3), \tau_{111}(T_+)$ parametrize such a structure on a pair of pants. When the pair of pants is glued along C, then two more shear parameters $\sigma_1(C) = u - 3v, \sigma_2(C) = u + 3v$ are involved.

PROPOSITION 5.4. *The convex real projective structures on T with the boundary curve being parabolic are parametrized by*

$$\sigma_1(B_1), \sigma_1(B_2), \sigma_1(B_3), \tau_{111}(T_+)$$

and $\sigma_1(C), \sigma_2(C)$ where C is a meridian curve so that $T \setminus C$ is a pair of pants.

When the boundary curve B becomes quasi-hyperbolic, it determines a codimension 1 boundary of $\mathcal{T}$.

PROPOSITION 5.5. *The space of convex real projective structures on T with the boundary curve being quasi-hyperbolic, is parametrized by*

$$\sigma_1(B_i), i = 1,2,3, \sigma_1(C), \sigma_2(C), \tau_{111}(T_\pm).$$

PROOF. From the proof of Theorem 5.3, if we set $B = A_1$, then

$$\sigma_1(B_2) = -\sigma_2(B_3)$$

and $\sigma_1(B_2) = \log s$ using $\lambda_2 = \lambda_3, \mu_2 = \mu_3$. Also we get

$$\sigma_1(B_1) = 2\log s + \sigma_2(B_1), \sigma_1(B_3) = 2\log s + \sigma_2(B_2).$$

Hence $\sigma_1(B_i), i = 1,2,3, \tau_{111}(T_\pm), \sigma_1(C), \sigma_2(C)$ parametrize the space. □

For some related calculations and materials, see [**5**].

Acknowledgments

The author thanks F. Bonahon for the conversations on projective structures and parametrizations during his visit to USC for several occasions. He also thanks S. Tillmann for his interests and pointing out relevant references.

References

[1] I. Adeboye and D. Cooper, *The area of convex projective surfaces and Fock-Goncharov coordinates*, J. Topol. Anal. **12** (2020), no. 2, 533–546, DOI 10.1142/S1793525319500560. MR4119114

[2] W. Ballmann, M. Gromov, and V. Schroeder, *Manifolds of nonpositive curvature*, Progress in Mathematics, vol. 61, Birkhäuser Boston, Inc., Boston, MA, 1985. MR823981

[3] F. Bonahon and G. Dreyer, *Parameterizing Hitchin components*, Duke Math. J. **163** (2014), no. 15, 2935–2975, DOI 10.1215/0012794-2838654. MR3285861

[4] F. Bonahon and I. Kim, *The Goldman and Fock-Goncharov coordinates for convex projective structures on surfaces*, Geom. Dedicata **192** (2018), 43–55, DOI 10.1007/s10711-017-0233-1. MR3749422

[5] A. Casella, D. Tate and S. Tillmann, *Moduli spaces of real projective structures on surfaces*, preprint.

[6] S. Choi and W. M. Goldman, *Convex real projective structures on closed surfaces are closed*, Proc. Amer. Math. Soc. **118** (1993), no. 2, 657–661, DOI 10.2307/2160352. MR1145415

[7] B. Colbois, C. Vernicos, and P. Verovic, *L'aire des triangles idéaux en géométrie de Hilbert* (French, with French summary), Enseign. Math. (2) **50** (2004), no. 3-4, 203–237. MR2116715

[8] V. V. Fock and A. B. Goncharov, *Moduli spaces of convex projective structures on surfaces*, Adv. Math. **208** (2007), no. 1, 249–273, DOI 10.1016/j.aim.2006.02.007. MR2304317

[9] P. B. Eberlein, *Geometry of nonpositively curved manifolds*, Chicago Lectures in Mathematics, University of Chicago Press, Chicago, IL, 1996. MR1441541

[10] P. Foulon and I. Kim, *Topological entropy and bulging deformation of real projective structures on surface*, submitted.

[11] W. M. Goldman and J. J. Millson, *Local rigidity of discrete groups acting on complex hyperbolic space*, Invent. Math. **88** (1987), no. 3, 495–520, DOI 10.1007/BF01391829. MR884798

[12] W. M. Goldman, *Convex real projective structures on compact surfaces*, J. Differential Geom. **31** (1990), no. 3, 791–845. MR1053346

[13] W. Goldman, *Bulging deformations of convex $\mathbb{RP}^2$-manifolds*, unpublished manuscript.

[14] O. Guichard and A. Wienhard, *Anosov representations: domains of discontinuity and applications*, Invent. Math. **190** (2012), no. 2, 357–438, DOI 10.1007/s00222-012-0382-7. MR2981818

[15] I. Kim and S. Kim, *Primitive stable representations in higher rank semisimple Lie groups*, to appear in Revista Matematica Complutense.

[16] I. Kim and J. Porti, *Uniformly Primitive stable representations in semisimple Lie groups*, in preparation.

[17] F. Labourie, *Anosov flows, surface groups and curves in projective space*, Invent. Math. **165** (2006), no. 1, 51–114, DOI 10.1007/s00222-005-0487-3. MR2221137

[18] L. Marquis, *Surface projective convexe de volume fini* (French, with English and French summaries), Ann. Inst. Fourier (Grenoble) **62** (2012), no. 1, 325–392. MR2986273

[19] Y. N. Minsky, *On dynamics of $Out(F_n)$ on* $\mathrm{PSL}_2(\mathbb{C})$ *characters*, Israel J. Math. **193** (2013), no. 1, 47–70, DOI 10.1007/s11856-012-0086-0. MR3038545

[20] T. Zhang, *The degeneration of convex $\mathbb{RP}^2$ structures on surfaces*, Proc. Lond. Math. Soc. (3) **111** (2015), no. 5, 967–1012, DOI 10.1112/plms/pdv051. MR3477227

[21] T. Zhang, *Degeneration of Hitchin representations along internal sequences*, Geom. Funct. Anal. **25** (2015), no. 5, 1588–1645, DOI 10.1007/s00039-015-0342-7. MR3426063

SCHOOL OF MATHEMATICS, KOREA INSTITUTE FOR ADVANCED STUDY, HEOGIRO 85, DONGDAEMEN-GU, SEOUL, 02455, KOREA

Email address: inkang@kias.re.kr

Contemporary Mathematics
Volume **760**, 2020
https://doi.org/10.1090/conm/760/15294

Examples of character varieties in characteristic p and ramification

Luisa Paoluzzi and Joan Porti

To Steve Boyer on the occasion of his 65th birthday

ABSTRACT. We study $\mathrm{SL}_2(\mathbb{F})$-character varieties of knots over algebraically closed fields $\mathbb{F}$. We give a sufficient condition in terms of the double branched cover of a 2-bridge knot (or, equivalently, of its Alexander polynomial) on the characteristic of $\mathbb{F}$, an odd prime, for the $\mathrm{SL}_2(\mathbb{F})$-character variety to present ramification phenomena. Finally we provide several explicit computations of character varieties to illustrate the result, exhibiting also other types of ramification.

1. Introduction

Character varieties have turned out to be powerful tools in the comprehension of 3-manifolds in general and knots in particular. In the case of knots, different invariants are related or connected to character varieties. This is, for instance, the case of the A-polynomial [**5**], and the algebraic and geometric properties of the excellent component of a hyperbolic knot [**10, 16, 23**]. In this paper we are interested in a far less explored type of invariants, namely the finite set of odd prime numbers p for which the character variety ramifies in characteristic p.

Recall that for a finitely presented group Γ its $\mathrm{SL}_2(\mathbb{C})$-character variety, denoted $X(\Gamma)$, is an algebraic set parametrising, roughly speaking, the conjugacy classes of representations of Γ into $\mathrm{SL}_2(\mathbb{C})$. The algebraic set $X(\Gamma)$ is determined by a finite set of polynomial equations with integer coefficients. If p is a prime number, one can consider the polynomial equations obtained from the previous ones by reducing their coefficients mod p. It follows from work by González–Acuña and Montesinos–Amilibia [**8**] that these new equations define the variety of characters of representations of Γ into $\mathrm{SL}_2(\mathbb{F})$, where $\mathbb{F}$ is an algebraically closed field of characteristic p, provided that p is odd.

It is not hard to see that for almost every p several features of the structure of the variety of characters of Γ over $\mathbb{F}$ coincide with those of $X(\Gamma)$ over $\mathbb{C}$: this is for instance the case of the number of irreducible components of the algebraic set and their dimensions. We shall say that p *ramifies* if there is a discrepancy between the behaviour of $X(\Gamma)$ and that of the character variety in characteristic

2010 *Mathematics Subject Classification.* Primary 57M25; Secondary 20C99, 57M50.
Key words and phrases. Character varieties, knots, double branched covers, ramification.
Partially supported by the Spanish Mineco through grant MTM2015–66165–P.

p. We are interested in understanding what types of ramification phenomena can appear in character varieties of knots (i.e. when Γ is the fundamental group of a knot exterior) and which primes ramify.

The existence of two different types of ramification phenomena were pointed out in previous papers by the authors. Both variations in the dimensions of irreducible components and in their number can occur: the former phenomenon appears for character varieties of orbifold structures of certain Montesinos knots [**21**], and the latter for knots admitting symmetries of order p [**22**]. In both cases, the appearance of ramification phenomena is related to the elementary fact that a matrix in $\mathrm{SL}_2(\mathbb{F})$ has order p if and only if it is parabolic (or unitary), i.e. has trace equal to 2.

The present paper is devoted to exhibiting examples of yet another type of ramification phenomena that we might describe as *order*-1 *ramification* in contrast to the ramification observed so far that we might call of *order* 0.

To make this discussion more concrete, let us consider an explicit example: the character variety of the figure-eight knot. It is well-known that this algebraic set can be defined by the equation $(x-2)(x^2+(1-t^2)x+t^2-1)=0$ where t represents the trace of a meridian, while x is the trace of the product of two generators of the group: a meridian and a conjugate of its inverse (see for instance [**8**, Section 6], with the variables $z=t^2-x$ and $y=t^2$). The character variety consists of two 1-dimensional irreducible components: the first one is the component of characters of abelian representations, while the second one is Thurston's excellent component containing the character of the lift to $\mathrm{SL}_2(\mathbb{C})$ of the hyperbolic holonomy of the knot. It is easy to convince oneself that, regardless of p, the $\mathrm{SL}_2(\mathbb{F})$-character variety always consists of two 1-dimensional irreducible components. However, although both components are smooth and intersect in precisely two points over $\mathbb{C}$, this is no longer the case for all p. Indeed, by computing the partial derivatives of the second component one has $2x-t^2+1$ and $-2t(x-1)$ and one realises that, for $p=5$, the second component is no more smooth at the point $t=0$ and $x=2$: the point is a cuspidal point and the two points of intersection between the two components collapse to this singular one in characteristic 5. Observe that if one considers PSL_2-characters instead of SL_2-characters, that is if we take $T=t^2$ as a coordinate instead of t, then the point of coordinates $T=0$ and $x=2$ is smooth even in characteristic 5 on each of the components, but the intersection between both components is transverse in characteristic zero or $p\neq 5$, and tangent in characteristic $p=5$.

DEFINITION 1.1. We say that $X(\Gamma)$ has a ramification of *order*-1 *type in characteristic* p if in the character variety in characteristic p either the type of intersection between two irreducible components changes (in particular, two transverse components become tangent), or the singularity type of a point changes (in particular, a singular point appears on a smooth component).

Note that ramifications of order 0 and 1 may appear at the same time if, for instance, an irreducible component splits into two and they meet at a point.

The following result gives a conceptual explanation of the reason why 5 is a ramified prime for the character variety of the figure-eight knot.

THEOREM 1.2. *Let K be a 2-bridge knot, $\Delta_K(t)$ its Alexander polynomial, and p a prime. We consider $X(\pi_1(K))$. If p divides $|\Delta_K(-1)|$, that is the order of the homology of the 2-fold branched cover of K, then p ramifies.*

The above is a special case of a more technical result for whose statement we need to introduce some notation first. Let K be a knot and M_n its n-fold cyclic branched cover. Let $\mu \in \pi_1(K)$ be an element representing a meridian of K. We have the following exact sequence of groups

$$1 \longrightarrow H \longrightarrow \pi_1(K)/ << \mu^4 >> \longrightarrow \mathbb{Z}/2\mathbb{Z} \longrightarrow 1.$$

where H can be seen as the orbifold fundamental group of the orbifold whose underlying topological space is M_2 and whose singular locus is the lift of K with order 2 singularity. The commutator subgroup H' of H is characteristic, so we can consider the quotient $\Gamma = (\pi_1(K)/ << \mu^4 >>)/H'$.

THEOREM 1.3. *Let K be a knot, $\Delta_K(t)$ its Alexander polynomial, and p a prime. We consider $X(\Gamma)$, where Γ is defined as above. If p divides $|\Delta_K(-1)|$, that is the order of $H_1(M_2)$, then p ramifies for $X(\Gamma)$.*

Remark that, for 2-bridge knots, Γ is a binary dihedral group, extension of $\mathbb{Z}/2\mathbb{Z}$ by $\pi_1(M_2) = H_1(M_2)$. In addition, $\mathrm{SL}_2(\mathbb{C})$-representations of Γ are lifts of representations of $\pi_1(K)/ << \mu^2 >>$ to $\mathrm{PSL}_2(\mathbb{C})$.

A natural question is to determine what ramification phenomena can appear in the situation of Theorem 1.3 and if the ramification is for $X(K)$ itself and not just for its algebraic subset $X(\Gamma)$. The easiest class of examples to consider is that of torus knots (see Section 7). Over $\mathbb{C}$ the character variety of a torus knot consists of a finite number of rational curves and the only ramification phenomena that occur are when several irreducible components collapse onto a single one. It follows that the character variety of the torus knot $T(a,b)$ ramifies for every odd prime p dividing ab, in particular whenever p divides $|\Delta_{T(a,b)}(-1)|$.

Although torus knots show that the ramification phenomena that can be observed in the situation of Theorem 1.3 is not always of the type seen for the figure-eight knot, one might expect that this latter behaviour is generic, at least for hyperbolic knots. The following corroborates this hypothesis.

PROPOSITION 1.4. *Let K be a hyperbolic 2-bridge knot with Alexander polynomial $\Delta_K(t)$. Let p be a divisor of $|\Delta_K(-1)|$. If all roots of Δ_K are simple, then the ramification phenomena occurring at p are of order* 1.

By [**9**] the hypothesis that all roots of $\Delta_K(t)$ are simple implies that the curve of abelian characters intersects transversely the other components of $X(K)$. In Lemma 4.1 we show that the intersection is no more transverse but tangent in characteristic p whenever p divides $|\Delta_K(-1)|$. Furthermore, the arguments of [**9**] can be used to argue that ramifications in the intersection of both components will appear at most for primes that divide the discriminant of $\Delta_K(t)$, provided this discriminant does not vanish (namely, all roots of $\Delta_K(t)$ are simple). See Remark 4.3. Notice that $\Delta_K(-1)$ divides the discriminant of $\Delta_K(t)$.

Incidentally, the techniques of the proof yield the following fact, which has been proved by Huynh, Le, and Tran in [**14**] using different methods.

PROPOSITION 1.5. *Let K be a hyperbolic 2-bridge knot with Alexander polynomial $\Delta_K(t)$. Let $\alpha = |\Delta_K(-1)|$. If α and $(\alpha-1)/2$ are prime numbers, then the character variety of K over $\mathbb{C}$ consists precisely of two irreducible components: the one whose characters correspond to abelian representations and Thurston's excellent one.*

In [**14**] the authors only assume α prime and $(\alpha-1)/2$ coprime with $(\beta-1)/2$, for the two-bridge knot β/α. We do not know whether these conditions can be further weakened. On the other hand, there are examples of hyperbolic 2-bridge knots such that neither α nor $(\alpha-1)/2$ are prime numbers and yet their character varieties consist of just two irreducible components.

To complete our analysis and to improve our understanding of ramification phenomena, we provide explicit computations of character varieties for some knots, notably for the tunnel number-one $P(2,2,3)$-pretzel knot 8_5 and for the π-hyperbolic knot 8_{18}. We also study character varieties for some 2-bridge knots. Among them, the knot 8_9 admits a ramified prime that does not come from the situation of Theorem 1.3. Quite interestingly, in this case we observe a new type of ramification, namely an irreducible component splitting into two irreducible ones.

Note that the aforementioned examples cover all possible geometries of double branched covers of hyperbolic knots, that is spherical, Seifert fibred, and hyperbolic. As for the case of hyperbolic knots with double branched covers admitting a non trivial JSJ-decomposition, their character varieties seem hard to compute because of the "large number" of generators of their groups (none of these is a tunnel number-one knot, for instance). Instead, we compute explicitly the character varieties of two tunnel number-one satellite knots, both with companion the trefoil knot. The first is a cable knot whose double branched cover is a graph manifold, while the second is a Whitehead double so the JSJ-decomposition of its double branched cover contains a hyperbolic piece. As a side remark, the double branched covers of these two knots are also the double branched covers of hyperbolic knots. Of course, it is not clear if there might be a relationship between the character varieties of two knots sharing the same double cover.

The paper is organised as follows. In Section 2 we recall basic facts about character varieties. The proofs of Theorems 1.2 and 1.3 follow from the discussion in Section 3, while Section 4 deals with the specificities of 2-bridge knots. This latter section also contains detailed computations of character varieties for some 2-bridge knots and the primes for which they ramify. The character varieties of the knots 8_5 and 8_{18} are studied in Sections 5 and 6, respectively. Section 7 is dedicated to character varieties of torus knots, while in Section 8 we discuss two examples of character varieties of satellite knots. Finally, some auxiliary computations are provided in the appendices.

The computations in this paper were performed by using Maple™.

2. Varieties of characters

In this section we recall relevant facts about character varieties and introduce some notation.

Let Γ be a finitely presented group and $\mathbb{F}$ an algebraically closed field. Given a representation $\rho\colon \Gamma \longrightarrow \mathrm{SL}_2(\mathbb{F})$, one can define a map $\chi_\rho\colon \Gamma \longrightarrow \mathbb{F}$ by composing with the trace function. Such map is called *the character of the representation* ρ. We will denote $X(\Gamma)_{\mathbb{F}}$ the set of all characters of representations of Γ. Note that each element γ of Γ induces an evaluation map $\tau_\gamma\colon X(\Gamma)_{\mathbb{F}} \longrightarrow \mathbb{F}$ defined as $\tau_\gamma(\chi)=\chi(\gamma)$.

For simplicity, when $\mathbb{F}=\mathbb{C}$ we will usually write $X(\Gamma)$ instead of $X(\Gamma)_{\mathbb{C}}$.

PROPOSITION 2.1 ([**6, 8**]). *The set of characters $X(\Gamma)$ is an affine algebraic set defined over $\mathbb{Z}$, which embeds in $\mathbb{C}^N$ with coordinate functions $(\tau_{\gamma_1},\dots,\tau_{\gamma_N})$ for some $\gamma_1,\dots,\gamma_N\in\Gamma$.*

The affine algebraic set $X(\Gamma)$ is called the *character variety* of Γ: it can be interpreted as the algebraic quotient of the variety of representation of Γ into $\mathrm{SL}_2(\mathbb{C})$ by the conjugacy action of $\mathrm{PSL}_2(\mathbb{C}) = \mathrm{SL}_2(\mathbb{C})/\mathcal{Z}(\mathrm{SL}_2(\mathbb{C}))$.

Note that the set $\{\gamma_1, \dots, \gamma_N\}$ in the above proposition can be chosen to contain a generating set of Γ. For Γ the fundamental group of a knot exterior, we will then assume that it always contains a representative of the meridian.

It is worth pointing out that the polynomial equations with integer coefficients of Proposition 2.1 only determine $X(\Gamma)$ as an algebraic set and not as a scheme.

A careful analysis of the arguments in [**8**] shows that Proposition 2.1 still holds if $\mathbb{C}$ is replaced by any algebraically closed field $\mathbb{F}$, provided that its characteristic is different from 2. Let $\mathbb{F}_p$ denote the field with p elements and $\overline{\mathbb{F}}_p$ its algebraic closure. We have:

PROPOSITION 2.2 ([**8**]). *Let $p > 2$ be an odd prime number. The set of characters $X(\Gamma)_{\overline{\mathbb{F}}_p}$ associated to representations of Γ into $\mathrm{SL}_2(\overline{\mathbb{F}}_p)$ is an algebraic set which embeds in ${\overline{\mathbb{F}}_p}^N$ with coordinate functions $(\tau_{\gamma_1}, \dots, \tau_{\gamma_N})$ for the same $\gamma_1, \dots, \gamma_N \in \Gamma$ seen in Proposition 2.1. Moreover, $X(\Gamma)_{\overline{\mathbb{F}}_p}$ is defined by the reductions mod p of the polynomials with coefficients in $\mathbb{Z}$ which define $X(\Gamma)$.*

The key observation here is that, as in the case of complex numbers, the algebraic set $X(\Gamma)_{\overline{\mathbb{F}}_p}$ can be again interpreted as the algebraic quotient of the variety of representation of Γ into $\mathrm{SL}_2(\overline{\mathbb{F}}_p)$ by the conjugacy action of $\mathrm{PSL}_2(\overline{\mathbb{F}}_p) = \mathrm{SL}_2(\overline{\mathbb{F}}_p)/\mathcal{Z}(\mathrm{SL}_2(\overline{\mathbb{F}}_p))$.

We will be mainly interested in the situation where Γ is the fundamental group $\pi_1(\mathbf{S}^3 \setminus K)$ of the exterior of a knot K. In this case we shall write $X(K)_{\mathbb{F}}$ or $X(K)$ instead of $X(\pi_1(\mathbf{S}^3 \setminus K))_{\mathbb{F}}$ or $X(\pi_1(\mathbf{S}^3 \setminus K))$ respectively.

Since the abelianisation of a knot group is infinite cyclic, for each element A of $\mathrm{SL}_2(\mathbb{F})$ there is precisely one abelian representation of $\pi_1(\mathbf{S}^3 \setminus K)$ defined by sending a meridian of K to A. The set of abelian representations projects onto a one-dimensional, rational component, of $X(K)_{\mathbb{F}}$.

If K is a hyperbolic knot, $\pi_1(\mathbf{S}^3 \setminus K)$ admits a hyperbolic holonomy representation into $\mathrm{PSL}_2(\mathbb{C})$. It is not hard to see that such representation lifts to a representation $\pi_1(\mathbf{S}^3 \setminus K) \longrightarrow \mathrm{SL}_2(\mathbb{C})$. It was proved by Thurston that the character of this representation is a smooth point on a one-dimensional irreducible component of $X(K)$, known as the *canonical* or *excellent component* of K. More generally, Thurston proved that every irreducible component of $X(K)$ has dimension at least one [**13, 26**].

3. Characters vanishing at the meridian

Consider the map

$$t = \tau_\mu \colon X(K) \to \mathbb{C},$$

sending a character to the trace of the meridian. We are interested in its fiber at 0:

$$Y := t^{-1}(0) \subset X(K).$$

Note that if we choose the trace of the meridian to be one of the coordinates of $X(K)$, the map is just a projection onto a coordinate and the fibre the intersection of $X(K)$ with a hyperplane. The condition $t(\chi_\rho) = 0$ means that χ_ρ is the character of a representation ρ that maps μ to a matrix A such that $A^2 = -\mathrm{Id}$. Consider the induced representation $\hat\rho$ obtained by composing ρ with the natural quotient $\mathrm{SL}_2(\mathbb{C}) \longrightarrow \mathrm{PSL}_2(\mathbb{C})$: $\hat\rho$ restricts to a representation $\hat\rho'$ of the double branched covering, M_2, of K.

We decompose the set $Y = t^{-1}(0)$ into three (possibly empty) disjoint subsets, according to the behaviour of $\hat\rho'$

$$Y = Y^{triv}(K) \sqcup Y^{ab}(K) \sqcup Y^{nab}(K),$$

where:

- $\chi \in Y^{triv}(K)$ if there is a representation ρ with character χ such that $\hat\rho'$ is trivial;
- $\chi \in Y^{ab}(K)$ if there is a representation ρ with character χ such that $\hat\rho'$ is abelian, but no such abelian representation is trivial;
- $Y^{nab}(K)$ if every representation ρ with character χ is such that $\hat\rho'$ is not abelian.

Note that $Y^{triv}(K) = X^{ab}(K) \cap Y$ always consists of a single point, the character of the abelian representation in $t^{-1}(0)$.

In characteristic zero, $Y^{ab}(K)$ is the set of binary dihedral representations and also the set of metabelian (and non-abelian) representations (see [**15, 18**]). As the cardinality of $H_1(M_2, \mathbb{Z})$ is $|\Delta_K(-1)|$, an odd integer, the cardinality of $Y^{ab}(K)$ is

$$|Y^{ab}(K)| = \frac{|\Delta_K(-1)| - 1}{2}.$$

Thus $Y^{ab}(K) = \emptyset$ if and only if $\Delta_K(-1) = \pm 1$.

Note that the subset $Y^{triv}(K) \cup Y^{ab}(K)$ is a subvariety of $Y(K)$ and thus of $X(K)$. Indeed, this set corresponds to the fibre $t^{-1}(0)$ of $X(\Gamma)$, where Γ is the quotient of $\pi_1(\mathbf{S}^3 \setminus K)$ defined in the introduction.

Before describing the reduction mod p of $Y^{ab}(K)$ we need a lemma on finite order elements of $\mathrm{PSL}_2(\mathbb{K})$. Its proof is a straightforward consequence of the structure of $\mathrm{PSL}_2(\mathbb{F}_{p^k})$ (see [**7**]).

LEMMA 3.1. *Let $\mathbb{F}$ be an algebraically closed field of characteristic $p \geq 3$ and $A \in \mathrm{PSL}_2(\mathbb{F})$ an element of finite order. Then either* $\mathrm{order}(A)$ *is coprime with p or* $\mathrm{order}(A) = p$. *In addition:*

- *When* $\mathrm{order}(A)$ *is coprime with p, then A is diagonalisable.*
- *When* $\mathrm{order}(A) = p$, *then A is conjugate to a matrix of the form* $\pm\left(\begin{smallmatrix} 1 & * \\ 0 & 1 \end{smallmatrix}\right)$.

COROLLARY 3.2. *For any prime p, write $\Delta_K(-1) = \pm p^k m$, with $m \geq 1$ an integer coprime with p. Then the reduction mod p of $Y^{ab}(K)$ has cardinality $\frac{m-1}{2}$.*

The previous corollary is an example of collapse when reducing mod p. Notice that when $m = 1$, i.e. when $\Delta_K(-1) = \pm p^k$, the reduction mod p of $Y^{ab}(K)$ collapses to a single point, the abelian character in $Y^{triv}(K)$. Note that, in characteristic p, this is now also the character of a reducible dihedral representation.

Regarding $Y^{nab}(K)$, we are interested in the following cases for K:

(1) K is a two bridge knot. Then $Y^{nab}(K) = \emptyset$ because M_2 is a lens space.
(2) K is a Montesinos knot. Then, as M_2 is Seifert fibred, $Y^{nab}(K)$ consists of representations whose projections into $\mathrm{PSL}_2(\mathbb{C})$ factor through the base 2-orbifold of the Seifert fibration. In particular for pretzel knots $Y^{nab}(K)$ is finite, because the variety of characters of triangle groups has finite cardinality.

(3) For π-hyperbolic knots $Y^{nab}(K)$ contains the lift of the holonomy of the orbifold with cone angle π. This is an isolated point of $Y(K)$ (by Weil local rigidity), but there may be more characters in $Y^{nab}(K)$.

4. Two-bridge knots

Let K be a 2-bridge knot. Recall that K is determined by its 2-*bridge notation*, a rational number of the form β/α, with α odd: one can recover a four-plat description of K from a continued fraction expansion for β/α, and the resulting knot does not depend on the chosen expansion. Moreover the 2-fold branched cover M_2 of K is precisely the lens space $L(\alpha,\beta)$ (see [**3**, Ch 12]).

The fundamental group $\pi_1(\mathbf{S}^3\setminus K)$ of a 2-bridge knot K admits a presentation of the form $\langle a,b \mid aw=wb\rangle$, with a and b meridians and w a word in the generators, a and b, and their inverses.

Since a and b are conjugate, the character variety $X(K)$ of K is a plane curve with coordinates $t=\tau_a=\tau_b$ and $x=\tau_{ab^{-1}}$:

$$\{(t,x)\in\mathbb{C}^2 \mid P(t,x)(x-2)=0\},$$

where $x=2$ is the set of characters of abelian representations, and its complement consists of characters of irreducible representations.

Notice that P is an even polynomial of the variable t (i.e. a polynomial in t^2): there is a natural action of $H^1(\mathbf{S}^3\setminus K;\mathbb{Z}/2)\cong\mathbb{Z}/2$ on the variety of characters, that maps a point with coordinates (t,x) to $(-t,x)$. By taking $T=t^2$ as a new variable, this defines the variety of $\mathrm{PSL}_2(\mathbb{C})$-characters [**8**,**10**].

Because of their simple structure, character varieties of 2-bridge knots have been widely studied, see for instance [**4**,**10**,**16**,**19**,**20**], see also [**12**] for 2-bridge links.

Here, again, we are interested in $Y(K)$. We have:

$$P(0,x)=\Phi_\alpha(x), \qquad \text{where } \Phi_\alpha(\lambda+\frac{1}{\lambda})=\frac{\lambda^\alpha-1}{\lambda-1}\lambda^{\frac{1-\alpha}{2}}.$$

Namely, the factors of Φ_α define the intersection with $\mathbb{R}$ of the corresponding cyclotomic extension.

LEMMA 4.1. *For $p\mid\alpha$, write $\alpha=p^r\alpha'$, with $\alpha'\in\mathbb{Z}$ coprime with p.*

(1) *$P(0,x)=0$ consists of $\frac{\alpha-1}{2}$ smooth points in $\mathbb{C}$.*
(2) *The reduction mod p of $P(0,x)=0$ consists of $\frac{\alpha'-1}{2}$ points in $\overline{\mathbb{F}}_p$.*
(3) *If $p^r\neq 3$, then the intersection between $P(t,x)=0$ and $x=2$ mod p is not transverse.*

PROOF. The assertions on the cardinality are proved in the previous corollary. Smoothness is a consequence of the fact that the roots of Φ_α are simple. It is also proved in [**2**, Lemmas 4.4 and 4.5].

In terms of the polynomial Φ_α we have:

$$\Phi_\alpha(x)\equiv(x-2)^{\frac{p^r-1}{2}}(\Phi_{\alpha'}(x))^{p^r} \mod p$$

For assertion (3) use that

$$P(t,x)=\Phi_\alpha(x)+t^2Q(t^2,x)$$

and the previous congruence of $\Phi_\alpha(x)$ mod p. Notice that we require $p^r\neq 3$ so that the exponent of $(x-2)$ is $\frac{p^r-1}{2}>1$. □

REMARK 4.2. One may ask whether this order-1 ramification phenomenon is a consequence of an order-0 ramification phenomenon that occurs at the level of the character variety $X(K)$, namely that different irreducible components of $\{(t,x)\in\mathbb{C}^2 \mid P(t,x)=0\}$ get identified together.

More specifically, for $\alpha = p$ we may ask whether $P(t,x)$ can be a power mod p. As the characteristic p is odd, if $P(t,x)$ were a power mod p, then it would be the power of an even polynomial in t. However in [**10**, Proposition 5.2] it is shown that the degree of $P(t,x)$ in t^2 is strictly less than the degree of $P(0,x)$, so this can never be the case.

REMARK 4.3. Let D be the discriminant of $\Delta_K(t)$. Assume that $D\neq 0$. Then, in characteristic zero, all roots of $\Delta_K(t)$ are simple, and by [**9**] the intersection of the abelian component $x=2$ with $P(t,x)=0$ is everywhere transverse. In characteristic p, if $p\nmid D$, then the intersection is still transverse, in particular the points $(t,2)$ are smooth points of $P(t,x)=0$. When $p\mid D$, new singularities or tangencies with $x=2$ may appear at those points. Notice also that $\Delta_K(-1)\mid D$.

In some instances, we may also count how many components $X(K)$ has:

LEMMA 4.4. *If α is prime, then $P(t,x)$ is irreducible over $\mathbb{Q}$. If in addition $\frac{\alpha-1}{2}$ is also prime, then it is irreducible over $\mathbb{C}$.*

PROOF. The curve of characters $X(K)$ has no ideal points with $t=0$, by [**1**, Lemma A.1.1], hence every irreducible component of $X(K)$ intersects the line $t=0$ in at least one point. Thus, to prove that P is irreducible (over $\mathbb{Q}$ or over $\mathbb{C}$) we need to check that all points in

$$Y^{ab}(K)=\{(0,x)\mid P(0,x)=0\}$$

are contained in the same irreducible component (recall that here $Y^{nab}(K)=\emptyset$). As $P(0,x)=\Phi_\alpha(x)$ is irreducible over $\mathbb{Q}$, because α is prime, $P=0$ has only one component over $\mathbb{Q}$, so P is irreducible over $\mathbb{Q}$.

For the second assertion, we notice first that at least two points of $Y^{ab}(K)$ lie in the same component over $\mathbb{C}$. Those points correspond to the lifts of the spherical holonomy of the orbifold with cone angle π. More precisely, the universal covering of $SO(4)$ is $SU(2)\times SU(2)$, and the lift of the holonomy yields two different characters in $Y^{ab}(K)$, one for the projection to each factor $SU(2)$, see for instance [**11**]. In addition, there is a curve of smooth points that joins both characters [**1,24,25,27**], coming from holonomies of spherical and Euclidean cone structures, hence they lie in the same irreducible component.

Once we know that two points of $Y^{ab}(K)$ lie in the same component, we consider the action of the Galois group. If P decomposes in $\mathbb{C}$, it decomposes with coefficients in a number field $\mathbb{K}$, that we may assume to contain the real cyclotomic field (containing the solutions of $\Phi_\alpha(z)$). The Galois group $\mathrm{Gal}(\mathbb{K}\mid\mathbb{Q})$ acts on $P=0$ by preserving or permuting irreducible components, in particular if $P=0$ contains more than one irreducible component, then $Y^{ab}(K)$ has a nontrivial partition induced by the components of $P=0$. As $\mathrm{Gal}(\mathbb{K}\mid\mathbb{Q})$ acts transitively on $Y^{ab}(K)$, all subsets of this partition have the same cardinality, which must be either 1 or $(\alpha-1)/2$, because $(\alpha-1)/2$ is prime. It cannot be 1 because we showed in the previous paragraph that the cardinality is at least 2. Hence it is $(\alpha-1)/2$ and there is a single component. □

The reader should compare this result with [**19**] where a condition is provided for the character variety of a 2-bridge knot to have at least two irreducible components, besides the abelian one.

In the following we provide some examples of character varieties of 2-bridge knots, for different values of α, notably α a prime number, a power of a prime number, and a product of different primes, in order to cover potentially different kinds of behaviour. In the first three instances, we also compute all ramified primes.

EXAMPLE: $6_1 = 7/9$

From the presentation

$$\pi_1(\mathbf{S}^3 \setminus 6_1) = \langle \alpha, \beta \mid \beta\alpha^{-1}\beta\alpha\beta^{-1}\alpha^{-1}\beta\alpha\beta^{-1} = \alpha^{-1}\beta\alpha\beta^{-1}\alpha^{-1}\beta\alpha\beta^{-1}\alpha \rangle,$$

using coordinates

$$t = \mathrm{tr}_\alpha = \mathrm{tr}_\beta, \qquad x = \mathrm{tr}_{\alpha\beta^{-1}},$$

and taking traces of the terms in the relation, we easily compute $P(t,x)$:

$$t^4x^2 - 3\,t^4x - 2\,t^2x^3 + 2\,t^4 + 2\,t^2x^2 + x^4 + 5\,t^2x + x^3 - 4\,t^2 - 3\,x^2 - 2\,x + 1$$

Even if 9 is not prime, $P(t,x)$ is irreducible over $\mathbb{C}$, as we will see in a while. We give here a reason why it is irreducible over $\mathbb{Q}$. We reproduce the argument in the proof of Lemma 4.4: setting $t = 0$, we have

$$P(0,x) = \Phi_9(x) = \Phi_3(x)\psi(x),$$

where the solution of $\Phi_3(x) = x + 1 = 0$ is -1, twice the real part of the primitive roots of unity of order 3, and the solutions of $\psi(x) = x^3 - 3x + 1 = 0$ are twice the real parts of primitive roots of unity of order 9, see (4.1). Thus, as there is no ideal point with $t = 0$ [**1**, Lemma A.1.1], to prove irreducibly over $\mathbb{Q}$ we must show that there are two points in the same component, with coordinates $(0,-1)$ and $(0,x_2)$, with $\psi(x_2) = 0$. Following the argument of Lemma 4.4, those points are provided by the lift to $\mathrm{SU}(2) \times \mathrm{SU}(2)$ of the holonomy of the spherical orbifold. Up to conjugation, the lift of the holonomy of $\pi_1(M_2)$ is a cyclic group generated by

$$\left(\begin{pmatrix} e^{i\,\theta_1/2} & 0 \\ 0 & e^{i\,\theta_1/2} \end{pmatrix}, \begin{pmatrix} e^{i\,\theta_2/2} & 0 \\ 0 & e^{i\,\theta_2/2} \end{pmatrix} \right)$$

By the description of the action of $\mathrm{SU}(2) \times \mathrm{SU}(2)$ on $\mathrm{SU}(2) \cong S^3$ [**17**], and as $M_2 \cong L(9,7)$, θ_1 and θ_2 must satisfy

$$\theta_2 - \theta_1 = \frac{1}{9}2\pi, \qquad \theta_1 + \theta_2 = \frac{7}{9}2\pi.$$

Hence

$$\theta_1 = \frac{1}{3}2\pi \qquad \theta_2 = \frac{2}{9}2\pi,$$

and $x_1 = 2\cos\theta_1 = -1$ and $x_2 = 2\cos\theta_2$ are the values we are looking for.

Observe also that the action of the Galois group permutes the points $(0,x_2)$ with $x_2^3 - 3x_2 + 1 = 0$, while fixing the point $(0,-1)$, so there is a single component over $\mathbb{C}$.

As just observed, for $t = 0$ we have:

$$P(0,x) = \Phi_3(x)\psi(x) = (x+1)\left(x^3 - 3\,x + 1\right).$$

According to Theorem 1.2, $Y(K)$ collapses to a single point mod 3, because

$$P(0,x) \equiv (x-2)^4 \mod 3.$$

Note that the curve of irreducible characters meets the line of abelian ones in two points with multiplicity 2: these are computed as the solutions of the equation $0 = P(t,2) = -2t^2+9$. The two components also have an intersection at infinity, also of multiplicity 2. The two (finite) intersections collapse to a single one in characteristic $p = 3$ (and in this characteristic only), and the two components become tangent. The latter assertion is easily seen by rewriting $P(t,x) = t^4[(x-2)^2 + (x-2)] + t^2[2(x-2)^3 - 10(x-2)^2 + 11(x-2) + 42] + (x+1)[(x-2)^3 + 6(x-2)^2 + 9(x-2) + 3]$ so that one also observes that the component is not smooth at this point in characteristic $p = 3$.

We want to show that $p = 3$ is the only ramified prime for the character variety of the knot 6_1. Clearly, for no prime the dimension of the variety can decrease or increase and it will always consists of at least two irreducible components. We will prove that the component defined by $P(t,x) = 0$ is smooth over $\mathbb{C}$, and so is the corresponding component in any characteristic $p > 3$.

The partial derivatives are

$$\partial_t P(t,x) = 2t[t^2(2x^2 - 6x + 4) - 2x^3 + 2x^2 + 5x - 4]$$

and

$$\partial_x P(t,x) = t^4(2x-3) + t^2(-6x^2 + 4x + 5) + 4x^3 + 3x^2 - 6x - 2.$$

Assume first that $t = 0$. In this case we need to find the common solutions of $P(0,x) = 0$ and $\partial_x P(0,x) = 0$: the discriminant of $P(0,x)$ is 3^6 so these two polynomials have a common root only if $p = 3$ and we already know what happens in that case.

We can thus assume that $t \neq 0$. Consider now the polynomial $(2x^2 - 6x + 4)$: its roots are 1 and 2. A computation shows that $\partial_t P(t,1)$ and $\partial_t P(t,2)$ cannot be zero in characteristic 0 or odd.

We can thus multiply $\partial_x P(t,x)$ times $(2x^2 - 6x + 4)^2$ in order to eliminate t, and we get $\partial_x P(t,x)(2x^2 - 6x + 4)^2 = 3x^2 - 4x$.

Assume $x = 0$. In this case $t^2 = 1$ and $P(1,0) = -1$, so this case cannot arise.

If $p = 3$, then again we must have $x = 0$ and the preceding reasoning applies.

We can thus assume that $p \neq 3$ and $x = 4/3$ and $t^2 = 10/9$ and again $P(4/3, 10/9)$ can never be 0 for any p.

We now will show that the variety consists of precisely two irreducible components in every characteristic ($\neq 2$). Obviously the abelian component, being a line, can never split into more than one component. It is thus enough to prove that $P(t,x)$ is always irreducible. Assume by contradiction that the curve corresponding to $P(t,x) = 0$ is not irreducible. Its irreducible components must intersect in the projective closure in some singular points, since we are considering algebraically closed fields. Similarly, every irreducible component must meet the line at infinity. We now observe that our curve intersects the line at infinity in two points, one with (homogeneous) coordinates equal to $t = 1$ and $x = 0$ and a second one with coordinates $t = 0$ and $x = 1$. The first point has multiplicity four and is smooth with tangent the line at infinity in any characteristic $\neq 2$. The second one has multiplicity two and is a singular point, again with tangent the line at infinity. Since the first point is smooth, it must belong to a single irreducible component. If $p \neq 3$ this irreducible component must also contain the second point at infinity, which is the only singular point of the curve. Because of the multiplicities of intersection, this irreducible component must be a curve of degree at least five. It follows that if there were a second component it should be a line, passing through the second point and tangent to the line at infinity. This is however absurd since the component cannot be the line at infinity. If $p = 3$, the component containing the smooth point at infinity might have degree 4, since there are two singular points. We consider then its intersection with the line of equation $x = 2$. Because of its degree, this component meets the line at the singular point $(t = 0, x = 2)$. If the multiplicity of intersection were 4, then the component could not intersect the line elsewhere and thus it could not intersect the remaining irreducible component(s) since it could not pass through the second singular point which is also on this line. This shows that the component must pass through the second point at infinity even in this case and the conclusion follows as in the previous one.

Example: $4_1 = 3/5$

We already discussed the character variety of the figure-eight knot in the introduction. It is well-known that it consists of just two irreducible components: the abelian one and Thurston's excellent one. This is also a consequence of Proposition 1.5 since both $\alpha = 5$ and $(\alpha - 1)/2 = 2$ are prime numbers.

The same type of arguments and computations used in the previous example show that in this case, too, there is a single odd prime for which the character variety ramifies, that is $p = 5$. Recall that in this case we have

$$P(t, x) = (x^2 + (1 - t^2)x + t^2 - 1)$$

while the partial derivatives are

$$\partial_t P(t, x) = 2t[x - 1]$$

and

$$\partial_x P(t, x) = 2x - t^2 + 1.$$

We see that singular points can only occur for $t = 0$ or $x = 1$. The latter case, is impossible as $P(t, 1) = 1$ for any t. In the former case we must have $2x + 1 = 0$ or equivalently $x = -1/2$ since we are assuming $p \neq 2$. This gives $P(t, x) = -5/4$, so we have singular points only for $p = 5$. This is precisely the situation described by Theorem 1.2.

To see that the reduction mod p of the polynomial $P(t, x)$ is irreducible for all $p \geq 3$, one observes that the points at infinity of $P(t, x)$ are both smooth. As a consequence, $P(t, x)$ can only split if $p = 5$ and at most into two components. Since the degrees of the two irreducible components must be 2 and 1, the component of degree one is necessarily one of the tangents at the singular point $(t = 0, x = 1)$, however the two tangents do not meet the line at infinity at the same points as $P(t, x)$. The computational details are left to the reader.

Example: $7_4 = 11/15$

The character variety of this double-twist knot was computed and studied in [**4**] by Chu who was interested in understanding intersections between different, non abelian, irreducible components. Indeed, in this case the variety has two irreducible components, besides the abelian one, as shown by [**16**, **19**].

Rewriting the equations in [**4**] using our notation, we have

$$P(t,x) = (-1+2x^2+x^3-x^2t^2)(1+4x-4x^2-x^3+x^4-2xt^2+3x^2t^2-x^3t^2).$$

Theorem 1.2 tells us that $p=3$ and $p=5$ are ramified primes for this variety. We wish to show that these are the only primes that ramify.

We start by studying each irreducible component separately. We begin with

$$P_1(t,x) = -1+2x^2+x^3-x^2t^2$$

whose partial derivatives are

$$\partial_t P_1(t,x) = -2tx^2$$

and

$$\partial_x P_1(t,x) = -2xt^2+4x+3x^2.$$

We see that both derivatives are 0 if $x=0$ but $P_1(t,0)=-1$, so we can exclude the value $x=0$. It follows that we must have $t=0$ and $4+3x=0$. If $p=3$ there is no solution, so we can assume $p\neq 3$ and $x=-4/3$. We get $P_1(0,-4/3)=-5/27$ so, as expected, $p=5$ is a ramified prime. Writing $P_1(t,x)$ as $-t^2(x-2)^2-4t^2(x-2)-4t^2+(x-2)^3+8(x-2)^2+20(x-2)+50$, one sees that, in characteristic $p=5$, the tangent at the singular point $t=0$, $x=-4/3\equiv 2$ has equation $t^2-2(x-2)^2=0$, where coefficients are thought mod 5.

To understand whether this component splits in some characteristic, we now analyse its points at infinity. There are two such points, both intersecting the line at infinity with multiplicity 2. One of them, with homogeneous coordinates $t=0$ and $x=1$, is smooth with tangent the line at infinity, while the other, with homogeneous coordinates $t=1$ and $x=0$, is singular with tangent of equation $-x^2=0$. Consider now the irreducible component containing the smooth point at infinity. Since the curve only meets the line of equation $x=0$ at infinity, such irreducible component passes through both points at infinity and has degree at least 3. If it had degree 3, the second irreducible component would have degree one and would have to coincide with the tangent at the singular point, that is the line of equation $x=0$. This is clearly impossible, so this curve is irreducible in all characteristics.

We pass now to the second component

$$P_2(t,x) = 1+4x-4x^2-x^3+x^4-2xt^2+3x^2t^2-x^3t^2$$

whose partial derivatives are

$$\partial_t P_2(t,x) = 2t(-2x+3x^2-x^3)$$

and

$$\partial_x P_2(t,x) = 4-8x-3x^2+4x^3+t^2(-2+6x-3x^2).$$

From the first derivative, we see that either $x\in\{0,1,2\}$ or $t=0$. For each value of x in $\{0,1,2\}$ we get $P_2(t,x)=1$, so we must have $t=0$. Computing the resultant of the polynomials $1+4x-4x^2-x^3+x^4$ and $4-8x-3x^2+4x^3$ one obtains $3^2 5^3$: as expected the curve has singular points in characteristics $p=3$ and $p=5$.

As for the previous component, we show that this component as well is always irreducible. As before we consider the points at infinity of the curve: again we have two points. The first one has homogeneous coordinates $t=0$ and $x=1$, it is smooth with tangent the line at infinity, and the multiplicity of intersection between the curve and the line at infinity at this point is 2. The second point has homogeneous coordinates $t = 1$ and $x = 0$, it is a singular point with tangent of equation $x^3 = 0$ and multiplicity of intersection 3. The very same argument seen in the previous case, shows that the irreducible component containing the smooth point at infinity must pass through the second one as well and have degree at least 3. If the curve splits in two or more irreducible components, the component of degree at most two must pass through the singular point at infinity with tangent of equation $x = 0$. As a consequence such component can only be the line $x = 0$, with multiplicity one or two. This is however impossible.

We turn now our attention to the points of intersection between the two components. Since x cannot be equal to 0, we can use $P_1(t, x)$ to eliminate t^2 from $P_2(t, x)$. We obtain the polynomial $x^2 - 2x + 2$ that always has two roots in any characteristic different from 2 (cfr. [**4**]).

EXAMPLE: $8_9 = 7/25$

From the presentation

$$\pi_1(\mathbf{S}^3 \setminus 8_9) = \langle \alpha, \beta \mid \beta\alpha^{-1}\beta\alpha\beta^{-1}\alpha^{-1}\beta\alpha\beta^{-1} = \alpha^{-1}\beta\alpha\beta^{-1}\alpha^{-1}\beta\alpha\beta^{-1}\alpha \rangle$$

by taking traces and using the coordinates $t = \mathrm{tr}_\alpha = \mathrm{tr}_\beta$, $x = \mathrm{tr}_{\alpha\beta^{-1}}$ with the help of symbolic software we get

$$P(t,x) = t^{18}x^3 - 6\,t^{18}x^2 - 9\,t^{16}x^4 + 12\,t^{18}x + 47\,t^{16}x^3 + 36\,t^{14}x^5 - 8\,t^{18} - 66\,t^{16}x^2 - 160\,t^{14}x^4 - 84\,t^{12}x^6 - 12\,t^{16}x + 115\,t^{14}x^3 + 308\,t^{12}x^5 + 126\,t^{10}x^7 + 56\,t^{16} + 273\,t^{14}x^2 + 35\,t^{12}x^4 - 364\,t^{10}x^6 - 126\,t^8x^8 - 232\,t^{14}x - 928\,t^{12}x^3 - 441\,t^{10}x^5 + 266\,t^8x^7 + 84\,t^6x^9 - 140\,t^{14} + 187\,t^{12}x^2 + 1515\,t^{10}x^4 + 763\,t^8x^6 - 112\,t^6x^8 - 36\,t^4x^{10} + 756\,t^{12}x + 630\,t^{10}x^3 - 1362\,t^8x^5 - 679\,t^6x^7 + 20\,t^4x^9 + 9\,t^2x^{11} + 124\,t^{12} - 1648\,t^{10}x^2 - 1755\,t^8x^4 + 647\,t^6x^6 + 345\,t^4x^8 + 2\,t^2x^{10} - x^{12} - 725\,t^{10}x + 1796\,t^8x^3 + 1964\,t^6x^5 - 108\,t^4x^7 - 95\,t^2x^9 - x^{11} + 42\,t^{10} + 1744\,t^8x^2 - 924\,t^6x^4 - 1163\,t^4x^6 - 27\,t^2x^8 + 11\,x^{10} - 93\,t^8x - 2206\,t^6x^3 + 76\,t^4x^5 + 358\,t^2x^7 + 10\,x^9 - 118\,t^8 - 29\,t^6x^2 + 1544\,t^4x^4 + 120\,t^2x^6 - 45\,x^8 + 422\,t^6x + 225\,t^4x^3 - 565\,t^2x^5 - 36\,x^7 + 25\,t^6 - 560\,t^4x^2 - 201\,t^2x^4 + 84\,x^6 - 85\,t^4x + 326\,t^2x^3 + 56\,x^5 + 25\,t^4 + 95\,t^2x^2 - 70\,x^4 - 46\,xt^2 - 35\,x^3 - 6\,t^2 + 21\,x^2 + 6\,x - 1$$

It intersects $t = 0$ at

$$(x^2 + x - 1)\left(x^{10} - 10\,x^8 + 35\,x^6 + x^5 - 50\,x^4 - 5\,x^3 + 25\,x^2 + 5\,x - 1\right) \tag{4.1}$$

Which corresponds to roots of order 5 and 25, respectively. This polynomial is congruent to

$$(x-2)^{12} \mod 5$$

Thus the whole $Y(K)$ collapses to a single point when reducing mod 5.

Unexpectedly, $P(t,x)$ factors non trivially mod 7 (but not mod 5):

$$\begin{aligned}P(t,x) \equiv &(t^{12}x^2 + 3\,t^{12}x + t^{10}x^3 + 4\,t^{12} + 6\,t^{10}x^2 + t^8x^4 + 6\,t^{10}x + 2\,t^8x^3 + t^6x^5 \\ &+ 5\,t^{10} + 6\,t^8x^2 + 5\,t^6x^4 + t^4x^6 + 6\,t^8x + 4\,t^6x^3 + t^4x^5 + t^2x^7 + 2\,t^8 \\ &+ t^6x^2 + 4\,t^2x^6 + x^8 + 4\,t^6x + 2\,t^4x^3 + t^2x^5 + 5\,t^6 + 4\,t^4x^2 + t^4x \\ &+ 4\,t^4 + 4\,t^2x^2 + 4\,x^4 + 5\,xt^2 + 4\,x^3 + 5\,t^2 + 5\,x^2 + 2\,x + 6) \\ &\times (t^6x + 5\,t^6 + 4\,t^4x^2 + 3\,t^4x + 3\,t^2x^3 + 6\,t^4 + 6\,x^4 + 4\,xt^2 + 6\,x^3 \\ &+ 4\,t^2 + 4\,x^2 + 3\,x + 1) \qquad \mod 7\end{aligned}$$

Using symbolic software again, we may check that in characteristic zero $P(t,x) = 0$ has no singular points. In fact, besides characteristic $p = 5$ and $p = 7$, in characteristic $p = 23$ one finds that $(t,x) = (2,\pm 6)$ is a singular point of the variety. Notice that the discriminant of the Alexander polynomial is $13225 = 5^2\,23^2$, which, by Remark 4.3, implies that there are singular points with $x = 2$ in characteristic p at most for $p = 5$ and $p = 23$.

Example: 11/23

In this case the order of the homology of the double branched cover is a prime number and moreover we have that $11 = (23-1)/2$ is also prime. We mainly study this twist knot because its character variety appears as a subvariety of the Whitehead double we study in Section 8. Indeed, there is a degree-one map from the Whitehead double to this knot obtained by sending the trefoil companion onto the trivial knot. Of course the map induces a surjection of fundamental groups and so an injection of character varieties.

We give here a presentation obtained from a presentation of the Whitehead link 5_1^2 by surgery on one of its two components. This will become handy when studying the Whitehead double.

We have for the Whitehead link

$$\pi_1(\mathbf{S}^3 \setminus 5_1^2) = \langle \alpha, b \mid \alpha b^{-1}\alpha^{-1} b\alpha^{-1} b^{-1}\alpha = b^{-1}\alpha b^{-1}\alpha^{-1} b\alpha^{-1} b^{-1}\alpha b\rangle,$$

and for the twist knot

$$\pi_1(\mathbf{S}^3 \setminus 11/23) = \langle \alpha, \beta \mid u = \beta^{-1}\alpha, b = u^6, u = \alpha b^{-1}\alpha^{-1} b\alpha^{-1} b^{-1}\alpha b, u = b^{-1}ub\rangle,$$

where the group of the twist knot is obtained by killing by Dehn surgery the slope u^6b^{-1}, where $u = \alpha b^{-1}\alpha^{-1}b\alpha^{-1}b^{-1}\alpha b$, so that the relation of the Whitehead link group becomes $u = b^{-1}ub$ which is redundant in the quotient group; finally one can recover a standard 2-bridge presentation for the twist knot by observing that $\beta = \alpha u^{-1}$ is conjugate to α hence a meridian that can be chosen as a generator of the group.

Hence we take coordinates

$$t = \mathrm{tr}_\alpha = \mathrm{tr}_\beta, \qquad x = \mathrm{tr}_{\alpha^{-1}\beta},$$

and by taking traces on the equality

$$\alpha^{-1}(\beta\alpha^{-1})^5\beta\alpha(\beta^{-1}\alpha)^5\beta = (\beta\alpha^{-1})^5(\beta\alpha)(\beta^{-1}\alpha)^5$$

we get

$$(x-2)P(t,x) = 0$$

where

$$\begin{aligned}P(t,x) = t^2x^{10} - t^2x^9 - x^{11} - 8\,t^2x^8 - x^{10} + 7\,t^2x^7 + 10\,x^9 + 22\,t^2x^6 + 9\,x^8 \\ - 16\,t^2x^5 - 36\,x^7 - 24\,t^2x^4 - 28\,x^6 + 13\,t^2x^3 + 56\,x^5 \\ + 9\,t^2x^2 + 35\,x^4 - 3\,t^2x - 35\,x^3 - 15\,x^2 + 6\,x + 1.\end{aligned}$$

with

$$P(0,x) \equiv -(x-2)^{11} \mod 23$$

An analysis similar to those seen for the first three examples shows that $p = 23$ is the only prime that ramifies for this knot.

5. A pretzel example

For a Montesinos knot K that is not a 2-bridge one, $Y^{nab}(K) \neq \emptyset$, as it contains irreducible representations of the double branched cover M_2. Since M_2 is Seifert fibered, those representations from $\pi_1(M_2)$ to $\mathrm{PSL}_2(\mathbb{C})$ map the fibre to the identity (the centre of $\mathrm{PSL}_2(\mathbb{C})$ is trivial), hence they are irreducible representations of the 2-orbifold, the space of fibres. For a Montesinos knot with k tangles, this will yield components in $Y^{nab}(K)$ of dimension up to $k-3$, see for instance [**21**]. For a pretzel knot, we have $k = 3$.

Example: $8_5 = P(3,3,2)$

Here $\Delta_K(-1) = 21$, hence $|Y^{ab}(K)| = 10$. If we restrict these characters to M_2, they are characters of ten cyclic representations $H_1(M_2,\mathbb{Z}) \to \mathrm{PSL}_2(\mathbb{C})$, one of them has order 3, three have order 7, and six have order 21 (the cardinality is one half of Euler's ϕ).

Since K is the pretzel $(2,3,3)$, elements in $Y^{nab}(K)$ are characters of representations in SU(2).

We will consider reductions mod p for $p = 3, 7$.

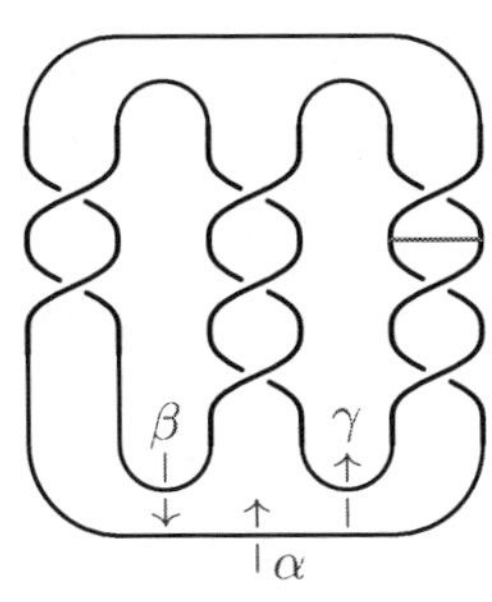

FIGURE 1. The knot 8_5. The generators α, β and γ of its fundamental group and a tunnel.

The group of 8_5 is generated by three meridians, α, β and γ in Figure 1. Since it has tunnel number one, the group can also be generated by two elements. Those are γ and $a = \gamma^{-1}\alpha^{-1}(\beta\gamma)^2$, see Appendix A. Hence we choose coordinates:

$$t = \mathrm{tr}_\gamma$$
$$x = \mathrm{tr}_a$$
$$y = \mathrm{tr}_{\gamma a}$$

With the help of symbolic software, there are 5 components:

(1) Abelian component

$$\begin{aligned} x &= t^2 - 2 \\ y &= t^3 - 3t, \end{aligned}$$

because x and y correspond to traces of elements whose abelianisations are 2 and 3 respectively.

(2) A "trefoil" component:

$$y = x - 1 = 0$$

namely, a component that looks like the variety of irreducible characters of the trefoil knot and which is induced by an epimorphism onto the trefoil group.

(3) The canonical component.

$$\begin{aligned} tx^3 - t^2y - ty^2 - x^2y + 2\,tx + 2\,yx - t - y &= 0, \\ t^2yx + y^2tx - 2\,tx^2 - ty^2 - x^2y - y^3 - tx + yx + t + 2\,y &= 0, \\ ty^3x + tx^2y - y^2x^2 - y^4 - 2\,txy - 2\,x^3 - xy^2 - ty + 2\,x^2 + 3\,y^2 + 4\,x - 2 &= 0. \end{aligned}$$

(4) An exotic component:

$$x + 1 = ty + y^2 - 2 = 0$$

(5) A second exotic component:

$$\begin{aligned} t^2x^2y - tx^3 - 2\,y^2tx + x^2y + y^3 + 2\,tx - t - 3\,y &= 0, \\ t^3yx + t^2xy^2 + ty^3x - t^3y - t^2x^2 - 2\,t^2y^2 - tx^2y - ty^3 - y^2x^2 & \\ -y^4 + t^2 + 4\,ty + 4\,y^2 &= 0, \\ tx^5y + tx^4y - x^6 - x^4y^2 + 2\,t^2xy^2 - 4\,tyx^3 + 2\,ty^3x - x^5 - x^3y^2 & \\ -t^3y - t^2y^2 - 7\,tx^2y - 2\,ty^3 + 6\,x^4 + y^2x^2 - 2\,y^4 + t^2x + 6\,x^3 & \\ +3\,xy^2 + t^2 + 8\,ty - 8\,x^2 + 7\,y^2 - 8\,x - 1 &= 0 \end{aligned}$$

Next we describe the intersection of each component with $t = 0$:

(1) Abelian: $x = -2$, $y = 0$. This is the point in $Y^{triv}(K)$.
(2) Trefoil: $y = 0$, $x = 1$. This point lies in $Y^{ab}(K)$ (it corresponds to a representation $\hat{\rho}'$ of order 3).
(3) Canonical: it is the union of two sets:
   - $y = x^3 - x^2 - 2x + 1 = 0$ that lies in $Y^{ab}(K)$ (3 points of corresponding to representations $\hat{\rho}'$ of order 7), and
   - $x - 1 = y^2 - 2 = 0$, two points in $Y^{nab}(K)$.
(4) Exotic: $x + 1 = y^2 - 2 = 0$, two points in $Y^{nab}(K)$.
(5) Second exotic: $y = -x^6 - x^5 + 6\,x^4 + 6\,x^3 - 8\,x^2 - 8\,x - 1 = 0$. Six points in $Y^{ab}(K)$ (corresponding to representations $\hat{\rho}'$ of order 21).

Notice that the 4 points in $Y^{nab}(K)$ correspond to two non conjugate representations in $\mathrm{PSL}_2(\mathbb{C})$ (x is already a variable of the characters in $\mathrm{PSL}_2(\mathbb{C})$ but y is not).

This knot is π-spherical, so the lift of its holonomy in $\mathrm{SU}(2) \times \mathrm{SU}(2)$ projects to the tetrahedral group in one $\mathrm{SU}(2)$ factor (for the basis 2-orbifold), and to the binary dihedral group of order 28 in the other $\mathrm{SU}(2)$ factor. This dihedral group in one of the factors corresponds to an action of $\mathbf{S}^1$ that is preserved up to orientation, yielding the orbifold Seifert fibration.

Regarding the ramifications, we obviously have the collapses of $Y^{ab}(K)$ mod 3 and mod 7:

- When reducing mod 3, the point corresponding to a represenatation $\hat{\rho}'$ of order 3 becomes trivial (in $Y^{triv}(K)$). The 6 points corresponding to representations $\hat{\rho}'$ of order 21 are identified to some of the 3 points of order 7.
- When reducing mod 7, the points corresponding to representations $\hat{\rho}'$ of order 7 become also trivial (in $Y^{triv}(K)$). The 6 points corresponding to representations $\hat{\rho}'$ of order 21 become the point corresponding to the representation $\hat{\rho}'$ of order 3.

6. A π-hyperbolic knot

The knot 8_{18} is π-hyperbolic, that is its double branched cover is hyperbolic. Here the canonical component splits into two irreducible components in characteristic 3.

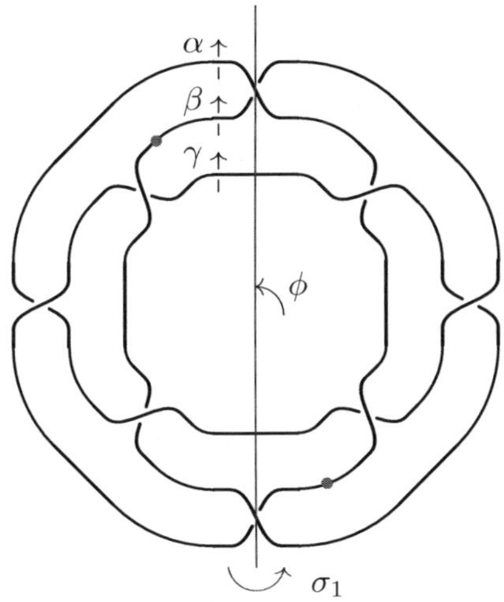

FIGURE 2. The knot 8_{18}. The period ϕ, the axis of the strong inversion σ_1 and the fixed points of the orientation reversing involution σ_2 marked as thick red dots

The knot 8_{18} has a period ϕ, corresponding to the rotation of order four in Figure 2. The symmetry group of the knot is generated by the rotation together with two involutions σ_1 and σ_2. The orientation preserving involution σ_1 is a rotation of angle π about an axis represented as a blue line in Figure 2: it is a strong inversion of the knot. The orientation-reversing involution σ_2 is reflection with two fixed points belonging to the knot (after a suitable conformal change of coordinates of $\mathbf{S}^3 = \mathbb{R}^3 \cup \{\infty\}$ sending one of the two fixed points to ∞ and the other to the origin, σ_2 can be seen as the linear map with matrix minus the identity).

Here $|\Delta_K(-1)| = 45$, and $H_1(M_2) = \mathbb{Z}/5 \oplus \mathbb{Z}/3 \oplus \mathbb{Z}/3$.

We sketch the computation of $X(8_{18})$ in Appendix B, here we just describe it. We use the following coordinates for $X(8_{18})$:

$$\begin{aligned} t &= tr_\alpha = tr_\beta = tr_\gamma \\ x &= tr_{\alpha\beta^{-1}} \\ y &= tr_{\beta^{-1}\gamma} \\ z &= tr_{\alpha\gamma} \\ w &= tr_{\alpha\beta^{-1}\gamma} \end{aligned}$$

We next list the 11 components of $X(8_{18})$.

Abelian component:

$$\langle -t + w, x - 2, y - 2, -t^2 + z + 2 \rangle \tag{6.1}$$

The symmetries act trivially here. When $t = 0$ we get the character that is trivial on M_2, the double branched covering.

The canonical component:

(6.2)
$$\langle tw + y^2 + \left(-t^2 - 1\right) y + 1, -t^2 + y + z, 5\, t^2 - 4\, tw + w^2 - 6, -t^2 + x + y - 1 \rangle$$

It can also be written as

$$\begin{aligned} tw + y^2 + \left(-t^2 - 1\right) y + 1 &= 0 \\ z &= t^2 - y \\ x &= z + 1 \\ 5\, t^2 - 4\, tw + w^2 &= 6 \end{aligned}$$

It is a double branched covering of the curve described in [**12**] obtained by considering only the variables t and y (we can get rid of x and z, the projection is the elimination of w).

This component is fixed pointwise by the period ϕ and by σ_1. The involution σ_2 preserves the component but permutes the holonomy of the complete structure with its complex conjugate.

The intersection with $t = 0$ yields precisely the 4 lifts of the hyperbolic holonomy of the orbifold ($t = 0$, $y = (1 \pm \sqrt{-3})/2$, $w = \pm\sqrt{6}$). The 4 points correspond to complex conjugation ($y = (1 \pm \sqrt{-3})/2$) and a choice of sign ($w = \pm\sqrt{6}$) as the abelianisation of the orbifold group is $\mathbb{Z}/2$ (it only concerns the variable w because $t = 0$). When reducing mod 3, these 4 points become a single point.

It should be noticed that the last equation

$$5\, t^2 - 4\, tw + w^2 = 6$$

factors mod 3 as the product of two lineal equations, hence this component splits into 2 components when reducing mod 3.

Four "trefoil-like" components: There are four components that look like the variety of characters of the trefoil:

$$\langle -1+z, -t+w, x-2, -t^2+y+1\rangle \tag{6.3}$$

$$\langle -1+z, -t+w, y-2, -t^2+x+1\rangle \tag{6.4}$$

$$\langle -1+z, -t^3+2\,t+w, -t^2+x+1, -t^2+y+1\rangle \tag{6.5}$$

$$\langle -t^3+2\,t+w, -t^2+x+1, -t^2+y+1, -t^2+z+2\rangle \tag{6.6}$$

They come from a surjection from $\pi_1(\mathbf{S}^3\setminus 8_{18})$ to the group of the trefoil. Namely, if we add the relation $\alpha=\beta$ then we get the group of the trefoil, and this corresponds to the component (6.3).

For computational purposes, it is useful to give a parametrisation of these components:

$$\begin{array}{llllr} x=2, & y=t^2-1, & z=1, & w=t, & (6.3)\\ x=t^2-1, & y=2, & z=1, & w=t, & (6.4)\\ x=t^2-1, & y=t^2-1, & z=1, & w=t^3-2t, & (6.5)\\ x=t^2-1, & y=t^2-1, & z=t^2-2, & w=t^3-2t, & (6.6) \end{array}$$

The period (the symmetry of order 4) permutes (6.3) with (6.5), and (6.4) with (6.6). The square of the period fixes point-wise these components.

The involution σ_1 preserves (6.3) and (6.5) and permutes (6.4) with (6.6).

The involution σ_2 permutes (6.3) and (6.4) and preserves the components (6.5) and (6.6). Thus the group of symmetries acts transitively on these components.

The intersection of the 4 components is the point $x=y=2$, $z=1$, $w=t=\pm\sqrt{3}$, which also lies in the abelian component (the discriminant of the Alexander polynomial vanishes).

When $t=0$, we get the conjugacy classes of 4 representations of order 3 of M_2, as $H_1(M_2)\cong\mathbb{Z}/3\oplus\mathbb{Z}/3\oplus\mathbb{Z}/5$.

The figure eight knot component: This component looks like the character variety of the figure eight knot and comes from a surjection of fundamental groups. The figure eight knot is in fact the quotient of the 8_{18} by ϕ^2, the square of the period. The ideal is:

$$\langle -t+w, y^2+\left(-t^2+1\right)y+t^2-1, -t^2+y+z, -t^2+x+y+1\rangle \tag{6.7}$$

It can also be presented as

$$\begin{aligned} w&=t\\ (y-1)t^2&=y^2+y-1\\ z&=t^2-y\\ x\,y&=x+y \end{aligned}$$

It is point-wise invariant by the period and by the involution σ_1 and preserved by σ_2, but σ_2 permutes x and y. In particular σ_2 swaps the holonomy of the figure eight knot with its complex conjugate.

When $t=0$, $y^2+y-1=0$ corresponds to the two conjugacy classes of nontrivial representations of $\mathbb{Z}/5$.

Four more components: When $t = 0$, there are still 16 conjugacy classes of representations of M_2 of order 15. They belong to 4 components. The first one is:

$$(6.8)\quad \langle x^2 + (-t^2+1)\,x + t^2 - 1,\\ (t^2-1)\,x + y - z - t^4 + 3\,t^2 - 1,\\ (-t^2+1)\,xz + (t^2+1)\,x + z^2 + (t^4 - 4\,t^2 + 1)\,z - t^4 + 3\,t^2 - 1,\\ - t^3 + xt - zt + 3\,t + w\rangle$$

The second one is:

$$(6.9)\quad \langle y^2 + (-t^2+1)\,y + t^2 - 1,\\ w - xt + (-t^3 + 2\,t)\,y + t^5 - 4\,t^3 + 4\,t,\\ x^2 + (t^2-1)\,xy + (-t^4 + 2\,t^2 - 1)\,x + (-t^4 + 2\,t^2 + 1)\,y + t^6 - 4\,t^4 + 4\,t^2 - 1,\\ xy + (-t^2+1)\,x + (-t^4 + 3\,t^2 - 1)\,y + z + t^6 - 5\,t^4 + 6\,t^2 - 1\rangle$$

The third one

$$(6.10)\quad \langle z^2 + (-t^2-1)\,z + 2\,t^2 - 1,\\ w + xzt - 2\,xt + (-t^5 + 4\,t^3 - 4\,t)\,z + 2\,t^5 - 7\,t^3 + 5\,t,\\ x^2 + (-t^2+1)\,xz + (t^2-1)\,x + (t^4 - 2\,t^2 - 1)\,z - 2\,t^4 + 5\,t^2 - 1,\\ xz - x + y + (-t^4 + 3\,t^2 - 1)\,z + 2\,t^4 - 6\,t^2 + 1\rangle$$

The last component is:

$$(6.11)\quad \langle t^2 - y - z,\\ - zx^2 + xy^2 + xyz + xz^2 + x^2 - 4\,xy - xz - 2\,y^2 - yz - z^2 + 5\,y + 2\,z - 1,\\ x^2yz - x^2y - zx^2 - 4\,xyz - xz^2 + 4\,xy + 6\,xz + 3\,yz + z^2 - x - 2\,y - 5\,z,\\ - x^2z^2 + xz^3 + 2\,zx^2 + yz^2 - z^3 - x^2 + xy - 4\,xz - 2\,yz + x - 2\,y + 4\,z + 1,\\ z^2yx - 4\,xyz - 2\,xz^2 + y^2z - yz^2 + 3\,xy + 6\,xz - 2\,y^2 + 2\,yz + 2\,z^2 - 2\,x + 2\,y - 5\,z - 1,\\ - x^2z^2 + z^2y^2 + 2\,z^3y + z^4 + 2\,zx^2 - 2\,xyz + xz^2 - y^2z - 7\,yz^2 - 5\,z^3 - x^2 + 3\,xy + xz\\ - 2\,y^2 + 5\,yz + 6\,z^2 - 2\,x + 4\,y - z - 1,\\ 2\,x^2z^2 + zy^3 - 3\,z^3y - 2\,z^4 - 4\,zx^2 + 3\,xyz - 3\,xz^2 - 2\,y^3 - 8\,y^2z + 6\,yz^2 + 9\,z^3 + 2\,x^2 - 5\,xy\\ + 16\,y^2 + 10\,yz - 5\,z^2 + 5\,x - 22\,y - 10\,z + 4,\\ xz^2t - ty^2z - 2\,z^2yt - z^3t - 2\,xzt + 2\,ty^2 + 10\,zyt + 5\,z^2t - 12\,ty - 10\,zt + 7\,t + w\rangle$$

Each component intersects $t = 0$ in 4 points, corresponding to representations of M_2 or order 15. The action of the symmetry group on the components is:

	σ_1			σ_2			ϕ	
(6.8)	$\mapsto$	(6.8)	(6.8)	$\mapsto$	(6.9)	(6.8)	$\mapsto$	(6.11)
(6.9)	$\mapsto$	(6.10)	(6.9)	$\mapsto$	(6.8)	(6.9)	$\mapsto$	(6.10)
(6.10)	$\mapsto$	(6.9)	(6.10)	$\mapsto$	(6.11)	(6.10)	$\mapsto$	(6.9)
(6.11)	$\mapsto$	(6.11)	(6.11)	$\mapsto$	(6.10)	(6.11)	$\mapsto$	(6.8)

So the four components are equivalent.

Characters of $Y(8_{18}) = Y^{triv}(8_{18}) \cup Y^{ab}(8_{18}) \cup Y^{nab}(8_{18})$:

- $Y^{triv}(8_{18})$ has a single point, with coordinates $t = w = 0$, $x = y = 2$, and $z = -2$.
- $Y^{ab}(8_{18})$ has $\frac{|\Delta(-1)|-1}{2} = 22$ points. When looking at the order of the induced representation of $\pi_1(M_2)$ in $\mathrm{PSL}_2(\mathbb{C})$, they are distributed as follows: the 2 points corresponding to representations $\hat{\rho}'$ of order 5 lie in the figure eight knot component, the 4 points corresponding to representations $\hat{\rho}'$ of order 3 lie in trefoil components, one per component, and the 16 points corresponding to representations $\hat{\rho}'$ of order 15 on the exotic components (4 per component). In characteristic 3, $Y^{ab}(8_{18})$ has $(5-1)/2 = 2$ points, in characteristic 5, $(9-1)/2 = 4$ points.
- $Y^{nab}(8_{18})$ has 4 points, all of them in the canonical component: two lifts of the orbifold holonomy and their complex conjugate.

When reducing mod 3: The 4 points of $Y^{ab}(8_{18})$ in the trefoil components collapse to $Y^{triv}(8_{18})$. The 2 points in the figure eight knot component remain as different points. On each exotic component, the 4 points become (the same) 2 points, that are also the 2 points for the figure eight component. The four points in $Y^{nab}(8_{18})$ become a single one, but different from the previous ones.

When reducing mod 5: The 4 points of $Y^{ab}(8_{18})$ in the trefoil components remain as 4 different points. The four points of $Y^{ab}(8_{18})$ in each exotic component become a single one, in fact the same as a point in one of the trefoil components, and points in different exotic components go to different trefoil components. The 2 points in the figure eight component collapse to $Y^{triv}(8_{18})$. The 4 points in $Y^{nab}(8_{18})$ remain as different points.

7. Torus Knots

For $m \leq n \in \mathbb{N}$ coprime, the (m,n)-torus knot is denoted by $T(m,n)$. Since

$$\pi_1(\mathbf{S}^3 \setminus T(m,n)) \cong \langle a, b \mid a^m = b^n \rangle,$$

irreducible representations map the central element $a^m = b^n$ to the centre $\{\pm \mathrm{Id}\}$ (and only to $-\mathrm{Id}$ for $m = 2$). Thus, besides the abelian one, the components of the variety of characters are determined by the conjugacy classes of elements in SL_2 to which a and b are mapped. Therefore, when p divides either m or n, the reduction mod p ramifies, so that some components get identified. To describe precisely this ramification, we introduce the following polynomials:

Definition 7.1. For $k \in \mathbb{Z}$, $k \geq 1$, we define $\Phi_k, \Psi_k \in \mathbb{Z}[x]$ to be the polynomials determined by the condition:

$$\Phi_k(\lambda + \frac{1}{\lambda}) = \begin{cases} \dfrac{\lambda^k - 1}{\lambda - 1}\lambda^{(-k+1)/2} & k \text{ odd} \\ \dfrac{\lambda^k - 1}{\lambda^2 - 1}\lambda^{-k/2+1} & k \text{ even} \end{cases}$$

and

$$\Psi_k(\lambda + \frac{1}{\lambda}) = \begin{cases} \dfrac{\lambda^k + 1}{\lambda + 1}\lambda^{(-k+1)/2} & k \text{ odd} \\ (\lambda^k + 1)\lambda^{-k/2} & k \text{ even} \end{cases}$$

for every $\lambda \in \mathbb{C}^*$.

REMARK 7.2. These polynomials satisfy the following property: for $A \in \mathrm{SL}_2(\mathbb{K})$ with $\mathbb{K}$ an algebraically closed field of characteristic 0 or coprime with k:

- $\Phi_k(\mathrm{tr}(A)) = 0$ if and only if $A^k = \mathrm{Id}$ and $A \neq \pm \mathrm{Id}$;
- $\Psi_k(\mathrm{tr}(A)) = 0$ if and only if $A^k = -\mathrm{Id}$ and $A \neq -\mathrm{Id}$.

This remark follows easily from viewing λ in Definition 7.1 as an eigenvalue of a matrix with determinant 1. The following formulae are going to be useful later:

$$\Psi_{2k}(\mathrm{tr}(A)) = \mathrm{tr}(A^k) \qquad \forall A \in \mathrm{SL}_2 \tag{7.1}$$

$$\Psi_{2k}(x) - 2 = \begin{cases} (x-2)\Phi_k(x)^2 & k \text{ odd} \\ (x^2-4)\Phi_k(x)^2 & k \text{ even} \end{cases} \tag{7.2}$$

$$\Psi_{2k}(x) + 2 = \begin{cases} (x+2)\Psi_k(x)^2 & k \text{ odd} \\ \Psi_k(x)^2 & k \text{ even} \end{cases} \tag{7.3}$$

The proofs of these formulae and other properties of Φ_k and Ψ_k are provided in Appendix C.

We describe now $X(T(m,n))$. Using coordinates

$$\begin{aligned} x &= \mathrm{tr}_a \\ y &= \mathrm{tr}_b \\ z &= \mathrm{tr}_{ab^{-1}} \end{aligned}$$

we have

$$X(T(2,n)) = X^{ab}(T(2,n)) \cup \{(x,y,z) \in \mathbb{C}^3 \mid x = \Psi_n(y) = 0\},$$

and, for $m > 2$,

$$\begin{aligned} X(T(m,n)) = X^{ab}(T(m,n)) &\cup \{(x,y,z) \in \mathbb{C}^3 \mid \Phi_m(x) = \Phi_n(y) = 0\} \\ &\cup \{(x,y,z) \in \mathbb{C}^3 \mid \Psi_m(x) = \Psi_n(y) = 0\} \end{aligned}$$

In both cases there are $(m-1)(n-1)/2 + 1$ components, including $X^{ab}(T(m,n))$.

The components other than $X^{ab}(T(m,n))$ are lines in the coordinates (x,y,z), as they are defined by $x = x_0$ and $y = y_0$, for some values $x_0, y_0 \in \mathbb{C}$.

To understand ramifications we use the following lemma, which is straightforward keeping in mind that the map sending each element of a field of characteristic p to its pth power is a morphisms of the field:

LEMMA 7.3. *For a prime $p > 2$, if $p \mid k$ and $k = p^r k'$ with r maximal we have:*

$$\Phi_k(u) \equiv \begin{cases} \Phi_{k'}(u)^{p^r}(u-2)^{(p^r-1)/2} & \text{for } k \text{ odd,} \\ \Phi_{k'}(u)^{p^r}(u^2-4)^{(p^r-1)/2} & \text{for } k \text{ even,} \end{cases} \mod p,$$

and

$$\Psi_k(u) \equiv \begin{cases} \Psi_{k'}(u)^{p^r}(u+2)^{(p^r-1)/2} & \text{for } k \text{ odd.} \\ \Psi_{k'}(u)^{p^r} & \text{for } k \text{ even,} \end{cases} \mod p.$$

In the statement of the lemma, notice that $\Phi_1 = \Psi_1 = 1$.

Notice that in characteristic p an element has trace ± 2 precisely when its p-th power is equal to $\pm \mathrm{Id}$. In our situation, when x or y is $\pm 2 \mod p$, the representation can still be irreducible. Thus:

COROLLARY 7.4. *Let $\mathbb{F}$ be an algebraically closed field of characteristic $p > 2$. If $p \mid n$, then the $(m-1)(n-1)/2$ components of $X(T(m,n))$ containing irreducible characters collapse to $(m-1)(n'+1)/2$ components in $X(T(m,n))_{\mathbb{F}}$, all of them containing irreducible characters, where n' is coprime with p satisfying $n = p^r n'$.*

More specifically, among the $(m-1)(n-1)/2$ components present in characteristic zero, $(m-1)(n-p^r)/2$ collapse to $(m-1)(n'-1)/2$ components in groups of p^r, and $(m-1)(p^r-1)/2$ collapse to $(m-1)/2$ components (that contain irreducible characters, as z may take arbitrary values).

The components with irreducible characters are pairwise disjoint, but we may ask how they intersect $X^{ab}(T(m,n))$.

LEMMA 7.5. *In characteristic 0 or p coprime with m and n, $X^{ab}(T(m,n))$ intersects each other irreducible component of $X(T(m,n))$ transversely in two distinct points.*

When reducing mod p, if $p \mid n$ then $X^{ab}(T(m,n))$ intersects the components with $y = \pm 2$ tangentially at single point.

PROOF. A necessary and sufficient condition for a representation to be abelian or reducible is that the trace of the commutator of the generators a and b is 2; namely, that the coordinates satisfy

$$x^2 + y^2 + z^2 - xyz - 4 = 0.$$

The discriminant in z of this equation is $(x-2)(x+2)(y-2)(y+2)$. This implies that the intersection with a line given by $x = x_0$ and $y = y_0$ is transverse and consists precisely of two points if and only if $x_0 \neq \pm 2$ and $y_0 \neq \pm 2$. This holds in characteristic 0 or coprime with m and n.

When $p \mid n$, if $y_0 \neq \pm 2$ then the discriminant does not vanish and the previous discussion applies. So assume $y_0 = 2$. Notice that the abelianisation $\pi_1(\mathbf{S}^3 \setminus T(m,n)) \to \mathbb{Z}$ maps a to n and b to m. Hence, for any element $\gamma \in \pi_1(\mathbf{S}^3 \setminus T(m,n))$ whose abelianisation generates $\mathbb{Z}$, if $\tau = \operatorname{tr}_\gamma$, then by (7.1) $X^{ab}(T(m,n))$ is parametrised by:

$$x = \operatorname{tr}_{\gamma^n} = \Psi_{2n}(\tau), \quad y = \operatorname{tr}_{\gamma^m} = \Psi_{2m}(\tau), \quad z = \operatorname{tr}_{\gamma^{-m+n}} = \Psi_{2|m-n|}(\tau).$$

We are going to show that the derivatives of x and y with respect to τ vanish at any intersection point with coordinates $(x_0, 2, z_0)$ (this yields tangency with the component defined by $x = x_0$ and $y = y_0$). Since $p \mid n$, by Lemma 7.3 $x = \Psi_{2n}(\tau) \cong (\Psi_{2n}(\tau))^p$ and its derivative $\mod p$ with respect to τ vanishes for any τ. For the derivative of y, write $y = 2 + (\Psi_{2m}(\tau) - 2)$ and use (7.2):

$$y(\tau) = 2 + \begin{cases} (\tau - 2)\Phi_m(\tau)^2 & \text{for } k \text{ odd,} \\ (\tau^2 - 4)\Phi_m(\tau)^2 & \text{for } k \text{ even.} \end{cases}$$

As $x_0 \neq \pm 2 \mod p$, then $\tau_0 \neq \pm 2$, hence $\Phi_m(\tau_0) = 0$ and the derivative of y with respect to τ at $\tau = \tau_0$ vanishes.

When $y = -2$, the proof is similar, using (7.3). □

8. Satellite knots

We will study in detail two examples, both with companion the trefoil knot. Recall from the section on torus knots that the fundamental group of the trefoil knot 3_1 is

$$\pi_1(\mathbf{S}^3 \setminus T(2,3)) \cong \langle \beta, \gamma \mid \beta^2 = \gamma^3 \rangle,$$

where the fibre of the Seifert fibration is $\beta^2=\gamma^3$ and $\gamma^{-1}\beta$ represents a meridian.

From this presentation, the character variety of the trefoil has coordinates $(\mathrm{tr}_\beta,\mathrm{tr}_\gamma,\mathrm{tr}_{\gamma^{-1}\beta})$. It has two components. Both components correspond to characters of representations of quotients of the group: the characters of the first component correspond to representations where the centre is sent to minus the identity

$$X^{irr}(3_1)=\{(\mathrm{tr}_\beta,\mathrm{tr}_\gamma,\mathrm{tr}_{\gamma^{-1}\beta})\in\mathbb{C}^3\mid \mathrm{tr}_\beta=0,\ \mathrm{tr}_\gamma=1\},$$

the characters of the second component correspond to abelian representations

$$X^{ab}(3_1)=\{(\mathrm{tr}_\beta,\mathrm{tr}_\gamma,\mathrm{tr}_{\gamma^{-1}\beta})\in\mathbb{C}^3\mid \mathrm{tr}_\beta=\mathrm{tr}^3_{\gamma^{-1}\beta}-3\,\mathrm{tr}_{\gamma^{-1}\beta},\ \mathrm{tr}_\gamma=\mathrm{tr}^2_{\gamma^{-1}\beta}-2\},$$

as $\gamma^{-1}\beta$ represents a meridian. Because every representation of the satellite knots we will be considering induces a representation of the trefoil knot, one way to determine the representations, and hence the characters, of the satellite knots is to establish which representations of the patterns (in our case the $T(2,4)$ torus link, and the Whitehead link) can be glued along the common boundary to representations of the trefoil knot to give global representations.

EXAMPLE: A CABLE KNOT K_c

This knot exterior is obtained by gluing the exterior of the trefoil knot to the exterior of the $T(2,4)$-torus link in such a way that the fibre of the trefoil knot is identified to the meridian of one of the components of $T(2,4)$ while the meridian of the trefoil knot is glued to the longitude: this ensures that the resulting knot has tunnel number one. A presentation for the torus link is

$$\pi_1(\mathbf{S}^3\setminus T(2,4))\cong\langle\alpha,b\mid b\alpha b^{-1}\alpha=\alpha b^{-1}\alpha b\rangle.$$

Here α and b represent meridians of the two components of $T(2,4)$. One easily observes that the longitude associated to b is $\alpha b^{-1}\alpha b$, so that the exterior of the cable knot has the following presentation:

$$\pi_1(\mathbf{S}^3\setminus K_c)\cong\langle\alpha,\beta\mid b=\beta^2,\gamma^{-1}\beta=\alpha b^{-1}\alpha b,\beta^2=\gamma^3,b\alpha b^{-1}\alpha=\alpha b^{-1}\alpha b\rangle.$$

The variety of characters of $T(2,4)$ has coordinates $(\mathrm{tr}_\alpha,\mathrm{tr}_b,\mathrm{tr}_{\alpha b^{-1}})$ and two components. One that maps the centre to the identity:

$$X^{irr}(T(2,4))=\{(\mathrm{tr}_\alpha,\mathrm{tr}_b,\mathrm{tr}_{\alpha b^{-1}})\in\mathbb{C}^3\mid \mathrm{tr}_{\alpha b^{-1}}=0\}$$

and the abelian component

$$X^{ab}(T(2,4))=\{(\mathrm{tr}_\alpha,\mathrm{tr}_b,\mathrm{tr}_{\alpha b^{-1}})\in\mathbb{C}^3\mid \mathrm{tr}^2_\alpha+\mathrm{tr}^2_b+\mathrm{tr}^2_{\alpha b^{-1}}-\mathrm{tr}_\alpha\mathrm{tr}_b\mathrm{tr}_{\alpha b^{-1}}-4=0\}.$$

To compute $X(K_c)$ we use coordinates

$$t=\mathrm{tr}_\alpha,\ z=\mathrm{tr}_\beta,\ x=\mathrm{tr}_{\alpha\beta^{-1}}\,.$$

We start with $X(3_1)$ and $X(T(2,4))$ and we add the relations:

$$\begin{aligned}
&\mathrm{tr}_b=\mathrm{tr}_{\beta^2}=z^2-2\\
&\mathrm{tr}_{\alpha b^{-1}}=\mathrm{tr}_{\alpha\beta^{-2}}=z\,x-t\\
&\mathrm{tr}_\gamma=\mathrm{tr}_{\beta^{-1}\alpha^{-1}\beta^2\alpha^{-1}}=tz^2x-t^2z-z^3-zx^2+tx+3\,z\\
&\mathrm{tr}_{\beta\gamma^{-1}}=\mathrm{tr}_{\alpha b^{-1}\alpha b}=tz^3x-t^2z^2-z^4-z^2x^2+t^2+4\,z^2-2
\end{aligned}\tag{8.1}$$

We assume first that the restriction to $\pi_1(\mathbf{S}^3\setminus 3_1)$ lies in $X^{irr}(3_1)$. This imposes the conditions $z=0$ and $\mathrm{tr}_\gamma=1$, which in (8.1) yield

$$\mathrm{tr}_b=-2,\quad \mathrm{tr}_{\alpha b^{-1}}=-t,\quad 1=t\,x,\quad \mathrm{tr}_{\beta\gamma^{-1}}=t^2-2$$

These equalities are incompatible with $t_{\alpha b^{-1}} = 0$, hence the intersection with $X^{irr}(T(2,4))$ is empty. On the other hand, these equalities imply that

$$\mathrm{tr}_\alpha^2 + \mathrm{tr}_b^2 + \mathrm{tr}_{\alpha b^{-1}}^2 - \mathrm{tr}_\alpha \,\mathrm{tr}_b \,\mathrm{tr}_{\alpha b^{-1}} - 4 = 0,$$

that defines $X^{ab}(T(2,4))$. Thus we get a first component

$$X_1(K_c) = \{(t,z,x) \in \mathbb{C}^3 \mid z = 0,\ t\,x = 1\}.$$

Next we suppose that the restriction to $\pi_1(\mathbf{S}^3 \setminus 3_1)$ lies in $X^{ab}(3_1)$, namely it is abelian. Therefore the corresponding representations factor through a Dehn filling on $T(2,4)$. One way to compute the Dehn filling is to add a generator η and the relations

$$\beta = \eta^3, \qquad \gamma = \eta^2,$$

as those are the relations that abelianise the trefoil. The quotient of $\pi_1(\mathbf{S}^3 \setminus K_c)$ by these relations is

$$\pi_1(\mathbf{S}^3 \setminus K_c)/\langle \beta = \eta^3, \gamma = \eta^2 \rangle \cong \langle \alpha, \eta \mid (\alpha^{-1}\eta^6)^2 = \eta^{11} \rangle.$$

Thus we get the group of the torus knot $T(2,11)$, and therefore

$$X(K_c) \cong X_1(K_c) \cup X(T(2,11)).$$

The six components of $X(T(2,11))$ in the coordinates (t,z,x) become:

$$X_2(K_c) = \{(t,z,x) \in \mathbb{C}^3 \mid t = x\,z, \quad z^5 - z^4 - 4\,z^3 + 3\,z^2 + 3\,z - 1 = 0\},$$

for the five curves of irreducible representations, and

$$X^{ab}(K_c) = \{(t,z,x) \in \mathbb{C}^3 \mid x = t^5 - 5\,t^3 + 5\,t,\ z = t^6 - 6\,t^4 + 9\,t^2 - 2\}.$$

This last component is the abelian one. Notice that the polynomial equation says that x equals the trace of α^5 and z that of α^6. Also notice that the five components in $X_2(K_c)$ collapse to a single one mod 11, because

$$\left(z^5 - z^4 - 4\,z^3 + 3\,z^2 + 3\,z - 1\right) \equiv (z-2)^5 \mod 11.$$

Example: a Whitehead double K_w

In this example, the cable space $\mathbf{S}^3 \setminus T(2,4)$ is replaced with the exterior of the Whitehead link, with presentation

$$\pi_1(\mathbf{S}^3 \setminus 5_1^2) = \langle \alpha, b \mid \alpha b^{-1}\alpha^{-1}b\alpha^{-1}b^{-1}\alpha = b^{-1}\alpha b^{-1}\alpha^{-1}b\alpha^{-1}b^{-1}\alpha b \rangle,$$

where α and b are meridians, see Figure 3. The gluing is as in the previous example, thus giving the following presentation

$$\pi_1(\mathbf{S}^3 \setminus K_w) \cong \langle \alpha, \beta \mid b = \beta^2, \gamma^{-1}\beta = \alpha b^{-1}\alpha^{-1}b\alpha^{-1}b^{-1}\alpha b, \beta^2 = \gamma^3,$$
$$\alpha b^{-1}\alpha^{-1}b\alpha^{-1}b^{-1}\alpha = b^{-1}\alpha b^{-1}\alpha^{-1}b\alpha^{-1}b^{-1}\alpha b \rangle.$$

Claim 8.1. *There is no character in $X(K_w)$ whose restriction to the trefoil lies in $X^{irr}(3_1)$.*

Proof. Assume there is such a character χ_ρ. The gluing requires that a meridian of the Whitehead link, b, is glued to the fibre of the trefoil. Hence, $\rho(b) = -\mathrm{Id}$, i.e. it is trivial in $\mathrm{PSL}_2(\mathbb{C})$. We notice that the image by ρ of the longitude that commutes with b must also be trivial in $\mathrm{PSL}_2(\mathbb{C})$, because each component of the

Whitehead link is a trivial knot and the linking number between the two components vanishes. As there is no representation in $X^{irr}(3_1)$ whose restriction to the peripheral subgroup is trivial in $\mathrm{PSL}_2(\mathbb{C})$, we get a contradiction. □

Hence all characters in $X(K_w)$ restrict to abelian characters in $X^{ab}(3_1)$. Therefore, we obtain the same representations we get by replacing the trefoil knot exterior by a solid torus. In other terms, these representations correspond to representations of the knot obtained by Dehn filling one of the two components of the Whitehead link $(5_1^2)_r$:

$$X(K_w) \cong X((5_1^2)_r),$$

where $r \in \mathbb{Q}$ is the slope of the filling. To determine r, we add the generator η and the relations

$$\beta = \eta^3, \qquad \gamma = \eta^2,$$

to the presentation of $\pi_1(\mathbf{S}^3 \setminus K_w)$ as in the previous example, because those are precisely the abelianisation relations of the trefoil. In particular η is a longitude that commutes with b (i.e. they are in the same component and in the same choice of peripheral group in the conjugacy class). Thus, we get the presentation

$$\pi_1(\mathbf{S}^3 \setminus (5_1^2)_r) = \langle \alpha, b, \gamma \mid b = \gamma^6, \gamma = \alpha b^{-1} \alpha^{-1} b \alpha^{-1} b^{-1} \alpha b, b\gamma = \gamma b \rangle.$$

Hence, the filling meridian is $b\gamma^{-6}$, i.e. the filling slope is $r = -1/6$.

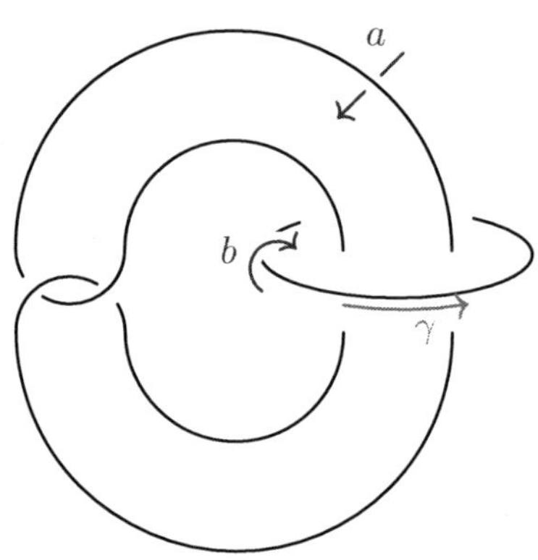

FIGURE 3. The meridians a and b and the longitude γ.

To understand the manifold obtained from the Dehn filling with slope $b\gamma^{-6}$, the standard procedure is to cut open along a disc bounded by the component of the link with meridian b, apply 6 turns, glue back again, so that the new filling slope becomes $1/0 = \infty$. This leads to a twist knot with twelve half-twists, that would not be alternating, and simplifies to 11 half-twists as in Figure 4. This is the two-bridge knot 11/23. The variety of characters of this two-bridge knot has been discussed in Section 4. It ramifies for $p = 23$.

Appendix A. Computations for the knot 8_5

From Figure 1, using Wirtinger presentation we compute:

$$\pi_1(\mathbf{S}^3 \setminus 8_5) \cong \langle \alpha, \beta, \gamma \mid [\alpha^{-1}, \gamma^{-1}]\alpha[\beta^{-1}, \alpha] = \gamma,\ [\beta^{-1}, \alpha]\beta[\gamma, \beta] = \gamma \rangle$$

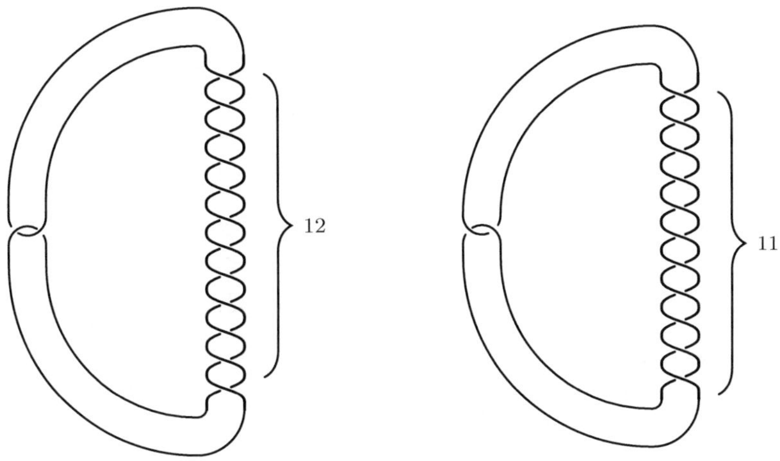

FIGURE 4. The projection on the left has 12 half-twists but it is not alternating, it simplifies to the projection on the right, which is a diagram of the 2-bridge knot 11/23.

Next we perform elementary changes to obtain a presentation with two generators. Firstly we introduce the elements $w = \alpha\gamma$ and $z = \beta\gamma$ and we eliminate α and β:

$$\pi_1(\mathbf{S}^3 \setminus 8_5) \cong \langle \gamma, w, z \mid w^{-1}\gamma^{-1}w^2z^{-1}w\gamma^{-1}zw^{-1} = z^{-2}w\gamma^{-1}zw^{-1}z^2\gamma^{-1}z^{-1} = 1\rangle$$

Setting $a = w^{-1}z^2$ and eliminating w we get:

$$\pi_1(\mathbf{S}^3 \setminus 8_5) \cong \langle \gamma, a, z \mid az^{-2}\gamma^{-1}z^2a^{-1}z^2a^{-1}z(a^{-1}\gamma^{-1}za)z^{-2} = a^{-1}\gamma^{-1}za\gamma^{-1} = 1\rangle$$

The second relation allows to write $z = \gamma a\gamma a^{-1}$ and to simplify the bracket in the first one:

$$az^{-2}\gamma^{-1}z^2a^{-1}z^2a^{-1}z\gamma z^{-2} = 1.$$

From this relation we get

$$az^{-2}\gamma^{-1}z^2a^{-1} = z^2\gamma^{-1}z^{-1}az^{-2},$$

which tells that $\gamma^{-1}z^{-1}a$ is conjugate to γ^{-1}. Taking traces it yields the equation

$$t^2x^2y + t\left(-x^3 - 2\,xy^2 + 2\,x - 1\right) + x^2y + y^3 - 3\,y = 0 \tag{A.1}$$

Next taking traces on the relation

$$\gamma^{-1}z^2a^{-1}z^2a^{-1}z\gamma = z^2az^2$$

gives

$$(t^2x^3y^2 - 2\,tx^4y - 2\,tx^2y^3 - y^2xt^2 + x^5 + 2\,x^3y^2 + xy^4 + 6\,tx^2y + ty^3 - txy - 5\,x^3 - 5\,xy^2 - 2\,ty + x^2 + y^2 + 5\,x - 2) \times (t^3x^4y^3 - 3\,t^2x^5y^2 - 3\,t^2x^3y^4 - y^3x^2t^3 + 3\,tx^6y + 6\,tx^4y^3 + 3\,tx^2y^5 + 9\,t^2x^3y^2 + 2\,t^2xy^4 - x^7 - 3\,x^5y^2 - 3\,x^3y^4 - xy^6 - 15\,tx^4y - 16\,tx^2y^3 - ty^5 - 4\,y^2xt^2 + 7\,x^5 + 14\,x^3y^2 + 7\,xy^4 + 18\,tx^2y + 4\,ty^3 - 14\,x^3 - 14\,xy^2 - 3\,ty + 7\,x + 1) = 0 \tag{A.2}$$

We consider the ideal generated by the polynomials in (A.1) and (A.2). We then use symbolic software to look for its prime decomposition (namely, the decomposition of its radical) and we get the five components we described. This gives an algebraic set that contains $X(8_5)$. We check that it is precisely $X(8_5)$ exploiting the fact that each component intersects $t = 0$ in one of the points described in Section 5.

The equations are invariant by the symmetry $(t, x, y) \mapsto (-t, x, -y)$, because t and y are traces of elements whose projections to $\mathbb{Z}/2\mathbb{Z}$ are nontrivial, and x is the trace of an element that projects to zero in $\mathbb{Z}/2\mathbb{Z}$.

Appendix B. Computations for the knot 8_{18}

To obtain a presentation of its fundamental group, we start with three meridians α, β and γ in Figure 2. We apply the map induced by the period $\phi_*^{\pm1}$ on these generators, which is equivalent to moving right and left in Figure 5.

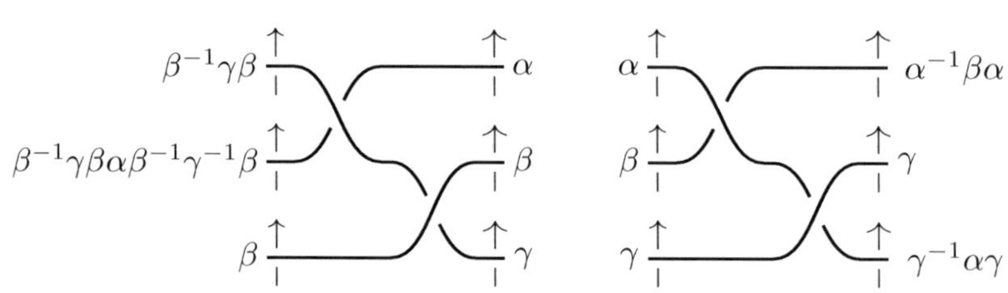

FIGURE 5. The generators α, β and γ and their image by $\phi_*^{\pm1}$

If we iterate this process once in each direction, we obtain the same elements in the fundamental group, yielding to relations. More precisely, the upper and lower strands yield

$$\beta^{-1}\gamma\beta\alpha^{-1}\beta^{-1}\gamma^{-1}\beta\gamma\beta\alpha\beta^{-1}\gamma^{-1}\beta = \alpha^{-1}\beta^{-1}\alpha\gamma\alpha^{-1}\beta\alpha$$
$$\beta^{-1}\gamma\beta\alpha\beta^{-1}\gamma^{-1}\beta = \gamma\alpha\gamma^{-1}\alpha^{-1}\beta\alpha\gamma\alpha^{-1}\gamma^{-1}$$

Since one relation is redundant, we skip the relation of the middle strand, for it is longer. As they are written, these relations establish that the meridians α, β and γ are conjugate, so that by taking traces we would get the equality $t = t$. Instead, by modifying them slightly, we derive other equivalent relations, from which we compute and equate traces:

$$\beta\gamma\alpha\gamma^{-1}\alpha^{-1}\beta\alpha\gamma = \gamma\beta\alpha\beta^{-1}\gamma^{-1}\beta\gamma\alpha \tag{B.1}$$

$$\gamma\beta\alpha^{-1}\beta^{-1}\gamma^{-1}\beta\gamma\beta\alpha\beta^{-1} = \beta\alpha^{-1}\beta^{-1}\alpha\gamma\alpha^{-1}\beta\alpha\beta^{-1}\gamma \tag{B.2}$$

$$\gamma^{-1}\beta\gamma\alpha\gamma^{-1}\alpha^{-1}\beta\alpha\gamma = \beta\alpha\beta^{-1}\gamma^{-1}\beta\gamma\alpha \tag{B.3}$$

$$\gamma\alpha\gamma^{-1}\alpha^{-1}\beta\alpha\gamma\alpha^{-1} = \beta^{-1}\gamma\beta\alpha\beta^{-1}\gamma^{-1}\beta\gamma \tag{B.4}$$

Taking traces to (B.1) and writing them as polynomials in our coordinates yields:

$$\begin{aligned}
&-t^6z + t^6 + t^4xz + 2\,t^4yz + t^4z^2 - 2\,t^4y - t^3wz - 2\,t^2xyz - t^2xz^2\\
&\quad - t^2y^2z - t^2yz^2 - t^4 + t^3w - t^2x^2 + t^2xz + t^2y^2 + twxz + twyz + twz^2\\
&\quad + xy^2z + t^2y - 2\,t^2z - twy - twz - w^2z + x^2y + yz^2 - z^3 + t^2 - y + 3\,z = 0
\end{aligned}$$

We do the same for (B.2):

$$\begin{aligned}
&t^6x + t^5wx - t^4x^2y - t^6 - t^5w + t^4w^2 - t^4x^2 - t^4xy - 2\,t^4xz - t^3wx^2 - 3\,t^3wxy\\
&-t^3wxz + t^2x^3y + 2\,t^2x^2y^2 + t^2x^2yz + t^4x + 2\,t^4y + t^4z + 3\,t^3wx + t^3wy + t^2w^2x - t^2w^2y\\
&- t^2x^2y + 2\,t^2x^2z + 2\,t^2xyz + t^2xz^2 - twx^2y + twx^2z + 2\,twxy^2 - x^3yz - x^2y^3 + t^4\\
&- t^3w - 3\,t^2w^2 - t^2xz - t^2y^2 - t^2yz + 3\,twxy - 2\,twxz + 2\,twyz - x^2y^2 - x^2z^2\\
&- 2\,xy^2z - 2\,t^2x - t^2y - twx + twz + x^2y - yz^2 - t^2 + tw + xz + y + 2 = 0
\end{aligned}$$

For (B.3):

$$\left(t^2 - y - z\right)\left(t^3z - t^3 - t^2w - txz + xt + ty - zt + wx\right) = 0$$

And for (B.4):

$$\begin{aligned}
-\left(t^2 - y - z\right)(t^4z - 2\,t^4 - t^2xz - t^2z^2 + 2\,t^2x - twy + twz + xy^2 - xyz&\\
+ xz^2 + 3\,t^2 - tw + yz - 2\,x - 1) = 0&
\end{aligned}$$

According to [**8**] for instance, since the presentation of the group has three generators, there is an additional polynomial relation between the coordinates of the character variety. In our situation, since $t = tr_\alpha = tr_\beta = tr_\gamma$ the equation reads:

$$w^2 + (t^3 - t\,x - t\,y - t\,z)w - t^2x - t^2y - t^2z + xyz + 3\,t^2 + x^2 + y^2 + z^2 - 4 = 0 \tag{B.5}$$

It turns out that these five polynomials still do not generate the ideal. To see that, we consider again the action of the period ϕ. The induced action on the characters is given by $\chi \mapsto \chi \circ \phi_*^{-1}$, where ϕ_* is the action on the fundamental group, By Figure 5 (right) ϕ_*^{-1} is given by

$$\begin{array}{ccc}
\alpha & & \alpha^{-1}\beta\alpha\\
\beta & \mapsto & \gamma\\
\gamma & & \gamma^{-1}\alpha\gamma
\end{array}$$

We consider the ideal generated by taking the trace in Equations (B.1), (B.2), (B.3) and (B.4), their image by ϕ_*, and Equation (B.5), thus a total of 9 polynomials. We then use symbolic software to look for its prime decomposition (namely, the decomposition of its radical) and we get the eleven components we described. This gives an algebraic set that contains $X(8_{18})$. We check that it is precisely $X(8_{18})$ because each component intersects $t = 0$ in one of the points described in Section 3.

We now describe the action of the involutions σ_1 and σ_2 on the variety of characters.

From Figure 2 the action of σ_1 on the fundamental group is

$$\begin{array}{ccc} \alpha & & \alpha^{-1}\beta^{-1}\alpha \\ \beta & \mapsto & \alpha^{-1}\beta^{-1}\alpha^{-1}\beta\alpha \\ \gamma & & \alpha^{-1}\beta^{-1}\gamma^{-1}\beta\alpha \end{array}$$

Up to inner homomorphisms, it can be simplified as:

$$\begin{array}{ccc} \alpha & & \beta^{-1} \\ \beta & \mapsto & \alpha^{-1} \\ \gamma & & \gamma^{-1} \end{array}$$

Hence $\sigma_1^*\colon X(8_{18}) \to X(8_{18})$ writes as:

$$\begin{pmatrix} t \\ x \\ y \\ z \\ w \end{pmatrix} \mapsto \begin{pmatrix} t \\ x \\ t^2 - z \\ t^2 - y \\ t^3 - t\,y - t\,z + w \end{pmatrix}$$

From Figure 2, on the fundamental group σ_2 acts as

$$\begin{array}{ccc} \alpha & & \gamma \\ \beta & \mapsto & \gamma\beta\gamma^{-1} \\ \gamma & & \gamma\beta\alpha\beta^{-1}\gamma^{-1} \end{array}$$

and on the variety of characters

$$\begin{pmatrix} t \\ x \\ y \\ z \\ w \end{pmatrix} \mapsto \begin{pmatrix} t \\ y \\ x \\ t^2 + t\,w - x\,y - z \\ w \end{pmatrix}$$

The map induced by the period ϕ on the variety of characters is

$$\begin{pmatrix} t \\ x \\ y \\ z \\ w \end{pmatrix} \mapsto \begin{pmatrix} t \\ -t^4 + t^2x + t^2y + zt^2 + t^2 - tw - xz - y \\ t^2 - z \\ -t^4z + t^4 + t^2xz + yzt^2 + z^2t^2 - t^2x - t^2y - zwt - xz^2 + tw - yz + x \\ -t^5 + t^3x + t^3y + zt^3 + 2\,t^3 - wt^2 - xzt - xt - 2\,ty - z \end{pmatrix}$$

From this we may derive the description of the action of the symmetry group on the variety of characters given in Section 6.

Appendix C. The polynomials Φ_k and Ψ_k

Here we provide the proof of some of the results stated in Section 7 for the polynomials Φ_k and Ψ_k of Definition 7.1. First of all we rewrite Definition 7.1 to make computations simpler. The polynomials are determined by the conditions

$$\Phi_k(\lambda + \lambda^{-1}) = \begin{cases} \dfrac{\lambda^{k/2} - \lambda^{-k/2}}{\lambda^{1/2} - \lambda^{-1/2}} & k \text{ odd} \\[2ex] \dfrac{\lambda^{k/2} - \lambda^{-k/2}}{\lambda - \lambda^{-1}} & k \text{ even} \end{cases}$$

and

$$\Psi_k(\lambda+\lambda^{-1}) = \begin{cases} \dfrac{\lambda^{k/2}+\lambda^{-k/2}}{\lambda^{1/2}+\lambda^{-1/2}} & k \text{ odd} \\ (\lambda^{k/2}+\lambda^{-k/2}) & k \text{ even} \end{cases}$$

for every $\lambda \in \mathbb{C}^*$. As $\Psi_{2k}(\lambda+\lambda^{-1}) = \lambda^k + \lambda^{-k}$, by viewing $\lambda^{\pm 1}$ as the eigenvalues of a matrix, (7.1) is clear.

Now we give the proof of (7.2), but we skip the proof of (7.3) as it is analogous.

PROOF OF (7.2). Set $u = \lambda + \lambda^{-1}$. For every k we have

$$\Psi_{2k}(u) - 2 = \lambda^k + \lambda^{-k} - 2 = (\lambda^{k/2} - \lambda^{-k/2})^2 \tag{C.1}$$

When k is odd, (C.1) gives:

$$\Psi_{2k}(u) - 2 = (\lambda^{1/2} - \lambda^{-1/2})^2 \left(\frac{\lambda^{k/2} - \lambda^{-k/2}}{\lambda^{1/2} - \lambda^{-1/2}} \right)^2 = (u-2)\Phi_k(u)^2.$$

When k is even:

$$\Psi_{2k}(u) - 2 = (\lambda - \lambda^{-1})^2 \left(\frac{\lambda^{k/2} - \lambda^{-k/2}}{\lambda - \lambda^{-1}} \right)^2 = (u^2-4)\Phi_k(u)^2,$$

which concludes the proof of (7.2). □

PROOF OF LEMMA 7.3. Let $k = p^r k'$ be as in the statement of the lemma. The main idea is to use the identity

$$\lambda^{k/2} \pm \lambda^{-k/2} \equiv (\lambda^{k'/2} \pm \lambda^{-k'/2})^{p^r} \mod p$$

and to adapt it to each case. For instance for k odd:

$$\begin{aligned} \Phi_k(u) = \frac{\lambda^{k/2} - \lambda^{-k/2}}{\lambda^{1/2} - \lambda^{-1/2}} &\equiv \frac{(\lambda^{k'/2} - \lambda^{-k'/2})^{p^r}}{(\lambda^{1/2} - \lambda^{-1/2})^{p^r}} (\lambda^{1/2} - \lambda^{-1/2})^{p^r - 1} \\ &= \Phi_{k'}(u)^{p^r} (u-2)^{(p^r-1)/2}. \end{aligned}$$

A similar trick applies to Φ_k for k even and to Ψ_k for k either even or odd. □

Other properties are $\deg(\Phi_k) = \deg(\Psi_k) = (k-1)/2$ for k odd, and $\deg(\Phi_k) = k/2 - 1$ and $\deg(\Psi_k) = k/2$ for k even.

These polynomials are easy to compute if we take into account the recursive relations, as polynomials in $\mathbb{Z}[x]$:

$$\Phi_k = x\Phi_{k-2} - \Phi_{k-4} \qquad \text{and} \qquad \Psi_k = x\Psi_{k-2} - \Psi_{k-4},$$

or:

$$\begin{cases} \Phi_{2k} = \Phi_{2k-1} - \Phi_{2k-2} \\ \Phi_{2k+1} = (x+2)\Phi_{2k} - \Phi_{2k-1} \end{cases} \qquad \text{and} \qquad \begin{cases} \Psi_{2k} = (x+2)\Psi_{2k-1} - \Psi_{2k-2} \\ \Psi_{2k+1} = \Psi_{2k} - \Psi_{2k-1} \end{cases}$$

with $\Phi_1 = \Phi_2 = \Psi_1 = 1$ and $\Psi_2 = x$. Here are the polynomials for k up to 10:

$$
\begin{aligned}
\Phi_1 &= 1 & \Psi_1 &= 1\\
\Phi_2 &= 1 & \Psi_2 &= x\\
\Phi_3 &= x+1 & \Psi_3 &= x-1\\
\Phi_4 &= x & \Psi_4 &= x^2-2\\
\Phi_5 &= x^2+x-1 & \Psi_5 &= x^2-x-1\\
\Phi_6 &= x^2-1 & \Psi_6 &= x^3-3x\\
\Phi_7 &= x^3+x^2-2\,x-1 & \Psi_7 &= x^3-x^2-2\,x+1\\
\Phi_8 &= x^3-2\,x & \Psi_8 &= x^4-4\,x^2+2\\
\Phi_9 &= x^4+x^3-3\,x^2-2\,x+1 & \Psi_9 &= x^4-x^3-3\,x^2+2\,x+1\\
\Phi_{10} &= x^4-3\,x^2+1 & \Psi_{10} &= x^5-5\,x^3+5\,x
\end{aligned}
$$

Acknowledgements

The authors are indebted to Alan Reid for suggesting the exploration of order-1 type ramification. L. Paoluzzi is also thankful to UAB for hospitality during her visit when the contents of the present paper were originally discussed.

References

[1] M. Boileau and J. Porti, *Geometrization of 3-orbifolds of cyclic type* (English, with English and French summaries), Astérisque **272** (2001), 208. Appendix A by Michael Heusener and Porti. MR1844891

[2] S. Boyer and X. Zhang, *Finite Dehn surgery on knots*, J. Amer. Math. Soc. **9** (1996), no. 4, 1005–1050, DOI 10.1090/S0894-0347-96-00201-9. MR1333293

[3] G. Burde and H. Zieschang, *Knots*, 2nd ed., De Gruyter Studies in Mathematics, vol. 5, Walter de Gruyter & Co., Berlin, 2003. MR1959408

[4] M. Chu, *Detecting essential surfaces as intersections in the character variety*, Algebr. Geom. Topol. **17** (2017), no. 5, 2893–2914, DOI 10.2140/agt.2017.17.2893. MR3704247

[5] D. Cooper, M. Culler, H. Gillet, D. D. Long, and P. B. Shalen, *Plane curves associated to character varieties of 3-manifolds*, Invent. Math. **118** (1994), no. 1, 47–84, DOI 10.1007/BF01231526. MR1288467

[6] M. Culler and P. B. Shalen, *Varieties of group representations and splittings of 3-manifolds*, Ann. of Math. (2) **117** (1983), no. 1, 109–146, DOI 10.2307/2006973. MR683804

[7] L. E. Dickson, *Linear groups: With an exposition of the Galois field theory*, with an introduction by W. Magnus, Dover Publications, Inc., New York, 1958. MR0104735

[8] F. González-Acuña and J. M. Montesinos-Amilibia, *On the character variety of group representations in* $\mathrm{SL}(2,\mathbf{C})$ *and* $\mathrm{PSL}(2,\mathbf{C})$, Math. Z. **214** (1993), no. 4, 627–652, DOI 10.1007/BF02572429. MR1248117

[9] M. Heusener, J. Porti, and E. Suárez Peiró, *Deformations of reducible representations of 3-manifold groups into* $\mathrm{SL}_2(\mathbf{C})$, J. Reine Angew. Math. **530** (2001), 191–227, DOI 10.1515/crll.2001.003. MR1807271

[10] H. M. Hilden, M. T. Lozano, and J. M. Montesinos-Amilibia, *On the arithmetic 2-bridge knots and link orbifolds and a new knot invariant*, J. Knot Theory Ramifications **4** (1995), no. 1, 81–114, DOI 10.1142/S0218216595000053. MR1321291

[11] H. M. Hilden, M. T. Lozano, and J. M. Montesinos-Amilibia, *On volumes and Chern-Simons invariants of geometric 3-manifolds*, J. Math. Sci. Univ. Tokyo **3** (1996), no. 3, 723–744. MR1432115

[12] H. M. Hilden, M. T. Lozano, and J. M. Montesinos-Amilibia, *On the character variety of periodic knots and links*, Math. Proc. Cambridge Philos. Soc. **129** (2000), no. 3, 477–490, DOI 10.1017/S0305004100004679. MR1780499

[13] M. Kapovich, *Hyperbolic manifolds and discrete groups*, Progress in Mathematics, vol. 183, Birkhäuser Boston, Inc., Boston, MA, 2001. MR1792613

[14] T. T. Q. Le and A. T. Tran, *On the AJ conjecture for knots*, Indiana Univ. Math. J. **64** (2015), no. 4, 1103–1151, DOI 10.1512/iumj.2015.64.5602. With an appendix written jointly with Vu Q. Huynh. MR3385787

[15] X. S. Lin, *Representations of knot groups and twisted Alexander polynomials*, Acta Math. Sin. (Engl. Ser.) **17** (2001), no. 3, 361–380, DOI 10.1007/s101140100122. MR1852950

[16] M. L. Macasieb, K. L. Petersen, and R. M. van Luijk, *On character varieties of two-bridge knot groups*, Proc. Lond. Math. Soc. (3) **103** (2011), no. 3, 473–507, DOI 10.1112/plms/pdr003. MR2827003

[17] J. M. Montesinos, *Classical tessellations and three-manifolds*, Universitext, Springer-Verlag, Berlin, 1987. MR915761

[18] F. Nagasato, *Finiteness of a section of the* $\mathrm{SL}(2,\mathbb{C})$*-character variety of the knot group*, Kobe J. Math. **24** (2007), no. 2, 125–136. MR2488756

[19] T. Ohtsuki, *Ideal points and incompressible surfaces in two-bridge knot complements*, J. Math. Soc. Japan **46** (1994), no. 1, 51–87, DOI 10.2969/jmsj/04610051. MR1248091

[20] T. Ohtsuki, R. Riley, and M. Sakuma, *Epimorphisms between 2-bridge link groups*, The Zieschang Gedenkschrift, Geom. Topol. Monogr., vol. 14, Geom. Topol. Publ., Coventry, 2008, pp. 417–450, DOI 10.2140/gtm.2008.14.417. MR2484712

[21] L. Paoluzzi and J. Porti, *Non-standard components of the character variety for a family of Montesinos knots*, Proc. Lond. Math. Soc. (3) **107** (2013), no. 3, 655–679, DOI 10.1112/plms/pds094. MR3100780

[22] L. Paoluzzi and J. Porti, *Invariant character varieties of hyperbolic knots with symmetries*, Math. Proc. Cambridge Philos. Soc. **165** (2018), no. 2, 193–208, DOI 10.1017/S0305004117000391. MR3833997

[23] Kathleen L. Petersen and Alan W. Reid, *Gonality and genus of canonical components of character varieties*, `arXiv:1408.3665`.

[24] J. Porti, *Spherical cone structures on 2-bridge knots and links*, Kobe J. Math. **21** (2004), no. 1-2, 61–70. MR2140602

[25] J. Porti and H. Weiss, *Deforming Euclidean cone 3-manifolds*, Geom. Topol. **11** (2007), 1507–1538, DOI 10.2140/gt.2007.11.1507. MR2326950

[26] William P. Thurston, *The Geometry and Topology of Three-Manifolds*, (Electronic version from 2002 available at `http://www.msri.org/publications/books/gt3m/`), 1980.

[27] H. Weiss, *Local rigidity of 3-dimensional cone-manifolds*, J. Differential Geom. **71** (2005), no. 3, 437–506. MR2198808

Aix-Marseille Univ, CNRS, Centrale Marseille, I2M, UMR 7373, 13453 Marseille, France

Email address: `luisa.paoluzzi@univ-amu.fr`

Departament de Matemàtiques, Universitat Autònoma de Barcelona, 08193 Cerdanyola del Vallès, Spain, and Centre de Recerca Matemàtica

Email address: `porti@mat.uab.cat`

Contemporary Mathematics
Volume **760**, 2020
https://doi.org/10.1090/conm/760/15295

Gonality and genus of canonical components of character varieties

Kathleen L. Petersen and Alan W. Reid

ABSTRACT. Let M be a finite volume hyperbolic 3-manifold with two cusps, and X_0 a canonical component of the $\mathrm{SL}_2(\mathbb{C})$ character variety of any hyperbolic filling of a single cusp of M. We show that the gonality of X_0 is bounded above independent of the filling, and obtain bounds for the genus of X_0 and the degree of a canonical component of the associated A-polynomial curve. We explicitly compute the gonality for several examples, including the double twist knots.

1. Introduction

Let M be a complete, orientable finite volume hyperbolic 3-manifold with cusps. The $\mathrm{SL}_2(\mathbb{C})$ character variety of M, denoted by $X(M)$, is a complex algebraic set arising as an algebro-geometric quotient of the representation variety $R(M) = \{\rho : \pi_1(M) \to \mathrm{SL}_2(\mathbb{C})\}$ (see §4 for more details). Thurston showed that any irreducible component of a character variety that contains the character of a discrete faithful representation has complex dimension equal to the number of cusps of M. Such components are called *canonical or geometric components* and are denoted $X_0(M)$. We will use the term canonical component.

Character varieties have been fundamental tools in studying the topology of M (we refer the reader to [**24**] for more details about the applications of the character variety to the topology of 3-manifolds), and canonical components carry a wealth of topological information about M, including containing subvarieties associated to Dehn fillings of M.

When M has exactly one cusp, any canonical component is a complex curve. The aim of this paper is to study how some of the natural invariants of these complex curves correspond to the underlying manifold M. In particular, we will concentrate on invariants of families of curves arising from Dehn filling on a 2-cusped hyperbolic manifold. This has been studied already in the the context of genus; for example for the families of twist knots and tunnel number one once-punctured torus bundles which both arise from Dehn fillings of the Whitehead link complement (see [**14**] and [**1**]).

2020 *Mathematics Subject Classification.* Primary 57M27; Secondary 14H51, 57M50.

Key words and phrases. character variety, cusped hyperbolic 3-manifold, gonality.

This work was partially supported by grants from the Simons Foundation (#209226 and #430077 to the first author), and by the National Science Foundation (to the second author).

In this paper we focus on how the *gonality* varies on families of curves arising from Dehn filling on a 2-cusped hyperbolic manifold. Recall that the gonality of a curve X, $\gamma(X)$, is the lowest degree of a rational map from X to $\mathbb{C}$.

Our main result about gonality is the following.

THEOREM 1.1. *Let M be a complete, orientable, finite volume hyperbolic 3-manifold with two cusps. Then there exists a constant c depending only on M such that for any hyperbolic manifold N obtained by filling one cusp of M, and for any canonical component X_0 of $X(N)$ we have*

$$\gamma(X_0) \leq c.$$

The figure-8 knot complement and its so-called sister manifold are both integral surgery of one component of the Whitehead link complement. Their canonical components are well-known to be an elliptic curve, and a rational curve, respectively. These have gonality 2 and 1. As such, there can be different values of $\gamma(X_0)$ appearing in Theorem 1.1 as the X_0 vary according to the Dehn fillings of M.

Unlike genus, gonality is not a topological invariant of curves, but rather is intimately connected to the geometry of the curve. There are connections between gonality and genus, most notably the Brill-Noether theorem which gives an upper bound for gonality in terms of genus (see § 2.2 and § 5) but in some sense these are orthogonal invariants. For example, all hyperelliptic curves all have gonality two, but can have arbitrarily high genus. Moreover, for $g > 2$, there are curves of genus g of different gonality. We refer the reader to § 2 for precise definitions.

Our techniques can also be used to obtain information about the genera (denoted by g) and degree (denoted by d) of related varieties as we now discuss. This needs some additional notation.

Denote by ∂M the boundary of the compact manifold obtained from M by truncating the cusps of M. Assume that M has two cusps and fix a framing of each component of ∂M, denoted by $\{m_1, l_1\}$ and $\{m_2, l_2\}$, which we often refer to as meridional and longitudinal parameters for the ith cusp.

The inclusion map from $\pi_1(\partial M)$ to $\pi_1(M)$ induces a map from $X(M)$ to $X(\partial M)$. Let $A(M)$ denote the image of this map and let $A_0(M)$ denote the image of a canonical component; such a component is also referred to as a canonical component. The variety $A(M)$ sits naturally in $\mathbb{C}^4(m_1, l_1, m_2, l_2)$. For $r = p/q \neq 0$ in lowest terms, we define the naive height of r to be $h(r) = \max\{|p|, |q|\}$ with $h(0) = h(\infty) = 1$. Let $M(-, r)$ denote the manifold obtained by $r = p/q$ filling ofthe second cusp of M.

THEOREM 1.2. *Let M be a complete, orientable, finite volume hyperbolic 3-manifold with two cusps. Then there exists a constant c depending only on M such that for any hyperbolic manifold $N = M(-, r)$ obtained by filling one cusp of M, and for any canonical component A_0 of $A(N)$ we have*

$$g(A_0) \leq c \cdot h(r)^2.$$

Using work of Dunfield [**7**] (see also §4) we deduce the following.

COROLLARY 1.3. *Let M be a complete, orientable, finite volume hyperbolic 3-manifold with two cusps. Then there exists a constant c depending only on M such that for any hyperbolic manifold $N = M(-, r)$ obtained by filling one cusp of M satisfying $|H^1(N; \mathbb{Z}/2\mathbb{Z})| = 2$ and any canonical component X_0 of $X(N)$ we have*

$$g(X_0) \leq c \cdot h(r)^2.$$

Unlike the genus or gonality of a curve, the degree of a variety is inherently tied to its embedding in ambient space. We obtain upper bounds for the degree of a 'natural model' of $A_0(M(-,r))$.

THEOREM 1.4. *Let M be a complete, orientable, finite volume hyperbolic 3-manifold with two cusps. Then there exists a constant c depending only on M such that for any hyperbolic manifold $N = M(-,r)$ obtained by filling one cusp of M, and for any canonical component A_0 of $A(N)$ we have*

$$d(A_0) \leq c \cdot h(r).$$

Work of Hoste and Shanahan [**11**] demonstrates that the degree bounds in Theorem 1.4 are sharp.

In contrast with the main results of this paper, we also show that by filling two cusps of a three cusped manifold, we can construct examples of knots (so-called double twist knots) for which the gonality of the canonical component can be made arbitrarily large (see §11.2 for details). Existence of families of curves with large gonality has been the subject of much recent research, including connections between the existence of towers of curves whose gonality goes to infinity and expander graphs [**8**].

We also remark on the gonalities of the character varieties of the once-punctured torus bundles of tunnel number one [**1**]. These manifolds are all integral surgeries on one cusp of the Whitehead link and the corresponding gonalities are all equal to one or two. Both the double twist knot examples and the once-punctured torus bundles of tunnel number one examples illustrate linear growth in the genus of the character varieties. We do not know if a quadratic bound such as in Theorem 1.2 is optimal.

We complete the introduction by sketching the proof of Theorem 1.1 . With $A_0(M) \subset \mathbb{C}^4$ as above, it follows that each cusp of M determines a rational map from $A_0(M)$ to $\mathbb{C}^2$ (denoted by ϖ_1 and ϖ_2). For simplicity we consider only the case when each of these maps has dense image in $\mathbb{C}^2$ (this is typically the case). In particular, it follows that the projection map ϖ_1 defines a degree $d > 0$ map between these surfaces. We then deduce from this that for all but finitely many 1-cusped hyperbolic manifolds N obtained by Dehn filling on the second cusp of M, the gonality of $A_0(N)$ is bounded by the degree of ϖ_1.

2. Algebraic Geometry and Preliminary Lemmas

2.1. Preliminaries. We follow Shafarevich [**23**] and Hartshorne [**9**]. The 'natural models' of the character varieties of interest are complex affine algebraic sets, which typically have singularities at infinity. We will rely on an analysis of rational maps between possibly singular affine sets. Such a map $f : X \dashrightarrow Y$ behaves (on most of X and $f(Y)$) as a combination of a very nice map and a map with controlled bad parts. We now present some algebraic geometry to make this precise.

Let $\mathbb{C}[T]$ be the polynomial ring with coefficients in $\mathbb{C}$ in the variables $T_1, \ldots, T_n$. A (Zariski) closed subset of $\mathbb{C}^n$ is a subset $X \subset \mathbb{C}^n$ consisting of the common vanishing set of a finite number of polynomials in $\mathbb{C}[T]$. We let $\mathfrak{U}_X$ denote the *ideal of* X, the ideal in $\mathbb{C}[T]$ consisting of all polynomials which vanish on X. If I is an ideal in $\mathbb{C}[T]$ we let $V(I)$ denote the *vanishing set* of I in $\mathbb{C}^n$, and if $f \in \mathbb{C}[T]$ let $V(f)$ be the vanishing set of f in $\mathbb{C}^n$. The *coordinate ring* of X is $\mathbb{C}[X] = \mathbb{C}[T]/\mathfrak{U}_X$. This ring is the ring of all polynomials on X, up to the equivalence of being equal on X;

adding any polynomial in $\mathfrak{U}_X$ to a polynomial f does not change the value of f on X. If a closed set X is irreducible then the field of fractions of the coordinate ring $\mathbb{C}[X]$ is the *function field* or *field of rational functions* of X; it is denoted $\mathbb{C}(X)$.

A function f defined on X with values in $\mathbb{C}$ is *regular* if there exists a polynomial $F(T)$ with coefficients in $\mathbb{C}$ such that $f(x) = F(x)$ for all $x \in X$. A map $f : X \to Y$ is *regular* if there exist n regular functions $f_1, \dots, f_n$ on X such that $f(x) = (f_1(x), \dots, f_n(x))$ for all $x \in X$. Any regular map $f : X \to Y$ defines a pullback homomorphism of $\mathbb{C}$-algebras $f^* : \mathbb{C}[Y] \to \mathbb{C}[X]$ as follows. If $u : Y \to Z$ we define $v : X \to Z$ by $v(x) = u(f(x))$. We define f^* by $f^*(u) = v$. A *rational map* $f : X \dashrightarrow Y \subset \mathbb{C}^m$ is an m-tuple of rational functions $f_1, \dots, f_m \in \mathbb{C}(X)$ such that, for all points $x \in X$ at which all the f_i are defined, $f(x) = (f_1(x), \dots, f_m(x)) \in Y$. We define the pullback f^* as above. A rational map $f : X \dashrightarrow Y$ is called *dominant* if $f(X)$ is (Zariski) dense in Y. A rational map $f : X \dashrightarrow Y$ is called *birational* if f has a rational inverse, and in this case we say that X and Y are birational. A regular map $f : X \to Y$ is *finite* if $\mathbb{C}[X]$ is integral over $\mathbb{C}[Y]$. These definitions naturally extend to the projective setting. A *morphism* $\varphi : X \to Y$ is a continuous map such that for every open $V \subset Y$ and every regular function $f : Y \to \mathbb{C}$ the function $f \circ \varphi : \varphi^{-1}(Y) \to \mathbb{C}$ is regular.

Character varieties are naturally affine sets. It is useful to consider an embedding of such an affine set into projective space. Given an affine variety X, a *projective completion* of X is the Zariski closure of an embedding of X in some projective space. A *smooth projective model* of X is a smooth projective variety birational to a projective completion of X. Such a model is unique up to isomorphism if X is a curve. The *degree* of any projective curve X, denoted $d(X)$ is the maximum number of points of intersection of X with a hyperplane not containing any component of X. This depends on the specific embedding of X.

The gonality of a curve is, in the most basic sense, the degree of a map from that curve to $\mathbb{C}^1$. Intuitively the degree of a map is the number of preimages of 'most points' in the image. Let X and Y be irreducible varieties of the same dimension and $f : X \dashrightarrow Y$ a dominant rational map. The degree of the field extension $f^*(\mathbb{C}(Y)) \subset \mathbb{C}(X)$ is finite and is called the *degree* of f. That is,

$$\deg f = [\mathbb{C}(X) : f^*(\mathbb{C}(Y))].$$

This is well defined for rational and regular maps.

We are now in a position to see that dominant rational maps between possibly singular curves can be decomposed in a pleasing manner.

PROPOSITION 2.1. *Let X be a possibly singular irreducible affine variety of dimension n and Y an algebraic set of dimension n. Assume that $g : X \dashrightarrow Y$ is a dense rational map.*

(1) *$\deg(g) = d$ is finite.*
(2) *For all but finitely many co-dimension one subvarieties W of X,*

$$\deg(g\mid_W) \leq d.$$

(3) *If $n \leq 2$ then $g^{-1}(y)$ consists of at most d points except for perhaps finitely many $y \in Y$. In X there are finitely many irreducible sets of co-dimension one such that if x is not contained in the union of these sets then $g^{-1}(g(x))$ contains only finitely many points.*

PROOF. By Hironaka [**10**] one can resolve the singularities of a variety by a proper birational map (which has degree one). Therefore, the variety X is birational to a smooth projective variety X', and similarly Y is birational to a smooth projective Y'. The rational map $g : X \dashrightarrow Y$ induces a rational map $g' : X' \dashrightarrow Y'$ of the same degree by pre and post composition with the birational inverses,

$$X \overset{\alpha}{\dashrightarrow} X' \overset{g'}{\rightarrow} Y' \overset{\beta}{\dashrightarrow} Y.$$

By Hironaka's resolution of indeterminacies [**10**], g' can be resolved into a regular map by a sequence of blowups. Therefore we can take g' to be a proper morphism. By Stein Factorization (see [**9**] Corollary 11.5), a proper morphism factors as a proper morphism with connected fibers followed by a finite map. It follows that the degree is finite. By the Weak Factorization Theorem [**28**] these birational maps (since they are over $\mathbb{C}$) can be decomposed as a series of blow-ups and blow-downs over smooth complete varieties.

A blow up or blow down is defined for all points outside of a codimension one set. Therefore, outside of a finite union of codimension one subvarieties, the map g is the composition of isomorphisms, a proper morphism with connected fibers, and a finite map. If $W \subset X$ is a codimension one subvariety not in this set, then the degree of $g\mid_W$ is at most the degree of g.

If X is a curve, then part 3 follows from the observation that in this case, Stein Factorization implies that g' is a finite map. If X is a surface, then it suffices to show the content of the lemma for X and Y smooth projective surfaces, and $g : X \to Y$ a dense morphism. By Stein Factorization such a g can be decomposed into a map with connected fibers followed by a finite map. For the finite map, the number of pre-images of a point is bounded above by the degree. Since the maps are between surfaces, the map with connected fibers must be birational, so part 3 follows. □

LEMMA 2.2. *Let $f : X \dashrightarrow Y$ be a dominant rational map of affine algebraic sets. If X is irreducible, then Y is irreducible. (That is, the closure of $f(X)$ is irreducible.)*

PROOF. By Proposition 2.1 it suffices to assume that f is regular. The map f induces a map $f^* : \mathbb{C}[Y] \to \mathbb{C}[X]$ of coordinate rings. If $\mathfrak{U}_X \subset \mathbb{C}[T]$ and $\mathfrak{U}_Y \subset \mathbb{C}[S]$ then $\mathbb{C}[X] = \mathbb{C}[T]/\mathfrak{U}_X$ and $\mathbb{C}[Y] = \mathbb{C}[S]/\mathfrak{U}_Y$. Since $f(X)$ is dense in Y, it follows that f^* is an isomorphic inclusion (see [**23**] page 31). An algebraic set is irreducible if and only if the vanishing ideal is prime. Therefore $\mathfrak{U}_X$ is prime and $\mathbb{C}[X]$ has no zero divisors. We conclude that $\mathbb{C}[Y]$ has no zero divisors, and therefore $\mathfrak{U}_Y$ is prime and Y is irreducible.

□

2.2. Gonality. The algebraic definition of gonality is as follows.

DEFINITION 2.3. The *gonality* of a curve X in $\mathbb{P}^n$ is

$$\gamma(X) = \min\{[\mathbb{C}(X) : \mathbb{C}(x)] : x \in \mathbb{C}(X) - \mathbb{C}\}.$$

Since $\mathbb{C}(X)$ is the function field of X and consists of the rational functions on X, we can reformulate the definition; the gonality of X is the minimal degree of a dominant rational map from X to $\mathbb{P}^1$. Therefore, gonality is a birational invariant; two varieties which are birational (but not necessarily isomorphic) have equal gonality. If X is an affine curve, X is birational to a smooth projective model and as such the gonality of X and X' are equal, and this is the minimal degree of a dominant rational map from X to $\mathbb{C}$.

By Noether's Normalization theorem, if X is complex projective variety of dimension n, then there is a linear subspace disjoint from X such that the projection onto that subspace is finite to one and surjective. This is also true for affine varieties; if Z is an affine variety of dimension n there is a finite degree map from Z onto a dense subset of $\mathbb{C}^n$. Therefore, gonality is always finite.

REMARK 2.4. Gonality of an affine variety is sometimes defined as the minimal degree of a regular map to $\mathbb{C}$ (or of a finite map to $\mathbb{C}$). This (regular) gonality is easily seen to be greater than or equal to the (rational) gonality. In fact, there are examples where these two notions differ. The affine curve determined by the equation $xy = 1$ is not isomorphic to $\mathbb{C}$ and so the (regular) gonality is not one. One can see that there is a regular degree two map onto $\mathbb{C}$, by projecting onto the line $y = -x$. Therefore, the (regular) gonality is two. Projection onto the x coordinate is a dominant map to $\mathbb{C}$ as the image is $\mathbb{C} - \{0\}$ and has degree one; the (rational) gonality is one.

In the projective setting the gonality defined by using rational, regular, or finite maps are all equal, and are therefore equal to the (rational) gonality of a corresponding affine variety. If there is a dominant rational map $g : X \dashrightarrow Y$ between affine (possibly singular) curves then this induces a finite dominant morphism f between smooth projective models X' and Y' for X and Y, where the degree of g and the degree of f are the same.

Gonality can also be defined using the language of Riemann surfaces instead of complex curves. Here the definitions involving rational or regular maps correspond to meromorphic or holomorphic functions from the Riemann surface to $\mathbb{P}^1$. Similarly, one can define the gonality in terms of the minimal degree of a branched or unbranched covering of $\mathbb{P}^1$ by the Riemann surface.

Hyperelliptic curves are curves defined by an equation of the form $y^2 = f(x)$ where f is a separable polynomial of degree greater than four. Such a curve has gonality equal to two, realized by the map $(x, y) \mapsto (x, y^2)$. However, the genus is $\lfloor (d-1)/2 \rfloor$ where d is the degree of f.

Nonetheless, the Brill-Noether theorem gives an upper bound for gonality in terms of genus. In particular it states that if a curve has gonality γ and genus g then

$$\gamma \leq \lfloor \frac{g+3}{2} \rfloor.$$

(See § 5 for more discussion.)

2.3. Key lemmas. The following lemma will be essential for our gonality bounds.

LEMMA 2.5. *Let $g : X \dashrightarrow Y$ be a degree d dominant rational map of irreducible (possibly singular) affine or projective curves. Then*

$$\gamma(Y) \leq \gamma(X) \leq d \cdot \gamma(Y).$$

PROOF. The second inequality is clear. To show the first, we follow [**21**] Proposition A.1, including the proof for completeness. Let $f \in \mathbb{C}(X)$ be a degree d map realizing the gonality of X. Let $p(T) \in \mathbb{C}(Y)[T]$ be the characteristic polynomial of f when f is viewed as an element in the extension $\mathbb{C}(X)$ of $\mathbb{C}(Y)$. Let F be a finite extension of $\mathbb{C}(Y)$ such that $p(T)$ factors into the product of r monomials $(T - f_i)$, with $f_i \in F$.

Viewed as a function in F, the degree of f is $d[F : \mathbb{C}(X)]$. This is also the degree of the f_i since they are all in the same $\mathrm{Aut}(F/\mathbb{C}(Y))$ orbit. The polar divisors of a coefficient of p viewed in F is at most the sum of the polar divisors of the f_i. (These polar divisors are just the effective divisors $(f_i)_\infty$.)

Therefore, each coefficient has degree at most $dr[F : \mathbb{C}(X)] = d[F : \mathbb{C}(Y)]$ as a function in F, and therefore degree at most d as a function in $\mathbb{C}(Y)$. Since f is non-constant, at least one of these coefficients is non-constant. Therefore the gonality of Y is at most d. □

We now introduce some notation. Commonly $\alpha(x) \in \Theta(\beta(x))$ is written to mean that there are positive constants c_1 and c_2 independent of x such that

$$c_1\beta(x) \leq \alpha(x) \leq c_2\beta(x).$$

Notice that if $\alpha(x) \in \Theta(\beta(x))$ then also

$$c_2^{-1}\alpha(x) \leq \beta(x) \leq c_1^{-1}\alpha(x)$$

so $\beta(x) \in \Theta(\alpha(x))$. This defines an equivalence relation.

DEFINITION 2.6. For positive quantities $\alpha = \alpha(x)$ and $\beta = \beta(x)$ we write $\alpha \sim \beta$ if $\alpha \in \Theta(\beta)$.

We will use $\sim$ in conjunction with quantities such as $\gamma(X_0(M(-, r)))$ to mean that the constants do not depend on r but only on M and the framing of ∂M. Note that $\alpha(x) \sim 1$ means that there is a positive constant c independent of x such that $\alpha(x) \leq c$.

We now compute the gonality of the curves given by $x^p - y^q$ and $x^p y^q - 1$ in $\mathbb{C}^2(x, y)$. First we establish some notation for these sets.

DEFINITION 2.7. Let p and q be relatively prime integers with $q > 0$. With $r = p/q$, let

$$\mathcal{C}(r) = V(x^{|p|}y^q - 1) \cup V(x^{|p|} - y^q)$$

in $\mathbb{C}^2(x, y)$. Let

$$\mathcal{H}(r) = V(x^{|p|}y^q - 1) \cup V(x^{|p|} - y^q)$$

in $\mathbb{C}^4$ where x and y are among the four coordinates, so $\mathcal{H}(r)$ is a vanishing set in $\mathbb{C}^4(x, y, x', y')$, for example. (Therefore, $\mathcal{H}(r)$ and $\mathcal{C}(r)$ are each the union of two hypersurfaces unless $p = 0$.)

LEMMA 2.8. *Assume that p and q are relatively prime integers. Then the gonality of each irreducible component of $\mathcal{C}(r) \subset \mathbb{C}^2(x, y)$ is one.*

PROOF. If either p or q is zero, the polynomial is $x = 1$ or $y = 1$ and is therefore naturally a $\mathbb{C}^1$ and has gonality one. Now assume that $p, q > 0$. We will compute the gonality of $V(x^p y^q - 1)$ since this vanishing set is birationally equivalent to the vanishing set of $x^p y^q - 1$ under the map $(x, y) \mapsto (x, y^{-1})$.

Since p and q are relatively prime, there are integers a and b such that $ap + bq = 1$. Let $\varphi : \mathbb{C}^2(x, y) \dashrightarrow \mathbb{C}^2(x', y')$ be the rational map given by

$$(x, y) \mapsto (x^b y^{-a}, x^p y^q) = (x', y').$$

This has an inverse φ^{-1} given by

$$(x', y') \mapsto ((x')^q (y')^a, (x')^{-p} (y')^b).$$

The map φ is birational, so $\gamma(V(x^py^q-1))$ is equal to the gonality of the image of $V(x^py^q-1)$ under φ. This image is dense in $y'=1$. As $y'=1$ naturally defines a $\mathbb{C}^1$, the gonality of $V(y'-1)$ and $V(x^py^q-1)$ are both one.

□

2.4. Gonality and projection. We will now prove a few lemmas which will be essential in bounding the gonalities we are interested in. In § 4 we will relate the situation below to that of the character varieties of interest. Recall the definition of $\mathcal{C}(r)$ and $\mathcal{H}(r)$ from Definition 2.7. By Lemma 2.8 the gonality of (each irreducible component of) $\mathcal{C}(r)\subset\mathbb{C}^2$ is equal to one.

REMARK 2.9. If a curve C is reducible, by $\gamma(C)$ we mean the maximal gonality of any irreducible curve component of C.

DEFINITION 2.10. Define the maps ϖ_1 and ϖ_2 to be the projection maps

$$\varpi_i:\mathbb{C}^4(x_1,y_1,x_2,y_2)\to\mathbb{C}^2(x_i,y_i)$$

for $i=1,2$.

By Definition 2.7, with $\mathcal{H}(r)\subset\mathbb{C}^4(x_1,y_1,x_2,y_2)$ and $\mathcal{C}(r)\subset\mathbb{C}^2(x_2,y_2)$ then $\mathcal{C}(r)=\varpi_2(\mathcal{H}(r))$.

LEMMA 2.11. *Let V be an irreducible surface in $\mathbb{C}^4(x_1,y_1,x_2,y_2)$ such that $\dim_{\mathbb{C}}\varpi_i(V)>0$. Then for all but finitely many r, $V\cap\mathcal{H}(r)$ consists of a finite union of curves.*

PROOF. The dimension of V is two, and the dimension of $\mathcal{H}(r)$ is three. Therefore, each irreducible component of their intersection has dimension at least one (see [**9**] Section I.7). As a result, the dimension of any irreducible component of $V\cap\mathcal{H}(r)$ is one, or two.

Assume there are infinitely many r such that $V\cap\mathcal{H}(r)$ contains an irreducible component of dimension two. The set $\mathcal{H}(r)$ is of the form $\mathbb{C}^2\times\mathcal{C}(r)$, and in $\mathbb{C}^2(x_2,y_2)$ any two distinct $\mathcal{C}(r)$ intersect in only finitely many points. We conclude that $V\subset\mathbb{C}^2\times P$ for some point P. Therefore, $\varpi_2(V)=P$ has dimension zero, which does not occur by hypothesis. It follows that $V\cap\mathcal{H}(r)$ consists of a finite union of curves for all but finitely many r.

□

The algebraic sets whose gonalities we will bound are of the form $\varpi_1(V\cap\mathcal{H}(r))$. We will show that the dimension $\varpi_i(V)$ is either one or two for our sets. The proof of the main theorem will consist of three cases, depending on the dimension of $\varpi_i(V)$. When the dimension of $\varpi_i(V)$ is two for $i=1,2$, we will prove our gonality bounds by bounding the gonality of $\varpi_2(V\cap\mathcal{H}(r))$, using this to bound the gonality of $V\cap\mathcal{H}(r)$, and finally, using this to bound the gonality of $\varpi_1(V\cap\mathcal{H}(r))$. The following lemmas will be essential for our bounds.

LEMMA 2.12. *Let V be an irreducible surface in $\mathbb{C}^4(x_1,y_1,x_2,y_2)$ such that $\dim_{\mathbb{C}}\varpi_i(V)>0$. Assume that $V\cap\mathcal{H}(r)$ is a finite union of curves and $\varpi_2(V)$ is dense in $\mathbb{C}^2$. Then for any irreducible curve component C of $V\cap\mathcal{H}(r)$*

$$\gamma(C)\sim 1$$

where the implied constant depends only on V.

PROOF. By Proposition 2.1 the image $\varpi_2(V)$ consists of all of $\mathbb{C}^2(x_2, y_2)$ except perhaps a finite union of curves and points. The image $\varpi_2(\mathcal{H}(r)) = \mathcal{C}(r) \subset \mathbb{C}^2(x_2, y_2)$ and since two distinct $\mathcal{C}(r)$ intersect in only a finite set of points, we conclude that $\varpi_2(V \cap \mathcal{H}(r))$ maps onto a dense subset of $\mathcal{C}(r)$ for all but perhaps finitely many r. Therefore, it suffices to show the bound for these r where $\varpi_2(V \cap \mathcal{H}(r))$ is dense in $\mathcal{C}(r)$.

By Proposition 2.1, the degree $d = \deg(\varpi_2|_V)$ is finite, and for all but a finite number of points $\{P_m\}$ in the image of $\varpi_2|_V$, the number of pre-images of any point is bounded by d. For some such P_m, the pre-image of P_m is contained in a finite union of curves $\{C_{m,n}\}$ in V. The set $V \cap \mathcal{H}(r)$ is a union of curves and each of these curves intersects a $C_{m,n}$ in a finite collection of points unless $C_{m,n}$ coincides with a curve component of $V \cap \mathcal{H}(r)$.

Let D be an irreducible curve component of $V \cap \mathcal{H}(r)$. If $C_{m,n}$ coincides with D then as there are only finitely many $C_{m,n}$ the maximal gonality of any such D is bounded by a constant depending only on V. If $C_{m,n}$ does not coincide with D then the degree of $\varpi_2|_D$ is bounded by d, which is finite and independent of r. By Lemma 2.5 we conclude that $\gamma(D) \sim \gamma(\mathcal{C}(r))$. By Lemma 2.8, $\gamma(\mathcal{C}(r)) = 1$, which concludes the proof. □

LEMMA 2.13. *Let V be an irreducible surface in $\mathbb{C}^4(x_1, y_1, x_2, y_2)$ such that $\dim_{\mathbb{C}} \varpi_i(V) > 0$. Assume that $V \cap \mathcal{H}(r)$ is a finite union of curves and $\varpi_2(V)$ is a curve in $\mathbb{C}^2$. Then for any irreducible curve component C of $V \cap \mathcal{H}(r)$*

$$\gamma(C) \sim 1$$

where the implied constant depends only on V.

PROOF. The projection $\varpi_2(V \cap \mathcal{H}(r))$ is a finite collection of points unless a curve component of $\mathcal{C}(r)$ is contained in the closure of $\varpi_2(V) \subset \mathbb{C}^2(x_2, y_2)$. This can only occur for finitely many r as $\varpi_2(V)$ is a curve. For all r in this finite set, $\gamma(V \cap \mathcal{H}(r))$ is bounded.

It now suffices to consider those r such that $\varpi_2(V \cap \mathcal{H}(r))$ is a finite collection of points. For these r, any curve in $V \cap \mathcal{H}(r)$ must be of the form $D \times P$ where $D \subset \mathbb{C}^2(x_1, y_1)$ is a curve and $P \subset \mathbb{C}^2(x_2, y_2)$ is a point. (If $(1, 1)$ is in $\varpi_2(V)$ then it is possible that P are the same for all r since $(1, 1)$ is contained in each D.)

It suffices to universally bound the gonality of one such $D_r \times P_r$ as there are only finitely many irreducible components of each $V \cap \mathcal{H}(r)$. Call such a point $P_r = (a_r, b_r)$. We use the notation established in § 2.1. Let $\{\varphi_i\}$ be a generating set for $\mathfrak{U}_V$ where $\varphi_i = \varphi_i(x_1, y_1, x_2, y_2)$. A generating set for $\mathfrak{U}_{D_r \times P_r}$ is contained in $\{\varphi^*_{r,i}\}$ where $\varphi^*_{r,i} = \varphi_i(x_1, y_1, a_r, b_r)$. The degrees of $\varphi_i|_{x_1}$ and $\varphi_i|_{y_1}$ provide an upper bound for the degrees of $\varphi^*_{r,i}$. These upper bounds are independent of r. It follows that the degree d_r of $D_r \times P_r$ is bounded by some d which is independent of r.

The genus of $D_r \times P_r$ equals the genus of D_r as they are isomorphic. Moreover, the degree of D_r equals the degree of $D_r \times P_r$ as seen by the form of the $\varphi^*_{r,i}$. Therefore, we identify $D_r \times P_r$ with the plane curve $D_r \subset \mathbb{C}^2(x_1, y_1)$. By the genus degree formula the genus g_r of D_r satisfies

$$g_r \leq \tfrac{1}{2}(d_r - 1)(d_r - 2).$$

(There is equality if the plane curve is non-singular.) The Brill-Noether bound states that for $\gamma_r = \gamma(D_r)$,

$$\gamma_r \leq \lfloor \frac{g_r + 3}{2} \rfloor.$$

Therefore

$$\gamma_r \leq \lfloor \frac{g_r + 3}{2} \rfloor \leq \frac{1}{4}(d_r - 1)(d_r - 2) + 2.$$

The lemma follows as $d_r \leq d$ and d is independent of r. □

3. Dehn filling and character varieties

3.1. Assume that M is a finite volume hyperbolic 3-manifold whose boundary consists of exactly k torus cusps. Fix a framing for each cusp; let T_i be the ith connected component of ∂M and let μ_i and λ_i be an oriented choice of meridian and longitude for T_i. For $r_i = p_i/q_i \in \mathbb{Q} \cup \{1/0\}$, $i = 1, \ldots, k$, written in lowest terms (with q non-negative), let $M(r_1, \ldots, r_k)$ denote the manifold obtained by performing r_i Dehn filling on the ith cusp of M. We write $r_i = -$ to indicate that the ith cusp remains complete and is not filled. By Thurston's Hyperbolic Dehn Surgery Theorem ([**25**] Theorem 5.8.2), for all but finitely many values of r in $\mathbb{Q} \cup \{1/0\}$ the manifold obtained by r filling of one cusp of M is hyperbolic.

We fix a finite presentation for $\Gamma \cong \pi_1(M)$, with $M_i = [\mu_i]$, $L_i = [\lambda_i]$ amongst the generators for $i = 1, \ldots, k$. If Γ has presentation

$$\Gamma = \langle \gamma_1, \ldots, \gamma_n : \varrho_1, \ldots \varrho_l \rangle$$

then by van Kampen's theorem, a presentation for

$$\Gamma(r_1, \ldots, r_k) \cong \pi_1(M(r_1, \ldots, r_k))$$

is given by

$$\Gamma(r_1, \ldots, r_k) = \langle \gamma_1, \ldots, \gamma_n : \varrho_1, \ldots \varrho_l, w_1, \ldots, w_k \rangle$$

where $w_i = M_i^p L_i^q$ unless $r = p/q = -$ in which case it is empty.

We state the next lemma for convenience. The proof is clear.

LEMMA 3.1. *Let M be a finite volume two-cusped hyperbolic 3-manifold. Then*

$$|H^1(M(-, r); \mathbb{Z}/2\mathbb{Z})| \leq |H^1(M; \mathbb{Z}/2\mathbb{Z})|$$

for all $r \in \mathbb{Q} \cup \{1/0\}$.

3.2. We will now define various varieties associated to the fundamental group of a 3-manifold M (see [**26**].) For a finitely generated group Γ, and a homomorphism $\rho : \Gamma \to \mathrm{SL}_2(\mathbb{C})$, the set

$$\mathrm{Hom}(\Gamma, \mathrm{SL}_2(\mathbb{C})) = \{\rho : \Gamma \to \mathrm{SL}_2(\mathbb{C})\}$$

naturally carries the structure of an affine algebraic set defined over $\mathbb{C}$. This is called the *representation variety*, $R(\Gamma)$. Given an element $\gamma \in \Gamma$ we define the trace function $I_\gamma : R(\Gamma) \to \mathbb{C}$ as $I_\gamma(\rho) = \mathrm{tr}(\rho(\gamma))$. Similarly, given a representation $\rho \in R(\Gamma)$ we define the character of ρ to be the function $\chi_\rho : \Gamma \to \mathbb{C}$ given by $\chi_\rho(\gamma) = \mathrm{tr}(\rho(\gamma))$. The $\mathrm{SL}_2(\mathbb{C})$-*character variety* of Γ is the set of all characters,

$$X(\Gamma) = \{\chi_\rho : \rho \in R(\Gamma)\}.$$

Such a set is called a *natural model* for the character variety, as opposed to a set isomorphic to one obtained in this manner. The functions $t_\gamma : R(\Gamma) \to X(\Gamma)$ defined by $t_\gamma(\rho) = \chi_\rho(\gamma)$ are regular maps. The coordinate ring of $X(\Gamma)$ is $T \otimes \mathbb{C}$

where T is the subring of all regular functions from $R(\Gamma)$ to $\mathbb{C}$ generated by 1 and the t_γ functions.

If Γ is isomorphic to Γ', then the sets $R(\Gamma)$ and $R(\Gamma')$ are isomorphic, as are $X(\Gamma)$ and $X(\Gamma')$. We write $R(M)$ or $X(M)$ to denote a specific $R(\pi_1(M))$ or $X(\pi_1(M))$, respectively, which is well-defined up to isomorphism. The inclusion homomorphism $i : \pi_1(\partial M) \to \pi_1(M)$ induces a map $i^* : X(M) \to X(\pi_1(\partial M)) := X(\partial M)$ by $\chi_\rho \mapsto \chi_{\rho|_{\pi_1(\partial M)}}$. We write $A(M)$ to denote the closure of the (positive dimensional components of the) image of this map.

The representation variety $R(M)$ naturally sits in $\mathbb{C}^{4n}(g_{11}, g_{12}, g_{13}, g_{14}, \ldots, g_{n4})$ where g_{i1}, g_{i2}, g_{i3} and g_{i4} are the four entries of the 2×2 matrix $\rho(\gamma_i)$. Fix a generating set for $\pi_1(\partial M)$, $\{M_1, L_1, \ldots, M_k, L_k\}$ where each pair $\{M_i, L_i\}$ generates $\pi_1(T_i)$, the fundamental group a connected component of ∂M. Let $m_i^{\pm 1}$ be the eigenvalues of $\rho(M_i)$ and $l_i^{\pm 1}$ be the eigenvalues of $\rho(L_i)$.

Define the following functions on $\mathbb{C}(g_{11}, \ldots, g_{n4}, m_1^{\pm 1}, l_1^{\pm 1}, \ldots, m_k^{\pm 1}, l_k^{\pm 1})$

$$I_{M_i} - (m_i + m_i^{-1}), \quad I_{L_i} - (l_i + l_i^{-1}), \quad I_{M_i L_i} - (m_i l_i + m_i^{-1} l_i^{-1}).$$

The extended representation variety, $R_E(M)$ is the algebraic set in $\mathbb{C}^{4n} \times (\mathbb{C}^*)^{2k}$ cut out by $\mathfrak{U}_{R(M)}$ and the equations above for $i = 1, \ldots, k$. The extended character variety $X_E(M)$ is defined similarly. The *eigenvalue variety*, $E(M)$, is the closure of the image of $R_E(M)$ under the projection to the coordinates $m_1, l_1, \ldots, m_k, l_k$. It naturally sits in $(\mathbb{C}^*)^{2k}(m_1^{\pm 1}, l_1^{\pm 1}, \ldots, m_k^{\pm 1}, l_k^{\pm 1})$. We have the following commuting diagram where the maps p_1, p_2, and p_3 are all finite to one of the same degree.

$$\begin{array}{ccccc} R_E(M) & \xrightarrow{t_E} & X_E(M) & \xrightarrow{i_E^*} & E(M) \\ {\scriptstyle p_1}\downarrow & & {\scriptstyle p_2}\downarrow & & {\scriptstyle p_3}\downarrow \\ R(M) & \xrightarrow[t]{} & X(M) & \xrightarrow[i^*]{} & A(M) \end{array}$$

An irreducible component of $R(M)$ containing a discrete and faithful representation is denoted $R_0(M)$, and the image in $X(M)$ is denoted $X_0(M)$ and called a *canonical component*. For a given M there can be more than one canonical component, as there are multiple lifts of a discrete faithful representation to $\mathrm{SL}_2(\mathbb{C})$. We call an irreducible component of $A(M)$ a canonical component and write $A_0(M)$ if it is the image of a canonical component in $X(M)$. This definition also extends to the extended varieties.

We can define the $\mathrm{PSL}_2(\mathbb{C})$ representation and character varieties in a similar way (see [**3**]). We denote the $\mathrm{PSL}_2(\mathbb{C})$ character variety of M by $Y(M)$ and a canonical component by $Y_0(M)$. We can form the restriction variety $B(M)$ analogous to $A(M)$ as above (and also form $B_0(M)$ when appropriate). The group $\mathrm{PSL}_2(\mathbb{C})$ is isomorphic to $\mathrm{Isom}^+(\mathbb{H}^3)$, the group of orientation preserving isometries of hyperbolic half space. Since the index $[\mathrm{Isom}(\mathbb{H}^3) : \mathrm{Isom}^+(\mathbb{H}^3)] = 2$, for M a finite volume hyperbolic 3-manifold with at least one cusp there are two $\mathrm{PSL}_2(\mathbb{C})$ characters corresponding to discrete faithful representations of $\pi_1(M)$. These correspond to different orientations of M. The authors know of no examples where these two characters lie on different irreducible components of $Y(M)$.

The coordinate ring of the quotient is $T_e \otimes \mathbb{C}$ where T_e is the subring of T consisting of all elements invariant under the action of μ_2. Culler [**6**] (Corollary 2.3) showed that a faithful representation of any discrete torsion-free subgroup to $\mathrm{PSL}_2(\mathbb{C})$ lifts to $\mathrm{SL}_2(\mathbb{C})$, and moreover (Theorem 4.1) that a representation $\rho : \Gamma \to$

$\mathrm{PSL}_2(\mathbb{C})$ lifts to a representation to $\mathrm{SL}_2(\mathbb{C})$ if and only if every representation in the path component containing ρ lifts.

Thurston's Dehn Surgery Theorem asserts that any neighborhood of the character of a discrete faithful representation in $Y(M)$ contains all but finitely many Dehn filling characters (see [**25**] Theorem 5.8.2 and [**2**] Theorem E.5.1 and Remark E.5.2).

Since a discrete faithful character is always a simple point of $Y(M)$ [**22**] it follows that $Y_0(M(-,r))$ is contained in some $Y_0(M)$ for all but finitely many r. Therefore, for all but finitely many r, $X_0(M(-,r))$ is contained in some canonical component of $X(M)$. There are $|H^1(\Gamma;\mathbb{Z}/2\mathbb{Z})|$ lifts of $\rho : \Gamma \to \mathrm{PSL}_2(\mathbb{C})$ to $\mathrm{SL}_2(\mathbb{C})$, so there are $2|H^1(\Gamma;\mathbb{Z}/2\mathbb{Z})|$ $\mathrm{SL}_2(\mathbb{C})$ characters corresponding to discrete and faithful representations.

In the light of Lemma 2.5 we now establish bounds on the degrees of maps between the varieties of interest.

LEMMA 3.2. *Let M be a finite volume hyperbolic 3-manifold with exactly two cusps. Assume that $M(-,r)$ is hyperbolic.*

(1) *$Y_0(M(-,r) \to B_0(M(-,r))$ has degree 1.*
(2) *$X_0(M(-,r)) \to A_0(M(-,r))$ has degree at most $\frac{1}{2}|H^1(M;\mathbb{Z}/2\mathbb{Z})|$.*
(3) *$E_0(M(-,r)) \to A_0(M(-,r))$ has degree 4.*
(4) *$X_0(M(-,r)) \to Y_0(M(-,r))$ has degree at most $|H^1(M;\mathbb{Z}/2\mathbb{Z})|$.*

These mappings are all dense.

PROOF. Part (1) follows directly from Dunfield [**7**] (Theorem 3.1), who showed that the mapping $i^* : Y_0(M(-,r)) \to B_0(M(-,r))$ is a birational map onto its image. Dunfield [**7**] (Corollary 3.2), also proved that the degree of the map $i^* : X_0(M(-,r)) \to A_0(M(-,r))$ is at most $\frac{1}{2}|H^1(M(-,r);\mathbb{Z}/2\mathbb{Z})|$. By Lemma 3.1 this is bounded by $\frac{1}{2}|H^1(M;\mathbb{Z}/2\mathbb{Z})|$. This establishes part (2). The mapping from $E_0(M(-,r))$ to $X_0(M(-,r))$ is given by the action of inverting both entries of the pair (m_1,l_1) to (m_1^{-1},l_1^{-1}). This has degree at most four, giving part (3). Part (4) follows from the observation that the number of lifts of a representation $\rho : \Gamma \to \mathrm{PSL}_2(\mathbb{C})$ to $\mathrm{SL}_2(\mathbb{C})$ is $|H^1(\Gamma;\mathbb{Z}/2\mathbb{Z})|$. Therefore the degree of the map is at most $|H^1(M(-,r);\mathbb{Z}/2\mathbb{Z})|$ which is bounded above by $|H^1(M;\mathbb{Z}/2\mathbb{Z})|$ by Lemma 3.1. since the maps commute.

The mappings must all have dense image in the target space since all maps considered are maps between curves and the image must have dimension one as the degrees of the maps are bounded. □

Given Lemma 3.2, and Definition 2.6 we record the following for convenience.

PROPOSITION 3.3. *Let M be a finite volume hyperbolic 3-manifold with exactly two cusps. Assume that $M(-,r)$ is hyperbolic. Define the set*

$$\mathfrak{S} = \{A_0(M(-,r)), B_0(M(-,r)), E_0(M(-,r)), X_0(M(-,r)), Y_0(M(-,r))\}.$$

For any $V, W \in \mathfrak{S}$, $\gamma(V) \sim \gamma(W)$.

PROOF. By Lemma 2.5 it suffices to show that for all but finitely many r that there is a dominant map from V to W or from W to V whose degree is independent of r. The lemma now follows from Lemma 3.2. □

4. Gonality in Dehn surgery space

In this section we restrict to the case where M is a finite volume hyperbolc 3-manifold with exactly two cusps. Before we proceed, we define some Chebyshev polynomials which will appear as traces.

DEFINITION 4.1. For any integer k let $f_k(x)$ be the kth Fibonacci polynomial, that is defined $f_0(x) = 0$, $f_1(x) = 1$ and for all other k, define $f_k(x)$ recursively by the relation

$$f_{k+1}(x) + f_{k-1}(x) = xf_k(x).$$

Define $g_k = f_k - f_{k-1}$, and $h_k(x) = f_{k+1} - f_{k-1}$.

REMARK 4.2. With $x = s + s^{-1}$ then

$$f_k(x) = \frac{s^k - s^{-k}}{s - s^{-1}}, \quad g_k(x) = \frac{s^k + s^{1-k}}{s+1}, \quad \text{and} \quad h_k(x) = s^k + s^{-k}.$$

The effect of $r = p/q$ Dehn filling of the second cusp of M on the fundamental group is that it introduces the group relation $M_2^p L_2^q = 1$. Therefore, for a representation ρ the relation introduces an additional matrix relation $\rho(M_2)^p \rho(L_2)^q = I_2$ where I_2 is the 2×2 identity matrix. Up to conjugation, $\rho(M_2)$ can be taken to be upper triangular, and since M_2 and L_2 commute, $\rho(L_2)$ is upper triangular as well. (If $\rho(M_2) = I_2$ we can conjugate so that $\rho(L_2)$ is upper triangular.) The character and eigenvalue varieties are not changed by conjugation. Therefore, we may take

$$\rho(M_2) = \begin{pmatrix} m & s \\ 0 & m^{-1} \end{pmatrix} \quad \text{and} \quad \rho(L_2) = \begin{pmatrix} l & t \\ 0 & l^{-1} \end{pmatrix}$$

where $m, l \in \mathbb{C}^*$ and $s, t \in \mathbb{C}$. The commuting condition ensures that

$$t(m - m^{-1}) = s(l - l^{-1}). \tag{1}$$

By the Cayley-Hamilton theorem,

$$\rho(M_2)^p = \begin{pmatrix} m^p & sf_p(m + m^{-1}) \\ 0 & m^{-p} \end{pmatrix} \quad \text{and} \quad \rho(L_2)^{-q} = \begin{pmatrix} l^{-q} & -tf_q(l + l^{-1}) \\ 0 & l^q \end{pmatrix}.$$

As $\rho(M_2)^p = \rho(L_2)^{-q}$, we conclude that

$$m^p l^q = 1 \quad \text{and} \quad sf_p(m + m^{-1}) = -tf_q(l + l^{-1}).$$

If neither m nor l is ± 1 then (1) implies that

$$s\frac{m^p - m^{-p}}{m - m^{-1}} = t\frac{m^p - m^{-p}}{l - l^{-1}}.$$

With $m^p l^q = 1$ the condition $sf_p(m+m^{-1}) = -tf_q(l+l^{-1})$ follows. Therefore, the relation $sf_p(m + m^{-1}) = -tf_q(l + l^{-1})$ follows from (1) and $m^p l^q = 1$ unless m or l is ± 1. If $m = \pm 1$ or $l = \pm 1$, an additional identity with s and t is required to deduce $sf_p(m+m^{-1}) = -tf_q(l+l^{-1})$ from the commuting condition and $m^p l^q = 1$. (The commuting condition (1) is part of the defining equations for $X(M)$ as M_i and L_i commute in the group.) In the next section, we will see how this additional relation manifests as an intersection in the character varieties. First, we will discuss the containment of these sets.

The set $X(\partial M)$ is contained in $\mathbb{C}^4(m_1, l_1, m_2, l_2)$ where the coordinates are distinguished eigenvalues of the meridians and longitudes of the cusps. Any irreducible component has dimension at most two, and the image of a canonical component has dimension two (See [**24**] and [**26**] Proposition 12). The set $A(M(-,r))$ is contained in $\mathbb{C}^2(m,l)$ where m and l are identified with m_1 and l_1, and any canonical component has dimension one.

For all but finitely many r, $Y_0(M(-,r))$ is contained in some $Y_0(M)$. Therefore for all but finitely many r, $X_0(M(-,r))$ is contained in some $X_0(M)$. There may be numerous lifts of $Y_0(M)$, each containing infinitely many $X_0(M(-,r))$. As $A_0(M) \subset \mathbb{C}^4$ and $A_0(M(-,r)) \subset \mathbb{C}^2$ it is not the case that $A_0(M(-,r)) \subset A_0(M)$. In Lemma 4.5 we will show the relationship between these two sets.

4.1. Filling relations. Recall the notation established in §2; if f is a polynomial, $V(f)$ indicates the vanishing set of (the ideal generated by) f in a specified $\mathbb{C}^n$. We will make extensive use of the sets and polynomials from Definition 2.7. We will write $V(m_2^{\pm p} l_2^q - 1)$ for $V(m_2^{|p|} l_2^q - 1) \cup V(m_2^{|p|} - l_2^q)$.

LEMMA 4.3. *Assume that $M(-,r)$ is hyperbolic. For all but finitely many r there is some $X_0(M)$ such that*

$$X_0(M(-,r)) \subset X_0(M) \cap \Big(V(m_2^{\pm p} l_2^q - 1)\Big).$$

PROOF. For all but perhaps finitely many r such that $M(-,r)$ is hyperbolic there is some $X_0(M)$ containing a given $X_0(M(-,r))$. Consider these $r = p/q$. The fundamental group $\pi_1(M(-,r))$ is the quotient of $\pi_1(M)$ corresponding to the addition of a single relation, $M_2^p L_2^q = 1$. Therefore, using a presentation for $\pi_1(M)$ where M_2 and L_2 are among the generators, $X(M(-,r))$ naturally is contained in $X(M)$ (since both use the generators for $\pi_1(M)$, and representations of $\pi_1(M(-,r))$ simply satisfy an additional relation). The trace of $\rho(M_2)^p$ equals the trace of $\rho(L_2)^{-q}$. By construction, these traces are $h_p(\chi_\rho(M_2))$ and $h_q(\chi_\rho(L_2))$ with $\chi_\rho(M_2) = m_2 + m_2^{-1}$ and $\chi_\rho(L_2) = l_2 + l_2^{-1}$. This shows that

$$X_0(M(-,r)) \subset X_0(M) \cap V\big(h_p(m_2 + m_2^{-1}) - h_q(l_2 + l_2^{-1})\big).$$

It suffices to show that

$$V\big(h_p(m_2 + m_2^{-1}) - h_q(l_2 + l_2^{-1})\big) \subset V(m_2^{\pm p} l_2^q - 1).$$

This is equivalent to showing that the solution set of $m_2^p + m_2^{-p} = l_2^q + l_2^{-q}$ is contained in the solution set of $m_2^{\pm p} l_2^q = 1$ in $\mathbb{C}^4(m_2, l_2)$. This follows directly from the observation that the solutions to an equation of the form $x + x^{-1} = y + y^{-1}$ are $x = y^{\pm 1}$.

□

REMARK 4.4. It is possible that other trace relations will specify the sign in the relation $m_2^{\pm p} l_2^q - 1$ but this is not necessarily the case.

We will assume that M is hyperbolic and $X_0(M(-,r)) \subset X_0(M)$. The sets $A_0(M(-,r))$ and $A_0(M)$ lie in different ambient spaces, but we have the following lemma which allows us to concretely relate $A_0(M(-,r))$ with $A(M)$. Since M is hyperbolic $A_0(M)$ is a surface, and for all but finitely many r, $A_0(M(-,r))$ is a curve. The image under the projection map, $\varpi_i(A_0(M))$, has dimension at most two. The following lemma will allow us to deduce that $\varpi_i(A_0(M))$ cannot be a point, and therefore is either a curve or is dense in $\mathbb{C}^2$. (The image is irreducible by Lemma 2.2.) Our analysis will depend upon the dimension of these images.

LEMMA 4.5. *Assume that $M(-,r)$ is hyperbolic. For all but finitely many r, $A_0(M) \cap \mathcal{H}(r)$ is a finite union of curves. The variety $A_0(M(-,r))$ is a curve component of the closure of $\varpi_1(A_0(M) \cap \mathcal{H}(r))$ for some $A_0(M)$.*

PROOF. The group $\pi_1(M(-,r))$ is a quotient of $\pi_1(M)$; the group presentation only has the additional relation that $M_2^p L_2^q = 1$, where $r = p/q$. Using the notation established earlier in this section, in the construction of $X(\partial M)$ and $A(M(-,r))$ the $(1,1)$ entry of $\rho(M_i)$ is identified with m_i, and the $(1,1)$ entry for $\rho(L_i)$ is identified with l_i. The manifestation of this group quotient on the character varieties is that $X_0(M(-,r))$ is contained in the intersection of some $X_0(M)$ and $V(m_2^{\pm p} l_2^q - 1)$, as shown in Lemma 4.3.

The epimorphism $\pi_1(M) \to \pi_1(M(-,r))$ surjects $\pi_1(\partial M)$ onto $\pi_1(\partial M(-,r))$ and this induces maps

$$X(M) \stackrel{i^*}{\to} X(\partial M) \stackrel{\varpi_1}{\to} X(\partial M(-,r)).$$

The set $X(\partial M)$ is dense in $\mathbb{C}^4(m_1, l_1, m_2, l_2)$ and the set $X(\partial M(-,r))$ is contained in $\mathbb{C}^2(m_1, l_1)$.

Since $X_0(M(-,r)) \subset X_0(M) \cap V(m_2^{\pm p} l_2^q - 1)$ by Lemma 4.3, it follows that

$$i_*(X_0(M(-,r))) \subset i_*(X_0(M)) \cap i_*(V(m_2^{\pm p} l_2^q - 1))).$$

Therefore, $i_*(X_0(M(-,r))) \subset A_0(M) \cap V(m_2^{\pm p} l_2^q - 1)$ in $\mathbb{C}^4$. Because $A_0(M(-,r))$ is the closure of $\varpi_1(i_*(X_0(M(-,r))))$, by projecting we conclude that $A_0(M(-,r))$ is contained in (the closure of) $\varpi_1(A_0(M) \cap V(m_2^{\pm p} l_2^q - 1))$, which is $\varpi_1(A_0(M) \cap \mathcal{H}(r))$.

To show that $A_0(M) \cap \mathcal{H}(r)$ is a finite union of curves, by Lemma 2.11 it suffices to show $\dim_{\mathbb{C}}(\varpi_i(A_0(M))) > 0$ for $i = 1, 2$. Since $M(-,r)$ is hyperbolic and has exactly one cusp $A_0(M(-,r))$ is a curve. From the above,

$$A_0(M(-,r)) \subset \varpi_1(A_0(M) \cap \mathcal{H}(r)) \subset \varpi_1(A_0(M))$$

which implies that $\dim_{\mathbb{C}} \varpi_1(A_0(M)) > 0$. By symmetry (filling the other cusp), $\dim_{\mathbb{C}} \varpi_2(A_0(M)) > 0$ as well.

□

4.2. Gonality relations. To control gonality we will analyze the dimension of $\varpi_1(A_0(M))$. The authors expect that the dimension of $\varpi_1(A_0(M))$ is usually two, and comment on a connection between this dimension and geometric isolation below. Our first lemma shows that if $\dim \varpi_1(A_0(M)) = 1$ then there are only finitely many $A_0(M(-,r))$. We conclude that in this situation, the gonality of any $A_0(M(-,r))$ is bounded independently of r.

LEMMA 4.6. *Assume that $\dim \varpi_1(A_0(M)) = 1$. For all but finitely many r such that $A_0(M(-,r)) \subset \varpi_1(A_0(M))$, all of the curves $A_0(M(-,r))$ are identical.*

PROOF. Lemma 4.5 implies that for all but finitely many r, $A_0(M(-,r))$ is contained in $\varpi_1(A_0(M) \cap \mathcal{H}(r))$. For all such r, $A_0(M(-,r))$ is a curve and since the dimension of $\varpi_1(A_0(M))$ is one, it follows that they must be the same curve. □

REMARK 4.7. A similar statement applies to $B_0(M)$. If there is a single canonical component $Y_0(M)$ then Lemma 4.6 shows that for all but finitely many r the curves $B_0(M(-,r))$ are the same curve.

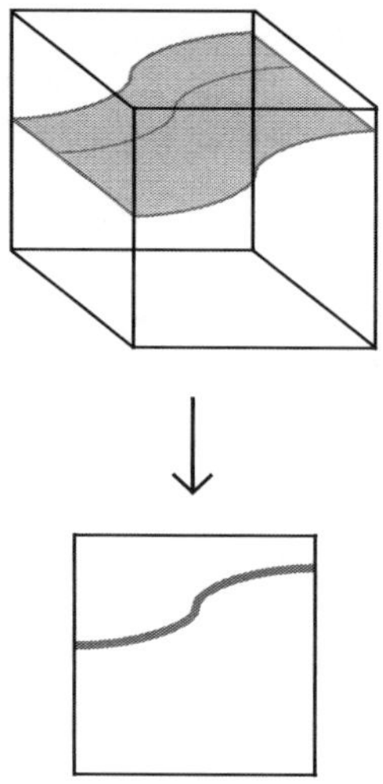

FIGURE 1. The set $A_0(M) \subset \mathbb{C}^4(m_1, l_1, m_2, l_2)$ with the intersection $A_0(M) \cap \mathcal{H}(r)$ and the projection $\varpi_1(A_0(M)) \subset \mathbb{C}^2(m_1, l_1)$ when this projection has dimension 1. The set $A_0(M(r)) \subset \varpi_1(A_0(M))$.

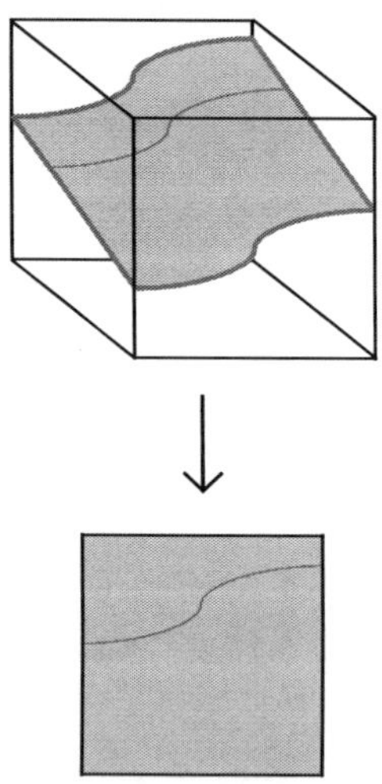

FIGURE 2. The set $A_0(M) \subset \mathbb{C}^4(m_1, l_1, m_2, l_2)$ with the intersection $A_0(M) \cap \mathcal{H}(r)$ and the projection $\varpi_1(A_0(M)) \subset \mathbb{C}^2(m_1, l_1)$ when $\dim \varpi_1(A_0(M)) = 2$. The set $A_0(M(r)) \subset \varpi_1(A_0(M))$.

Now we deal with the case that $\dim \varpi_1(A_0(M)) = 2$. In order to bound the gonality of $A_0(M(-, r))$ it suffices to bound the gonality of $A_0(M) \cap \mathcal{H}(r)$.

LEMMA 4.8. *Assume that* $\dim \varpi_1(A_0(M)) = 2$. *Then for all but finitely many* r *such that* $A_0(M(-, r)) \subset \varpi_1(A_0(M))$ *the set* $A_0(M) \cap \mathcal{H}(r)$ *is a finite union of curves. Additionally, for any irreducible curve component* C *of* $A_0(M) \cap \mathcal{H}(r)$ *with*

$A_0(M(-,r))$ *contained in the closure of* $\varpi_1(C)$

$$\gamma(C) \sim \gamma(A_0(M(-,r))).$$

PROOF. It suffices to prove the lemma for a single irreducible component $A_0(M)$ whose projection contains infinitely many $A_0(M(-,r))$. By Lemma 2.5 it suffices to show that with the exception of finitely many r, the degree of ϖ_1 restricted to $A_0(M) \cap \mathcal{H}(r)$ is independent of r.

We are assuming that $\varpi_1 : A_0(M) \to \mathbb{C}^2$ is dense. By Proposition 2.1 this is a dense degree d map between surfaces, for some d. Moreover, for all but finitely many points in the image of $\varpi_1(A_0(M))$ the pre image of a point consists of at most d points in $A_0(M)$. The image $\varpi_1(A_0(M))$ is $\mathbb{C}^2$ with the possible exception of finitely many curves and points. By Lemma 4.5 except for finitely many points, $A_0(M(-,r))$ is contained in the image of $\varpi_1(A_0(M))$.

Therefore, for all but finitely many points P on any curve $A_0(M(-,r))$, the number of pre-images of P under ϖ_1 is at most d. As a result, if $C \subset \big(A_0(M) \cap \mathcal{H}(r)\big)$ is an irreducible component such that $A_0(M(-,r))$ is contained in the closure of $\varpi_1(C)$, the map $\varpi_1 \mid_C$ is degree at most d. Since d independent of r, we have $\gamma(C) \sim \gamma(A_0(M(-,r)))$. □

4.3. Proof of Theorem 1.1.

PROOF OF THEOREM 1.1. From Proposition 3.3, it suffices to bound the gonality of $A_0(M(-,r))$. For all but finitely many r, $X_0(M(-,r))$ is contained in some $X_0(M)$. With Lemma 4.5, we conclude that for all but finitely many r, $A_0(M(-,r))$ is contained in some $\varpi_1(A_0(M))$. As there are only finitely many $A_0(M)$, it suffices to bound the gonality of all r such that $A_0(M(-,r))$ is contained in a fixed $\varpi_1(A_0(M))$.

The proof will follow from cases, depending on the possible dimensions of $\varpi_i(A_0(M))$. We will use the fact, proven in Lemma 4.5, that for all but finitely many r the set $A_0(M) \cap \mathcal{H}(r)$ is a finite union of curves, so that for such an r there is a curve component $C = C(r)$ of $A_0(M) \cap \mathcal{H}(r)$ such that $A_0(M(-,r))$ is contained in $\varpi_1(C)$.

First, assume that $\dim \varpi_1(A_0(M)) = \dim \varpi_2(A_0(M)) = 2$. In this case, $A_0(M)$ is a surface in $\mathbb{C}^4$ whose image under both ϖ_1 and ϖ_2 is dense in $\mathbb{C}^2$. By Lemma 4.8 since $\dim \varpi_2(A_0(M)) = 2$, for all but finitely many r, there is an irreducible curve component $C = C(r)$ of $A_0(M) \cap \mathcal{H}(r)$ such that $A_0(M(-,r))$ is contained in the closure of $\varpi_1(C)$. Moreover, $\gamma(C) \sim \gamma(A_0(M(-,r)))$. Therefore it suffices to bound the gonality of any curve component C of $A_0(M) \cap \mathcal{H}$. By Lemma 2.12 since $C \subset A_0(M) \cap \mathcal{H}(r)$ and $\varpi_2(A_0(M))$ is dense in $\mathbb{C}^2$ it follows that $\gamma(C) \sim 1$. By the transitivity of $\sim$ we conclude that $\gamma(A_0(M(-,r))) \sim 1$. (In this case, for all but finitely many r, the implied upper bounds for the gonality are as follows. From the projection ϖ_1 we have $\gamma(A_0(M(-,r))) \leq \gamma(C)$ and from ϖ_2 we have $\gamma(C) \leq \deg \varpi_2$, so that $\gamma(A_0(M(-,r))) \leq \deg \varpi_2$.)

Now we consider the case where $\dim \varpi_1(A_0(M)) = 1$. In this case $\varpi_1(A_0(M))$ is a curve; let D denote its closure. By Lemma 4.5 for all but finitely many r we have $A_0(M(-,r)) \subset D$. In this case, as r varies the the $A_0(M(-,r))$ are only a finite number of different curves. As a result, $\gamma(A_0(M(-,r))$ is bounded by the maximum gonality of this finite set of curves and is independent of r.

Now, assume that $\dim \varpi_1(A_0(M)) = 2$ and $\dim \varpi_2(A_0(M)) = 1$. Let C be an irreducible curve component of $A_0(M) \cap \mathcal{H}(r)$ with $A_0(M(-,r))$ contained in the closure of $\varpi_1(C)$. By Lemma 4.8, since ϖ_1 is dense in $\mathbb{C}^2$,

$$\gamma(C) \sim \gamma(A_0(M(-,r))).$$

Therefore, it suffices to bound $\gamma(C)$. By Lemma 2.13, $\gamma(C) \sim 1$. (In this case, for all but finitely many r, the implied upper bounds for the gonality are as follows. From the projection ϖ_1 we have $\gamma(A_0(M(-,r))) \leq \gamma(C)$. From Lemma 2.13 the bound for $\gamma(C)$ in this case follows from the Brill Noether theorem and the genus degree formula as the degrees of the defining equations for any curve C is bounded independent of r. □

REMARK 4.9. The proof of Theorem 1.1 implies that if $\dim \varpi_1(A_0(M)) = 1$, then $\varpi_1(A_0(M))$ is an irreducible curve and the curve $A_0(M(-,r))$ is the same for all r such that $\varpi_1(A_0(M))$ contains $A_0(M(-,r))$. A similar statement is true for $B_0(M(-,r))$. As there are only two characters of discrete faithful representations in $Y_0(M)$, there are at most two $B_0(M)$ and at most two projections.

It is possible that all $A_0(M(-,r))$ are the same if $\dim \varpi_2(A_0(M)) = 1$. This will occur if $\varpi_2(A_0(M) \cap \mathcal{H}(r))$ contains one of the common intersection points of the $\mathcal{C}(r) = \varpi_2(\mathcal{H}(r))$.

REMARK 4.10. By Proposition 3.3 the bounds in Theorem 1.1 translate to bounds for other sets as well. We obtain an analogous statement for each of the sets

$$A_0(M(-,r)), B_0(M(-,r)), E_0(M(-,r)), X_0(M(-,r)), Y_0(M(-,r)).$$

For a given r such that $M(-,r)$ is hyperbolic, the correspondence between two such sets is established by a mapping between the sets. Even though this is ambiguous, any lift can be traced back to Y_0 so there is a correspondence between different lifts.

REMARK 4.11. The authors do not know of any examples where the dimensions of $\varpi_1(A_0(M))$ and $\varpi_2(A_0(M))$ differ. When the dimensions are 1, then $A_0(M) \cong C_1 \times C_2$, with C_i is a curve in $\mathbb{C}^2(m_i, l_i)$. In this case, l_1 depends only on m_1 and l_2 depends only on m_2. Hence, it follows from [**17**] Theorem 4.3 that the first cusp is geometrically isolated from the second (and vice versa).

5. Genus and degree

The gonality of a curve is intimately related to both the (geometric) genus of the curve, and the degree of the curve. Like gonality, the genus is a birational invariant. The degree of a curve depends on the embedding of the curve. Assume that C is a curve with a specified embedding into some ambient space. Let g, d, and γ be the genus, degree, and gonality of C, respectively. The Brill-Noether theorem relates the genus and the gonality of a curve. It states that

$$\gamma \leq \lfloor \frac{g+3}{2} \rfloor.$$

The genus degree formula (the adjunction formula) states that if C is a smooth plane curve then

$$g = \tfrac{1}{2}(d-1)(d-2).$$

Each singularity of order s reduces the right hand side by $\frac{1}{2}s(s-1)$.

With Theorem 1.1 we can establish bounds for the genus. We have shown that the gonality is bounded. We now consider the variety $A_0(M(-,r))$ which naturally is contained in $\mathbb{C}^2(m,l)$.

THEOREM 1.4. Let M be a finite volume hyperbolic 3-manifold with two cusps. If $M(-,r)$ is hyperbolic, then there is a positive constant c_1 depending only on M and the framing of the second cusp such that

$$d(A_0(M(-,r))) \leq c_1 \cdot h(r).$$

If one of the projections $\varpi_i(A_0(M))$ is a curve, then there is a positive constant c_2 depending only on M and the framing of the second cusp, such that

$$d(A_0(M(-,r))) \leq c_2.$$

PROOF. First, we will bound the degree of (any irreducible component of) the intersection of a surface X in $\mathbb{C}^4(x_1,y_1,x_2,y_2)$ with $\mathcal{H}(r)$. As the dimension of $\mathcal{H}(r)$ is three and the dimension of X is two, the condition that they intersect in a curve implies that the intersection is proper. It is a consequence of Bézout's theorem (see [**9**] Theorem I,7.7) that with $r = p/q$, since $\deg \mathcal{H}(r) = \max\{|p|,|q|\}$, the degree of the intersection is at most

$$\deg(X \cap \mathcal{H}(r)) \leq \deg X \cdot \deg \mathcal{H}(r) = \deg X \cdot \max\{|p|,|q|\}.$$

For any variety $W \subset \mathbb{C}^4$ the degree of $\varpi_1(W)$ is at most the degree of W. To see this, assume that W intersects a complimentary hyperplane H in the point P. Then $\varpi_1(W)$ intersects $\varpi_1(H)$ in $\varpi_1(P)$. It suffices to see that we can take H so that $\varpi_1(H)$ is a complimentary hyperplane to $\varpi_1(W)$ and that this respects multiplicities.

Therefore, with $X = A_0(M)$ by Lemma 4.5, $A_0(M(-,r))$ is contained in the closure of $\varpi_1(A_0(M) \cap \mathcal{H}(r))$. We conclude that

$$\deg A_0(M(-,r)) \leq \deg(\varpi_1(A_0(M) \cap \mathcal{H}(r))) \leq \deg(A_0(M) \cap \mathcal{H}(r))$$

and by Bézout's theorem

$$\deg(A_0(M) \cap \mathcal{H}(r)) \leq \deg A_0(M) \max\{|p|,|q|\}.$$

This implies the first assertion.

Now assume that $\varpi_2(A_0(M))$ is a curve. Let $\{\varphi_i\}$ be a generating set for $\mathfrak{U}_{A_0(M)}$ where $\varphi_i = \varphi_i(x_1,y_1,x_2,y_2)$. Then $\mathfrak{U}_{A_0(M)\cap\mathcal{H}(r)}$ is generated by $\{\varphi_i(x_1,y_1,a_r,b_r)\}$ as in the proof of Lemma 2.13. The degree of $A_0(M) \cap \mathcal{H}(r)$ is bounded by the degrees of these equations. The map ϖ_1 is a projection. It follows that degree of the the image $\varpi_1(A_0(M) \cap \mathcal{H}(r))$ is bounded by the degree (in x_1 and y_1) of the $\varpi_i(x_1,y_1,x_2,y_2)$.

If $\varpi_1(A_0(M))$ is a curve, the result follows from Lemma 4.6. □

We can bound the genus using the genus-degree formula. Specifically, we have the following.

THEOREM 1.2. Let M be a finite volume hyperbolic 3-manifold with two cusps. If $M(-,r)$ is hyperbolic, there is a positive constant c depending only on M and the framing of the second cusp such that

$$g(A_0(M(-,r))) \leq c \cdot h(r)^2.$$

PROOF. It suffices to consider curves of degree greater than two, since the genus of plane curves of smaller degree at most two is bounded. The theorem follows directly from the genus-degree formula which implies that if C is a plane curve of degree d then the genus of C is at most $\frac{1}{2}(d-1)(d-2)$, which is less than $\frac{1}{2}d^2$. By Theorem 1.4 there is a constant c such that $d \leq ch(r)$ so that the genus us bounded by $\frac{1}{2}c^2h(r)^2$.

□

REMARK 5.1. For $r = p/q$ with $pq \neq 0$ this is

$$g(A_0(M(-,r))) \leq c\max\{p^2, q^2\}.$$

This statement can also be made for $B_0(M(-,r))$. These genus bounds do not appear sharp. In § 7.2 we will show that the double twist knots the genus is linear in the number of twists in each twist region. Similarly, the genus bounds from [**1**] are linear in $h(r)$.

6. Newton polygons

Bounds for the genus and the gonality of a plane curve can be obtained from the associated Newton Polygon. Let $f \in \mathbb{C}[x^{\pm 1}, y^{\pm 1}]$ be an irreducible Laurent polynomial which defines a curve $V(f)$ in $(\mathbb{C} - \{0\})^2$. Let $\Delta(f)$ be the associated Newton polygon of f. In 1893, Baker related the genus of such a curve to the volume of the Newton Polygon.

THEOREM 6.1 (Baker). *The geometric genus of $V(f)$ is at most the number of lattice points in the interior of $\Delta(f)$.*

Recently, Castryck and Cools showed a similar relationship for the gonality of $V(f)$. A $\mathbb{Z}$-affine transformation is a map from $\mathbb{R}^2$ to $\mathbb{R}^2$ of the form $x \mapsto Ax + b$ where $A \in \mathrm{GL}_2(\mathbb{Z})$ and $b \in \mathbb{Z}^2$. We say that two lattice polygons Δ and Δ' are equivalent if there is a $\mathbb{Z}$-affine transformation φ such that $\varphi(\Delta) = \Delta'$. The lattice width of Δ is the smallest integer $s \geq 0$ such that there is a $\mathbb{Z}$-affine transformation φ such that $\varphi(\Delta)$ is contained in the horizontal strip $\{(x, y) \in \mathbb{R}^2 : 0 \leq y \leq s\}$.

THEOREM 6.2 (Castryck-Cools). *The gonality of $V(f)$ is at most the lattice width of $\Delta(f)$.*

Culler and Shalen [**5**] showed that if M has only one cusp, then $X_0(M)$ determines a norm on $H_1(\partial M, \mathbb{R})$. This norm encodes many topological properties of M including information about the Dehn fillings of M. This construction can be applied to other curve components of $X(M)$ yielding a semi norm in general. Culler and Shalen, and Boyer and Zhang [**3**] used these semi norms to study surfaces in 3-manifolds. Boyer and Zhang also extended this semi norm to $Y(M)$.

We observe the following about non-norm curve components.

THEOREM 6.3. *Let M be a compact, connected, irreducible, ∂-irreducible 3-manifold with boundary torus. Let $Y \subset Y(M)$ be a curve which is not a norm curve. Then Y is birational to $\mathbb{C}$.*

PROOF. By Boyer and Zhang [**3**] (Proposition 5.4.3), if M has one cusp any norm curve of $X(M)$ maps to a component of the A-polynomial whose Newton polygon is two-dimensional, and any curve that is not a norm curve maps to a component of the A-polynomial that is one-dimensional or zero-dimensional. Therefore, the Newton polygon of a non-norm curve is a line segment or a point. This has empty interior, so by Baker's theorem, or Castryck and Cools' Theorem, the geometric genus is zero. It follows that the curve is birational to $\mathbb{C}$. □

7. Examples

Theorem 1.1 shows that the gonality of $X_0(M(-,r))$ is bounded independent of r. In general, it is difficult to determine the exact gonality of such curves, as explicit computations of character varieties are difficult and only a few infinite families have been computed. In this section, we will compute the gonality of the character varieties associated to two families of manifolds which arise as Dehn surgeries. These examples will demonstrate that the linear bounds in Theorem 1.4 are sharp, and moreover that there are character varieties with arbitrarily large gonality. All of the filled manifolds in the following examples are knot complements in S^3. For these $M(-,r)$, $A_0(M(-,r))$ is birational to $X_0(M(-,r))$.

FIGURE 3. The once-punctured torus bundle of tunnel number one, M_n, is $-(n+2)$ surgery on one component of the Whitehead link, shown above.

7.1. Once-punctured torus bundles of tunnel number one. The family of once-punctured torus bundles of tunnel number one is an infinite family of one-cusped manifolds, indexed as $\{M_n\}$ for $n \in \mathbb{Z}$. These manifolds are hyperbolic for $|n| > 2$ and can be realized as integral fillings of one component of the Whitehead link complement, W; the manifold M_n is $W(-,-(n+2))$.

The character varieties of these manifolds were explicitly computed in [**1**]. For each hyperbolic M_n there is a single canonical component in the $\mathrm{SL}_2(\mathbb{C})$ character variety. The curves $X_0(M_n)$ are all hyperelliptic curves of genus $\lfloor\frac{1}{2}(|n-1|-1)\rfloor$ if $n \not\equiv 2 \pmod 4$ and $\lfloor\frac{1}{2}(|n-1|-3)\rfloor$ if $n \equiv 2 \pmod 4$. (The variety $X_0(M_3)$ associated to the figure-eight knot sister is a rational curve. The variety $X_0(M_{-3})$ associated to the figure-eight knot complement and the variety $X_0(M_6)$ are elliptic curves.) The discrepancy arising when $n \equiv 2 \pmod 4$ is due to the fact that in this case there are two components of irreducible representations, the canonical component and an affine line component.

For this family $r = p/q = (-(n+2))/1$; the gonality of $X_0(M_n)$ is bounded by 2, and the genus is increasing linearly in p. Note that Theorem 1.2 gives an upper bound of the form $c \cdot p^2$ for the genus. The canonical components of the

$\mathrm{PSL}_2(\mathbb{C})$ character variety are all birational to $\mathbb{C}^1$ and we conclude that here both the gonality and the genus are bounded by the constant one.

7.2. Double twist knots. The double twist knots are the knots $J(k,l)$ shown in the right hand side of Figure 4. When exactly one of k and l is odd, $J(k,l)$ is a knot. When k and l are even, these knots are the result of $-1/k$ surgery and $-1/l$ surgery on two components of the link shown in the left hand side of Figure 4 where k and l are the number of half twists in the boxes. When k and l are odd, this is a two component link and is the result of $-1/k$ surgery and $-1/l$ surgery on two components of a similar link with half twists added (see Figure 5); these links are dealt with in the next section. Since $J(k,l)$ is ambient isotopic to $J(l,k)$ we may assume that $l = 2n$. The knot complement of $J(k,l)$ is hyperbolic unless $|k|$ or $|l|$ is less than two, or $k = l = \pm 2$. The knots with $|k|$ or $|l|$ equal to two are the twist knots. These include the figure-eight knot, $J(2,-2) = J(-2,2) = J(2,3)$ and the trefoil $J(2,2) = J(-2,-2)$. We now restrict our attention to these hyperbolic knots.

Let $Y_0(k,l)$ denote $Y_0(S^3 - J(k,l))$, and $X_0(k,l)$ denote $X_0(S^3 - J(k,l))$. For a fixed k_0, by Theorem 1.1

$$\gamma(Y_0(k_0,l)) \sim c = c(k_0).$$

A similar statement is true if we fix l.

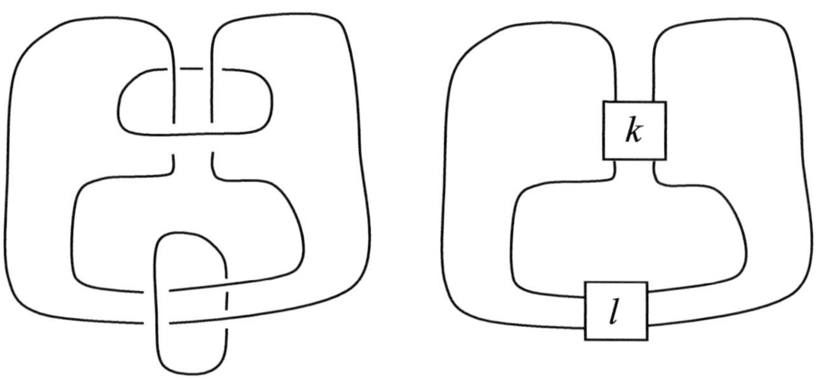

FIGURE 4. The $J(k,l)$ knot is the result of $-1/k$ and $-1/l$ surgery on the three component link pictured on the left.

The character varieties of these knot complements were computed in and analyzed in [**14**]. The component of the $\mathrm{PSL}_2(\mathbb{C})$ character variety of the double twist knot $S^3 - J(k,l)$ which contain irreducible representations is birationally equivalent to the curve defined by the following equations in $\mathbb{P}^1 \times \mathbb{P}^1$:

$$(*) \qquad \begin{aligned} g_{m'+1}(r)g_{n'}(t) &= g_{m'}(r)g_{n'+1}(t) \ \text{ if } \ k = 2m', l = 2n' \\ f_{m'+1}(r)g_{n'}(t) &= f_{m'}(r)g_{n'+1}(t) \ \text{ if } \ k = 2m'+1, l = 2n'. \end{aligned}$$

(The polynomials f_i and g_i were introduced in Definition 4.1.) These curves are smooth and irreducible in $\mathbb{P}^1 \times \mathbb{P}^1$ when $k \neq l$. When $k = l$ the canonical component is given by $r = t$. The genus can be computed from the bidegree of these polynomials. Let $m = \lfloor \frac{|k|}{2} \rfloor$ and $n = \lfloor \frac{|l|}{2} \rfloor$ and let $D(k,l)$ be the projective closure of the variety determined by the equations $(*)$ in $\mathbb{P}^1(r) \times \mathbb{P}^1(t)$. When $k \neq l$ the bidegree is (m, n). The genus of $X_0(k, l)$ is determined by studying the ramification of the map from $X_0(k, l)$ to $Y_0(k, l)$.

THEOREM 1 ([**14**] Theorems 6.2 and 6.5). *Let $J(k, l)$ be a hyperbolic with $m = \lfloor \frac{|k|}{2} \rfloor$. If $k \neq l$ the genus of $Y_0(k, l)$ is $(m-1)(n-1)$ and the genus of $X_0(k, l)$ is $3mn - m - n - b$ where*

$$b = \begin{cases} (-1)^{k+l} & \text{if} \quad (-1)^{k+l}kl > 0 \\ 0 & \text{otherwise} \end{cases}$$

The genus of $Y_0(k, k)$ is zero and the genus of $X_0(k, k)$ is $n - 1$.

We will compute the gonality of these curves by relating the gonality to the degree of the curves. In 1883 Noether [**18**] proved that a non-singular plane curve of degree $d \geq 2$ in $\mathbb{P}^2$ has gonality equal to $d - 1$, where d is the degree of the curve. (See [**16**] Theorem 2.3.1, page 73.) Certain well-behaved singularities are known to modify this relationship in an understood manner. The curves determined in [**14**] are smooth, but lie in $\mathbb{P}^1 \times \mathbb{P}^1$; the birational mapping to $\mathbb{P}^2$ introduces singularities in the curves. We now prove a $\mathbb{P}^1 \times \mathbb{P}^1$ version of Noether's theorem by understanding this mapping.

LEMMA 7.1. *Let C be a smooth irreducible curve in $\mathbb{P}^1 \times \mathbb{P}^1$ of bidegree (M, N) where $MN \neq 0$, then $\gamma(C) = \min\{M, N\}$.*

(If $M = 0$ or $N = 0$ then C is isomorphic to $\mathbb{P}^1$ and therefore has gonality one.)

PROOF. Consider an irreducible curve $C \subset \mathbb{P}^2$ of degree d. Let ν be the maximum multiplicity of a singular point of C. The projection map from the point with multiplicity ν to a line has degree $d - \nu$, so we conclude that $\gamma(C) \leq d - \nu$. Ohkouchi and Sakai [**19**] (Theorem 3) show that if C has at most two singular points (or infinitely near singular points), then $\gamma(C) = d - \nu$. (See [**15**] page 194 for a discussion of infinitely near points.)

We consider the birational mapping $\phi : \mathbb{P}^1 \times \mathbb{P}^1 \to \mathbb{P}^2$ given by

$$([x : w], [y : z]) \mapsto [wy : zx : xy].$$

This has birational inverse given by

$$\phi^{-1}([a : b : c]) = ([c : a], [c : b]).$$

Let D be a smooth curve of bidegree (M, N) in $\mathbb{P}^1 \times \mathbb{P}^1$ with $M \geq N$. It suffices to show that $\nu = M$, $deg(\phi(D)) = M + N$, and $\phi(D)$ has exactly two singular or infinitely near singular points; for then Ohkouchi and Sakai's result implies that $\gamma(\phi(D)) = d - \nu = N$. Since ϕ is birational, $\gamma(D) = \gamma(\phi(D))$ proving the result.

If D is a smooth curve of bidegree (M, N) then D is the vanishing set of a polynomial $f(x, y)$ after projectivizing in $\mathbb{P}^1 \times \mathbb{P}^1$. If $f(x, y) = \sum c_{n,m} x^n y^m$ the projectivization of f is

$$F(x, w, y, z) = \sum c_{n,m} x^n w^{N-n} y^m z^{M-m}.$$

When $x = 0$ (therefore $w = 1$) this reduces to $\sum c_{0,M} y^M z^{M-m}$. We conclude that there are M points on D such that $x = 0$. Similarly there are N points corresponding to $y = 0$ (and $z = 1$). When the polynomial has a constant term, then these points do not coincide. If f has no constant term, then $([0 : 1], [0 : 1]) \in D$. We will assume, by taking an isomorphic copy of D, if necessary, that $([0 : 1], [0 : 1]) \notin D$, so f has a constant term.

Under the birational mapping, $\phi([0:1],[y:z]) = [y:0:0] = [1:0:0]$ (if $y \neq 0$). Likewise, $\phi([x:w],[0:1]) = [0:x:0] = [0:1:0]$ (if $x \neq 0$). It follows that the N points above get identified, as do the M points. The image $C = \phi(D)$ in $\mathbb{P}^2$ has a singularity of order M and one of order N. It is straightforward to see that if $xy \neq 0$ then a neighborhood of the point $([x:w],[y:z])$ maps smoothly into $\mathbb{P}^2$. Therefore $\phi(D)$ has exactly two singular points, one of degree M and one of degree N.

The degree of C is given by the intersection number CH where H' is a hyperplane section of $\mathbb{P}^2$. Similarly, the degree of D is the intersection number DH where H is a hyperplane section of $\mathbb{P}^1 \times \mathbb{P}^1$. The divisor class group of $\mathbb{P}^1 \times \mathbb{P}^1$ is $Cl(\mathbb{P}^1 \times \mathbb{P}^1) = \mathbb{Z} \oplus \mathbb{Z}$. Let $E = \mathbb{P}^1 \times \{x\}$ and $F = \{x\} \times \mathbb{P}^1$. Then D is linearly equivalent to $ME+NF$, that is $D \sim ME+NF$. (Two divisors are linear equivalent if their difference is a principal divisor.) The curves of bidegree (M,N) correspond to the effective divisors of the class $ME+NF$. Similarly, $H \sim E+F$.

Therefore, since the degree of a divisor depends only on the linear equivalence class, $deg(D) = deg(ME+NF)$. Since the degree of D is the intersection number DH, and $D \sim ME+NF$ and $H \sim E+F$ it follows that the degree of D is the intersection number of $ME+NF$ and $E+F$, which is $M+N$. □

We now use this to determine the gonality of the curves $Y_0(k,l)$.

THEOREM 7.2. *Let $J(k,l)$ be a hyperbolic knot and let $m = \lfloor \frac{1}{2}|k| \rfloor$ and $n = \lfloor \frac{1}{2}|l| \rfloor$.*

If $k \neq l$ then $\gamma(Y_0(k,l)) = \min\{m,n\}$.
If $k = l$ then $\gamma(Y_0(k,l)) = 1$.

PROOF. The equations $(*)$ defining $Y_0(k,l)$ are irreducible and determine smooth curves in $\mathbb{P}^1 \times \mathbb{P}^1$ by [**14**]. These equations have bidegree (m,n) if $k \neq l$. By Lemma 7.1 if $nm \neq 0$ the gonality of $Y_0(k,l)$ is $\min\{m,n\}$. If $m,n = 0$ then $k = 0, \pm 1$ or $l = 0, \pm 1$ and $J(k,l)$ is not hyperbolic. It suffices to consider the case when $k = l = 2n'$. In this case, the canonical component for the $\mathrm{PSL}_2(\mathbb{C})$ character variety is isomorphic to $\mathbb{P}^1 \times \mathbb{P}^1$, given by $r = t$. We conclude that the gonality is one.

□

COROLLARY 7.3. *For all positive integers n there are infinitely many double twist knots K such that*

$$\gamma(Y_0(S^3 - K)) = n.$$

Now we compute the gonality of the canonical components of the $\mathrm{SL}_2(\mathbb{C})$ character varieties. By Lemma 2.5 since $X_0(k,l)$ is a degree 2 branched cover of $Y_0(k,l)$ we conclude that

$$\min\{m,n\} \leq \gamma(X_0(k,l)) \leq 2\min\{m,n\}.$$

We now explicitly show that the gonality equals this upper bound.

THEOREM 7.4. *Let $J(k,l)$ be a hyperbolic knot and let $m = \lfloor \frac{1}{2}|k| \rfloor$ and $n = \lfloor \frac{1}{2}|l| \rfloor$.*

If $k \neq l$ then $\gamma(X_0(k,l)) = 2\min\{m,n\}$.
If $k = l$ then $\gamma(X_0(k,l)) = 2$.

PROOF. We first consider the case when $k = l = 2n'$. By [**14**], a 'natural model' for $X_0(2n', 2n') \subset \mathbb{C}^2(x, r)$ is given by

$$x^2 = \frac{(r+2)f_{n'}(r)^2 - 1}{f_{n'}(r)^2}$$

and is birational to the hyperelliptic curve

$$w^2 = (r+2)f_{n'}(r)^2 - 1.$$

This has genus $n' - 1$ by [**1**]. If $J(k, k)$ is hyperbolic then $|n'| > 2$ and the gonality of $X_0(k, k)$ is two.

Now, assume that $k \neq l$ and $J(k, l)$ is hyperbolic. If m or n is one, then one can verify directly from the defining equations that $X_0(k, l)$ is an elliptic or hyperelliptic curve and therefore has gonality two. We will now assume that m and n are at least two.

Consider the following situation. Let X and Y be smooth curves of genus g_X and g_Y and let $f : X \to Y$ be a degree k simple morphism and $h : X \to \mathbb{P}^1$ a degree d morphism. (A simple morphism is one that does not factor through a nontrivial morphism.) The Castelnuovo-Severi inequality (see [**4**] Theorem 2.1) states that if

$$(*) \qquad d \leq \frac{g_X - kg_Y}{k - 1}$$

then h factors through f.

Let X be a smooth projective model for $X_0(k, l)$ and Y a smooth projective model for $Y_0(k, l)$. Since there is a degree two birational map from $X_0(k, l)$ to $Y_0(k, l)$, by Stein Factorization, there is a degree two morphism from X to Y.

From [**14**], we have a degree $k = 2$ mapping, with $g_X = 3mn - m - n - b$ (for an explicitly determined $b \in \{0, \pm 1\}$) and $g_Y = (m-1)(n-1)$. Therefore, the righthand side of the inequality is

$$g_X - 2g_Y = (3mn - m - n - b) - 2(m-1)(n-1) = mn + m + n - b - 2.$$

Since $b \in \{0, \pm 1\}$ the right-hand side of $(*)$ is at most $mn+m+n-1$. We conclude that any dense map from $X_0(k, l)$ to $\mathbb{P}^1$ of degree at most $mn + m + n - 1$ factors through a map from $Y_0(k, l)$ to $\mathbb{P}^1$. Let d be the minimal degree of a map from $X_0(k, l)$ to $\mathbb{P}^1$. Since the gonality of $Y_0(k, l)$ is $\min\{m, n\}$ we conclude that

$$\min\{m, n\} \leq d \leq 2\min\{m, n\}.$$

Since m and n are at least two,

$$d \leq 2\min\{m, n\} \leq mn + m + n - 1.$$

By the Castelnuovo-Severi inequality, the map factors through $Y_0(k, l)$. As the gonality of $Y_0(k, l)$ is $\min\{m, n\}$ by Lemma 7.2 we conclude that the smallest degree of a map from $X_0(k, l) \to \mathbb{P}^1$ is $2\min\{m, n\}$.

□

REMARK 7.5. The Brill-Noether theorem gives the bound $\gamma \leq \lfloor \frac{g+3}{2} \rfloor$ where g is the genus. For $Y_0(k, l)$ this inequality is

$$\min\{m, n\} = \gamma \leq \lfloor \tfrac{1}{2}(mn - m - n + 2) \rfloor$$

so the gonality is smaller than the Brill-Noether bound. For $k = l$ the genus of $Y_0(k, l)$ is zero and the gonality is one, meeting the Brill-Noether bound. The examples with $k \neq l$ show that for any N there is a double twist knot (a twist knot, in fact) such that $\frac{1}{2}(g+3)-\gamma > N$. That is, the discrepancy between the gonality and the Brill-Noether bound is unbounded in this family.

REMARK 7.6. Hoste and Shanahan have computed a recursive formula for the A-polynomials of the twist knots [**11**], the knots $J(\pm 2, n)$ and shown that these polynomials are irreducible. Their results show that the degree of the A-polynomials grow linearly in n. This shows that the linear bounds in Theorem 1.4 are sharp, since the twist knots are $-1/n$ surgery on one component of the Whitehead link.

7.3. Double Twist Links. Similar to the double twist knots, the double twist links are a family of two component links which are all two-bridge links. We will use the notation as above for the character varieties of these sets where k and l are the number of half twists in the boxes indicated below. These are hyperbolic when $|k|$ and $|l|$ are greater than one.

FIGURE 5. The two component link $J(k,l)$ is the result of $-1/k$ and $-1/l$ surgery on the four component link pictured on the left.

The $\mathrm{SL}_2(\mathbb{C})$ character varieties of these links were studied in [**20**], where natural and smooth models were determined. All irreducible components containing characters of irreducible representations are surfaces of the form $C\times\mathbb{C}$ where C is a curve. The genus of the curves C were determined. The degree of irrationality of a surface of the form $C\times\mathbb{C}$ is equal the gonality of C (see [**29**] (Prop 1) and [**27**]), and using the techniques established above, this was computed as well.

THEOREM 2 ([**20**] Theorems 1.3 and 1.4). *Let k and l be odd. If $k\neq l$ then the algebraic set $X_0(k,l)$ is birational to $C_0(k,l)\times\mathbb{C}$ where $C_0(k,l)$ is a curve of genus*

$$(\lfloor\tfrac{|k|}{2}\rfloor-1)(\lfloor\tfrac{|l|}{2}\rfloor-1)$$

and degree of irrationality

$$\min\{\lfloor\tfrac{|k|}{2}\rfloor,\lfloor\tfrac{|l|}{2}\rfloor\}.$$

If $k=l$ and $|l|\geq 3$ then $X_0(l,l)$ has the form $C_0(l,l)\times\mathbb{C}$ where $C_0(l,l)$ is a curve of genus zero and degree of irrationality one.

This produces an example where filling two cusps of a four cusped link produces a family of (two-cusped) manifolds where the degree of irrationality grows without bound.

8. Gonality, injectivity radius, and eigenvalues of the Laplacian

Since the smooth projective model of a (possibly singular) affine curve is a compact Riemann surface, it can also be viewed as a quotient of the Poincare disk by a discrete group of isometries of the Poincare disk, and as such it is natural to try to relate gonality to geometric and analytical properties associated to the hyperbolic metric; for example the injectivity radius, ϱ (i.e. half of the length of a shortest closed geodesic), or the first non-zero eigenvalue of the Laplacian, λ_1.

To that end, a famous result of Li and Yau [**13**] shows that for a compact Riemann surface of genus g, a bound for the gonality γ is given by:

$$\lambda_1(g-1) \leq \gamma$$

where γ is the gonality defined using regular maps. Since the (rational) gonality of a complex affine algebraic curve equals the (regular) gonality of a smooth projective completion, this bound applies to smooth projective models of character varieties. A Riemann surface of genus zero is rational and so has gonality one, and a Riemann surface of genus one is an elliptic curve and so has gonality two. For any higher genus, the above bound implies that

$$\lambda_1 \leq \frac{\gamma}{g-1}.$$

By Theorem 1.1 the (rational) gonality of $X_0(M(-,r))$ is bounded independent of r, and therefore so is the (regular) gonality, γ, of a smooth projective model. We conclude that λ_1 is bounded above independent of r for these curves. Moreover, if $\{r\}$ is sequence of rational numbers such that the genus of the character varieties grows, then since γ is bounded

$$\lambda_1 \leq \frac{\gamma}{g-1} \to 0.$$

For hyperbolic double twist knots the genera of $Y_0(k,l)$ and $X_0(k,l)$ was determined in [**14**]. (See § 7.2.) The genus of $X_0(k,l)$ is less than two only for the complements of the knots $J(2,-2)$ (the figure-eight) and $J(4,4)$; both have genus 1. The only hyperbolic double twist knots such that the genus of $Y_0(k,l)$ is less than two are $J(k,k)$, $J(2,-2)$, $J(2,-3)$ which have genus zero, and $J(4,\pm 5)$ which have genus one. When $k \neq l$, we obtain the following inequalities using Theorem 1

$$\lambda_1(Y_0(k,l)) \leq \frac{\min\{m,n\}}{(m-1)(n-1)}, \qquad \lambda_1(X_0(k,l)) \leq \frac{2\min\{m,n\}}{3mn-m-n-b}$$

where $X_0(k,l)$ and $Y_0(k,l)$ denote smooth projective models and $n = \lfloor \frac{1}{2}|k| \rfloor$ and $m = \lfloor \frac{1}{2}|l| \rfloor$. Both quantities $\to 0$ as m or $n \to \infty$. For the once-punctured torus bundles of tunnel number one, (see § 7.1) let $X_0(M_n)$ denote a smooth projective model. Then for $|n| > 4$

$$\lambda_1(X_0(M_n)) \leq \frac{2}{\lfloor \frac{1}{2}(|n-2|-2) \rfloor} \leq \frac{4}{|n-1|-3}$$

and we conclude that $\lambda_1 \to 0$ as $|n| \to \infty$.

More recently Hwang and To [**12**] show that the injectivity radius ϱ of a compact Riemann surface of genus $g \geq 2$ is bounded above by a function of gonality (defined for regular maps),

$$\varrho \leq 2\cosh^{-1}(\gamma).$$

With $X_0(M(-,r))$ denoting a smooth projective model, Theorem 1.1 implies that there is a constant c such that the gonality of these varieties is bounded above by c and therefore there is a constant d such that

$$\varrho(X_0(M(-,r))) \leq 2\cosh^{-1}(\gamma) = d.$$

It follows that the injectivity radius is bounded above for all $X_0(M(-,r))$, such that $g \neq 0, 1$.

By Theorem 7.4 and Theorem 7.2 for all hyperbolic double twist knots other than the figure-eight, if $k \neq l$

$$\varrho(X_0(k,l)) \leq 2\cosh^{-1}(2\min\{m,n\}), \quad \text{and} \quad \varrho(Y_0(k,l)) \leq 2\cosh^{-1}(\min\{m,n\}).$$

If $k = l$ then $\varrho(X_0(k,l)) \leq 2\cosh^{-1}(2)$. For all once-punctured torus bundles of tunnel number one with $|n| > 6$, $X_0(M_n)$ is a hyper elliptic curve. We conclude that for the corresponding smooth projective models the injectivity radius is bounded by $2\cosh^{-1}(2)$.

Motivated by the discussion here, we close with two questions. To state these define an infinite family of compact Riemann surfaces $\{\Sigma_j\}$ to be *an expander family* if the genera of $\Sigma_j \to \infty$ and there is a constant $C > 0$ for which $\lambda_1(\Sigma_j) > C$. One such family are congruence arithmetic surfaces.

QUESTION. (1) Does there exist an infinite family of 1-cusped hyperbolic 3-manifolds M_j for which $X_0(M_j)$ (resp. $Y_0(M_j)$) is an expander family?

(2) Does there exist an infinite family of 1-cusped hyperbolic 3-manifolds M_j for which $X_0(M_j)$ (resp. $Y_0(M_j)$) has arbitrarily large injectivity radius?

Acknowledgments

The first author would like to thank the University of Texas at Austin and the Max Planck Institut für Mathematik for their hospitality while working on this manuscript, and Martin Hils for helpful discussions. We would also like to thank two referees for helpful comments on a previous version of the paper that helped clarify some of the arguments.

References

[1] K. L. Baker and K. L. Petersen, *Character varieties of once-punctured torus bundles with tunnel number one*, Internat. J. Math. **24** (2013), no. 6, 1350048, 57, DOI 10.1142/S0129167X13500481. MR3078072

[2] R. Benedetti and C. Petronio, *Lectures on hyperbolic geometry*, Universitext, Springer-Verlag, Berlin, 1992. MR1219310

[3] S. Boyer and X. Zhang, *On Culler-Shalen seminorms and Dehn filling*, Ann. of Math. (2) **148** (1998), no. 3, 737–801, DOI 10.2307/121031. MR1670053

[4] Y. Choi and S. Kim, *Gonality and Clifford index of projective curves on ruled surfaces*, Proc. Amer. Math. Soc. **140** (2012), no. 2, 393–402, DOI 10.1090/S0002-9939-2011-10905-5. MR2846309

[5] M. Culler and P. B. Shalen, *Bounded, separating, incompressible surfaces in knot manifolds*, Invent. Math. **75** (1984), no. 3, 537–545, DOI 10.1007/BF01388642. MR735339

[6] M. Culler, *Lifting representations to covering groups*, Adv. in Math. **59** (1986), no. 1, 64–70, DOI 10.1016/0001-8708(86)90037-X. MR825087

[7] N. M. Dunfield, *Cyclic surgery, degrees of maps of character curves, and volume rigidity for hyperbolic manifolds*, Invent. Math. **136** (1999), no. 3, 623–657, DOI 10.1007/s002220050321. MR1695208

[8] J. S. Ellenberg, C. Hall, and E. Kowalski, *Expander graphs, gonality, and variation of Galois representations*, Duke Math. J. **161** (2012), no. 7, 1233–1275, DOI 10.1215/00127094-1593272. MR2922374

[9] R. Hartshorne, *Algebraic geometry*, Springer-Verlag, New York-Heidelberg, 1977. Graduate Texts in Mathematics, No. 52. MR0463157

[10] H. Hironaka, *Resolution of singularities of an algebraic variety over a field of characteristic zero. I, II*, Ann. of Math. (2) **79** (1964), 109–203; ibid. (2) **79** (1964), 205–326, DOI 10.2307/1970547. MR0199184

[11] J. Hoste and P. D. Shanahan, *A formula for the A-polynomial of twist knots*, J. Knot Theory Ramifications **13** (2004), no. 2, 193–209, DOI 10.1142/S0218216504003081. MR2047468

[12] J.-M. Hwang and W.-K. To, *Injectivity radius and gonality of a compact Riemann surface*, Amer. J. Math. **134** (2012), no. 1, 259–283, DOI 10.1353/ajm.2012.0007. MR2876146

[13] P. Li and S. T. Yau, *A new conformal invariant and its applications to the Willmore conjecture and the first eigenvalue of compact surfaces*, Invent. Math. **69** (1982), no. 2, 269–291, DOI 10.1007/BF01399507. MR674407

[14] M. L. Macasieb, K. L. Petersen, and R. M. van Luijk, *On character varieties of two-bridge knot groups*, Proc. Lond. Math. Soc. (3) **103** (2011), no. 3, 473–507, DOI 10.1112/plms/pdr003. MR2827003

[15] M. Miyanishi, *Algebraic geometry*, Translations of Mathematical Monographs, vol. 136, American Mathematical Society, Providence, RI, 1994. Translated from the 1990 Japanese original by the author. MR1284715

[16] M. Namba, *Families of meromorphic functions on compact Riemann surfaces*, Lecture Notes in Mathematics, vol. 767, Springer, Berlin, 1979. MR555241

[17] W. D. Neumann and A. W. Reid, *Rigidity of cusps in deformations of hyperbolic 3-orbifolds*, Math. Ann. **295** (1993), no. 2, 223–237, DOI 10.1007/BF01444885. MR1202390

[18] M. Noether, *Zur Grundlegung der Theorie der algebraischen Raumcurven* (German), J. Reine Angew. Math. **93** (1882), 271–318, DOI 10.1515/crll.1882.93.271. MR1579926

[19] M. Ohkouchi and F. Sakai, *The gonality of singular plane curves*, Tokyo J. Math. **27** (2004), no. 1, 137–147, DOI 10.3836/tjm/1244208480. MR2060080

[20] K. L. Petersen and A. T. Tran, *Character varieties of double twist links*, Algebr. Geom. Topol. **15** (2015), no. 6, 3569–3598, DOI 10.2140/agt.2015.15.3569. MR3450771

[21] B. Poonen, *Gonality of modular curves in characteristic p*, Math. Res. Lett. **14** (2007), no. 4, 691–701, DOI 10.4310/MRL.2007.v14.n4.a14. MR2335995

[22] J. Porti, *Torsion de Reidemeister pour les variétés hyperboliques* (French, with English and French summaries), Mem. Amer. Math. Soc. **128** (1997), no. 612, x+139, DOI 10.1090/memo/0612. MR1396960

[23] I. R. Shafarevich, *Basic algebraic geometry. 1*, 2nd ed., Springer-Verlag, Berlin, 1994. Varieties in projective space; Translated from the 1988 Russian edition and with notes by Miles Reid. MR1328833

[24] P. B. Shalen, *Representations of 3-manifold groups*, Handbook of geometric topology, North-Holland, Amsterdam, 2002, pp. 955–1044. MR1886685

[25] W.P. Thurston, *The geometry and topology of 3-manifolds*, Princeton University, Mathematics Department, 1978, available at `http://msri.org/publications/books/gt3m/`.

[26] S. Tillmann, *Boundary slopes and the logarithmic limit set*, Topology **44** (2005), no. 1, 203–216, DOI 10.1016/j.top.2004.07.002. MR2104008

[27] H.-o Tokunaga and H. Yoshihara, *Degree of irrationality of abelian surfaces*, J. Algebra **174** (1995), no. 3, 1111–1121, DOI 10.1006/jabr.1995.1170. MR1337188

[28] J. Włodarczyk, *Toroidal varieties and the weak factorization theorem*, Invent. Math. **154** (2003), no. 2, 223–331, DOI 10.1007/s00222-003-0305-8. MR2013783

[29] H. Yoshihara, *Degree of irrationality of a product of two elliptic curves*, Proc. Amer. Math. Soc. **124** (1996), no. 5, 1371–1375, DOI 10.1090/S0002-9939-96-03375-8. MR1327053

Department of Mathematics, Florida State University, 208 Love Building, 1017 Academic Way, Tallahassee, Florida 32306-4510
Email address: petersen@math.fsu.edu

Department of Mathematics, Rice University, 6100 Main Street, Houston, Texas 77005
Email address: alan.reid@rice.edu

Contemporary Mathematics
Volume **760**, 2020
https://doi.org/10.1090/conm/760/15296

Tropical varieties associated to ideal triangulations: The Whitehead link complement

Stephan Tillmann

ABSTRACT. This paper illustrates a computational approach to methods of Culler, Morgan and Shalen that uses ideal triangulations, spun-normal surfaces and tropical geometry. Certain affine algebraic sets associated to the Whitehead link complement as well as their logarithmic limit sets are computed. The projective solution space of spun-normal surface theory is related to the space of incompressible surfaces and to the unit ball of the Thurston norm. It is shown that all boundary curves of the Whitehead link complement are strongly detected by its character variety. The specific results obtained can be used to study the geometry and topology of the Whitehead link complement and its Dehn surgeries. The methods can be applied to any cusped hyperbolic 3-manifold of finite volume.

1. Introduction

The study of the topology and the geometry of 3-manifolds via affine algebraic sets, which are related to representations of the fundamental group into $\mathrm{SL}_2(\mathbb{C})$ and $\mathrm{PSL}_2(\mathbb{C})$, has led to many deep results about 3-manifolds including the proofs of the Smith conjecture [**27**], the cyclic surgery theorem [**12**] and the finite surgery theorem [**4**]. At the heart is a construction due to Culler and Shalen [**13**], which associates essential surfaces in the manifold to the finitely many ideal points of a curve in the character variety via actions on trees. Work of Morgan and Shalen [**28**] associates, more generally, transversely measured codimension-one laminations in the manifold to the ideal points of a component of the character variety.

For any ideal triangulation of a non-compact, orientable, topologically finite 3-manifold there is a so-called deformation variety [**39**], which is equipped with an algebraic map into (but possibly not onto) the character variety. The deformation variety is a version of Thurston's parameter space [**33**]; both depend on the choice of ideal triangulation and are therefore not unique (not even up to birational equivalence). The deformation variety is compactified in [**39**] using the *logarithmic limit set* due to Bergman [**1**], and it is shown that each point in the logarithmic limit set is associated with a transversely measured singular codimension-one foliation of M, formalising a relationship originally observed by Thurston. The logarithmic limit set of the deformation variety can be computed with the software package `gfan`, which resulted from work of Bogart, Jensen, Speyer, Sturmfels and Thomas [**3**].

2010 *Mathematics Subject Classification.* Primary 57M25, 57N10; Secondary 14A25.

The ramification of this is an algorithmic approach to the character variety techniques of Culler, Morgan and Shalen in general, and to computing boundary curves strongly detected by the character variety in particular.

This paper illustrates this algorithmic approach to the character variety techniques of Culler, Morgan and Shalen by applying it to the Whitehead link complement (with its standard triangulation). To give a complete and unified picture, all relevant varieties and the maps between them are computed for the Whitehead link complement, and the spun-normal surfaces and ideal points are analysed in detail. The resulting information recovers the computation of the space of incompressible surfaces by Floyd and Hatcher [**15**], as well as the boundary curve space computed by Lash [**25**] and Hoste and Shanahan [**20**]. It is shown that all non-compact spun-normal surfaces are essential and dual to ideal points of the character variety. However, it is also shown that there are *fake ideal points* of the deformation variety: there is a trivial, closed normal surface associated to an ideal point of the deformation variety of the Whitehead link complement.

The contents of this paper is as follows:

Section 2 (Character varieties, ideal points and essential surfaces). This section summarises background material pertaining to various varieties associated to representations of 3-manifold groups into $\mathrm{SL}_2(\mathbb{C})$ and $\mathrm{PSL}_2(\mathbb{C})$. It serves as a quick trip through Culler-Shalen theory and discusses Bergman's logarithmic limit set [**1**] and the eigenvalue variety [**37**]. The eigenvalue variety is a multi–cusped analogue of the A–polynomial of Cooper, Culler, Gillet, Long and Shalen [**9**].

Section 3 (The deformation variety and normal surfaces). This section describes the relationships between the deformation variety, the character and the eigenvalue varieties exhibited in [**39**]. The required elements of normal surface theory of an ideally triangulated 3-manifold are outlined, and the projective admissible solution space is defined and related to the logarithmic limit set of the deformation variety.

None of the material up to this point is new; it is mostly collated from [**1, 32, 37–39**] and included for convenience of the reader.

Section 4 (The Whitehead link). The ideal triangulation $\mathcal{T}_\mathcal{W}$ of the complement $\mathcal{W}$ of the Whitehead link that is used for all computations in this paper is described, and differences between the right-handed and the left-handed Whitehead link, as well as different choices of meridians and longitudes are discussed.

Section 5 (Deformation and Eigenvalue varieties). The deformation variety $\mathfrak{D}(\mathcal{T}_\mathcal{W})$ and the Dehn surgery component $\overline{\mathfrak{E}}_0(\mathcal{W})$ of the $\mathrm{PSL}_2(\mathbb{C})$–eigenvalue variety are computed. It turns out that $\mathfrak{D}(\mathcal{T}_\mathcal{W})$ and $\overline{\mathfrak{E}}_0(\mathcal{W})$ are birationally equivalent; this is shown via a variety $\overline{\mathfrak{C}}(\mathcal{W})$, which parameterises certain representations of $\pi_1(\mathcal{W})$ into $\mathrm{PSL}_2(\mathbb{C})$ and which can be viewed as a de-singularisation of $\mathfrak{D}(\mathcal{T}_\mathcal{W})$. These computations were used by Indurskis [**21**] to determine Culler-Shalen seminorms. The $\mathrm{SL}_2(\mathbb{C})$– and $\mathrm{PSL}_2(\mathbb{C})$–character varieties of $\mathcal{W}$, and in particular their respective Dehn surgery components $\mathfrak{X}_0(\mathcal{W})$ and $\overline{\mathfrak{X}}_0(\mathcal{W})$, are computed in Appendix C.

All maps constructed are summarised in below commutative diagram. The horizontal maps are birational isomorphisms, and all other maps are generically

4–to–1 and correspond to actions of the Klein four group. The bar in the notation indicates that a variety is associated to representations into $\mathrm{PSL}_2(\mathbb{C})$.

$$\begin{array}{ccccc}
 & & \mathfrak{C}_0(\mathcal{W}) & \xrightarrow{\ \mathrm{e}\ } & \mathfrak{E}_0(\mathcal{W}) \\
 & & \downarrow{\scriptstyle q_1} & & \downarrow{\scriptstyle q_3} \\
\mathfrak{D}(\mathcal{T}_{\mathcal{W}}) & \xrightarrow{\ r\ } & \overline{\mathfrak{C}}_0(\mathcal{W}) & \xrightarrow{\ \overline{\mathrm{e}}\ } & \overline{\mathfrak{E}}_0(\mathcal{W}) \\
 & \searrow{\scriptstyle \chi} & \downarrow & & \\
 & & \overline{\mathfrak{X}}_0(\mathcal{W}) & &
\end{array}$$

Section 6 (Embedded surfaces). The projective admissible solution space $\mathcal{PF}(\mathcal{T}_{\mathcal{W}})$ of spun-normal surface theory is computed, and a criterion due to Dunfield is used to determine which spun-normal surfaces are essential. The unit ball of the Thurston norm is derived using this information. Results by Walsh [**41**] and Kang and Rubinstein [**24**] on the normalisation of essential surfaces are used to determine the space $\mathcal{PL}(\mathcal{W})$ of all essential surfaces and the boundary curve space $\mathfrak{BC}(\mathcal{W})$. This computation is compared with work of Floyd and Hatcher [**15**], Lash [**25**] and Hoste and Shanahan [**20**].

Section 7 (Logarithmic limit sets). The logarithmic limit set $\mathfrak{D}_\infty(\mathcal{T}_{\mathcal{W}})$ of the deformation variety $\mathfrak{D}(\mathcal{T}_{\mathcal{W}})$ is homeomorphic to a closed subset of $\mathcal{PF}(\mathcal{T}_{\mathcal{W}})$ according to [**39**]. Both $\mathfrak{D}_\infty(\mathcal{T}_{\mathcal{W}})$ and the logarithmic limit set of the eigenvalue variety $\overline{\mathfrak{E}}_0(\mathcal{W})$ are determined. The computation in particular shows that all strict boundary curves of $\mathcal{W}$ are detected by the Dehn surgery component of the character variety. Since the boundary curves of fibres are detected by reducible characters, this shows that all boundary slopes of the Whitehead link are strongly detected. In addition, it is shown that $\mathfrak{D}_\infty(\mathcal{T}_{\mathcal{W}})$ has a fake ideal point.

2. Character varieties, ideal points and essential surfaces

This section summarises background material pertaining to various varieties associated to representations of 3-manifold groups into $\mathrm{SL}_2(\mathbb{C})$ and $\mathrm{PSL}_2(\mathbb{C})$. The first four subsections outline the bread and butter of Culler-Shalen theory; the reader is referred to Shalen [**32**] for details. This is followed up with a description of Bergman's logarithmic limit set [**1**] in §2.5, and the eigenvalue varieties [**37**] in §§2.6–2.7.

2.1. Character variety. Let Γ be a finitely generated group. The *representation variety* of Γ is the set $\mathfrak{R}(\Gamma) = \mathrm{Hom}(\Gamma, \mathrm{SL}_2(\mathbb{C}))$, which is regarded as an affine algebraic set (see [**13**]). Two representations are *equivalent* if they differ by an inner automorphism of $\mathrm{SL}_2(\mathbb{C})$. A representation is *irreducible* if the only subspaces of $\mathbb{C}^2$ invariant under its image are trivial. This is equivalent to saying that the representation cannot be conjugated to a representation by upper triangular matrices. Otherwise a representation is *reducible.*

Each $\gamma \in \Gamma$ defines a *trace function* $I_\gamma : \mathfrak{R}(\Gamma) \to \mathbb{C}$ by $I_\gamma(\rho) = \operatorname{tr}\rho(\gamma)$, which is an element of the coordinate ring $\mathbb{C}[\mathfrak{R}(\Gamma)]$. For each $\rho \in \mathfrak{R}(\Gamma)$, its *character* is the function $\chi_\rho : \Gamma \to \mathbb{C}$ defined by $\chi_\rho(\gamma) = \operatorname{tr}\rho(\gamma)$. Irreducible representations are determined by characters up to equivalence, and the reducible representations form a closed subset $\mathfrak{Red}(\Gamma)$ of $\mathfrak{R}(\Gamma)$. The collection of characters $\mathfrak{X}(\Gamma)$ can be regarded as an affine algebraic set, which is called the *character variety.* There is a regular map $\mathrm{t} : \mathfrak{R}(\Gamma) \to \mathfrak{X}(\Gamma)$ taking representations to characters. If Γ is the fundamental group of a topological space M, then $\mathfrak{R}(M)$ and $\mathfrak{X}(M)$ denote $\mathfrak{R}(\Gamma)$ and $\mathfrak{X}(\Gamma)$ respectively.

There also is a character variety arising from representations into $\mathrm{PSL}_2(\mathbb{C})$, see Boyer and Zhang [**5**], and the relevant objects are denoted by placing a bar over the previous notation. The natural map $\mathrm{q} : \mathfrak{X}(\Gamma) \to \overline{\mathfrak{X}}(\Gamma)$ is finite–to–one, but in general not onto. As with the $\mathrm{SL}_2(\mathbb{C})$–character variety, there is a surjection $\overline{\mathrm{t}} : \overline{\mathfrak{R}}(\Gamma) \to \overline{\mathfrak{X}}(\Gamma)$, which is constant on conjugacy classes, and with the property that if $\overline{\rho}$ is an irreducible representation, then $\overline{\mathrm{t}}^{-1}(\overline{\mathrm{t}}(\overline{\rho}))$ is the orbit of $\overline{\rho}$ under conjugation.

2.2. Dehn surgery component. Suppose M is the interior of an orientable, compact, connected 3-manifold $\overline{M}$ with non-empty boundary consisting of a pairwise disjoint union of tori. If M admits a complete hyperbolic structure of finite volume, then there is a discrete and faithful representation $\pi_1(M) \to \mathrm{PSL}_2(\mathbb{C})$. This representation is necessarily irreducible, as hyperbolic geometry otherwise implies that M has infinite volume. The following result (see [**32, 33**]) will be relied on:

THEOREM 1 (Thurston). *Let M be a complete hyperbolic 3-manifold of finite volume with h cusps, and let $\overline{\rho}_0 : \pi_1(M) \to PSL_2(\mathbb{C})$ be a discrete and faithful representation associated to the complete hyperbolic structure. Then $\overline{\rho}_0$ admits a lift ρ_0 into $SL_2(\mathbb{C})$ which is still discrete and faithful. The (unique) irreducible component $\mathfrak{X}_0$ in the $SL_2(\mathbb{C})$–character variety containing the character χ_0 of ρ_0 has (complex) dimension h.*

Furthermore, if $T_1, \ldots, T_h$ are the boundary tori of a compact core of M, and if γ_i is a non–trivial element in $\pi_1(M)$ which is carried by T_i, then $\chi_0(\gamma_i) = \pm 2$ and χ_0 is an isolated point of the set

$$\mathfrak{X}^* = \{\chi \in \mathfrak{X}_0 \mid I_{\gamma_1}^2 = \ldots = I_{\gamma_h}^2 = 4\}.$$

The respective irreducible components containing the so–called *complete representations* ρ_0 and $\overline{\rho}_0$ are denoted by $\mathfrak{R}_0(M)$ and $\overline{\mathfrak{R}}_0(M)$ respectively. In particular, $\mathrm{t}(\mathfrak{R}_0) = \mathfrak{X}_0$ and $\overline{\mathrm{t}}(\overline{\mathfrak{R}}_0) = \overline{\mathfrak{X}}_0$ are called the respective *Dehn surgery components* of the character varieties of M, since the holonomy representations of hyperbolic manifolds or orbifolds obtained by performing high order Dehn surgeries on M are near $\overline{\rho}_0$ (see [**33**]). We remark that Dehn surgery components are generally not unique, even if one considers representations into $\mathrm{PSL}_2(\mathbb{C})$ (see [**7**] for examples). The point is that there are two conjugacy classes of discrete and faithful representations into $\mathrm{PSL}_2(\mathbb{C})$. They correspond to the two different choices of orientation of M and are related by complex conjugation.

2.3. Essential surfaces. An *(embedded) surface* S in the topologically finite 3-manifold $M = \mathrm{int}(\overline{M})$ will always mean a 2–dimensional PL submanifold of M with the property that its closure $\overline{S}$ in $\overline{M}$ is *properly embedded* in $\overline{M}$, that is, a closed subset of $\overline{M}$ with $\partial\overline{S} = \overline{S} \cap \partial\overline{M}$. A surface S in M is said to be *essential* if its closure $\overline{S}$ is essential in $\overline{M}$ as described in the following definition:

DEFINITION. [**32**] *A surface $\overline{S}$ in a compact, irreducible, orientable 3-manifold $\overline{M}$ is said to be essential if it has the following five properties:*

(1) *$\overline{S}$ is bicollared;*
(2) *the inclusion $\pi_1(\overline{S}_i) \to \pi_1(\overline{M})$ is injective for every component $\overline{S}_i$ of $\overline{S}$;*
(3) *no component of $\overline{S}$ is a 2–sphere;*
(4) *no component of $\overline{S}$ is boundary parallel;*
(5) *$\overline{S}$ is nonempty.*

The boundary curves of $\overline{S}$, $\overline{S} \cap \partial\overline{M}$, are also called the *boundary curves* of S.

2.4. Ideal points, and valuations. In case that V is a 1–dimensional irreducible variety, there is a unique non–singular projective variety $\tilde{V}$ which is birationally equivalent to $\overline{J(V)}$. $\tilde{V}$ is called the *smooth projective completion* of V, and the *ideal points* of $\tilde{V}$ are the points of $\tilde{V}$ corresponding to $\overline{J(V)} - J(V)$ under the birational equivalence. Moreover, the function fields of V and $\tilde{V}$ are isomorphic.

Let $C \subset \mathfrak{X}(M)$ be a curve, i.e. a 1–dimensional irreducible subvariety, and denote its smooth projective completion by $\tilde{C}$. We will refer to the ideal points of $\tilde{C}$ also as ideal points of C. The function fields of C and $\tilde{C}$ are isomorphic. Denote them by K. Any ideal point ξ of $\tilde{C}$ determines a (normalised, discrete, rank 1) valuation ord_ξ of K, by

$$ord_\xi(f) = \begin{cases} k & \text{if } f \text{ has a zero of order } k \text{ at } \xi \\ \infty & \text{if } f = 0 \\ -k & \text{if } f \text{ has a pole of order } k \text{ at } \xi \end{cases}$$

Note that $ord_\xi(z) = 0$ for all non-zero constant functions $z \in \mathbb{C}$. In the language of algebraic geometry, the valuation ring $\{f \in K \mid ord_\xi(f) \geq 0\}$ of ord_ξ is the local ring at ξ.

Culler and Shalen [**13, 32**] associate essential surfaces in a 3-manifold M to *ideal points of curves* in the character variety $\mathfrak{X}(M)$. The ingredients are as follows:

(1) A curve in $\mathfrak{X}(M)$ yields a field F with a discrete valuation at each ideal point and a representation $\mathcal{P} : \pi_1(M) \to \mathrm{SL}_2(F)$.
(2) The group $\mathrm{SL}_2(F)$ acts on the associated Bruhat-Tits tree $\mathfrak{T}_v$ and using $\mathcal{P}$ we can pull this back to an action of $\pi_1(M)$ on $\mathfrak{T}_v$ that is non–trivial and without inversions.
(3) A non–trivial action of $\pi_1(M)$ without inversions on a trees gives essential surfaces in M via a construction due to Stallings.

2.5. Logarithmic limit set. Let

$$\mathbb{C}[X^{\pm}] = \mathbb{C}[X_1^{\pm 1}, \ldots, X_n^{\pm 1}].$$

Given an ideal $I \subset \mathbb{C}[X^{\pm}]$, let $V = V(I)$ be the corresponding subvariety of $(\mathbb{C} - \{0\})^n$. Bergman gives the following three descriptions of a *logarithmic limit set* in [**1**]:

1. The tropical variety: Define V_{val} as the set of n–tuples

$$(-v(X_1), \ldots, -v(X_n)) \tag{1}$$

as v runs over all real–valued valuations on $\mathbb{C}[X^{\pm}]/I$ satisfying $\sum v(X_i)^2 = 1$.

2. A geometric construction: Define the support $s(f)$ of an element $f \in \mathbb{C}[X^{\pm}]$ to be the set of points $\alpha = (\alpha_1, \ldots, \alpha_n) \in \mathbb{Z}^n$ such that $X^\alpha = X_1^{\alpha_1} \cdots X_n^{\alpha_n}$ occurs with non–zero coefficient in f. Then define V_{sph} to be the set of $\xi \in S^{n-1}$ such that for all non–zero $f \in I$, the maximum value of the dot product $\xi \cdot \alpha$ as α runs over $s(f)$ is assumed at least twice. Geometrically, this is:

$$V_{\text{sph}} = \bigcap_{0 \neq f \in I} \text{Sph}(\text{Newt}(f)),$$

where Newt(f) is the Newton polytope and Sph(Newt(f)) its spherical dual. The latter consists of all outward pointing unit normal vectors to the support planes of Newt(f) which meet Newt(f) in more than one point.

Bergman shows in [**1**] that $V_{\text{val}} = V_{\text{sph}}$.

3. The logarithmic limit set: Define $V_{\log}$ as the set of limit points on S^{n-1} of the set of real n–tuples in the interior of B^n:

$$\left\{ \frac{(\log|x_1|, \ldots, \log|x_n|)}{\sqrt{1 + \sum(\log|x_i|)^2}} \mid x \in V \right\}. \tag{2}$$

Bergman shows in [**1**] that we have $V_{\log} \subset V_{\text{val}} = V_{\text{sph}}$. It follows from work by Bieri and Groves [**2**] that all sets in fact coincide:

THEOREM 2 (Bergman, Bieri–Groves). *We have* $V_{\log} = V_{\text{val}} = V_{\text{sph}}$.

Let $V_\infty = V_{\log}$ be the logarithmic limit set of the variety V. The ideal generated by a set of polynomials $\{f_i\}_i$ shall be denoted by $I(f_i)_i$, and the variety generated by the ideal I is denoted by $V(I)$ or $V(f_i)_i$, if $I = I(f_i)_i$. The following fact completely determines the tropical variety of a principal ideal:

PROPOSITION 3. *Let* $f \in \mathbb{C}[X^\pm]$. *If* $f \neq 0$, *then* $V(f)_\infty = \text{Sph}(\text{Newt}(f))$.

In general, we have the following:

THEOREM 4 (Bieri-Groves Polyhedrality). *The logarithmic limit set* V_∞ *of an algebraic variety* V *in* $(\mathbb{C} - \{0\})^n$ *is a finite union of rational convex spherical polytopes, all having the same dimension, namely*

$$(\dim_{\mathbb{C}} V) - 1.$$

The following is of interest to applications:

LEMMA 5. [**37**] *Let* V *be an algebraic variety in* $(\mathbb{C} - \{0\})^m$, *and* $\xi \in V_\infty$ *be a point with rational coordinate ratios. Then there is a curve* C, *i.e. an irreducible subvariety of complex dimension one, in* V *such that* $\xi \in C_\infty$.

2.6. Eigenvalue variety. Let M be an orientable, irreducible, compact 3-manifold with non–empty boundary consisting of a disjoint union of h tori, denoted $T_1, \ldots, T_h$.

Given a presentation of $\pi_1(M)$ with a finite number, n, of generators, $\gamma_1, \ldots, \gamma_n$, introduce four affine coordinates (representing matrix entries) for each generator, which are denoted by g_{ij} for $i = 1, \ldots, n$ and $j = 1, 2, 3, 4$. View $\mathfrak{R}(M)$ as an affine algebraic set in $\mathbb{C}^{4n}$ defined by an ideal J in $\mathbb{C}[g_{11}, \ldots, g_{n4}]$. There are elements $I_\gamma \in \mathbb{C}[g_{11}, \ldots, g_{n4}]$ for each $\gamma \in \pi_1(M)$ such that $I_\gamma(\rho) = \text{tr}\,\rho(\gamma)$ for each $\rho \in \mathfrak{R}(M)$.

Identify $\mathcal{M}_i$ and $\mathcal{L}_i$ with generators of $\text{im}(\pi_1(T_i) \to \pi_1(M))$. Thus, $\mathcal{M}_i$ and $\mathcal{L}_i$ are words in the generators for $\pi_1(M)$.

In the ring $\mathbb{C}[g_{11},\ldots,g_{n4},m_1^{\pm 1},l_1^{\pm 1},\ldots,m_h^{\pm 1},l_h^{\pm 1}]$, we define the following polynomial equations:

$$I_{\mathcal{M}_i} = m_i + m_i^{-1}, \tag{3}$$

$$I_{\mathcal{L}_i} = l_i + l_i^{-1}, \tag{4}$$

$$I_{\mathcal{M}_i\mathcal{L}_i} = m_i l_i + m_i^{-1} l_i^{-1}, \tag{5}$$

for $i = 1,\ldots,h$. Let $\mathfrak{R}_E(M)$ be the variety in $\mathbb{C}^{4n} \times (\mathbb{C} - \{0\})^{2h}$ defined by J together with the above equations. For each $\rho \in \mathfrak{R}(M)$, the equations (3)–5 have a solution since commuting elements of $\mathrm{SL}_2(\mathbb{C})$ always have a common invariant subspace. The natural projection $p_1 : \mathfrak{R}_E(M) \to \mathfrak{R}(M)$ is therefore onto, and p_1 is a dominating map.

If $(a_1,\ldots,a_{4n}) \in \mathfrak{R}(M)$, then there is an action of $\mathbb{Z}_2^h$ on the resulting points

$$(a_1,\ldots,a_{4n},m_1,l_1,\ldots,m_h,l_h) \in \mathfrak{R}_E(M)$$

by inverting both entries of a tuple (m_i,l_i) to (m_i^{-1},l_i^{-1}). The group $\mathbb{Z}_2^h$ acts transitively on the fibres of p_1. The map p_1 is therefore finite–to–one with degree $\le 2^h$. The maximal degree is in particular achieved when the interior of M admits a complete hyperbolic structure of finite volume, since Theorem 1 implies that there are points in $\mathfrak{X}_0(M) - \cup_{i=1}^h \{I_{\mathcal{M}_i}^2 = 4\}$.

The *eigenvalue variety* $\mathfrak{E}(M)$ is the closure of the image of $\mathfrak{R}_E(M)$ under projection onto the coordinates $(m_1,\ldots,l_h)$. It is therefore defined by an ideal of the ring $\mathbb{C}[m_1^{\pm 1},l_1^{\pm 1},\ldots,m_h^{\pm 1},l_h^{\pm 1}]$ in $(\mathbb{C} - \{0\})^{2h}$.

Note that this construction factors through a variety $\mathfrak{X}_E(M)$, which is the character variety with its coordinate ring appropriately extended, since coordinates of $\mathfrak{X}(M)$ can be chosen such that I_γ (as a function on $\mathfrak{X}(M)$) is an element of the coordinate ring of $\mathfrak{X}(M)$ for each $\gamma \in \pi_1(M)$. Let $\mathrm{t}_E : \mathfrak{R}_E(M) \to \mathfrak{X}_E(M)$ be the natural quotient map, which is equal to $\mathrm{t} : \mathfrak{R}(M) \to \mathfrak{X}(M)$ on the first $4n$ coordinates, and equal to the identity on the remaining $2h$ coordinates.

There also is a restriction map $r : \mathfrak{X}(M) \to \mathfrak{X}(T_1) \times \ldots \times \mathfrak{X}(T_h)$, which arises from the inclusion homomorphisms $\pi_1(T_i) \to \pi_1(M)$. Therefore denote the map $\mathfrak{X}_E(M) \to \mathfrak{E}(M)$ by r_E. Denote the closure of the image of r by $\mathfrak{X}_\partial(M)$. There is the following commuting diagram of dominating maps:

$$\begin{array}{ccccc}
\mathfrak{R}_E(M) & \xrightarrow{\ \mathrm{t}_E\ } & \mathfrak{X}_E(M) & \xrightarrow{\ r_E\ } & \mathfrak{E}(M) \\
{\scriptstyle p_1}\downarrow & & {\scriptstyle p_2}\downarrow & & {\scriptstyle p_3}\downarrow \\
\mathfrak{R}(M) & \xrightarrow{\ \mathrm{t}\ } & \mathfrak{X}(M) & \xrightarrow{\ r\ } & \mathfrak{X}_\partial(M)
\end{array}$$

Note that the maps p_1, p_2, t and t_E have the property that every point in the closure of the image has a preimage, and that the maps p_1, p_2 and p_3 are all finite–to–one of the same degree.

Recall the construction by Culler and Shalen. Starting with a curve $C \subset \mathfrak{X}(M)$ and an irreducible component R_C in $\mathfrak{R}(M)$ such that $\mathrm{t}(R_C) = C$, one obtains the tautological representation $\mathcal{P} : \pi_1(M) \to SL_2(\mathbb{C}(R_C))$. Let R'_C be an irreducible component of $\mathfrak{R}_E(M)$ with the property that $p_1(R'_C) = R_C$. Then $\mathbb{C}(R'_C)$ is a finitely generated extension of $\mathbb{C}(R_C)$, and $\mathcal{P}$ may be thought of as a representation $\mathcal{P} : \pi_1(M) \to SL_2(\mathbb{C}(R'_C))$.

If $\overline{r_E\,\mathrm{t}_E(R'_C)}$ contains a curve E, then $\mathbb{C}(R'_C)$ is a finitely generated extension of $\mathbb{C}(E)$, and essential surfaces can be associated to each ideal point of E using $\mathcal{P}$.

Since the eigenvalue of at least one peripheral element blows up at an ideal point of E, the associated surfaces necessarily have non–empty boundary (see [**9**]).

Thus, if there is a closed essential surface associated to an ideal point of C, then either $\overline{r_E\,\mathrm{t}_E(R'_C)}$ is 0–dimensional, or there is a neighbourhood U of an ideal point ξ of R'_C such that there are (finite) points in $\overline{r_E\,\mathrm{t}_E(U)} - r_E\,\mathrm{t}_E(U)$. The later are called *holes in the eigenvalue variety*, examples of which are given in [**35**].

The action of $\mathbb{Z}_2^h$ on the eigenvalue variety $\mathfrak{E}(M)$ induces an action of $\mathbb{Z}_2^h$ on its logarithmic limit set $\mathfrak{E}_\infty(M) \subset S^{2h-1}$. The action is generated by taking $\xi = (x_1,\ldots,x_{2i-1},x_{2i},\ldots,x_{2h}) \in \mathfrak{E}_\infty(M)$ to $(x_1,\ldots,-x_{2i-1},-x_{2i},\ldots,x_{2h}) \in \mathfrak{E}_\infty(M)$ for $i=1,\ldots,h$. Factoring by these symmetries, a quotient of $\mathfrak{E}_\infty(M)$ is obtained in the space $\mathbb{R}P^{2h-1}/\mathbb{Z}_2^{h-1} \cong S^{2h-1}$. The quotient map extends to a map $\varphi : S^{2h-1} \to S^{2h-1}$ of spheres, which has degree 2^h. Let

$$P = \begin{pmatrix} 0 & 1 \\ -1 & 0 \end{pmatrix}$$

and let P_h be the block diagonal matrix with h copies of P on its diagonal. Then P_h is orthogonal, and its restriction to S^{2h-1} is a map of degree one. The main result of [**37**] is:

LEMMA 6 (Boundary slopes). *Let M be an orientable, irreducible, compact 3-manifold with non–empty boundary consisting of a disjoint union of h tori. If $\xi \in \mathfrak{E}_\infty(M)$ is a point with rational coordinate ratios, then there is an essential surface with boundary in M whose projectivised boundary curve coordinate is $\varphi(P_h\xi)$. Moreover, the image $\varphi(P_h\mathfrak{E}_\infty(M))$ is a closed subset of $\mathfrak{BC}(M)$.*

2.7. $\mathbf{PSL_2}(\mathbb{C})$–eigenvalue variety. Analogous to $\mathfrak{E}(M)$, one can define a $\mathrm{PSL}_2(\mathbb{C})$–eigenvalue variety $\overline{\mathfrak{E}}(M)$, since the function $I_\gamma^2 : \overline{\mathfrak{X}}(M) \to \mathbb{C}$ defined by $I_\gamma^2(\overline{\rho}) = (\mathrm{tr}\,\overline{\rho}(\gamma))^2$ is regular (i.e. polynomial) for all $\gamma \in \pi_1(M)$ (see [**5**]). Thus, a variety $\overline{\mathfrak{R}}_E(M)$ is constructed using the relations

$$\begin{aligned} I^2_{\mathcal{M}_i} &= M_i + 2 + M_i^{-1}, \\ I^2_{\mathcal{L}_i} &= L_i + 2 + L_i^{-1}, \\ I^2_{\mathcal{M}_i\mathcal{L}_i} &= M_iL_i + 2 + M_i^{-1}L_i^{-1}, \end{aligned}$$

for $i = 1,\ldots,h$.

Consider the representation of $\mathbb{Z}\oplus\mathbb{Z}$ into $\mathrm{PSL}_2(\mathbb{C})$ generated by the images of

$$\begin{pmatrix} i & 0 \\ 0 & -i \end{pmatrix} \quad \text{and} \quad \begin{pmatrix} 0 & 1 \\ -1 & 0 \end{pmatrix}. \tag{6}$$

In $\mathrm{PSL}_2(\mathbb{C})$, the image of this representation is isomorphic to $\mathbb{Z}_2\oplus\mathbb{Z}_2$, but the image of any lift to $\mathrm{SL}_2(\mathbb{C})$ is isomorphic to the quaternion group Q_8 (in Quaternion group notation). If $\overline{\rho} \in \overline{\mathfrak{R}}(M)$ restricts to such an irreducible abelian representation on a boundary torus T_i, then the equations $M_i+2+M_i^{-1} = 0$, $L_i+2+L_i^{-1} = 0$, $M_iL_i+2+M_i^{-1}L_i^{-1} = 0$ have no solution. In particular, the projection $\overline{p} : \overline{\mathfrak{R}}_E(M) \to \overline{\mathfrak{R}}(M)$ may not be onto.

The closure of the image of $\overline{\mathfrak{R}}_E(M)$ onto the coordinates $(M_1,L_1,\ldots,M_h,L_h)$ is denoted by $\overline{\mathfrak{E}}(M)$ and called the *$PSL_2(\mathbb{C})$-eigenvalue variety*. The relationship between ideal points of $\overline{\mathfrak{E}}(M)$ and strongly detected boundary curves is the same as for the $\mathrm{SL}_2(\mathbb{C})$–version, since the proof of Lemma 6 applies without change:

LEMMA 7. *Let M be an orientable, irreducible, compact 3-manifold with non–empty boundary consisting of a disjoint union of h tori. If $\xi \in \overline{\mathfrak{E}}_\infty(M)$ is a point with rational coordinate ratios, then there is an essential surface with boundary in M whose projectivised boundary curve coordinate is $\varphi(P_h\xi)$.*

3. The deformation variety and normal surfacess

This section summarises the approach to Culler-Morgan-Shalen theory using ideal triangulations, spun-normal surfaces and tropical geometry given in [**39**]. Throughout this section, M is the interior of an orientable, compact, connected 3-manifold with non-empty boundary consisting of a pairwise disjoint union of tori.

3.1. Ideal triangulations and shape parameters. An ideal (topological) triangulation $\mathcal{T}$ of M is a triple $\mathcal{T} = (\widetilde{\Delta}, \Phi, h)$, where

(1) $\widetilde{\Delta} = \bigcup_{k=1}^n \widetilde{\Delta}_k$ is a pairwise disjoint union of oriented standard Euclidean 3–simplices,
(2) Φ is a collection of orientation reversing Euclidean isometries between the 2–simplices in $\widetilde{\Delta}$; termed *face pairings*,
(3) $h\colon (\widetilde{\Delta} \setminus \widetilde{\Delta}^{(0)})/\Phi \to M$ is a homeomorphism, where the domain is given the quotient topology.

We write $p\colon (\widetilde{\Delta} \setminus \widetilde{\Delta}^{(0)}) \to M$ for the composition of the natural quotient map with h, and note that M is determined by $\mathcal{T}$.

The associated *pseudo-manifold* (or *end-compactification* of M) is $P = \widetilde{\Delta}/\Phi$. We denote the quotient map with the same letter, $p\colon \widetilde{\Delta} \to P$. Denote Σ^k the set of all (possibly singular) k–simplices in P. The degree $\deg(\sigma^k)$ of a k–simplex in Σ^k is the number of k–simplices in its preimage in $\widetilde{\Delta}$.

An *ideal k–simplex* is a k–simplex with its vertices removed. The vertices of the k–simplex are termed the *ideal vertices* of the ideal k–simplex. Similar for singular simplices. The standard terminology of (ideal) edges, (ideal) faces and (ideal) tetrahedra will be used for the (ideal) simplices of dimensions one, two, and three respectively.

Let Δ^3 be the standard 3–simplex with a chosen orientation. Suppose the edges Δ^3 are labeled by z, z' and z'' so that opposite edges have the same labeling and all labels are used. Then the cyclic order of z, z' and z'' viewed from each vertex depends only on the orientation of the 3–simplex. It follows that, up to orientation preserving symmetries, there are two possible labelings, and we fix one of these labelings.

Suppose $\Sigma^3 = \{\sigma_1, \ldots, \sigma_n\}$. Since M is orientable, the 3–simplices in Σ^3 may be oriented coherently. For each $\sigma_i \in \Sigma^3$, fix an orientation preserving simplicial map $f_i\colon \Delta^3 \to \sigma_i$. An Euler characteristic argument shows that $|\Sigma^3| = |\Sigma^1|$. Let $\Sigma^1 = \{e_1, \ldots, e_n\}$, and let $a_{ij}^{(k)}$ be the number of edges in $f_i^{-1}(e_j)$ which have label $z^{(k)}$.

For each $i \in \{1, \ldots, n\}$, define

$$p_i = z_i(1 - z_i'') - 1, \quad p_i' = z_i'(1 - z_i) - 1, \quad p_i'' = z_i''(1 - z_i') - 1, \tag{7}$$

and for each $j \in \{1, \ldots, n\}$, let

$$g_j = \prod_{i=1}^n z_i^{a_{ij}} (z_i')^{a_{ij}'} (z_i'')^{a_{ij}''} - 1. \tag{8}$$

Setting $p_i = p_i' = p_i'' = 0$ gives the *parameter equations*, and setting $g_j = 0$ gives the *hyperbolic gluing equations*. For a discussion and geometric interpretation of these equations, see [**30, 33**]. The parameter equations imply that $z_i^{(k)} \neq 0, 1$.

DEFINITION 8. *Suppose $\mathcal{T}$ is an ideal triangulation of M. The* deformation variety $\mathfrak{D}(\mathcal{T})$ *is the variety in $(\mathbb{C} - \{0\})^{3n}$ defined by the parameter equations and the hyperbolic gluing equations.*

3.2. Developing maps, characters and eigenvalues. The following facts can be found in [**42**, §5]; see also [**31**] for a more detailed discussion.

Given $Z \in \mathfrak{D}(\mathcal{T})$, each ideal tetrahedron in M has edge labels which can be lifted equivariantly to $\widetilde{M}$. Following [**42**], we define a continuous map $\text{dev}_Z\colon \widetilde{M} \to \mathbb{H}^3$, which maps every ideal tetrahedron σ in $\widetilde{M}$ to an ideal hyperbolic 3–simplex $\Delta(\sigma)$, such that the labels carried forward to the edges of $\Delta(\sigma)$ correspond to the shape parameters of $\Delta(\sigma)$ determined by its hyperbolic structure; see [**33**] for the geometry of hyperbolic ideal tetrahedra. If each shape parameter has positive imaginary part, then the associated hyperbolic ideal tetrahedron is positively oriented and dev_Z is a developing map for a (possibly incomplete) hyperbolic structure on M.

For each $Z \in \mathfrak{D}(\mathcal{T})$, the Yoshida map dev_Z can be used to define a representation $\rho_Z\colon \pi_1(M) \to \text{PSL}_2(\mathbb{C})$ as follows (see [**42**]). A representation into $\text{PSL}_2(\mathbb{C})$ is an action on $\mathbb{H}^3$, and this is the unique representation which makes dev_Z $\pi_1(M)$–equivariant: $\text{dev}_Z(\gamma \cdot x) = \rho_Z(\gamma)\,\text{dev}_Z(x)$ for all $x \in \widetilde{M}$, $\gamma \in \pi_1(M)$. Thus, ρ_Z is well–defined up to conjugation, since it only depends upon the choice of the embedding of the initial tetrahedron σ. This yields a well–defined map $\chi_{\mathcal{T}}\colon \mathfrak{D}(\mathcal{T}) \to \overline{\mathfrak{X}}(N)$ from the deformation variety to the $\text{PSL}_2(\mathbb{C})$–character variety. It is implicit in [**29**] that $\chi_{\mathcal{T}}$ is algebraic; see [**8**] for details using a faithful representation of $\text{PSL}_2(\mathbb{C}) \to SL(3, \mathbb{C})$. Note that the image of each peripheral subgroup under ρ_Z has at least one fixed point on the sphere at infinity. We summarise this discussion in the following lemma:

LEMMA 9. *Let M be the interior of a compact, connected, orientable 3-manifold with non-empty boundary consisting of a pairwise disjoint union of tori, and let $\mathcal{T}$ be an ideal triangulation of M. For each $Z \in \mathfrak{D}(\mathcal{T})$, there exists a representation $\overline{\rho}_Z\colon \pi_1(M) \to PSL_2(\mathbb{C})$ and a $\overline{\rho}_Z$–equivariant, continuous map $D_Z\colon \widehat{P} \to \overline{\mathbb{H}}^3$ with the property that for each ideal tetrahedron $\sigma \subset \widetilde{M}$, the image $D_Z(\sigma)$ is a hyperbolic ideal tetrahedron with edge invariants determined by Z. Moreover, $\overline{\rho}_Z$ is well-defined up to conjugacy and the well-defined map*

$$\chi_{\mathcal{T}}\colon \mathfrak{D}(\mathcal{T}) \to \overline{\mathfrak{X}}(M)$$

is algebraic.

For cusped hyperbolic 3-manifolds, there are a number of special facts that are well-known to follow from work in [**30, 33**]. The following highlights the fact that the deformation variety facilitates the study of the Dehn surgery components in the character variety, i.e. the components containing the characters of discrete and faithful representations.

THEOREM 10. *Let M be an orientable, connected, cusped hyperbolic 3-manifold. Let $\mathcal{T}$ be an ideal triangulation of M with the property that all edges are essential. Then there exists $Z \in \mathfrak{D}(\mathcal{T})$, such that $\overline{\rho}_Z\colon \pi_1(M) \to PSL_2(\mathbb{C})$ is a discrete and* *faithful representation. Moreover, the whole Dehn surgery component* $\overline{\mathfrak{X}}_0(M)$ *containing the character of* $\overline{\rho}_Z$ *is in the image of* $\chi_{\mathcal{T}}$.

Denote C a compact core of M; this is obtained by removing a small open regular neighbourhood from each vertex in P, such that C inherits a decomposition into truncated tetrahedra. Each boundary torus T_i, $i = 1, \ldots, h$, of C inherits a triangulation $\mathcal{T}_i$ induced by $\mathcal{T}$. Let γ be a closed simplicial path on T_i. In [**30**], the *holonomy* $\mu(\gamma)$ is defined as $(-1)^{|\gamma|}$ times the product of the moduli of the triangle vertices touching γ on the right, where $|\gamma|$ is the number of 1–simplices of γ, and the moduli asise from the corresponding edge labels.

At $Z \in \mathfrak{D}(\mathcal{T})$, evaluating $\mu(\gamma)$ gives a complex number $\mu_Z(\gamma) \in \mathbb{C} \setminus \{0\}$. It is stated in [**30, 33**] that $\mu_Z(\gamma)$ is the square of an eigenvalue; one has:

$$(\operatorname{tr} \overline{\rho}_Z(\gamma))^2 = \mu_Z(\gamma) + 2 + \mu_Z(\gamma)^{-1}.$$

This can be seen by putting a common fixed point of $\overline{\rho}_Z(\pi_1(T_i))$ at infinity in the upper–half space model, and writing $\overline{\rho}_Z(\gamma)$ as a product of Möbius transformations, each of which fixes an edge with one endpoint at infinity and takes one face of a tetrahedron to another.

Choose a basis $\{\mathcal{M}_i, \mathcal{L}_i\}$ of $\pi_1(T_i) \cong H_1(T_i)$ for each boundary torus T_i. Since $\mu_Z\colon \pi_1(T_i) \to \mathbb{C} \setminus \{0\}$ is a homomorphism for each $i = 1, \ldots, h$, there is a well–defined rational map:

$$\overline{\mathfrak{e}}\colon \mathfrak{D}(\mathcal{T}) \to (\mathbb{C} - \{0\})^{2h} \qquad \overline{\mathfrak{e}}(Z) = (\mu_Z(\mathcal{M}_1), \ldots, \mu_Z(\mathcal{L}_h)).$$

The closure of its image is contained in the $PSL_2(\mathbb{C})$*–eigenvalue variety* $\overline{\mathfrak{E}}(M)$ of [**37**].

3.3. Ideal points. Since $\mathfrak{D}(\mathcal{T})$ is a variety in $(\mathbb{C} \setminus \{0\})^{3n}$, Bergman's construction in [**1**] can be used to define its set of ideal points. Let

$$\begin{aligned} Z &= (z_1, z_1', z_1'', z_2, \ldots, z_n''), \\ \log|Z| &= (\log|z_1|, \ldots, \log|z_n''|), \\ u(Z) &= \frac{1}{\sqrt{1 + (\log|z_1|)^2 + \ldots + (\log|z_n''|)^2}}. \end{aligned}$$

The map $\mathfrak{D}(\mathcal{T}) \to B^{3n}$ defined by $Z \to u(Z)\log|Z|$ is continuous, and the *logarithmic limit set* $\mathfrak{D}_\infty(\mathcal{T})$ is the set of limit points on S^{3n-1} of its image. Thus, for each $\xi \in \mathfrak{D}_\infty(M)$ there is a sequence $\{Z_i\}$ in $\mathfrak{D}(M)$ such that

$$\lim_{i\to\infty} u(Z_i)\log|Z_i| = \xi.$$

The sequence $\{Z_i\}$ is said to *converge* to ξ, written $Z_i \to \xi$, and ξ is called an ideal point of $\mathfrak{D}(\mathcal{T})$. Whenever an edge invariant of a tetrahedron converges to one, the other two edge invariants "blow up". Thus, an ideal point of $\mathfrak{D}(\mathcal{T})$ is approached if and only if a tetrahedron degenerates.

Since the Riemann sphere is compact, there is a subsequence, also denoted by $\{Z_i\}$, with the property that each shape parameter converges in $\mathbb{C} \cup \{\infty\}$. In this case $\{Z_i\}$ is said to *strongly converge* to ξ. If ξ has rational coordinate ratios, a strongly convergent sequence may be chosen on a curve in $\mathfrak{D}(\mathcal{T})$ according to Lemma 5.

3.4. Normal surface theory. A *normal surface* in a (possibly ideal) triangulation $\mathcal{T}$ is a properly embedded surface which intersects each tetrahedron of $\mathcal{T}$ in a pairwise disjoint collection of *triangles* and *quadrilaterals*, as shown in Figure 1. These triangles and quadrilaterals are called *normal discs.* In an ideal triangulation of a non-compact 3-manifold, a normal surface may contain infinitely many triangles; such a surface is called *spun-normal* [**38**]. A normal surface may be disconnected or empty.

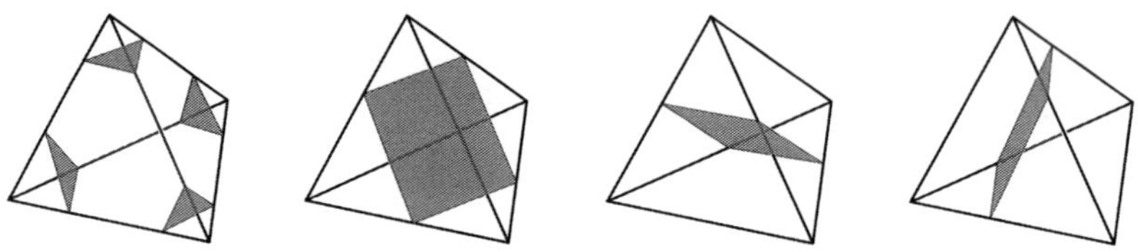

FIGURE 1. The seven types of normal disc in a tetrahedron.

We now describe an algebraic approach to normal surfaces. The key observation is that each normal surface contains finitely many quadrilateral discs, and is uniquely determined (up to normal isotopy) by these quadrilateral discs. Here a *normal isotopy* of M is an isotopy that keeps all simplices of all dimensions fixed. Let $\square$ denote the set of all normal isotopy classes of normal quadrilateral discs in $\mathcal{T}$. A normal quadrilateral disc is up to normal isotopy uniquely determined by a pair of opposite edges of a tetrahedron. Hence $|\square| = 3n$ where n is the number of tetrahedra in $\mathcal{T}$. These normal isotopy classes are called quadrilateral *types.*

We identify $\mathbb{R}^\square$ with $\mathbb{R}^{3n}$. Given a normal surface S, let $x(S) \in \mathbb{R}^\square = \mathbb{R}^{3n}$ denote the integer vector for which each coordinate $x(S)(q)$ counts the number of quadrilateral discs in S of type $q \in \square$. This *normal Q–coordinate* $x(S)$ satisfies the following two algebraic conditions.

First, $x(S)$ is admissible. A vector $x \in \mathbb{R}^\square$ is *admissible* if $x \geq 0$, and for each tetrahedron x is non-zero on at most one of its three quadrilateral types. This reflects the fact that an embedded surface cannot contain two different types of quadrilateral in the same tetrahedron.

Second, $x(S)$ satisfies a linear equation for each interior edge in M, termed a *Q–matching equation.* Intuitively, these equations arise from the fact that as one circumnavigates the earth, one crosses the equator from north to south as often as one crosses it from south to north. We now give the precise form of these equations. To simplify the discussion, we assume that M is oriented and all tetrahedra are given the induced orientation; see [**38**, Section 2.9] for details. These are easiest to describe with respect to a model for the neighbourhood of e in M, called its *abstract neighbourhood.*

The *degree of an edge* e in P, $\deg(e)$, is the number of 1–simplices in $\widetilde{\Delta}$ which map to e. Given the edge e in P, there is an associated *abstract neighbourhood* $B(e)$ of e. This is a ball triangulated by $\deg(e)$ 3–simplices, having a unique interior edge $\widetilde{e}$, and there is a well-defined simplicial quotient map $p_e\colon B(e) \to P$ taking $\widetilde{e}$ to e. This abstract neighbourhood is obtained as follows.

If e has at most one pre-image in each 3–simplex in $\widetilde{\Delta}$, then $B(e)$ is obtained as the quotient of the collection $\widetilde{\Delta}_e$ of all 3–simplices in $\widetilde{\Delta}$ containing a pre-image of e by the set Φ_e of all face pairings in Φ between faces containing a pre-image of

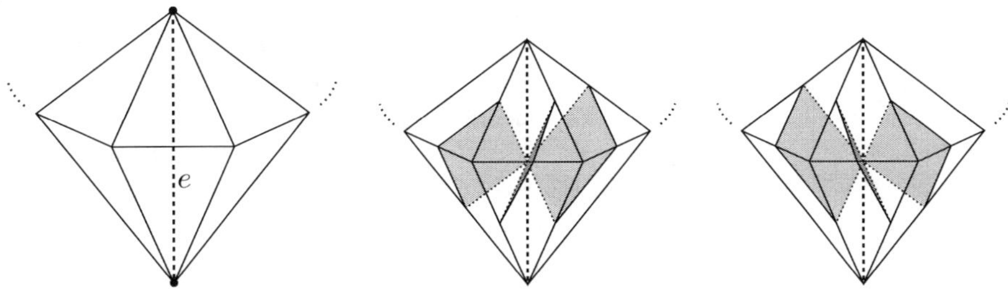

(a) The abstract neighbourhood $B(e)$ (b) Positive slope $+1$ (c) Negative slope -1

FIGURE 2. Slopes of quadrilaterals

e. There is an obvious quotient map $b_e\colon\ B(e) \to P$ which takes into account the remaining identifications on the boundary of $B(e)$.

If e has more than one pre-image in some 3–simplex, then multiple copies of this simplex are taken, one for each pre-image. The construction is modified accordingly, so that $B(e)$ again has a unique interior edge and there is a well defined quotient map $b_e\colon\ B(e) \to P$. Complete details can be found in [**38**], Section 2.3.

Let σ be a tetrahedron in $B(e)$. The boundary square of a normal quadrilateral of type q in σ meets the equator of $\partial B(e)$ if and only it has a vertex on e. In this case, it has a slope ± 1 of a well–defined sign on $\partial B(e)$ which is independent of the orientation of e. Refer to Figures 2(b) and 2(c), which show quadrilaterals with *positive* and *negative slopes* respectively.

Given a quadrilateral type q and an edge e, there is a *total weight* $\text{wt}_e(q)$ of q at e, which records the sum of all slopes of q at e (we sum because q might meet e more than once, if e appears as multiple edges of the same tetrahedron). If q has no corner on e, then we set $\text{wt}_e(q) = 0$. Given edge e in M, the Q–matching equation of e is then defined by $0 = \sum_{q\in\square} \text{wt}_e(q)\ x(q)$.

THEOREM 11. *For each $x \in \mathbb{R}^\square$ with the properties that x has integral coordinates, x is admissible and x satisfies the Q–matching equations, there is a (possibly non-compact) normal surface S such that $x = x(S)$. Moreover, S is unique up to normal isotopy and adding or removing vertex linking surfaces, i.e., normal surfaces consisting entirely of normal triangles.*

This is related to Hauptsatz 2 of [**18**]. For a proof of Theorem 11, see [**23**, Theorem 2.1] or [**38**, Theorem 2.4]. The set of all $x \in \mathbb{R}^\square$ with the property that (i) $x \geq 0$ and (ii) x satisfies the Q–matching equations is denoted $\mathcal{Q}(\mathcal{T})$. This naturally is a polyhedral cone, but the set of all admissible $x \in \mathbb{R}^\square$ typically meets $\mathcal{Q}(\mathcal{T})$ in a non-convex set $\mathcal{F}(\mathcal{T})$.

The *projectivised solution space* $\mathcal{PQ}(\mathcal{T})$ and the *projectivised admissible solution space* $\mathcal{PF}(\mathcal{T})$, are obtained as the quotient spaces under the multiplication action of the positive real numbers. These are usually identified via radial projection with compact subsets of either the standard simplex or a sphere centred at the origin. For the purpose of this paper, we view $\mathcal{PF}(\mathcal{T})$ as a compact subset of a sphere centred at the origin.

3.5. The relationship between gluing and matching equations. The deformation variety $\mathfrak{D}(\mathcal{T})$ is not defined by a principal ideal, hence its logarithmic limit set $\mathfrak{D}_\infty(\mathcal{T})$ is in general not directly determined by its defining equations (see [**37**] for details). However, it is contained in the intersection of the spherical duals of the Newton polytopes of its defining equations. The Newton polytope of a polynomial is the convex hull of its exponent vectors. The spherical dual of a polytope is the set of unit normal vectors to its supporting hyperplanes. For a polynomial p, we write $\mathrm{Sph}(p)$ for the spherical dual of the Newton polytope of p. See [**1, 37**] for details. From (7) and (8), we have

$$\mathfrak{D}_\infty(\mathcal{T}) \subseteq \mathfrak{D}_{\text{pre-}\infty}(\mathcal{T}) = \bigcap_{i=1}^{n} \big(\mathrm{Sph}(g_i) \cap \mathrm{Sph}(p_i) \cap \mathrm{Sph}(p'_i) \cap \mathrm{Sph}(p''_i)\big). \tag{9}$$

The set $\mathfrak{D}_{\text{pre-}\infty}(\mathcal{T})$ is termed a *tropical pre-variety.*

PROPOSITION 12 ([**39**], Proposition 3.1). *Let M be the interior of an orientable, connected, compact 3-manifold with non-empty boundary consisting of tori, and $\mathcal{T}$ be an ideal triangulation of M. The set $\mathfrak{D}_{pre\text{-}\infty}(\mathcal{T})$ is homeomorphic with the projective admissible solution space $\mathcal{PF}(\mathcal{T})$ of normal surface theory. In particular, $\mathfrak{D}_\infty(\mathcal{T})$ is homeomorphic with a closed subset of $\mathcal{PF}(\mathcal{T})$.*

The proof of the above proposition in [**39**] gives a canonical homeomorphism $N\colon \mathfrak{D}_\infty(\mathcal{T}) \to \mathcal{PF}(\mathcal{T})$ as follows. At its heart is the following simple relationship between gluing equations and matching equations. Recall the description of the hyperbolic gluing equation (8) of e_j:

$$1 = \prod_{i=1}^{n} z_i^{a_{ij}} (z'_i)^{a'_{ij}} (z''_i)^{a''_{ij}}.$$

Denoting $q_i^{(j)}$ the quadrilateral type separating the pair of edges with label $z_i^{(j)}$ in tetrahedron σ_i, it follows that the the Q–matching equation of e_j is:

$$0 = \sum_{i=1}^{n} (a''_{ij} - a'_{ij})q_i + (a_{ij} - a''_{ij})q'_i + (a'_{ij} - a_{ij})q''_i$$

(see [**38**], Section 2.9). Let

$$C_1 = \begin{pmatrix} 0 & 1 & -1 \\ -1 & 0 & 1 \\ 1 & -1 & 0 \end{pmatrix}, \tag{10}$$

and let C_n be the $(3n \times 3n)$ block diagonal matrix with n copies of C_1 on its diagonal. Hence letting A be the exponent matrix of the hyperbolic gluing equations and B the coefficient matrix of the Q–matching equations, we have $AC_n = B$. The desired natural homeomorphism is given by letting $N(\xi)$ be the unique normal Q–coordinate such that $\xi = C_n^T N(\xi)$ for a given $\xi \in \mathfrak{D}_\infty(\mathcal{T})$.

If ξ has rational coordinate ratios, one can associate a unique spun-normal surface $S(\xi)$ to it as follows. By assumption, there is $r > 0$ such that $rN(\xi)$ is an integer solution, and hence corresponds to a unique spun-normal surface without vertex linking components according to Theorem 11. One then requires r to be minimal with respect to the condition that the surface is 2–sided.

3.6. Essential surfaces and boundary curves. Morgan and Shalen [**28**] compactified the character variety of a 3-manifold by identifying ideal points with certain actions of $\pi_1(M)$ on $\mathbb{R}$–trees (given as points in an infinite dimensional space), and dual to these are codimension–one measured laminations in M. The work in [**39**] takes a different but related approach by compactifying the deformation variety with certain transversely measured singular codimension–one foliations (a finite union of convex rational polytopes), and dual to these are (possibly trivial) actions on $\mathbb{R}$–trees. The relationship between these compactifications is investigated in Section 6 of [**39**], with focus on the original consideration of ideal points of curves and surfaces dual to Bass–Serre trees due to Culler and Shalen [**13**]. A surface which can be reduced to an essential surface that is detected by the character variety in the sense of [**36**] will be called *weakly dual to an ideal point of a curve in the character variety.* The following result combines statements of Theorem 1.2 and Proposition 4.3 in [**39**]:

THEOREM 13. [**39**] *Let M be the interior of a compact, connected, orientable, irreducible 3-manifold with non-empty boundary consisting of a disjoint union of tori, and let $\mathcal{T}$ be an ideal triangulation of M. Let $\xi \in \mathfrak{D}_\infty(\mathcal{T})$ be a point with rational coordinate ratios, and assume that the spun-normal surface $S(\xi)$ is not closed.*

Then $S(\xi)$ is non-trivial and (weakly) dual to an ideal point of a curve in the character variety of M. Moreover, every essential surface detected by this ideal point has the same boundary curves as $S(\xi)$. In particular, the boundary curves of $S(\xi)$ are strongly detected.

It is an open question whether there is an ideal point of the deformation variety that is associated with a closed essential surface but with the property that any associated sequence of characters does not approach an ideal point of the character variety.

For a more general statement pertaining to all points in $\mathfrak{D}_\infty(\mathcal{T})$, and thus completing the description of its relationship with the Morgan–Shalen compactification, see Theorem 1.2 in [**39**].

4. The Whitehead link

The right–handed Whitehead link is shown in Figure 3(a). Throughout this paper, $\mathcal{W} = \mathcal{W}^+$ denotes its complement in S^3. This is a hyperbolic 3-manifold isometric with the same orientation to the manifold $m129$ in the census shipped with SnapPy. The complement of a left–handed Whitehead link is oppositely oriented, and denoted by $\mathcal{W}^-$.

(a) Right–handed Whitehead link (b) Orientations for peripheral elements

FIGURE 3

The standard triangulation $\mathcal{T}_{\mathcal{W}}$ with four ideal tetrahedra of $\mathcal{W}$ will be considered throughout; the triangulation is given in Figure 4 and Table 1, and is derived from SnapPy's manifold data (see Appendix A). We remark that $\mathcal{W}$ has three inequivalent minimal triangulations with four tetrahedra; they arise from the three diffent ways one can choose a "diagonal" in an octahedron; the triangulation used here is $m129$: #2 in Regina's cusped orientable manifold census. There are four tetrahedra in $\mathcal{T}_{\mathcal{W}}$, shown in Figure 4, and hence four edges, called red, green, black and blue. The black edge is labelled with one arrow, the blue with two, the green with three and the red with four. One of the cusps corresponds to the ideal endpoints of the red edge, and the other to the ideal endpoints of the green edge. The cusps are therefore referred to as the red cusp (also: cusp 0) and the green cusp (also: cusp 1) respectively. The face pairings are described in Table 1, where the face of tetrahedron σ_i opposite vertex j is denoted by $F_{i,j}$.

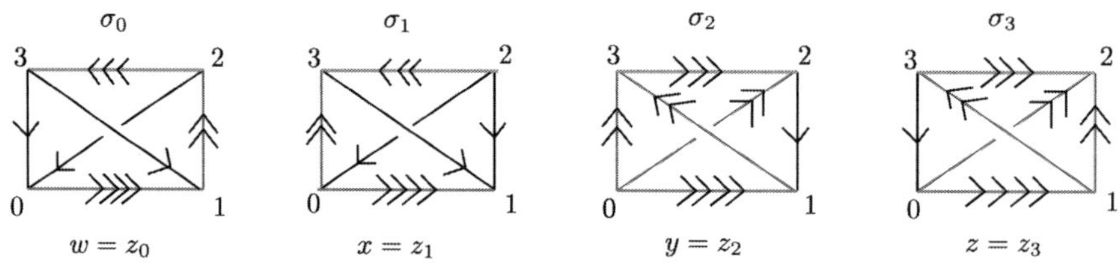

FIGURE 4. Triangulation $\mathcal{T}_{\mathcal{W}}$ of $\mathcal{W}$

$F_{0,0} \to F_{2,0}$	$F_{1,0} \to F_{0,1}$	$F_{2,0} \to F_{0,0}$	$F_{3,0} \to F_{2,1}$
$F_{0,1} \to F_{1,0}$	$F_{1,1} \to F_{3,1}$	$F_{2,1} \to F_{3,0}$	$F_{3,1} \to F_{1,1}$
$F_{0,2} \to F_{1,3}$	$F_{1,2} \to F_{2,3}$	$F_{2,2} \to F_{3,3}$	$F_{3,2} \to F_{0,3}$
$F_{0,3} \to F_{3,2}$	$F_{1,3} \to F_{0,2}$	$F_{2,3} \to F_{1,2}$	$F_{3,3} \to F_{2,2}$

TABLE 1. Face pairings

The convention for orientations of the peripheral elements is shown in Figure 3(b), where the right–hand rule is applied to a link projection and the induced triangulation of the cusp cross–section is viewed from outside the manifold.

The induced triangulations of the cusp cross–sections of $\mathcal{W}$ are given in Figure 5. There is a standard meridian, and two choices for longitudes are considered, which are termed *geometric* and *topological* and denoted by $\mathcal{L}^g$ and $\mathcal{L}^t$ respectively. The geometric longitudes are chosen by SnapPy for $m129$. They have the following property. Let $\mathcal{W}(p,q)$ be the manifold obtained from hyperbolic Dehn filling on one of the cusps. The coefficients $(p,q) \in \mathbb{R}^2$ are called *exceptional* if $\mathcal{W}(p,q)$ does not admit a complete hyperbolic metric. If the longitude is geometric, then the set of exceptional coefficients is contained in the rectangle with vertices $\pm(2,1)$, $\pm(2,-1)$ (see [**19**]). If a null–homologous longitude is chosen, the set of exceptional coefficients is contained in a parallelogram with vertices $\pm(-4,1)$, $\pm(0,1)$ for $\mathcal{W}^-$ (see [**29**]), and $\pm(4,1)$, $\pm(0,1)$ for $\mathcal{W}^+$. The natural linear maps between the Dehn surgery coefficients with respect to the different peripheral systems are given in Table 2.

$\{\mathcal{M},\mathcal{L}^t\}$ for $\mathcal{W}^-$	$\{\mathcal{M},\mathcal{L}^g\}$ for $\mathcal{W}^+$ and $\mathcal{W}^-$	$\{\mathcal{M},\mathcal{L}^t\}$ for $\mathcal{W}^+$
$(p+2q,-q)$	(p,q)	$(p+2q,q)$

TABLE 2. Surgery coefficients with respect to the peripheral systems

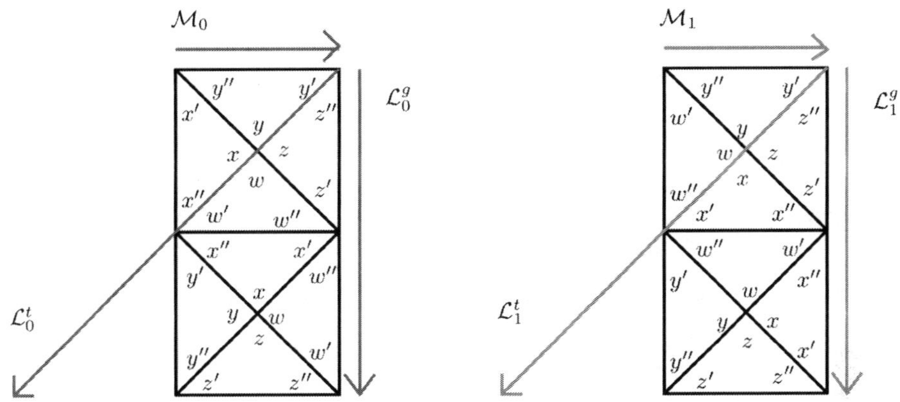

FIGURE 5. Triangulation of cusp cross–section and peripheral elements

5. Affine algebraic sets

This section determines the various affine algebraic sets associated with the Whitehead link complement, as well as the maps between them. Differences between result below and results in Neumann-Reid [**29**] arise from the fact that the latter use a left-handed link and have a different convention regarding shape parameters.

5.1. Deformation variety. The shape parameters are assigned in Figure 4, and the following hyperbolic gluing equations can be read off from this figure:

(red & green) $\quad 1 = wxyz$

(black) $\quad 1 = w'x'y'z'(w'')^2(x'')^2$

(blue) $\quad 1 = w'x'y'z'(y'')^2(z'')^2.$

The deformation variety $\mathfrak{D}(\mathcal{T}_\mathcal{W})$ is defined by these equations and the parameter relations:

$$(11)\quad 1 = w(1-w''),\quad 1 = x(1-x''),\quad 1 = y(1-y''),\quad 1 = z(1-z''),$$

$$(12)\quad 1 = w'(1-w),\quad 1 = x'(1-x),\quad 1 = y'(1-y),\quad 1 = z'(1-z),$$

$$(13)\quad 1 = w''(1-w'),\quad 1 = x''(1-x'),\quad 1 = y''(1-y'),\quad 1 = z''(1-z').$$

Using the fact that equations of the form $ww'w'' = -1$ result from the parameter relations, it is not hard to see that $\mathfrak{D}(\mathcal{T}_\mathcal{W})$ can be defined by the parameter relations in (11) and (12) together with the following two equations:

$$(14)\qquad 1 = wxyz \qquad \text{and} \qquad w''x'' = y''z''.$$

The variety $\mathfrak{D}(\mathcal{T}_\mathcal{W})$ is mapped into $(\mathbb{C}-\{0\})^4$ by the projection $\varphi(w, w', \ldots, z'') = (w, x, y, z)$. The closure of the resulting image is a variety defined by:

$$0 = 1 - wxyz, \tag{15}$$

$$0 = wx(1-y)(1-z) - (1-w)(1-x)yz. \tag{16}$$

This variety is the *parameter space* of **[30, 33]**, and is denoted by $\mathfrak{D}'(\mathcal{T}_\mathcal{W})$. For any $w, x, y, z \in \mathbb{C} - \{0, 1\}$ subject to (15) and (16), there is a unique point on $\mathfrak{D}(\mathcal{T}_\mathcal{W})$. Thus, $\mathfrak{D}(\mathcal{T}_\mathcal{W})$ and $\mathfrak{D}'(\mathcal{T}_\mathcal{W})$ are birationally equivalent, and the inverse map:

$$\begin{aligned}&\varphi^{-1}(w, x, y, z)\\ &= \left(w, \frac{1}{1-w}, \frac{w-1}{w}, x, \frac{1}{1-x}, \frac{x-1}{x}, y, \frac{1}{1-y}, \frac{y-1}{y}, z, \frac{1}{1-z}, \frac{z-1}{z}\right)\end{aligned}$$

is not regular at the intersection of $\mathfrak{D}'(\mathcal{T}_\mathcal{W})$ with the collection of hyperplanes

$$\{w = 1\} \cup \{x = 1\} \cup \{y = 1\} \cup \{z = 1\}$$

in $(\mathbb{C}-\{0\})^4$. If $\mathfrak{D}'(\mathcal{T}_\mathcal{W})$ intersects any one of these hyperplanes, then it intersects either exactly two or four of them.

One of the variables, say w, can be eliminated from the system of equations (15) & (16), and hence there is a map $\varphi' : \mathfrak{D}'(\mathcal{W}) \to (\mathbb{C}-\{0\})^3$ with the closure of its image defined by a single irreducible equation:

$$0 = g(x, y, z) = x - xy - xz + yz - xy^2z^2 + x^2y^2z^2. \tag{17}$$

Again, φ' is a birational isomorphism onto its image, and this in particular shows that $\mathfrak{D}(\mathcal{T}_\mathcal{W})$ is irreducible. Whence $\mathfrak{D}(\mathcal{T}_\mathcal{W})$ is an irreducible variety in $(\mathbb{C}-\{0\})^{12}$ defined by the 10 equations in (11), (12) and (14), and so:

PROPOSITION 14. *The deformation variety $\mathfrak{D}(\mathcal{T}_\mathcal{W})$ is a complete intersection variety of dimension two.*

5.2. Symmetries. There are symmetries in the defining equations of the deformation variety $\mathfrak{D}(\mathcal{T}_\mathcal{W})$, which descend to symmetries in (15) and (16). Consider the following involutions:

$$\tau_1(w, x, y, z) = (z, y, x, w) \quad \text{and} \quad \tau_2(w, x, y, z) = (y, z, w, x), \tag{18}$$

$$\tau_3(w, x, y, z) = (x, w, z, y), \tag{19}$$

$$\tau_4(w, x, y, z) = (w, x, z, y) \quad \text{and} \quad \tau_5(w, x, y, z) = (x, w, y, z). \tag{20}$$

Then $\tau_1\tau_2 = \tau_3 = \tau_4\tau_5$, and each of the pairs (18) and (20) generates a Klein four group. The elements $\tau_6 = \tau_1\tau_4$ and $\tau_7 = \tau_2\tau_4$ have order four, and the group D_4 generated by all involutions is a dihedral group. For any $p \in \mathfrak{D}'(\mathcal{T}_\mathcal{W})$ and any $\tau \in D_4$, $\tau p \in \mathfrak{D}'(\mathcal{T}_\mathcal{W})$, and if $\varphi^{-1} : \mathfrak{D}'(\mathcal{T}_\mathcal{W}) \to \mathfrak{D}(\mathcal{T}_\mathcal{W})$ is regular at p, then it is regular at τp.

5.3. Holonomies. Each cusp cross section inherits a triangulation induced by $\mathcal{T}_\mathcal{W}$. Let γ be a closed simplicial path on a cusp cross section. In **[30]**, the *derivative of the holonomy* $H'(\gamma)$ is defined as $(-1)^{|\gamma|}$ times the product of the moduli of the triangle vertices touching γ on the right, where $|\gamma|$ is the number of 1–simplices of γ, and the moduli asise from the corresponding edge labels.

The derivatives of the holonomies can this be read off from Figure 5, and simplify to:

$$H'(\mathcal{M}_0) = \frac{z}{w'y''} = \frac{x'z''}{y}, \quad H'(\mathcal{L}_0^t) = x^2y^2, \quad H'(\mathcal{L}_0^g) = \frac{(w'')^2}{(y'')^2} = \frac{(z'')^2}{(x'')^2}, \tag{21}$$

$$H'(\mathcal{M}_1) = \frac{w'z''}{y} = \frac{x}{w''z'}, \quad H'(\mathcal{L}_1^t) = w^2y^2, \quad H'(\mathcal{L}_1^g) = \frac{(z'')^2}{(w'')^2} = \frac{(x'')^2}{(y'')^2}. \tag{22}$$

The completeness equations for the holonomies yield that the complete hyperbolic structure is attained at $(w, x, y, z) = (i, i, i, i)$. The action of D_4 on the holonomies is described in Table 3, where $(m_i, l_i) = (H'(\mathcal{M}_i), H'(\mathcal{L}_i^t))$. The relationship between elements of D_4 and elements of the symmetry group of $\mathcal{W}$ can be deduced from this table.

	τ_1	τ_2	τ_3	τ_4	τ_5	τ_6	τ_7
(m_0, l_0)	(m_0, l_0)	(m_0^{-1}, l_0^{-1})	(m_0^{-1}, l_0^{-1})	(m_1^{-1}, l_1^{-1})	(m_1, l_1)	(m_1, l_1)	(m_1^{-1}, l_1^{-1})
(m_1, l_1)	(m_1^{-1}, l_1^{-1})	(m_1, l_1)	(m_1^{-1}, l_1^{-1})	(m_0^{-1}, l_0^{-1})	(m_0, l_0)	(m_0^{-1}, l_0^{-1})	(m_0, l_0)

TABLE 3. Action of D_4 on holonomies

5.4. Fundamental group. An abstract presentation of $\pi_1(\mathcal{W})$ can be computed from the triangulation using the set-up in the next section:

$$\pi_1(\mathcal{W}) = \langle \mathfrak{a}, \mathfrak{b}, \mathfrak{c}, \mathfrak{d} \mid \mathfrak{d} = \mathfrak{a}\mathfrak{b}, \mathfrak{b}\mathfrak{c}\mathfrak{d} = \mathfrak{c}\mathfrak{d}\mathfrak{c}, \mathfrak{a}\mathfrak{c} = \mathfrak{b}\mathfrak{a} \rangle,$$

and the peripheral elements are the following words in the generators:

$$\mathcal{M}_0 = \mathfrak{b}^{-1}, \qquad \mathcal{L}_0^g = \mathfrak{a}\mathfrak{d}^{-1}\mathfrak{c}^{-1}, \qquad \mathcal{L}_0^t = \mathcal{M}_0^{-2}\mathcal{L}_0^g,$$
$$\mathcal{M}_1 = \mathfrak{d}, \qquad \mathcal{L}_1^g = \mathfrak{a}^2\mathfrak{b}\mathfrak{c}, \qquad \mathcal{L}_1^t = \mathcal{M}_1^{-2}\mathcal{L}_1^g.$$

Thus, $\pi_1(\mathcal{W})$ can be generated by two meridians, and one obtains a single relation:

$$\begin{aligned} \pi_1(\mathcal{W}) = \langle \mathcal{M}_0, \mathcal{M}_1 \mid & \mathcal{M}_1\mathcal{M}_0\mathcal{M}_1\mathcal{M}_0^{-1}\mathcal{M}_1^{-1}\mathcal{M}_0^{-1}\mathcal{M}_1\mathcal{M}_0 \\ & = \mathcal{M}_0\mathcal{M}_1\mathcal{M}_0^{-1}\mathcal{M}_1^{-1}\mathcal{M}_0^{-1}\mathcal{M}_1\mathcal{M}_0\mathcal{M}_1 \rangle \end{aligned} \tag{23}$$

5.5. Developing map. Given a point $Z = (w, w', \ldots, z'') \in \mathfrak{D}(\mathcal{T}_{\mathcal{W}})$, there are (1) a continuous map $\mathrm{dev}_Z \colon \widetilde{\mathcal{W}} \to \mathbb{H}^3$ such that each ideal tetrahedron in $\widetilde{\mathcal{W}}$ (with the lifted ideal triangulation) is mapped to an ideal hyperbolic tetrahedron of the specified shape, and (2) a unique representation $\overline{\rho}_Z \colon \pi_1(\mathcal{W}) \to \mathrm{PSL}_2(\mathbb{C})$ which makes dev_Z $\pi_1(\mathcal{W})$–equivariant (see Section 2.5 in [**39**]). We start with an embedding of (a lift of) σ_0 with vertices at the points $0, 1, \infty, w$, and develop around the edge $[0, \infty]$ according to the parameters given by Z. This results in one lift of each ideal tetrahedron to $\mathbb{H}^3$. The resulting ideal vertices of the tetrahedra are indicated in Figure 6. Some tetrahedra may intersect or be "inverted". If all tetrahedra are positively oriented, then the resulting fundamental domain is an ideal octahedron.

Indeed, one obtains a combinatorial ideal octahedron, if one only applies the face pairings shown in Table 4. The Whitehead link complement is then obtained by applying the remaining face pairings, which match the boundary faces of the ideal octahedron.

$F_{0,2} \to F_{1,3}$	$F_{1,2} \to F_{2,3}$	$F_{2,2} \to F_{3,3}$	$F_{3,2} \to F_{0,3}$
$F_{0,3} \to F_{3,2}$	$F_{1,3} \to F_{0,2}$	$F_{2,3} \to F_{1,2}$	$F_{3,3} \to F_{2,2}$

TABLE 4. Face pairings used for fundamental domain

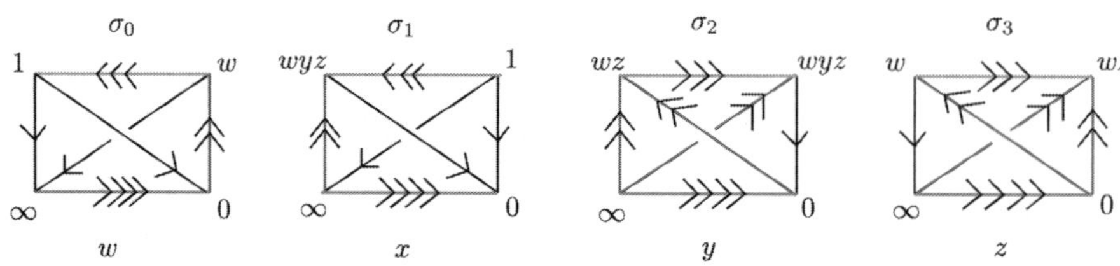

FIGURE 6. Developing map

Applied to the developed ideal tetrahedra in $\mathbb{H}^3$, the remaining face pairings are:

$$\rho_Z(\mathfrak{a})\colon\ [0,1,wyz] \to [\infty, w, 1], \qquad \rho_Z(\mathfrak{b})\colon\ [w,\infty,wz] \to [1,\infty,wyz],$$
$$\rho_Z(\mathfrak{c})\colon\ [0,1,w] \to [0,wyz,wz], \qquad \rho_Z(\mathfrak{c}\mathfrak{d})\colon\ [w,0,wz] \to [wz,\infty,wyz],$$

This determines a representation of $\pi_1(\mathcal{W})$ into $\mathrm{PSL}_2(\mathbb{C})$:

$$\rho_Z(\mathfrak{a}) = \sqrt{\frac{w'}{x'}}\begin{pmatrix} w+\frac{x'}{w'} & -\frac{x'}{w'} \\ 1 & 0 \end{pmatrix}, \qquad \rho_Z(\mathfrak{b}) = \sqrt{\frac{x'z''}{y}}\begin{pmatrix} \frac{y}{x'z''} & 1-x''z' \\ 0 & 1 \end{pmatrix},$$

$$\rho_Z(\mathfrak{c}) = \sqrt{\frac{w''y'}{x}}\begin{pmatrix} 1 & 0 \\ xw'(1-wy) & \frac{x}{w''y'} \end{pmatrix}, \quad \rho_Z(\mathfrak{d}) = \sqrt{\frac{y}{w'z''}}\begin{pmatrix} (1-wx)w' & \frac{y''}{w''} \\ \frac{w'}{x'} & \frac{1-xz}{w''} \end{pmatrix}.$$

We note:

$$\operatorname{tr}\rho_Z(\mathfrak{b}) = \sqrt{\frac{x'z''}{y}} + \sqrt{\frac{y}{x'z''}} = \sqrt{H'(\mathcal{M}_0)} + \frac{1}{\sqrt{H'(\mathcal{M}_0)}}.$$

A calculation using the identity $wxyz = 1$ and the standard identities for shape parameters (such as $ww'w'' = -1$ and $w'(1-w) = 1$) gives the following (where we have shown every second step):

$$\begin{aligned}\operatorname{tr}\rho_Z(\mathfrak{d}) &= \sqrt{\frac{y}{w'z''}}\left(w' - ww'x + \frac{1}{w''} - \frac{xz}{w''}\right) = \sqrt{\frac{y}{w'z''}}\left(1 + \frac{w'(z-1)}{yz}\right) \\ &= \sqrt{\frac{y}{w'z''}}\left(1 + \frac{w'z''}{y}\right) = \sqrt{H'(\mathcal{M}_1)} + \frac{1}{\sqrt{H'(\mathcal{M}_1)}}.\end{aligned}$$

This algebraically confirms the consistency of the choice of meridians (up to orientation) in §5.3 and §5.4.

At the complete structure, the peripheral elements have to be parabolic. This forces:

$$\mathfrak{a}_0 = \begin{pmatrix} 1+i & -1 \\ 1 & 0 \end{pmatrix}, \qquad \mathfrak{b}_0 = \begin{pmatrix} 1 & 1-i \\ 0 & 1 \end{pmatrix},$$
$$\mathfrak{c}_0 = \begin{pmatrix} 1 & 0 \\ -1+i & 1 \end{pmatrix}, \qquad \mathfrak{d}_0 = \begin{pmatrix} 1+i & 1 \\ 1 & 1-i \end{pmatrix}.$$

There is a well–defined map $\chi = \chi_{\mathcal{T}_W} \colon \mathfrak{D}(\mathcal{T}_W) \to \overline{\mathfrak{X}}(\mathcal{W})$, and it is shown in the next section that χ has degree 4. This corresponds to the observation that the manifolds $\mathcal{W}(p_0, q_0, p_1, q_1)$, $\mathcal{W}(-p_0, -q_0, p_1, q_1)$, $\mathcal{W}(p_0, q_0, -p_1, -q_1)$, $\mathcal{W}(-p_0, -q_0, -p_1, -q_1)$ are geometrically distinguished (they "spiral in different directions into the cusps"), whilst they have isomorphic fundamental groups.

5.6. Face pairings. A variety $\mathfrak{C}(\mathcal{W})$ that parameterises representations into $\mathrm{SL}_2(\mathbb{C})$ is computed as follows. Putting:

$$\rho(\mathcal{M}_0) = \begin{pmatrix} s & c \\ 0 & s^{-1} \end{pmatrix} \text{ and } \rho(\mathcal{M}_1) = \begin{pmatrix} u & 0 \\ 1 & u^{-1} \end{pmatrix}, \tag{24}$$

one obtains a single equation using (23):

$$\begin{aligned} 0 &= f(s,u,c) \\ &= (s - s^{-1})(u - u^{-1}) + c(s^{-2}u^{-2} - u^{-2} - s^{-2} + 4 - s^2 - u^2 + s^2u^2) \\ &\quad + c^2(2s^{-1}u^{-1} - su^{-1} - s^{-1}u + 2su) + c^3 \in \mathbb{C}[s^{\pm 1}, u^{\pm 1}, c]. \end{aligned} \tag{25}$$

Then $\mathfrak{C}(\mathcal{W})$ is an irreducible hypersurface in $(\mathbb{C}-\{0\})^2 \times \mathbb{C}$, and its intersection with $c = 0$ is the collection of lines $\{s^2 = 1, c = 0\} \cup \{u^2 = 1, c = 0\}$, which parametrises reducible representations. Moreover, $\mathfrak{C}(\mathcal{W})$ is a cover of the Dehn surgery component $\mathfrak{X}_0(\mathcal{W})$ of the character variety, since any irreducible representation of $\pi_1(\mathcal{W})$ into $\mathrm{SL}_2(\mathbb{C})$ is conjugate to an element of $\mathfrak{C}(\mathcal{W})$, and it is a 4–to–1 branched cover of $\mathfrak{X}_0(\mathcal{W})$ since $\mathfrak{C}(\mathcal{W})$ is not contained in the union of hypersurfaces $s^2 = 1$ and $u^2 = 1$.

If $f(s,u,c) = 0$, then $f(-s,u,-c) = f(s,-u,-c) = f(-s,-u,c) = 0$, and the four solutions correspond to the action of $\mathrm{Hom}(\pi_1(\mathcal{W}), \mathbb{Z}_2)$ on $\mathfrak{C}(\mathcal{W})$. To obtain a description of the corresponding quotient map $\mathfrak{C}(\mathcal{W}) \to \overline{\mathfrak{C}}(\mathcal{W})$, and hence a parametrisation of the associated variety $\overline{\mathfrak{C}}(\mathcal{W})$ of representations into $\mathrm{PSL}_2(\mathbb{C})$, note that (24) can be adjusted by a conjugation and rewritten in the form:

$$\rho(\mathcal{M}_0) = \frac{1}{s}\begin{pmatrix} s^2 & csu \\ 0 & 1 \end{pmatrix} \quad \text{and} \quad \rho(\mathcal{M}_1) = \frac{1}{u}\begin{pmatrix} u^2 & 0 \\ 1 & 1 \end{pmatrix}, \tag{26}$$

which is still subject to $0 = f(s,u,c)$. The map $\mathrm{q}_1 : (\mathbb{C}-\{0\})^2 \times \mathbb{C} \to (\mathbb{C}-\{0\})^2 \times \mathbb{C}$ defined by $\mathrm{q}_1(s,u,c) = (s^2, u^2, csu)$ can be identified with the natural quotient map $\mathfrak{C}(\mathcal{W}) \to \overline{\mathfrak{C}}(\mathcal{W})$, and the defining equation for $\overline{\mathfrak{C}}(\mathcal{W})$ can be derived from this relationship. Thus, $\overline{\mathfrak{C}}(\mathcal{W})$ can be viewed as a variety of representations into $GL_2(\mathbb{C})$:

$$\overline{\rho}_{GL}(\mathcal{M}_0) = \begin{pmatrix} \overline{s} & d \\ 0 & 1 \end{pmatrix} \quad \text{and} \quad \overline{\rho}_{GL}(\mathcal{M}_1) = \begin{pmatrix} \overline{u} & 0 \\ 1 & 1 \end{pmatrix} \quad \text{subject to} \tag{27}$$

$$\begin{aligned} 0 &= \overline{su}(\overline{s} - 1)(\overline{u} - 1) + d(1 - \overline{s} - \overline{u} + 4\overline{su} - \overline{s}^2\overline{u} - \overline{su}^2 + \overline{s}^2\overline{u}^2) \\ &\quad + d^2(2 - \overline{s} - \overline{u} + 2\overline{su}) + d^3 \in \mathbb{C}[\overline{s}^{\pm 1}, \overline{u}^{\pm 1}, d]. \end{aligned} \tag{28}$$

For each $\overline{\rho}_{GL}$ as above, there is a unique $\mathrm{PSL}_2(\mathbb{C})$–representation $\overline{\rho}$ such that $\overline{\rho}_{GL}$ and $\overline{\rho}$ are identical as representations into the group of projective transformations of $\mathbb{C}P^1$. Hence there is a bijection between the set of all of the above $GL_2(\mathbb{C})$–representations of $\pi_1(\mathcal{W})$ with $d \neq 0$ and the set of all irreducible $\mathrm{PSL}_2(\mathbb{C})$–representations of $\pi_1(\mathcal{W})$ (up to conjugation) which lift to $\mathrm{SL}_2(\mathbb{C})$.

LEMMA 15. *There is a birational isomorphism* $r : \mathfrak{D}(\mathcal{T}_W) \to \overline{\mathfrak{C}}(\mathcal{W})$.

PROOF. To construct the map $r : \mathfrak{D}(\mathcal{T}_\mathcal{W}) \to \overline{\mathfrak{C}}(\mathcal{W})$, conjugate the face pairings $\rho_Z(\mathfrak{a}), \ldots, \rho_Z(\mathfrak{d})$ by a *suitably chosen* matrix A to obtain a form analogous to (26), and then adjust the resulting representation by multiplication:

$$\overline{\rho}'_Z(\mathcal{M}_0) := \sqrt{\frac{x}{w''y'}}\ A\ \rho_Z(\mathfrak{b}^{-1})\ A^{-1} = \begin{pmatrix} \frac{w''y'}{x} & z-1 \\ 0 & 1 \end{pmatrix}$$

$$\text{and} \quad \overline{\rho}'_Z(\mathcal{M}_1) := \sqrt{\frac{w''z'}{x}}\ A\ \rho_Z(\mathfrak{d})\ A^{-1} = \begin{pmatrix} \frac{x}{w''z'} & 0 \\ 1 & 1 \end{pmatrix}.$$

It can be verified that $\overline{\rho}'_Z$ defines a $GL_2(\mathbb{C})$–representation for each $Z \in \mathfrak{D}(\mathcal{T}_\mathcal{W})$. This induces a map $r : \mathfrak{D}(\mathcal{T}_\mathcal{W}) \to \overline{\mathfrak{C}}(\mathcal{W})$ defined by:

$$r(w, w', \ldots, z'') = \left(\frac{w''y'}{x}, \frac{x}{w''z'}, z-1 \right). \tag{29}$$

An elementary calculation shows that this map is 1–1 and that its image is dense in $\overline{\mathfrak{C}}(\mathcal{W})$. A rational inverse is given by the following map $\overline{\mathfrak{C}}(\mathcal{W}) \to \mathfrak{D}'(\mathcal{T}_\mathcal{W})$:

$$(\overline{s}, \overline{u}, d) \to \left(\frac{d + d^2 - d\overline{s} + \overline{su} + d\overline{su}}{(1+d)(d+\overline{su})}, \frac{\overline{su}}{d + d^2 - d\overline{s} + \overline{su} + d\overline{su}}, 1 + \frac{d}{\overline{su}}, 1+d \right), \tag{30}$$

which composed with $\varphi : \mathfrak{D}'(\mathcal{T}_\mathcal{W}) \to \mathfrak{D}(\mathcal{T}_\mathcal{W})$ gives a map $r^{-1} : \overline{\mathfrak{C}}(\mathcal{W}) \to \mathfrak{D}(\mathcal{T}_\mathcal{W})$. Composition of the maps r and r^{-1} induces the identity on both $\mathfrak{D}(\mathcal{T}_\mathcal{W})$ and $\overline{\mathfrak{C}}(\mathcal{W})$, and this proves the lemma. □

The only singularity of $\mathfrak{D}(\mathcal{T}_\mathcal{W})$ is at infinity; at the ideal point where all of w, x, y, z tend to one, and the only singularity of $\mathfrak{D}'(\mathcal{W})$ is at the corresponding point $(w, x, y, z) = (1, 1, 1, 1)$. The variety $\overline{\mathfrak{C}}(\mathcal{W})$ is an irreducible variety without singularities, and the above proof shows that it parametrises developing maps when thought of as a variety in $(\mathbb{C} - \{0\})^3$. It is shown in Subsection B.3 that the intersection of $\overline{\mathfrak{C}}(\mathcal{W})$ with $c = 0$ corresponds to the ideal point of $\mathfrak{D}(\mathcal{T}_\mathcal{W})$ where $w, x, y, z \to 1$. Thus, it is a natural de-singularisation of $\mathfrak{D}(\mathcal{T}_\mathcal{W})$.

COROLLARY 16. *χ: $\mathfrak{D}(\mathcal{T}_\mathcal{W}) \to \overline{\mathfrak{X}}_0(\mathcal{W})$ is generically 4–to–1 and onto.*

PROOF. Note that χ: $\mathfrak{D}(\mathcal{T}_\mathcal{W}) \to \overline{\mathfrak{X}}(\mathcal{W})$ factors through $\overline{\mathfrak{C}}(\mathcal{W})$, and hence it is enough to show that the map $\overline{\mathfrak{C}}(\mathcal{W}) \to \overline{\mathfrak{X}}_0(\mathcal{W})$ is generically 4–to–1 and onto. Firstly, $\overline{\mathfrak{C}}(\mathcal{W})$ is irreducible, and hence its image in $\overline{\mathfrak{X}}(\mathcal{W})$ is irreducible. Secondly, the above lemma implies that $\overline{\mathfrak{C}}(\mathcal{W})$ contains a discrete and faithful representation, and hence maps to the Dehn surgery component. Since there are $\overline{\rho} \in \overline{\mathfrak{C}}(\mathcal{W})$ such that $(\operatorname{tr} \overline{\rho}(\mathcal{M}_0))^2 \neq 4 \neq (\operatorname{tr} \overline{\rho}(\mathcal{M}_1))^2$, there are generically four elements of each conjugacy class of representations contained in $\overline{\mathfrak{C}}(\mathcal{W})$. Hence, the degree of χ is four. □

5.7. Eigenvalue maps. As in [**37**], denote the respective eigenvalue varieties by $\mathfrak{E}(\mathcal{W})$ and $\overline{\mathfrak{E}}(\mathcal{W})$. The aim of this section is to show that the subvariety $\overline{\mathfrak{E}}_0(\mathcal{W})$ corresponding to $\overline{\mathfrak{X}}_0(\mathcal{W})$ is birationally equivalent to $\mathfrak{D}(\mathcal{T}_\mathcal{W})$. The following lemma establishes suitable affine coordinates and maps needed for this.

LEMMA 17. *There are a quotient map* q_3: $\mathfrak{C}(\mathcal{W}) \to \overline{\mathfrak{C}}(\mathcal{W})$ *corresponding to the action of* $\operatorname{Hom}(\pi_1(\mathcal{W}), \mathbb{Z}_2)$, *and* eigenvalue maps e: $\mathfrak{C}(\mathcal{W}) \to \mathfrak{E}(\mathcal{W})$ *and* $\overline{\mathrm{e}}$: $\overline{\mathfrak{C}}(\mathcal{W}) \to$

$\overline{\mathfrak{E}}(\mathcal{W})$ *such that the following diagram commutes:*

$$\begin{array}{ccc} \mathfrak{C}(\mathcal{W}) & \xrightarrow{\ \mathrm{e}\ } & \mathfrak{E}_0(\mathcal{W}) \\ \mathrm{q}_1 \downarrow & & \downarrow \mathrm{q}_3 \\ \overline{\mathfrak{C}}(\mathcal{W}) & \xrightarrow{\ \overline{\mathrm{e}}\ } & \overline{\mathfrak{E}}_0(\mathcal{W}) \end{array}$$

PROOF. Let $\vartheta_\gamma\colon \mathfrak{C}(\mathcal{W}) \to \mathbb{C}$ be the holomorphic map which takes ρ to the upper left entry of $\rho(\gamma)$. Then define $\mathrm{e}\colon \mathfrak{C}(\mathcal{W}) \to \mathfrak{E}(\mathcal{W})$ to be the map

$$\mathrm{e}(\rho) = (\vartheta_{\mathcal{M}_0}(\rho), \vartheta_{\mathcal{L}_0^t}(\rho), \vartheta_{\mathcal{M}_1}(\rho), \vartheta_{\mathcal{L}_1^t}(\rho)). \tag{31}$$

Since every $\rho \in \mathfrak{C}(\mathcal{W})$ is triangular on the peripheral subgroups, it follows that this map is well–defined. Denote the affine coordinates of $\mathfrak{E}(\mathcal{W})$ by (s,t,u,v), so they correspond to eigenvalues of $\mathcal{M}_0$, $\mathcal{L}_0^t$, $\mathcal{M}_1$ and $\mathcal{L}_1^t$.

Denote the map which takes $\overline{\rho}_{GL}$ to the upper left entry of $\overline{\rho}_{GL}(\gamma)$ by ϑ_γ as well, and let $\overline{\mathrm{e}}\colon \overline{\mathfrak{C}}(\mathcal{W}) \to \overline{\mathfrak{E}}(\mathcal{W})$ be the map

$$\overline{\mathrm{e}}(\overline{\rho}_{GL}) = (\vartheta_{\mathcal{M}_0}(\overline{\rho}_{GL}), \vartheta_{\mathcal{L}_0^t}(\overline{\rho}_{GL}), \vartheta_{\mathcal{M}_1}(\overline{\rho}_{GL}), \vartheta_{\mathcal{L}_1^t}(\overline{\rho}_{GL})). \tag{32}$$

To verify that this map has the right range, note that the upper left entries of $\overline{\rho}_{GL}(\mathcal{M}_0)$ and $\overline{\rho}_{GL}(\mathcal{M}_1)$ are the squares of the eigenvalues of matrices representing the associated $\mathrm{PSL}_2(\mathbb{C})$–representation. The longitudes are the following words in the meridians:

$$\mathcal{L}_0^t = \mathcal{M}_0^{-1}\mathcal{M}_1\mathcal{M}_0\mathcal{M}_1^{-1}\mathcal{M}_0^{-1}\mathcal{M}_1^{-1}\mathcal{M}_0\mathcal{M}_1 \tag{33}$$

$$\mathcal{L}_1^t = \mathcal{M}_1^{-1}\mathcal{M}_0\mathcal{M}_1\mathcal{M}_0^{-1}\mathcal{M}_1^{-1}\mathcal{M}_0^{-1}\mathcal{M}_1\mathcal{M}_0, \tag{34}$$

and hence $\overline{\rho}_{GL}(\mathcal{L}_i^t) = \overline{\rho}(\mathcal{L}_i^t)$ for any $\overline{\rho}_{GL}$ and its corresponding unique $\mathrm{PSL}_2(\mathbb{C})$–representation $\overline{\rho}$. In particular, $\vartheta_{\mathcal{L}_i^t}(\overline{\rho}_{GL})$ is an eigenvalue of $\overline{\rho}_{GL}(\mathcal{L}_i^t) = \overline{\rho}(\mathcal{L}_i^t)$ and it is independent of the choice of the signs of matrices representing $\overline{\rho}(\mathcal{M}_0)$ and $\overline{\rho}(\mathcal{M}_1)$. It follows that $\overline{\mathfrak{E}}(\mathcal{W})$ can be given the affine coordinates $(\overline{s}, t, \overline{u}, v)$.

The natural quotient map which makes the above diagram commute is therefore $\mathrm{q}_3\colon \mathfrak{E}(\mathcal{W}) \to \overline{\mathfrak{E}}(\mathcal{W})$ defined by $\mathrm{q}_3(s,t,u,v) = (s^2, t, u^2, v)$. This map clearly corresponds to the action of $\mathrm{Hom}(\pi_1(\mathcal{W}), \mathbb{Z}_2)$ on $\mathfrak{E}(\mathcal{W})$.

Since any irreducible $\mathrm{SL}_2(\mathbb{C})$–representation is conjugate to a representation in $\mathfrak{C}(\mathcal{W})$, the closure of the image of e is the component of $\mathfrak{E}(\mathcal{W})$ corresponding to the Dehn surgery component $\mathfrak{X}_0(\mathcal{W})$. It follows that the closure of the image of $\overline{\mathrm{e}}$ corresponds to $\overline{\mathfrak{X}}_0(\mathcal{W})$. In particular the composite mapping $\Psi = \overline{\mathrm{e}} \circ r\colon \mathfrak{D}(\mathcal{T}_\mathcal{W}) \to \overline{\mathfrak{E}}_0(\mathcal{W})$ is onto. □

From the face pairing $\rho_Z\colon \pi_1(M) \to \mathrm{PSL}_2(\mathbb{C})$, one computes $\vartheta_{\mathcal{M}_0}(\overline{\rho}_Z) = \frac{x'z''}{y}$, $\vartheta_{\mathcal{L}_0^t}(\overline{\rho}_Z) = xy$, $\vartheta_{\mathcal{M}_1}(\overline{\rho}_Z) = \frac{w'z''}{y}$ and $\vartheta_{\mathcal{L}_1^t}(\overline{\rho}_Z) = wy$. The map $\Psi\colon \mathfrak{D}(\mathcal{T}_\mathcal{W}) \to \overline{\mathfrak{E}}(\mathcal{W})$ with respect to the chosen coordinates is therefore given by

$$\Psi(w,w',w'',x,x',x'',y,y',y'',z,z',z'') = (\frac{x'z''}{y}, xy, \frac{w'z''}{y}, wy) \tag{35}$$

We have $\Psi\varphi^{-1}(i,i,i,i) = (1,-1,1,-1)$ at the complete structure.

LEMMA 18. *The map* $\Psi\colon \mathfrak{D}(\mathcal{T}_\mathcal{W}) \to \overline{\mathfrak{E}}_0(\mathcal{W})$ *is a birational isomorphism.*

PROOF. Since Ψ is a regular map and $\mathfrak{D}(\mathcal{T}_\mathcal{W})$ is irreducible, it follows that the closure of its image, which has been identified as $\overline{\mathfrak{E}}_0(\mathcal{W})$, is irreducible. Thus,

$\Psi\colon \mathfrak{D}(\mathcal{T}_\mathcal{W}) \to \overline{\mathfrak{E}}_0(\mathcal{W})$ is a regular map of irreducible varieties. It remains to show that it has degree one.

Assume that $\Psi\varphi^{-1}(w_0, x_0, y_0, z_0) = \Psi\varphi^{-1}(w_1, x_1, y_1, z_1)$, where (w_i, x_i, y_i, z_i) are two regular points of φ on $\mathfrak{D}'(\mathcal{T}_\mathcal{W})$. An elementary calculation shows that the points are either identical or we have $w_0 = x_0$ and $w_1 = x_1$. Thus, all points on which Ψ does not have degree one are contained on the hypersurface $w = x$. Since for any $w \in \mathbb{C} - \{0, \pm 1\}$, the point $\varphi^{-1}(w, -w^{-1}, w, -w^{-1})$ is contained in $\mathfrak{D}(\mathcal{T}_\mathcal{W})$, $\mathfrak{D}(\mathcal{T}_\mathcal{W})$ is not contained in this hypersurface. □

An inverse $\overline{\mathfrak{E}}_0(\mathcal{W}) \to \mathfrak{D}(\mathcal{T}_\mathcal{W})$ taking $(\overline{s}, t, \overline{u}, v) \to (w, w', \ldots, z'')$ is determined by the following map $\overline{\mathfrak{E}}_0(\mathcal{W}) \to \mathfrak{D}'(\mathcal{W})$ which can be computed from (35):

$$(\overline{s}, t, \overline{u}, v) \to \left(\frac{v(\overline{s}-\overline{u})}{\overline{s}t-\overline{u}v}, \frac{t(\overline{s}-\overline{u})}{\overline{s}t-\overline{u}v}, \frac{\overline{s}t-\overline{u}v}{\overline{s}-\overline{u}}, \frac{\overline{s}t-\overline{u}v}{tv(\overline{s}-\overline{u})} \right).$$

This map is not regular on a 1–dimensional subvariety of $\overline{\mathfrak{E}}_0(\mathcal{W})$, which is defined by the following three equations:

$$\overline{s} = \overline{u}, \qquad t = v, \qquad 0 = \overline{s} - t + \overline{s}t + \overline{s}^2 t - \overline{s}^2 t^2 - \overline{s}^3 t^2 + \overline{s}^4 t^2 - \overline{s}^3 t^3. \tag{36}$$

See Subsection B.1 for a computation of the eigenvalue varieties and this subvariety.

6. Embedded surfaces

This section gives a complete description of the space $\mathcal{PL}(\mathcal{W})$ of all essential surfaces in the Whitehead link complement, and deduces the associated boundary curve space $\mathfrak{BC}(\mathcal{W})$ and the unit ball of the Thurston norm. This is compared with previous work of Floyd and Hatcher [**15**], Lash [**25**] and Hoste and Shanahan [**20**].

The main tools used to analyse incompressible surfaces in the projective admissible solution space $\mathcal{PF}(\mathcal{T}_\mathcal{W})$ of spun-normal surface theory are a criterion due to Dunfield that determines which spun-normal surfaces are essential, and results by Walsh [**41**] and Kang and Rubinstein [**24**] on the normalisation of essential surfaces.

6.1. Complete description of the projective admissible solution space. The convention for quadrilateral coordinates for orientable manifolds in [**38**] is used. Table 5 indicates the position of the quadrilaterals in the tetrahedra. For reference, we also add the shape parameters, where we write $w = z_0$, $x = z_1$, $y = z_2$, and $z = z_3$. The notation ij/kl means that the particular quadrilateral type separates the vertices i and j from the vertices k and l.

Quadrilateral	q_i	q'_i	q''_i
Separates	01/23	03/12	02/13
Edge label	z_i	z'_i	z''_i

TABLE 5. Quadrilateral types

The Q–matching equations can be worked out directly from the triangulation or by using their relationship with the gluing equations described in §3.5. There are the following three equations:

$$\text{(red \& green)} \qquad 0 = q_0' - q_0'' + q_1' - q_1'' + q_2' - q_2'' + q_3' - q_3'',$$
$$\text{(black)} \qquad 0 = q_0 - 2q_0' + q_0'' + q_1 - 2q_1' + q_1'' - q_2 + q_2'' - q_3 + q_3''$$
$$\text{(blue)} \qquad 1 = -q_0 + q_0'' - q_1 + q_1'' + q_2 - 2q_2' + q_2'' + q_3 - 2q_3' + q_3''$$

These are equivalent to the following two, which correspond to (14):

$$0 = q_0' - q_0'' + q_1' - q_1'' + q_2' - q_2'' + q_3' - q_3'', \tag{37}$$
$$0 = q_0 - q_0' + q_1 - q_1' - q_2 + q_2' - q_3 + q_3'. \tag{38}$$

To obtain an admissible solution, one now sets two quadrilateral coordinates from each tetrahedron equal to zero and solves the above equations subject to this constraint. Hence the set $\mathcal{PF}(\mathcal{T}_\mathcal{W})$ can be computed by solving $3^4 = 81$ systems of linear equations. This is automated by feeding the triangulation to `Regina`. Alternatively, given its description as a tropical pre-variety in Proposition 12, one can use `gfan` to compute $\mathcal{PF}(\mathcal{T}_\mathcal{W})$ from the defining equation of $\mathfrak{D}(\mathcal{T}_\mathcal{W})$ using the command `tropical_intersection`.

The set $\mathcal{PF}(\mathcal{T}_\mathcal{W})$ is a finite union of convex spherical polytopes in $S^{11} \subset \mathbb{R}^{12}$. It turns out that there are 28 geodesic arcs and one geodesic quadrilateral, spanned by a total of 20 vertices. These are indicated in Figure 7, where geodesic arcs are represented as PL arcs.

The vertices are denote by $V_1, \ldots, V_{20}$, and called *vertex solutions*. These correspond to extremal solutions to the linear equations. Each vertex solution V_i is rescaled by a positive real to obtain a (minimal) integer solution, and the resulting normal surface is denoted by F_i. Hence the normal Q–coordinate $N(F_i)$ is a multiple of V_i. These normal surfaces are described in Table 6 and the information given in the table is as follows:

First, the normal Q–coordinate $N(F_i)$ is given, then the topological type of F_i, where T_1 stands for a once–punctured torus, S_3 for a thrice–punctured sphere, R_2 for a twice–punctured $\mathbb{R}P^2$. The abbreviations K_i and T_i will be used in subsequent figures for an i–punctured Klein bottle and an i–punctured torus respectively, and G_2 denotes a (closed) genus two surface.

The column *class* specifies the equivalence class (defined in Subsection 6.3) that the projective normal Q–coordinate V_i of F_i belongs to.

The column *∂–curves* encodes the number of boundary components on the respective cusps; if there are i boundary components on the (red) cusp 0, and j on the (green) cusp 1, this is written as i/j.

Last, the corresponding boundary curves are given as they are computed from the chosen (oriented, topological) peripheral system. The boundary curves are determined by the signed intersection numbers of the peripheral elements with the spun-normal surface. We here summarise how this is computed; the details can be found in [**39**, §4.2] and [**38**, §3.1]. Let γ be a closed simplicial path on a cusp cross section with respect to the triangulation induced by $\mathcal{T}$. The Q–modulus of a vertex with label z_i is $q_i'' - q_i'$, of a vertex with label z_i' is $q_i - q_i''$, and of a vertex with label z_i'' is $q_i' - q_i$. The linear functional $\nu(\gamma)$ is defined to be the sum of the Q–moduli of all vertices of triangles touching γ to the right. For our generators, we obtain

Vertex	q_0	q_0'	q_0''	q_1	q_1'	q_1''	q_2	q_2'	q_2''	q_3	q_3'	q_3''	type	class	∂–curves	$\nu(\mathcal{L}_0^t)$,	$-\nu(\mathcal{M}_0)$,	$\nu(\mathcal{L}_1^t)$,	$-\nu(\mathcal{M}_1)$
1	1	0	0	0	0	0	1	0	0	0	0	0	T_1	N_1'	0/1	0,	0,	0,	−1
2	0	0	0	1	0	0	1	0	0	0	0	0	T_1	N_1	1/0	0,	−1,	0,	0
3	0	0	0	1	0	0	0	0	0	1	0	0	T_1	N_1'	0/1	0,	0,	0,	1
4	1	0	0	0	0	0	0	0	0	1	0	0	T_1	N_1	1/0	0,	1,	0,	0
5	1	0	0	0	1	0	0	0	0	0	0	1	S_3	N_2'	2/1	−2,	0,	0,	−1
6	0	0	0	0	0	1	1	0	0	0	1	0	S_3	N_2'	2/1	2,	0,	0,	−1
7	0	0	1	0	0	0	1	0	0	0	1	0	S_3	N_2	1/2	0,	−1,	2,	0
8	0	1	0	1	0	0	0	0	0	0	0	1	S_3	N_2	1/2	0,	−1,	−2,	0
9	0	1	0	1	0	0	0	0	1	0	0	0	S_3	N_2'	2/1	2,	0,	0,	1
10	0	0	1	0	0	0	0	1	0	1	0	0	S_3	N_2'	2/1	−2,	0,	0,	1
11	0	0	0	0	0	1	0	1	0	1	0	0	S_3	N_2	1/2	0,	1,	−2,	0
12	1	0	0	0	1	0	0	0	1	0	0	0	S_3	N_2	1/2	0,	1,	2,	0
13	0	0	0	0	1	0	0	1	0	0	0	2	R_2	N_3'	1/1	−4,	−1,	−2,	−1
14	0	1	0	0	0	2	0	0	0	0	1	0	R_2	N_3'	1/1	4,	1,	−2,	−1
15	0	0	2	0	1	0	0	0	0	0	1	0	R_2	N_3	1/1	−2,	−1,	4,	1
16	0	1	0	0	0	0	0	1	0	0	0	2	R_2	N_3	1/1	−2,	−1,	−4,	−1
17	0	1	0	0	0	0	0	0	2	0	1	0	R_2	N_3'	1/1	4,	1,	2,	1
18	0	0	2	0	1	0	0	1	0	0	0	0	R_2	N_3'	1/1	−4,	−1,	2,	1
19	0	1	0	0	0	2	0	1	0	0	0	0	R_2	N_3	1/1	2,	1,	−4,	−1
20	0	0	0	0	1	0	0	0	2	0	1	0	R_2	N_3	1/1	2,	1,	4,	1

Angle	α_0	α_0'	α_0''	α_1	α_1'	α_1''	α_2	α_2'	α_2''	α_3	α_3'	α_3''	dual to vertex surfaces numbers
α^+	0	0	π	π	0	0	0	0	π	π	0	0	1, 5, 6, 13, 14, 16, 19
α^-	0	π	0	π	0	0	0	π	0	π	0	0	1, 5, 6, 7, 12, 15, 20
β^+	π	0	0	0	0	π	0	0	π	π	0	0	2, 7, 8, 13, 15, 16, 18
β^-	π	0	0	0	π	0	0	π	0	π	0	0	2, 6, 7, 8, 9, 14, 17
γ^+	π	0	0	0	0	π	π	0	0	0	0	π	3, 9, 10, 15, 17, 18, 20
γ^-	π	0	0	0	π	0	π	0	0	0	π	0	3, 8, 9, 10, 11, 16, 19
δ^+	0	0	π	π	0	0	π	0	0	0	0	π	4, 11, 12, 14, 17, 19, 20
δ^-	0	π	0	π	0	0	π	0	0	0	π	0	4, 5, 10, 11, 12, 13, 18

TABLE 6. Minimal representatives for vertex solutions and some dual angle structures

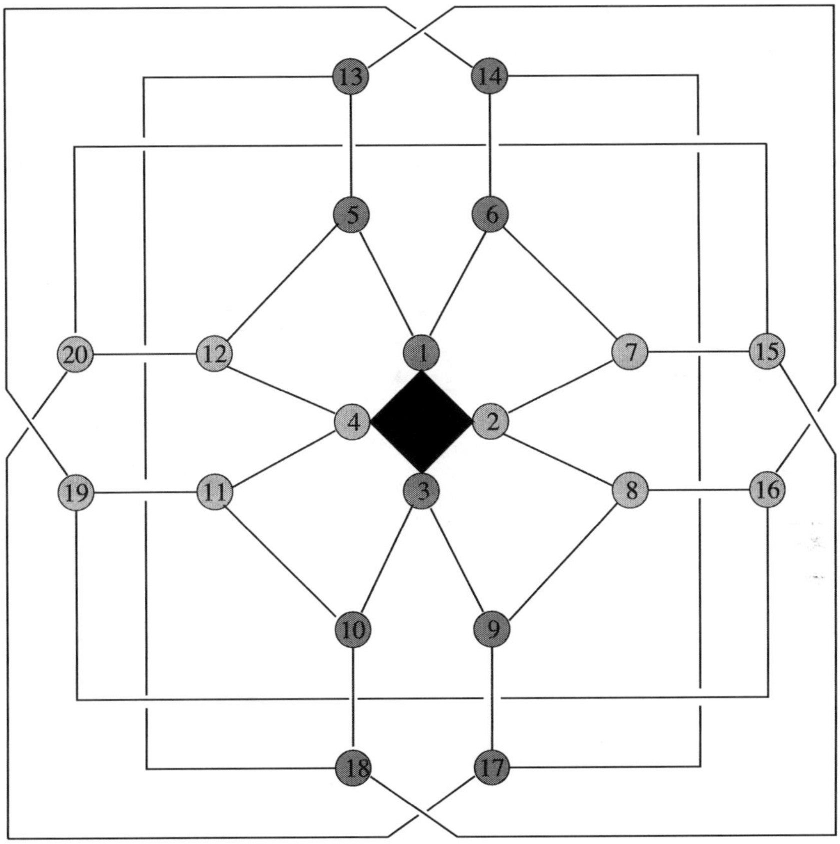

FIGURE 7. Projective admissible solution space $\mathcal{PF}(\mathcal{T}_\mathcal{W})$: The labels of the nodes correspond to the vertex solutions; the shown PL arcs between them can be realised as geodesics in S^{11} with pairwise disjoint interior, and the quadrilateral as a convex spherical quadrilateral.

the following linear functionals:

$$\nu(\mathcal{M}_0) = q_1 - q_1'' + q_2' - q_2'' - q_3 + q_3' \qquad \nu(\mathcal{L}_0^t) = -2q_1' + 2q_1'' - 2q_2' + 2q_2''$$
$$\nu(\mathcal{M}_1) = q_0 - q_0' - q_1' + q_1'' - q_3 + q_3'' \qquad \nu(\mathcal{L}_1^t) = -2q_0' + 2q_0'' - 2q_2' + 2q_2''$$

The signs in the table respect the transverse orientations, which are relevant when the boundary curves of surfaces corresponding to linear combinations of the $N(F_i)$ are computed.

6.2. Some spun-normal surfaces. Using the explicit description of normal surfaces, one can work out the topological type and the position of normal surfaces in the manifold. The pictures of the gluing pattern of some surfaces are shown in Figure 8, where the quadrilaterals and finitely many triangles are used to obtain compact surfaces, along whose boundary components infinite normal annuli have to be attached. The boundary components are drawn in the colour of the corresponding cusp. Triangle coordinates are labelled by numbering the vertices of the four tetrahedra from 0 to 15.

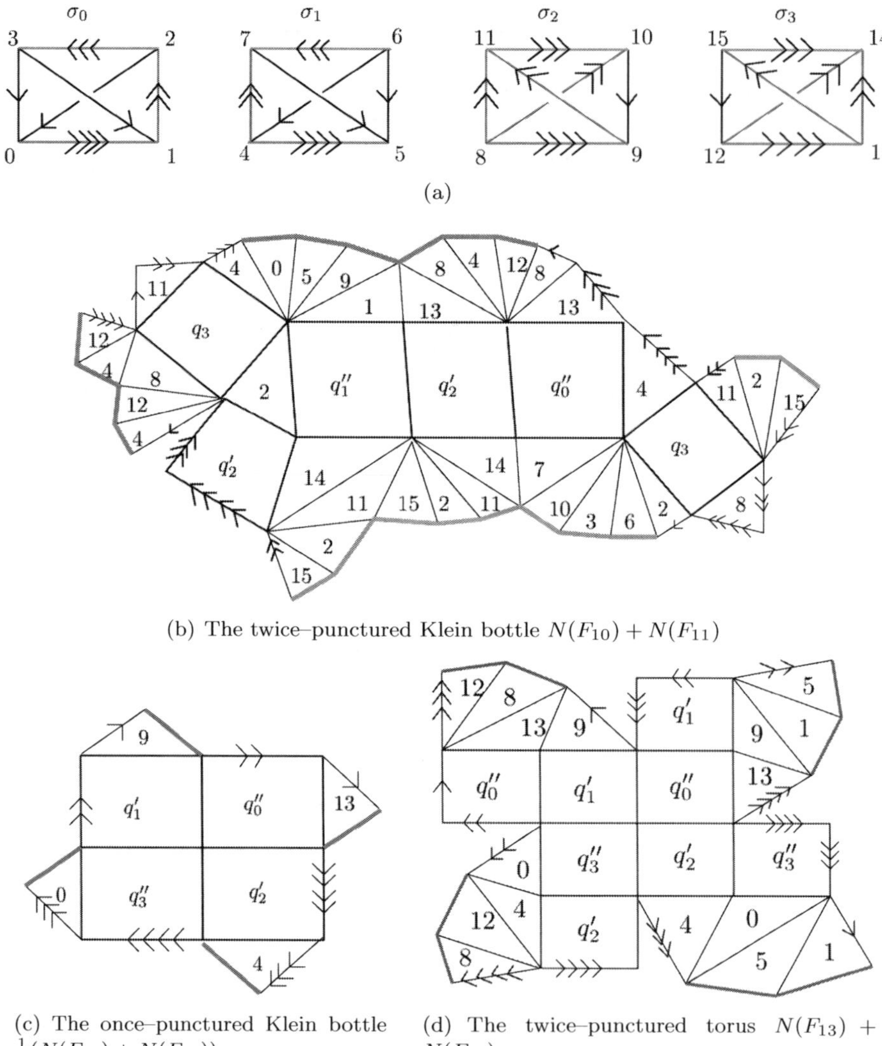

(a)

(b) The twice–punctured Klein bottle $N(F_{10}) + N(F_{11})$

(c) The once–punctured Klein bottle $\frac{1}{2}(N(F_{13}) + N(F_{18}))$

(d) The twice–punctured torus $N(F_{13}) + N(F_{18})$

FIGURE 8. Normal surfaces in the Whitehead link complement

6.3. Equivalence classes. A homeomorphism $N\colon \mathfrak{D}_{\text{pre-}\infty}(\mathcal{W}) \to \mathcal{PF}(\mathcal{T}_\mathcal{W})$ is given in [**39**] between $\mathcal{PF}(\mathcal{T}_\mathcal{W})$ and the tropical pre-variety $\mathfrak{D}_{\text{pre-}\infty}(\mathcal{W})$ obtained from the canonical defining equations. There is an induced action of the group D_4 of symmetries of $\mathfrak{D}(\mathcal{W})$ on $\mathfrak{D}_{\text{pre-}\infty}(\mathcal{W})$. Since the elements of D_4 interchange coordinates, the induced action corresponds to interchanging coordinate triples of elements in $\mathfrak{D}_\infty(\mathcal{W})$. Moreover, there is an induced action of D_4 on $\mathcal{PF}(\mathcal{T}_\mathcal{W})$ via the homeomorphism N which again corresponds to interchanging coordinate triples of elements.

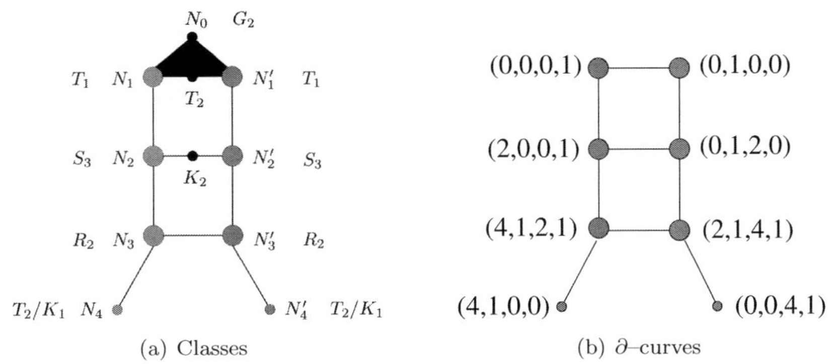

FIGURE 9. Surfaces in the Whitehead link complement

Indeed, D_4 can be identified with a group of symmetries of the triangulation. There is a Klein four group K_f, identified with $\langle \tau_1, \tau_2 \rangle$, which stabilises the cusps, and orbits of the action of K_f on $\mathcal{PF}(\mathcal{T_W})$ give six equivalence classes amongst the vertex solutions in $\mathcal{PF}(\mathcal{T_W})$:

$$N_1 = \{V_2, V_4\}, \qquad N_1' = \{V_1, V_3\},$$
$$N_2 = \{V_7, V_8, V_{11}, V_{12}\}, \qquad N_2' = \{V_5, V_6, V_9, V_{10}\},$$
$$N_3 = \{V_{15}, V_{16}, V_{19}, V_{20}\}, \qquad N_3' = \{V_{13}, V_{14}, V_{17}, V_{18}\}.$$

The classification up to isotopy is given in §6.5 and §6.6. The orbit under K_f corresponds to the different ways a surface can "spin into the cusps". In particular, the (two or four) normal coordinates in each of the classes N_k, N_k' correspond to isotopic surfaces.

Members of the classes N_k and N_k' are interchanged by symmetries interchanging the cusps. These symmetries correspond to the remaining elements of D_4. One can visualise the action of D_4 on $\mathcal{PF}(\mathcal{T_W})$ by considering the action of the dihedral group on Figure 7 induced by its standard action on a square. The quotient by the action of K_f is pictured in Figure 9(a), where the topological types of minimal representatives for some points are indicated.

Note that there are arcs in $\mathcal{PF}(\mathcal{T_W})$ connecting elements of N_3, e.g. the vertices V_{16} and V_{19}. The geometric sum $F_{16} + F_{19}$ is a twice–punctured torus. However, $\frac{1}{2}(N(F_{16}) + N(F_{19}))$ is also an admissible integer solution, and the corresponding normal surface is a once-punctured Klein bottle. The corresponding equivalence class (i.e. K_f orbit) is denoted by N_4. Similarly, for arcs in $\mathcal{PF}(\mathcal{T_W})$ joining elements of N_3' we obtain an equivalence class N_4' whose elements are midpoints of these arcs.

The surface determined by a minimal integer solution corresponding to the point $V_0 := \frac{1}{2}(V_1 + V_3) = \frac{1}{2}(V_2 + V_4)$ in the "centre" of $\mathcal{PF}(\mathcal{T_W})$ is a genus two surface. This point is fixed by all symmetries, and we denote its equivalence class by N_0. The quadrilateral spanned by V_1, V_2, V_3, V_4 is called the *centre square*.

The D_4 orbit of a point $P \in \mathcal{PF}(\mathcal{T_W})$ is now analysed. If $P = V_0$, then its equivalence class only contains one element. If P is contained in the square $[V_1, V_2, V_3, V_4]$ but not equal to its centre, then its D_4 orbit contains exactly four elements. Note that different elements in the square can have the same *∂–coordinate*

$$(\nu_P(\mathcal{L}_0), -\nu_P(\mathcal{M}_0), \nu_P(\mathcal{L}_1), -\nu_P(\mathcal{M}_1)).$$

This is true for instance for V_1 and $\frac{3}{4}V_1 + \frac{1}{4}V_3$. However, elements of the same D_4 orbit are distinguished by their ∂–coordinates. If P is contained in N_4 or N_4', or is the midpoint of an arc $[V, W]$ in $\mathcal{PF}(\mathcal{T_W})$ with $V \in N_i$, $W \in N_i'$ and $i = 2$ or 3, then its D_4 orbit contains exactly four elements. Moreover, the elements of the orbit are distinguished by their ∂–coordinates. If $P \in \mathcal{PF}(\mathcal{T_W})$ is not contained in any of the sets considered above, then its D_4 orbit contains exactly eight elements, and all these elements are distinguished by their ∂–coordinates.

Moreover, the elements of an equivalence class of a point in $\mathcal{PF}(\mathcal{T_W})$ have the same projectivised (i.e. unoriented) ∂–coordinate. Thus, the set of projectivised ∂–coordinates arising from $\mathcal{PF}(\mathcal{T_W})$ can be computed using the incidence structure amongst the equivalence classes, and the result is shown in Figure 9(b). This shows that each equivalence class is uniquely determined by its projectivised ∂–coordinate unless its elements are contained in the centre square of $\mathcal{PF}(\mathcal{T_W})$.

LEMMA 19. *An embedded spun-normal surface in $\mathcal{W}$ is uniquely determined by its* transversely oriented *boundary curves if its projectivised normal Q–coordinate is not contained in the centre square in $\mathcal{PF}(\mathcal{T_W})$.*

PROOF. Let S be an embedded spun-normal surface. Then there is a unique $\alpha > 0$ and a unique point $P \in \mathcal{PF}(\mathcal{T_W})$ such that $N(S) = \alpha P$. If P is not contained in the centre square, then there is no other point in $\mathcal{PF}(\mathcal{T_W})$ with the same ∂–coordinate. Thus, S is uniquely determined by

$$\begin{aligned}&\alpha(\nu_P(\mathcal{L}_0), -\nu_P(\mathcal{M}_0), \nu_P(\mathcal{L}_1), -\nu_P(\mathcal{M}_1))\\ =&(\nu_{N(S)}(\mathcal{L}_0), -\nu_{N(S)}(\mathcal{M}_0), \nu_{N(S)}(\mathcal{L}_1), -\nu_{N(S)}(\mathcal{M}_1)).\end{aligned}$$

This proves the lemma. □

6.4. Criteria for incompressibility. Dunfield and Garoufalidis [**14**] give a simple criterion for a vertex surface to be incompressible, namely that it be a vertex surface with non-empty boundary which has a quadrilateral in each tetrahedron. However, this criterion does not apply to any of the vertex normal surfaces in the given ideal triangulation of the Whitehead link complement.

A certificate for incompressibility can be given using a different result, which the author learned from Nathan Dunfield. This is only summarised here, and complete details will be given in [**40**]. A *semi-angle structure* is an assignment $q \mapsto \alpha(q)$ of a non-negative real number (called angle) to each quadrilateral type q in an ideal triangulation such that the sum of the angles associated to the three distinct quadrilateral types supported by each tetrahedron equals π, and the sum of all angles of the quadrilateral types facing an edge in the 3-manifold equals 2π. See [**26**] for a detailed discussion of this viewpoint, We view $\alpha \in [0, \pi]^\square = [0, \pi]^{3n}$, where $\square$ is the set of all isotopy classes of quadrilateral discs. We remark that if $\alpha \in (0, \pi)^\square$, then it is termed an *angle structure*. This can be viewed as a linear hyperbolic structure and we refer the reader to the excellent exposition in [**16**] for a history and applications of angle structures.

An embedded normal surface S has normal Q–coordinate $N(S) \in \mathbb{N}^\square = \mathbb{N}^{3n}$. Then S is said to be *dual to the semi-angle structure* α, if $\alpha \cdot N(S) = 0$, where the standard Euclidean inner product is taken. In other words, for each quadrilateral type q that has non-zero weight in S, we have $\alpha(q) = 0$.

THEOREM 20 (Dunfield). *Let M be the interior of a compact 3-manifold with boundary a non–empty disjoint union of tori. Let S be an embedded normal surface* *(possibly non–compact) in $(M;\mathcal{T})$ without any boundary parallel components. If S is dual to a semi-angle structure of $(M;\mathcal{T})$, then it is essential.*

Here is a sketch of the proof. Suppose S had a compression disc D. Then D can be put into a normal form relative to the triangulation and the normal surface, and inherits an induced semi-angle structure. The combinatorial Gauß-Bonnet formula then implies that D has non-positive Euler characteristic, a contradiction. Hence the surface is incompressible. The proof is concluded with an observation by Hatcher that incompressible implies boundary incompressible for an embedded surface with boundary only on the torus boundary components of a 3-manifold. A proof of Dunfield's theorem is given in [**40**].

The above theorem has the following simple corollary:

COROLLARY 21. *Let M be the interior of a compact 3-manifold with boundary a non–empty disjoint union of tori, $\mathcal{T}$ be an ideal triangulation of M, and $\alpha \in [0,\pi]^{\square}$ be a semi-angle structure. If S and F are compatible normal surfaces (possibly non–compact) that are both dual to α, then each 2–sided normal surface that is a Haken sum of S and F is essential.*

6.5. Essential normal surfaces. Listed in Table 5 are a number of semi-angle structures on $\mathcal{T}_\mathcal{W}$, as well as the vertex surfaces that are dual to these. In particular, each of the 20 vertex surfaces is essential. It follows from the information in the table and Corollary 21 that each 2–sided normal surface in $\mathcal{W}$ is essential except possibly those whose projectivised normal coordinates lie on the sides or the interior of the central square spanned by V_1, V_2, V_3, V_4.

Each surface F_1, F_2, F_2, F_4 is a once-punctured torus and can be given a natural orientation arising from the transverse orientation of its unique boundary curve. In this way, it represents an element $[F_i] \in H_2(\mathcal{W}^c, \partial\mathcal{W}^c;\mathbb{Z})$, where $\mathcal{W}^c$ is a compact core of $\mathcal{W}$. Note that $[F_3] = -[F_1]$ and $[F_4] = -[F_2]$. As in [**34**], one can now argue that the surfaces S_i are norm minimising, and that each normal surface with projectivised normal coordinate along the boundary of the central square is norm minimising, and hence essential. In particular, we have the following:

OBSERVATION 22. *The boundary of the central square naturally corresponds to the boundary of the unit ball of the Thurston norm.*

It is also shown in [**34**] that every essential surface homologous to a fibre is isotopic to a fibre. Whence every fibre arises as a normal surface with respect to $\mathcal{T}_\mathcal{W}$. This verifies a result of Kang and Rubinstein [**24**].

We now claim that every normal surface not in the boundary of the central square is not a fibre. This can be seen either by considering their boundary slopes, or by determining their image in homology. All classes are mapped to zero except for the surfaces along segments between thrice punctured spheres and once-punctured tori. These cannot be fibres since they only evaluate non-zero on one cusp. We have thus established:

OBSERVATION 23. *Every fibre in $\mathcal{W}$ normalises in precisely two ways. A spun-normal surface in $\mathcal{W}$ is a fibre if and only if its projectivised normal coordinate lies in the interior of an edge of the central square.*

It is also not difficult to describe the once–punctured tori corresponding to the vertices of the central square. The surfaces F_1 and F_3 have identical projectivised boundary slopes, and so do F_2 and F_4. Constructing the corresponding normal surfaces explicitly, one can see that for each of the solutions, there are four normal triangles which form an annulus on the surface and cap the quadrilaterals off at one of the cusps. The union of the two respective annuli on F_1 and F_3 is a boundary parallel torus. Similar for F_2 and F_4. It follows that each pair is obtained from a thrice punctured sphere that meets one of the boundary components in two meridians and the other in a longitude by tubing the two meridional boundary components together. The two choices of annulus give the two surfaces in each pair. However, these are in fact isotopic surfaces. The isotopy is indicated in Figure 10.

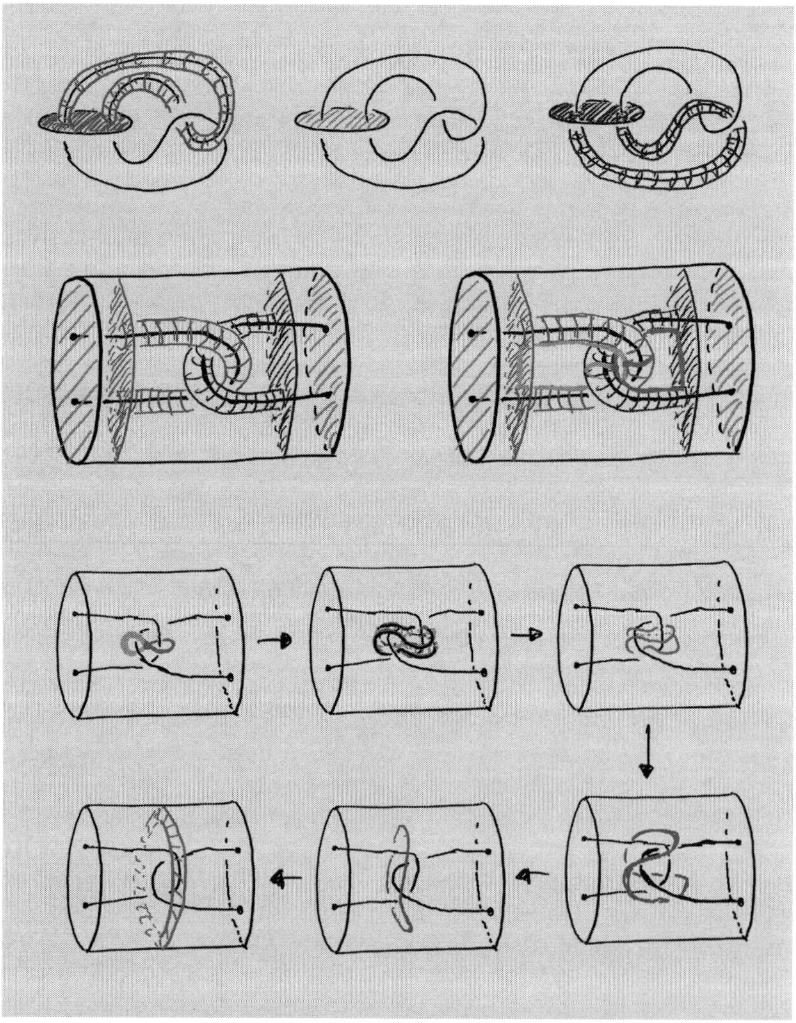

FIGURE 10. Isotopic once-punctured tori. The reader will find it a pleasant exercise in normal surface theory to show that the once-punctured tori normalise to the same normal surface in the complement of a *normal* thrice-punctured disc in $\mathcal{T}_\mathcal{W}$.

The central surface, G_2, is a closed genus two surface linking the red and green edge, and hence compressible. To one side, it compresses to a torus linking the red cusp, and to the other to a torus linking the green cusp. Its complements are therefore compression bodies. The surface G_2 is made of up four quadrilaterals and no triangles. Any surface F whose projectivised normal coordinate lies in the interior of the central square is a Haken sum of the form $F = n_1 S_i + n_2 S_j + n_3 G_2$, where S_i and S_j correspond to vertices of an edge of the central square. In particular, this has the same boundary slope as the surface $n_1 S_i + n_2 S_j$. From the local structure around the green and red edges, we see that each such surface is compressible, namely there are n_3 compression discs, and after the compressions we have a surface in same homology class as $n_1 S_i + n_2 S_j$ and with same Euler characteristic, and hence it is a fibre and in fact isotopic to this. At this point, we may record the following:

OBSERVATION 24. *A spun-normal surface in $\mathcal{W}$ is essential if and only if its projectivised normal coordinate does not lie in the interior of the central square.*

This completes the classification of projectivised essential normal surfaces up to normal isotopy.

6.6. Essential surfaces. It follows from the main result of Walsh [**41**] that every essential surface in $\mathcal{W}$ that is not a fibre or a semi-fibre can be normalised with respect to $\mathcal{T}_{\mathcal{W}}$. A semi-fibre is a separating surface that lifts to a fibre in a double cover. An inspection of the three double covers of $\mathcal{W}$ shows that there is no such surface in $\mathcal{W}$. It is well known to experts that every non-compact essential surface that is not a fibre can be put into spun-normal form by choosing, for each cusp on which the surface has a cusp, an appropriate direction of spinning. A proof of this was recently announced by Kang and Rubinstein [**24**]. It follows that the space $\mathcal{PL}(\mathcal{W})$ of projectivised classes of essential surfaces in $\mathcal{W}$ is precisely the orbit space of the action of K_f on the space of projectivised essential normal surfaces. This is shown in Figure 12(a). Here, the labels of the vertices are with reference to Figure 11, which is taken from an earlier version of [**15**], and the remaining surfaces can be determined from the symmetry interchanging the components.

Up to permuting the link components, the surfaces in Figure 11 have the following normalisations. The pair of once-punctured tori $\{S_7, S_8\}$ normalises to the pair $\{F_2, F_4\}$; the twice-punctured tori S_2 and S_3 normalise to the pairs $\{F_1 + F_4, F_2 + F_3\}$ and $\{F_1 + F_2, F_3 + F_4\}$; the thrice-punctured sphere S_5 normalises to the surfaces $\{F_7, F_8, F_{11}, F_{12}\}$; the twice-punctured projective plane S_4 normalises to the surfaces $\{F_{15}, F_{16}, F_{19}, F_{20}\}$; and the twice punctured torus S_6 normalises to the surfaces $\{F_{15} + F_{20}, F_{16} + F_{19}\}$.

The space of all incompressible surfaces in $\mathcal{W}$ was described by Floyd and Hatcher [**15**] using branched surfaces. They give an algorithm to compute a complex with rational barycentric coordinates that carries all essential surfaces that are *meridional incompressible.* The condition for an essential surface $S \subset S^3 \setminus L$ to be meridional incompressible is as follows. If there is a disc $D \subset S^3$ with $D \cap S = \partial D$ and D meeting L transversely in one point in the interior of D, then there is a disc $D' \subset S \cup L$ with $\partial D' = \partial D$, D' also meeting L transversely in one point. Geometrically, this corresponds to the existence of accidental parabolics. The complex carrying all meridional incompressible surfaces in the Whitehead link

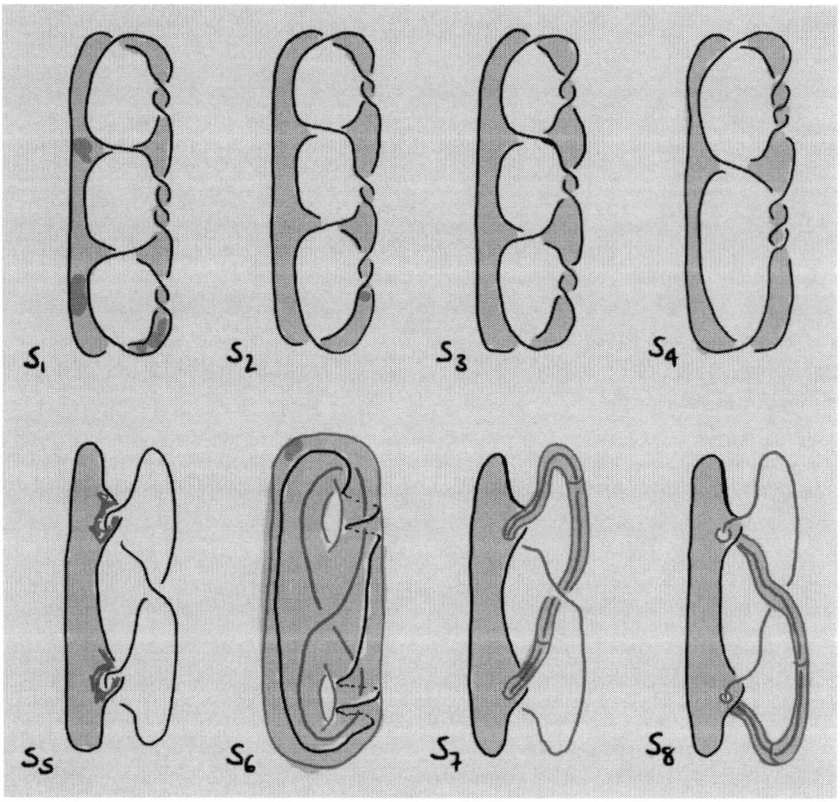

FIGURE 11. Representative surfaces as shown in a pre-print of [**15**].

complement is shown in Figure 12(b) — in [**15**, Figure 5.4] this is the 1–complex with label [2,1,2] (the meaning of this label is irrelevant here).

The vertices of the 1–complex in Figure 12(b) are labelled $1, 2, 3, 4, 4', 5, 5', 6, 6'$, and we denote the 1–simplex with vertices i and j by $[i, j]$. It follows from [**15**, Proposition 6.1] that the interior of each union $[5, 2] \cup [2, 5']$ and $[5, 3] \cup [3, 5']$ consists of fibres, and hence corresponds to a cone over a fibred face of the Thurston norm ball (after choosing coherent orientations of the surfaces representing the vertices).

An explanation of how to modify the algorithm of [**15**] to obtain the essential surfaces that do not satisfy the meridional incompressibility condition is given in §8 of [**15**]. The resulting complex carrying all essential surfaces in the Whitehead link complement according to [**15**] is the 2–complex shown in Figure 12(c). This does not agree with our argument in §6.5 that the space $\mathcal{PL}(\mathcal{W})$ of essential surfaces is as shown in Figure 12(a). A possible explanation is that the isotopy between the once-punctured tori S_7 and S_8 was missed in [**15**], because they are represented as non-isotopic surfaces in Figure 12(c).

6.7. Boundary curve space. The boundary curve space can be computed from the complex $\mathcal{PL}(\mathcal{W})$. The surfaces carried by the segment $[5, 3]$ in Figure 12(b) have the same boundary curves are those carried by $[5, 2]$. Similar for $[3, 5']$ and

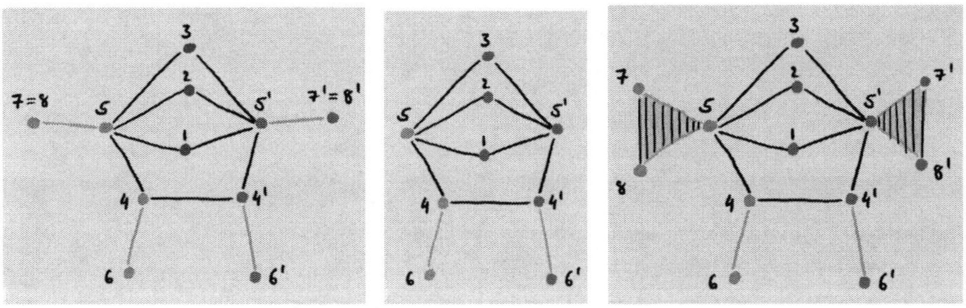

(a) $\mathcal{PL}(\mathcal{W})$: all essential surfaces (b) Meridional incompressible surfaces (c) Floyd-Hatcher

FIGURE 12. Spaces of essential surfaces in the Whitehead link complement: The complexes shown have rational barycentric coordinates; their bilateral symmetry corresponds to an involution exchanging the components of the Whitehead link; labels correspond to representative surfaces shown in Figure 11; and the vertical lines in the 2–simplices in 12(c) indicate the same projective isotopy class; the blue 2-simplices and 1-simplices correspond to surfaces obtained from disjoint unions, whilst the black 1-simplices arise from normal sums or branched surfaces.

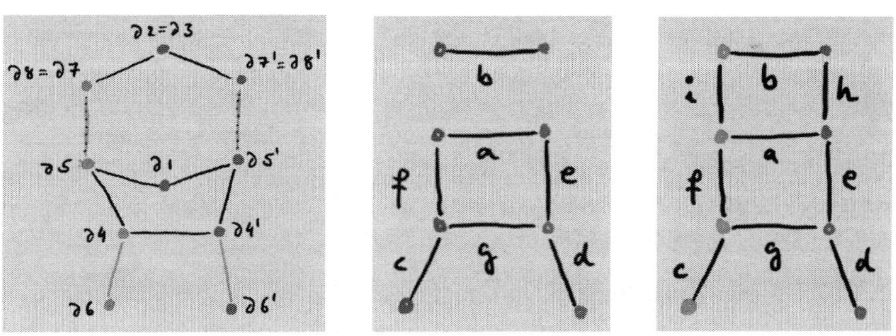

(a) Boundary curve space (b) meridional incompressible surfaces (c) all incompressible surfaces

FIGURE 13. Boundary curve space

$[2, 5']$. All other surfaces are uniquely determined by their boundary curves. Moreover, the projectivised curves match up as shown in Figure 13(a). The dotted blue segments correspond to boundary curves of surfaces that are not meridionally incompressible.

Using the notation of Lash [**25**], the boundary curve space is described explicitly as:

$$\begin{array}{ll} \{(-2,t,-2t,1) \mid t\in[0,\infty]\} & a \\ \cup\{(0,t,0,1) \mid t\in[0,\infty]\} & b \\ \cup\{(2t,t,4,1) \mid t\in(0,1)\} & c \\ \cup\{(4t,t,2,1) \mid t\in(1,\infty)\} & d \\ \cup\{(2t+2,t,2t,1) \mid t\in(0,1)\} & e \\ \cup\{(2,t,2t+2,1) \mid t\in(1,\infty)\} & f \\ \cup\{(-t+3,1,t+3,1) \mid t\in[-1,1\} & g \\ \cup\{(-2t,0,0,1) \mid t\in[0,1]\} & h \\ \cup\{(0,1,-2t,0) \mid t\in[0,1]\} & i \end{array}$$

Lash [**25**] only lists segments a–g and we include h and i to account for the surfaces that are not meridional incompressible. We also changed the sign of the meridian coordinates in [**25**] to account for a difference in conventions. We remark that Lash [**25**] and Hoste-Shanahan [**20**] appear to compute only the space of projectivised boundary slopes of essential surfaces that are meridional incompressible, following only the main algorithm of Floyd and Hatcher. This explains why their spaces are not connected.

7. Surfaces and boundary curves arising from degenerations

The logarithmic limit set of the deformation variety $\mathfrak{D}(\mathcal{T}_{\mathcal{W}})$ and the logarithmic limit set of the eigenvalue variety $\overline{\mathfrak{E}}_0(\mathcal{W})$ are determined in this section. Together with the results on essential surfaces of the previous section, these computations show that all strict boundary curves of $\mathcal{W}$ are detected by the Dehn surgery component of the character variety. Since the boundary curves of fibres are detected by reducible characters, this shows that all boundary slopes of the Whitehead link are strongly detected. In addition, it is shown that $\mathfrak{D}_\infty(\mathcal{T}_{\mathcal{W}})$ has a fake ideal point, and some interesting degenerations are analysed in detail.

7.1. The logarithmic limit sets. The tropical variety $\mathfrak{D}_\infty(\mathcal{W})$ is determined in Appendix D and shown in Figure 14(a). The above analysis of the rational maps and the normal surfaces together with the results concerning strongly detected boundary curves in [**39**] also determines the logarithmic limit set of $\overline{\mathfrak{E}}_0(\mathcal{W})$, which is shown in Figure 14(b). As discussed above, the isotopy relation is given by the action of the Klein four group K_f, and the respective quotient spaces are shown in Figure 15.

It follows from this calculation and Theorem 13 that all non-fibre essential surfaces in $\mathcal{W}$ are detected by the deformation variety and the character varieties. Moreover, no fibre is detected by the deformation variety nor the geometric component in the character variety. We also note that

(1) there is a *fake ideal point*, namely the centre V_0 of the central square, which corresponds to an ideal point of the deformation variety that maps to a finite point of the character variety and the associated surface compresses to two boundary parallel tori;

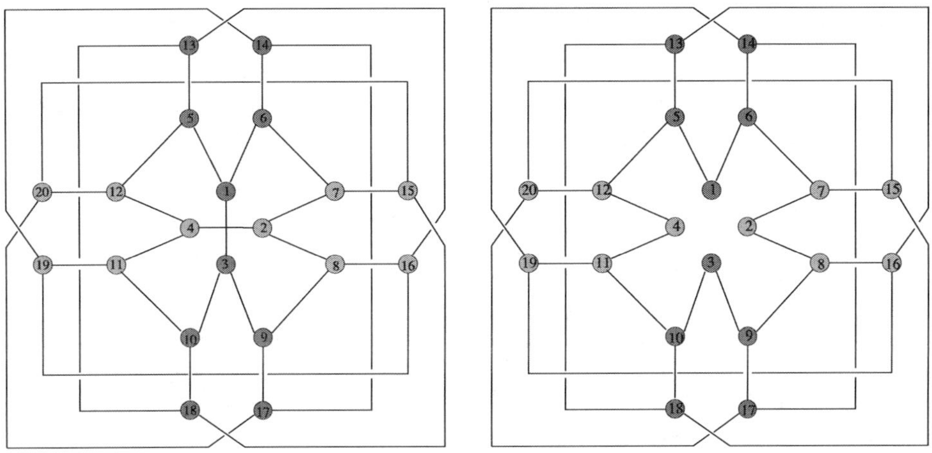

(a) Logarithmic limit set of $\mathfrak{D}(\mathcal{T}_{\mathcal{W}})$ (b) Logarithmic limit set of $\overline{\mathfrak{E}}_0(\mathcal{W})$

FIGURE 14. The logarithmic limit sets of $\mathfrak{D}(\mathcal{T}_{\mathcal{W}})$ and $\overline{\mathfrak{E}}_0(\mathcal{W})$: The labels of the nodes correspond to the vertex solutions; the shown PL arcs between them can be realised as geodesics in S^{11}.

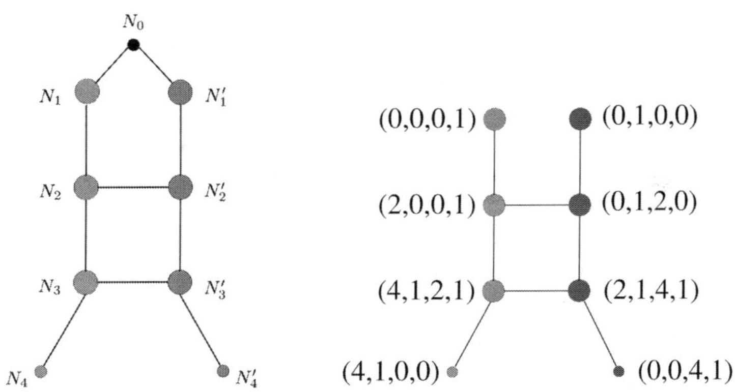

(a) Detected surfaces (b) Detected boundary curves

FIGURE 15. The logarithmic limit sets of $\mathfrak{D}(\mathcal{T}_{\mathcal{W}})$ and $\overline{\mathfrak{E}}_0(\mathcal{W})$ modulo $\mathbb{Z}_2^2$

(2) each ideal point that is in the interior of the segments $[V_1, V_0]$, $[V_2, V_0]$, $[V_3, V_0]$ or $[V_4, V_0]$ detects a surface that is not incompressible but can be compressed to an essential dual surface.

7.2. Detected slopes. The relationship between ideal points of the $\mathrm{PSL}_2(\mathbb{C})$–eigenvalue variety $\overline{\mathfrak{E}}(\mathcal{W})$ and ideal points of the deformation variety $\mathfrak{D}(\mathcal{W})$ has already been discussed, and it follows that the logarithmic limit set of $\overline{\mathfrak{E}}_0(\mathcal{W})$ is as pictured in Figure 14(b). In particular, all these slopes arise from sequences of irreducible representations. This was already shown for the slopes of meridional incompressible surfaces by Lash [**25**], and the present computation extends this to the boundary slopes of all essential surfaces that are not fibres.

OBSERVATION 25. *Let $S \subset \mathcal{W}$ be an essential surface that is not a fibre. Then the boundary slopes of S are strongly detected by a sequence of irreducible representations.*

It is well known that the boundary slopes of fibres are strongly detected by sequences of reducible representations. The slopes detected by both reducible and irreducible representations are precisely the slopes of the once-punctured tori in N_1 and N_1'. There are sequences of shape parameters which converge to ideal points corresponding to elements of N_1 and N_1' giving rise to sequences of irreducible representations in the character variety since the eigenvalue of a longitude is constant equal to -1 throughout the degeneration. There also are sequences of reducible representations detecting the same slopes:

$$\rho_m(\mathcal{M}_i) = \begin{pmatrix} 1 & 1 \\ 0 & 1 \end{pmatrix} \qquad \text{and} \qquad \rho_m(\mathcal{M}_j) = \begin{pmatrix} m & 0 \\ 0 & m^{-1} \end{pmatrix},$$

where $m \to 0$ or $m \to \infty$.

7.3. Some interesting degenerations. Recall that $\mathfrak{D}'(\mathcal{T}_{\mathcal{W}})$ is defined by

$$1 = wxyz \quad \text{and} \quad wx(1-y)(1-z) = (1-w)(1-x)yz, \tag{39}$$

and that for any $w, x, y, z \in \mathbb{C} - \{0, 1\}$ subject to these equations, there is a unique point on $\mathfrak{D}(\mathcal{W})$. We will now describe sequences of points in $\mathfrak{D}'(\mathcal{T}_{\mathcal{W}})$ which correspond to well–defined sequences in $\mathfrak{D}(\mathcal{W})$ approaching the desired ideal points.

1. The point $(w, x, y, z) = (w, -w^{-1}, w, -w^{-1})$ satisfies equations (39) for any $w \in \mathbb{C} - \{0\}$, and if w has positive imaginary part, it defines a positively oriented triangulation of $\mathcal{W}$, and in particular an incomplete hyperbolic structure on $\mathcal{W}$. This implies that the triangulation can be deformed through positively oriented tetrahedra from the complete structure where $w = i$ to the ideal points where $w \to 0$ or $w \to \pm 1$. These points correspond to V_1 and V_3 for $w \to 1$ and $w \to -1$ respectively, and to $\frac{1}{2}(V_{16} + V_{19})$ for $w \to 0$, since the growth rates of all parameters are equal.

We give one example, where we scale to integer coordinates. As $w \to 0$, the ideal point

$$\xi = (-1, 0, 1, 1, -1, 0, -1, 0, 1, 1, -1, 0)$$

is approached. The corresponding normal Q–coordinate is

$$N(\xi) = (0, 1, 0, 0, 0, 1, 0, 1, 0, 0, 0, 1),$$

$$\text{whilst} \quad \frac{1}{2}(N(S_{16}) + N(S_{19})) = (0, 1, 0, 0, 0, 1, 0, 1, 0, 0, 0, 1),$$

which we can rescale to $\frac{1}{2}(V_{16} + V_{19})$.

Moreover, one easily obtains the limiting eigenvalues. As $w \to 0$ there are the following power series expansions for the resulting eigenvalues:

$$\overline{s} = 1, \qquad\qquad t = -1,$$

$$\overline{u} = \frac{1+w}{w(1-w)} = \frac{1}{w} + 2\sum_{i=0}^{\infty} w^i, \qquad\qquad v = w^2.$$

The first two equations reflect the fact that one cusp is complete, and comparing the growth rates at the second cusp (remembering that $\overline{u}$ is the square of an eigenvalue, whilst v is an eigenvalue) gives a point in boundary curve space with coordinates $[0, 0, 4, 1]$.

2. For any $w \in \mathbb{C} - \{0, \pm 1\}$, the point

$$(w, x, y, z) = (w, \frac{1-w}{1+w}, -\frac{1+w}{1-w}, -w^{-1}) \in \mathfrak{D}'(\mathcal{T}_{\mathcal{W}})$$

gives a point on $\mathfrak{D}(\mathcal{W})$. Note that the associated triangulation involves either only flat tetrahedra or both positively and negatively oriented ones. One may use power series expansions to show that a detected surface for $w \to 0$ is S_8:

$$x = \frac{1-w}{1+w} = 1 + 2\sum_{i=1}^{\infty}(-1)^i w^i, \qquad y = -\frac{1+w}{1-w} = -1 - 2\sum_{i=1}^{\infty} w^i, \qquad z = -\frac{1}{w}.$$

Thus, as $w \to 0$, the ideal point

$$\xi = (-1, 0, 1, 0, 1, -1, 0, 0, 0, 1, -1, 0)$$

is approached. The corresponding normal Q–coordinate is

$$N(\xi) = (0, 1, 0, 1, 0, 0, 0, 0, 0, 0, 0, 1),$$

which coincides with the normal Q–coordinate of S_8, and may be rescaled to V_8.

3. Detecting elements of $N_3 \cup N_3'$ is a little more involved. Let

$$w = \frac{-1 - \epsilon^3 + \sqrt{5 - 4\epsilon + 4\epsilon^2 - 2\epsilon^3 + \epsilon^6}}{2(1+\epsilon^2)} \qquad x = \epsilon$$

$$y = -\epsilon \frac{2(1+\epsilon^2)}{-1 - \epsilon^3 + \sqrt{5 - 4\epsilon + 4\epsilon^2 - 2\epsilon^3 + \epsilon^6}} \qquad z = -\epsilon^{-2}$$

Formal substitution of these assignments in the defining equations of the deformation variety shows that whenever the expressions are defined for any small, non–zero ϵ, they determine a point of $\mathfrak{D}(\mathcal{W})$. Since w is differentiable at $\epsilon = 0$ and takes the value $w(0) = \frac{-1 \pm \sqrt{5}}{2}$, it follows that w and y have converging power series expansions at $\epsilon = 0$, and y has a zero of order one at $\epsilon = 0$. The first few terms are:

$$w(\epsilon) = \frac{-1 \pm \sqrt{5}}{2} - \frac{1}{\sqrt{5}}\epsilon + (1 - \frac{17}{5\sqrt{5}})\epsilon^2 + (-3 + \frac{123}{25\sqrt{5}})\epsilon^3 + \dots$$

$$y(\epsilon) = -\frac{2}{-1+\sqrt{5}}\epsilon - \frac{8}{\sqrt{5}(-1+\sqrt{5})^2}\epsilon^2 + \dots$$

Thus, as $w \to 0$, we approach

$$\xi = (0, 0, 0, -1, 0, 1, -1, 0, 1, 2, -2, 0).$$

A detected surface is therefore S_{13}.

4. Let $(w, x, y, z) = (w, w^{-1}, w^{-1}, w)$. Then $w \to 1$ approaches an ideal point corresponding to V_0. The chosen degeneration is through triangulations involving positively and negatively oriented tetrahedra, or triangulations which are entirely flat. In fact, there cannot be a degeneration through positively oriented tetrahedra to this ideal point, since the hyperbolic gluing equation $1 = wxyz$ would imply that throughout this degeneration $\arg(w) + \arg(x) + \arg(y) + \arg(z) = 2\pi$, whilst for each parameter the argument converges to zero.

Experimentation with SnapPy suggests that elements from all but the fourth set can be approached through degenerations only involving positively oriented tetrahedra. Using the projection of the right–handed Whitehead link, the following surgery coefficients can be approached through degenerations involving positively oriented triangulations:

- $(\infty), (4, 1)$. Four tetrahedra degenerate and the splitting surface is a 1–sided Klein bottle.
- $(\infty), (0, 1)$. Two tetrahedra degenerate and two become flat with parameters equal to -1. The splitting surface is a 2–sided torus.
- $(4, 0), (0, 2)$. Three tetrahedra degenerate, the remaining tetrahedron has shape $\frac{1}{2} + \frac{1}{2}i$. The limiting orbifold has volume approximately 0.9159. (SnapPy does not compute a splitting surface if all cusps have been filled.)
- $(8, 2), (4, 2)$. Three tetrahedra degenerate, the remaining tetrahedron has shape $1 + i$, and the limiting orbifold has volume approximately 0.9159. (Dito concerning the splitting surface.)

These examples will be put in a general framework in [**40**].

Appendices

Appendix A. Manifold data

The "gluing and completeness" data obtained from SnapPy is given in Table 7. The relationship to the shape parameters used here is: $w = z_1'$, $x = z_2''$, $y = z_3''$ and $z = z_4''$. From this, one can verify that the holonomies of the peripheral elements given by SnapPy are the inverses of the holonomies of the meridians and geometric longitudes given above, where SnapPy's cusp 0 corresponds to the green cusp (here: cusp 1), and cusp 1 to the red cusp (here: cusp 0).

	z_1	z_2	z_3	z_4	$1-z_1$	$1-z_2$	$1-z_3$	$1-z_4$	
$H'(\mathcal{M}_0)$	1	0	−1	0	−1	0	1	1	0
$H'(\mathcal{M}_1)$	0	0	0	1	1	−1	0	0	0
$H'(\mathcal{L}_0)$	1	0	0	0	0	1	−1	1	0
$H'(\mathcal{L}_1)$	−1	0	0	0	0	−1	−1	1	0
e_1	1	1	1	1	1	−2	0	0	−1
e_2	0	−1	−1	−1	−1	1	1	1	1
e_3	−1	1	1	1	1	0	−2	−2	−1
e_4	0	−1	−1	−1	−1	1	1	1	1

TABLE 7. SnapPy's gluing matrix for $m129$

Appendix B. Eigenvalue variety

B.1. Eigenvalue variety. Denote the component in $\mathfrak{E}(\mathcal{W})$ arising from reducible representations by $\mathfrak{E}^r(\mathcal{W})$. We have $\mathfrak{E}_0(\mathcal{W}) = \overline{\mathfrak{E}(\mathcal{W}) - \mathfrak{E}^r(\mathcal{W})}$, where the overline denotes the Zariski closure. With respect to the chosen affine coordinates, we have $\mathfrak{E}^r(\mathcal{W}) = \{t = v = 1\}$, which is 2–dimensional. $\mathfrak{E}_0(\mathcal{W})$ is 2–dimensional since it is the closure of the image of the eigenvalue map. From this, it also follows that the intersection of $\mathfrak{E}_0(\mathcal{W})$ and $\mathfrak{E}^r(\mathcal{W})$ is a union of two lines:

$$\mathfrak{E}^r(\mathcal{W}) \cap \mathfrak{E}_0(\mathcal{W}) = \{s = t = v = 1\} \cup \{t = u = v = 1\}.$$

The following calculations use well known elimination and extension theorems which can be found in [**10**]. Consider $\mathfrak{C}(\mathcal{W})$ and the eigenvalue map e (see equation (31)). The rational functions determining the eigenvalues of the longitudes are:

$$\begin{aligned} t =&\vartheta_{\mathcal{L}_0^t}(\rho) = s^{-2} - s^{-2}u^2 + u^2 + c(2s^{-1}u + s^{-3}u^{-1} - s^{-3}u) + c^2 s^{-2}, \\ v =&\vartheta_{\mathcal{L}_1^t}(\rho) = u^{-2} - u^{-2}s^2 + s^2 + c(2u^{-1}s + u^{-3}s^{-1} - u^{-3}s) + c^2 u^{-2}. \end{aligned}$$

Then $s^2t - u^2v = s^2 - u^2 + c(su^{-1} - s^{-1}u)$. Since $\mathfrak{C}(\mathcal{W})$ is not contained in the hyperplane $s^2 = u^2$, this shows that the map $\mathrm{e} : \mathfrak{C}(\mathcal{W}) \to \mathfrak{E}_0(\mathcal{W})$ has degree one, and also determines an inverse mapping:

$$(s, t, u, v) \to \left(s, u, \frac{s^2(t-1) + u^2(1-v)}{su^{-1} - s^{-1}u} \right). \tag{40}$$

Recall the defining equation (25) of $\mathfrak{C}(\mathcal{W})$:

$$\begin{aligned} f_1 = (s - s^{-1})(u - u^{-1}) &+ c(s^{-2}u^{-2} - u^{-2} - s^{-2} + 4 - s^2 - u^2 + s^2u^2) \\ &+ c^2(2s^{-1}u^{-1} - su^{-1} - s^{-1}u + 2su) + c^3. \end{aligned}$$

Two additional polynomials $f_2 = s^3ut + \dots$ and $f_3 = su^3v + \dots$ are obtained from the above expressions for t and v. The only variable to be eliminated is c, and the leading coefficients of c in f_1, f_2 and f_3 are monomials in $\mathbb{C}[s, u]$. The elimination is done using resultants (see [**10**] for details). Since s and u are units, it follows that the eigenvalue variety is defined by $Res(f_1, f_2, c) = Res(f_1, f_3, c) = Res(f_2, f_3, c) = 0$. Eliminating redundant factors from the resultants gives the following set of defining equations for $\mathfrak{E}_0(\mathcal{W})$:

$$\begin{aligned} h_1 = t &- s^2t + s^2t^2 - s^4t^2 - u^2 - 2s^2tu^2 + s^4tu^2 - t^2u^2 + 2s^2t^2u^2 \\ &+ s^4t^3u^2 + tu^4 - s^2tu^4 + s^2t^2u^4 - s^4t^2u^4, \\ h_2 = s^2 &- v - s^4v + u^2v + 2s^2u^2v + s^4u^2v - s^2u^4v + s^2v^2 - u^2v^2 \\ &- 2s^2u^2v^2 - s^4u^2v^2 + u^4v^2 + s^4u^4v^2 - s^2u^4v^3, \\ h_3 = s^4t &- s^6t - s^2tu^2 + s^4tu^2 + s^6t^2u^2 - s^2u^2v \\ &+ u^4v + s^2u^4v - 2s^4tu^4v - u^6v + s^2u^6v^2. \end{aligned}$$

The mapping (40) is not defined when $s^2 = u^2$. One may now compute the defining equations of the subvariety of $\mathfrak{E}_0(\mathcal{W})$ on which it is not defined:

$$s^2 = u^2, \qquad t = v, \qquad 0 = -t + s^2(1+t) + s^4t(1-t) - s^6t^2(1+t) + s^8t^2.$$

With a little more effort, one can compute the following inverse mappings defined on two open sets, which cover all but eight points of the eigenvalue variety:

$$\mathrm{e}_1^{-1} : \mathfrak{E}_0(\mathcal{W}) \to \mathfrak{C}(\mathcal{W}) \qquad (s, t, u, v) \to \left(s, u, \frac{(s^2 - v)(1 - u^2)}{su(1+v)} \right) \tag{41}$$

$$\mathrm{e}_2^{-1} : \mathfrak{E}_0(\mathcal{W}) \to \mathfrak{C}(\mathcal{W}) \qquad (s, t, u, v) \to \left(s, u, \frac{(u^2 - t)(1 - s^2)}{su(1+t)} \right) \tag{42}$$

The points corresponding to the complete structure are always singularities of the eigenvalue variety — here determined by the points where $t = v = -1$ and $s^2 = u^2 = 1$. The other points of $\mathfrak{E}_0(\mathcal{W})$ where neither of the above maps are defined are subject to $t = v = -1$ and $s^2 = u^2 = -1$.

The defining equations for the $\mathrm{PSL}_2(\mathbb{C})$–eigenvalue variety can be worked out from $\overline{\mathfrak{E}}(\mathcal{W})$ similarly to the above, or from the results of the previous subsection using Lemma 17. In particular, one has:

$$t = \vartheta_{\mathcal{L}_0^t}(\overline{\rho}_{GL}) = \overline{s}^{-1} - \overline{s}^{-1}\overline{u} + \overline{u} + d(2\overline{s}^{-1} + \overline{s}^{-2}\overline{u}^{-1} - \overline{s}^{-2}) + d^2\overline{s}^{-2}\overline{u}^{-1},$$
$$v = \vartheta_{\mathcal{L}_1^t}(\overline{\rho}_{GL}) = \overline{u}^{-1} - \overline{u}^{-1}\overline{s} + \overline{s} + d(2\overline{u}^{-1} + \overline{u}^{-2}\overline{s}^{-1} - \overline{u}^{-2}) + d^2\overline{u}^{-2}\overline{s}^{-1}.$$

Then $\overline{s}\overline{u}^2v - \overline{s}^2\overline{u}t = \overline{s}\overline{u}^2 - \overline{s}^2\overline{u} + d(\overline{u} - \overline{s})$ implies that the degree of the map $\overline{\mathfrak{C}}(\mathcal{W}) \to \overline{\mathfrak{E}}(\mathcal{W})$ is equal to one. It follows from the above discussion that there are only two points where neither of the following inverse maps is not defined:

$$\overline{\mathrm{e}}_1^{-1} : \overline{\mathfrak{E}}_0(\mathcal{W}) \to \overline{\mathfrak{C}}(\mathcal{W}) \qquad (\overline{s}, t, \overline{u}, v) \to \left(\overline{s}, \overline{u}, \frac{(\overline{s} - v)(1 - \overline{u})}{(1 + v)}\right) \tag{43}$$

$$\overline{\mathrm{e}}_2^{-1} : \overline{\mathfrak{E}}_0(\mathcal{W}) \to \overline{\mathfrak{C}}(\mathcal{W}) \qquad (\overline{s}, t, \overline{u}, v) \to \left(\overline{s}, \overline{u}, \frac{(\overline{u} - t)(1 - \overline{s})}{(1 + t)}\right). \tag{44}$$

The results of this and the previous subsections are summarised in the following lemma. This provides an alternative proof of Lemma 18.

LEMMA 26. *The varieties $\mathfrak{C}(\mathcal{W})$ and $\mathfrak{E}_0(\mathcal{W})$ are birationally equivalent, and so are the varieties $\overline{\mathfrak{C}}(\mathcal{W})$ and $\overline{\mathfrak{E}}_0(\mathcal{W})$.*

B.2. Dehn fillings on one cusp. If one of the cusps is assumed to be complete, then the resulting subvariety, which parametrises hyperbolic Dehn fillings on the other cusp, is defined by:

$$0 = 1 - t + 4\overline{s}t - \overline{s}^2t + \overline{s}^2t^2 \qquad \text{when } \overline{u} = 1, v = -1, \tag{45}$$
$$0 = 1 - v + 4\overline{u}v - \overline{u}^2v + \overline{u}^2v^2 \qquad \text{when } \overline{s} = 1, t = -1. \tag{46}$$

The boundary curves $(0,0,0,1)$, $(0,0,4,1)$ and $(0,1,0,0)$, $(4,1,0,0)$ are detected by these curves respectively. The above equations may be written as the following well–defined trace relations:

$$4 = \operatorname{tr}\overline{\rho}(\mathcal{M}_i^2) - \operatorname{tr}\overline{\rho}(\mathcal{M}_i^2\mathcal{L}_i^t) = \operatorname{tr}\overline{\rho}(\mathcal{M}_i^2) - \operatorname{tr}\overline{\rho}(\mathcal{L}_i^g) \text{ where } i \in \{0,1\}. \tag{47}$$

Curves in $\mathfrak{D}(\mathcal{T}_{\mathcal{W}})$ corresponding to these curves in the eigenvalue variety can readily be determined. Consider points in $\overline{\mathfrak{E}}_0(\mathcal{W})$ of the form $(1, -1, \overline{u}, v)$. The preimage under r of such a point is of the form $\varphi^{-1}(w, -w^{-1}, w, -w^{-1})$ for some $w \in \mathbb{C} - \{0, \pm 1\}$, and hence $H'(\mathcal{M}_1) = \frac{1+w}{w(1-w)}$ and $H'(\mathcal{L}_1^t) = w^2$.

B.3. Eigenvalues at $(\mathbf{w}, \mathbf{x}, \mathbf{y}, \mathbf{z}) = (\mathbf{1}, \mathbf{1}, \mathbf{1}, \mathbf{1})$. Using the implicit function theorem, the eigenvalues at the ideal point

$$\xi = (0, 1, -1, 0, 1, -1, 0, 1, -1, 0, 1, -1)$$

where $w, x, y, z \to 1$ can be computed. Note that this point maps to a singularity of $\mathfrak{D}'(\mathcal{T}_{\mathcal{W}})$. We use the projection $(w, x, y, z) \to (w, y, z)$, with image defined by:

$$0 = g(w, y, z) = w - wy - wz + yz - wy^2z^2 + w^2y^2z^2.$$

The holonomies of meridians take the form

$$H'(\mathcal{M}_0) = -yz\frac{1 - w}{1 - y}, \qquad H'(\mathcal{M}_1) = -\frac{1 - z}{yz(1 - w)},$$

and as ξ is approached, the eigenvalues of the longitudes converge to one. Since the growth rates of all parameters are equal, let

$$w = 1 + \epsilon, \qquad y = 1 - \overline{s}^{-1}\epsilon + \epsilon Y \qquad z = 1 - \overline{u}\epsilon,$$

and we are seeking $Y(\epsilon)$ such that $g(w(\epsilon), y(\epsilon), z(\epsilon)) = 0$ for all $\epsilon \in B_\delta(0)$ and $Y(0) = 0$ for some $\overline{s}, \overline{u} \in \mathbb{C} - \{0\}$. Substitution yields a polynomial $F(\epsilon, Y)$, and the implicit function theorem applies if $(\overline{s} - 1)(\overline{u} - 1) = 0$. We have

$$H'(\mathcal{M}_0) = \frac{(1 - \overline{u}\epsilon)(1 - \overline{s}^{-1}\epsilon + \epsilon Y)}{\overline{s}^{-1} + Y} \quad \text{and} \quad H'(\mathcal{M}_1) = \frac{\overline{u}}{(1 - \overline{u}\epsilon)(1 + \epsilon)}.$$

Thus, for any $\overline{s}, \overline{u} \in \mathbb{C} - \{0\}$, one can obtain the limiting eigenvalues $(1, 1, \overline{u}, 1)$ and $(\overline{s}, 1, 1, 1)$ as ξ is approached. These eigenvalues are precisely the eigenvalues of the reducible representations in $\overline{\mathfrak{E}}(\mathcal{W})$.

Appendix C. Character varieties

A complete description of the character varieties associated to $\mathcal{W}$ is now given.

Since $\mathfrak{E}(\mathcal{W})$ is irreducible, it follows that the $\mathrm{SL}_2(\mathbb{C})$–character variety is of the form $\mathfrak{X}(\mathcal{W}) = \mathfrak{X}_0(\mathcal{W}) \cup \mathfrak{X}^r(\mathcal{W})$, where $\mathfrak{X}_0(\mathcal{W})$ is the Dehn surgery component and $\mathfrak{X}^r(\mathcal{W})$ is the set of reducible characters. A defining equation for the $\mathrm{SL}_2(\mathbb{C})$–Dehn surgery component can be worked out from $\mathfrak{E}(\mathcal{W})$. Let $X = \operatorname{tr}\rho(\mathcal{M}_0), Y = \operatorname{tr}\rho(\mathcal{M}_1)$ and $Z = \operatorname{tr}\rho(\mathcal{M}_0\mathcal{M}_1)$, then $\mathfrak{X}_0(\mathcal{W})$ is defined by

$$0 = F(X, Y, Z) = XY + (2 - X^2 - Y^2)Z + XYZ^2 - Z^3. \tag{48}$$

The set of reducible characters is parametrised by $\operatorname{tr}\rho[\mathcal{M}_0, \mathcal{M}_1] = 2$, which is equivalent to:

$$4 = X^2 + Y^2 + Z^2 - XYZ.$$

The form of the relator in (23) implies that a $\mathrm{PSL}_2(\mathbb{C})$–representation lifts to $\mathrm{SL}_2(\mathbb{C})$ if and only if the relator is equal to the identity in $\mathrm{SL}_2(\mathbb{C})$ for any assignment of matrices representing the $\mathrm{PSL}_2(\mathbb{C})$–representation. The Dehn surgery component of the $\mathrm{PSL}_2(\mathbb{C})$–character variety can be worked out from (48). With $\overline{X} = X^2, \overline{Y} = Y^2$ and $\overline{Z} = Z^2$ one obtains:

$$\begin{aligned} 0 &= \overline{F}(\overline{X}, \overline{Y}, \overline{Z}) \\ &= -\overline{XY} + (4 - 4\overline{X} + \overline{X}^2 - 4\overline{Y} + \overline{Y}^2)\overline{Z} - (4 - 2\overline{X} - 2\overline{Y} + \overline{XY})\overline{Z}^2 + \overline{Z}^3. \end{aligned}$$

Note that $\overline{F}(X^2, Y^2, Z^2) = F(X, Y, Z)F(-X, Y, Z) = F(X, Y, Z)F(X, -Y, Z)$.

Computation of $\mathrm{PSL}_2(\mathbb{C})$–representations which do not lift to $\mathrm{SL}_2(\mathbb{C})$ reveals that there is only a finite set, parametrised by $(\overline{X}, \overline{Y}, \overline{Z}) = (0, 0, 2 \pm \sqrt{2})$. These points do not satisfy $\overline{F}$, and the corresponding representations are irreducible on the peripheral subgroups, with image isomorphic to $\mathbb{Z}_2 \oplus \mathbb{Z}_2$. In particular, they do not contribute any points to the $\mathrm{PSL}_2(\mathbb{C})$–eigenvalue variety. Thus, all boundary curves which are detected by $\overline{\mathfrak{E}}_0(\mathcal{W})$ are also detected by $\mathfrak{E}_0(\mathcal{W})$.

Appendix D. The Logarithmic limit set

The following calculation was first done by Debbie Yuster. The ideal formed by the hyperbolic gluing equations and parameter relations given in Section 5.1 is:

$$\begin{aligned} I =& \langle wxyz - 1, w'x'y'z'w''^2x''^2 - 1, w'x'y'z'y''^2z''^2 - 1, \\ & w - ww'' - 1, w' - ww' - 1, w'' - w''w' - 1, \\ & x - xx'' - 1, x' - x'x - 1, x'' - x''x' - 1, \\ & y - yy'' - 1, y' - y'y - 1, y'' - y''y' - 1, \\ & z - zz'' - 1, z' - z'z - 1, z'' - z''z' - 1\rangle. \end{aligned}$$

Solving the first three expressions for z, z', and z'' respectively gives:

$$z = \frac{1}{wxy}, \qquad z' = \frac{1}{w'x'y'w''^2x''^2}, \qquad z'' = \frac{w''x''}{y''}.$$

The sign in the last equation is determined by equations (14). Substitute these values back into the 15 ideal generators above. The first three generators become 0, and we obtain an ideal with 12 generators in the variables $w, w', w'', x, x', x'', y, y'$, and y''. The last three of these generators, respectively, are:

$$\begin{aligned} & \frac{1}{wxy} - \frac{w''x''}{wxyy''} - 1, \\ & \frac{1}{w'x'y'w''^2x''^2} - \frac{1}{w'x'y'w''^2x''^2wxy} - 1, \\ & \frac{w''x''}{y''} - \frac{1}{w''x''y''w'x'y'} - 1. \end{aligned}$$

We multiply each of these by the least common multiple of its denominators, thus clearing denominators. This gives the ideal:

$$\begin{aligned} J =& \langle w - ww'' - 1, w' - ww' - 1, w'' - w''w' - 1, x - xx'' - 1, x' - x'x - 1, \\ & x'' - x''x' - 1, y - yy'' - 1, y' - y'y - 1, y'' - y''y' - 1, y'' - w''x'' - wxyy'', \\ & w'w''^2x'x''^2y' - 1 - y''w''x''w'x'y', wxy - 1 - w'w''^2x'x''^2y'wxy\rangle. \end{aligned}$$

Since $ww'w'' + 1$, $xx'x'' + 1$, and $yy'y'' + 1$ are in J, we make the following substitutions:

$$w'' = \frac{-1}{ww'}, \qquad x'' = \frac{-1}{xx'}, \qquad y'' = \frac{-1}{yy'}$$

and clear denominators as before. After deleting repeated elements, we are left with four generators:

$$K = \langle ww' + 1 - w', xx' + 1 - x', yy' + 1 - y', yy' - w^2x^2yw'x' + ww'xx'\rangle.$$

We relabel the variables so they are alphabetically contiguous, starting from the beginning of the alphabet, as follows: $w \to a$, $w' \to b$, $x \to d$, $x' \to e$, $y \to c$, $y' \to f$.

We now perform a few operations to make computing the tropical variety easier. Namely, we homogenise the ideal by first homogenising the generators and then saturating with respect to the homogenising variable. For ideals I_0 and J_0, the saturation is the set of all elements f in the ambient ring such that $J_0^N f$ is contained in I_0, for some integer N. Saturating an ideal by a principal monomial ideal does not alter the tropical variety.

We homogenise using the variable g and call the resulting ideal L:

$$L = \langle ab + g^2 - bg, de + g^2 - eg, cf + g^2 - fg, cfg^5 - a^2d^2cbe + abdeg^3 \rangle.$$

We saturate L with respect to $\langle g \rangle$ and find minimal generators, using `Macaulay 2` [**17**]. The final ideal is:

$$\langle de - eg + g^2, cf - fg + g^2, ab - bg + g^2,$$
$$acd - ace + acg - bcd + bce - bcg - beg + bg^2 + cdg - ceg + cg^2 + eg^2 - fg^2 \rangle.$$

Next we run `Gfan`'s commands `tropical_starting_cone` and `tropical_traverse` on these generators, and we obtain a tropical variety isomorphic to the tropical variety of I with the isomorphism determined by the above substitutions. The result is shown in Figure 14(a).

Acknowledgments

The author thanks Debbie Yuster for the calculation in Appendix D, and the referee for suggestions that improved the exposition of this paper. This research has benefitted from the use of the following software packages: `SnapPy` by Marc Culler, Nathan Dunfield, Matthias Goerner and Jeff Weeks [**11**], `Regina` by Ben Burton [**6**], and `gfan` by Anders Jensen [**22**]. The author thanks Craig Hodgson, Diane Maclagan, Hyam Rubinstein, Saul Schleimer and Bernd Sturmfels for helpful discussions. Research of the author is currently supported by an Australian Research Council Future Fellowship (project number FT170100316).

References

[1] G. M. Bergman, *The logarithmic limit-set of an algebraic variety*, Trans. Amer. Math. Soc. **157** (1971), 459–469, DOI 10.2307/1995858. MR280489

[2] R. Bieri and J. R. J. Groves, *The geometry of the set of characters induced by valuations*, J. Reine Angew. Math. **347** (1984), 168–195. MR733052

[3] T. Bogart, A. N. Jensen, D. Speyer, B. Sturmfels, and R. R. Thomas, *Computing tropical varieties*, J. Symbolic Comput. **42** (2007), no. 1-2, 54–73, DOI 10.1016/j.jsc.2006.02.004. MR2284285

[4] S. Boyer and X. Zhang, *Finite Dehn surgery on knots*, J. Amer. Math. Soc. **9** (1996), no. 4, 1005–1050, DOI 10.1090/S0894-0347-96-00201-9. MR1333293

[5] S. Boyer and X. Zhang, *On Culler-Shalen seminorms and Dehn filling*, Ann. of Math. (2) **148** (1998), no. 3, 737–801, DOI 10.2307/121031. MR1670053

[6] Benjamin A. Burton, Ryan Budney, William Pettersson, et al., *Regina: Software for low-dimensional topology*, `http://regina-normal.github.io/`, 1999–2017.

[7] A. Casella, F. Luo, and S. Tillmann, *Pseudo-developing maps for ideal triangulations II: Positively oriented ideal triangulations of cone-manifolds*, Proc. Amer. Math. Soc. **145** (2017), no. 8, 3543–3560, DOI 10.1090/proc/13290. MR3652806

[8] A. A. Champanerkar, *A-polynomial and Bloch invariants of hyperbolic 3-manifolds*, ProQuest LLC, Ann Arbor, MI, 2003. Thesis (Ph.D.)–Columbia University. MR2704573

[9] D. Cooper, M. Culler, H. Gillet, D. D. Long, and P. B. Shalen, *Plane curves associated to character varieties of 3-manifolds*, Invent. Math. **118** (1994), no. 1, 47–84, DOI 10.1007/BF01231526. MR1288467

[10] D. A. Cox, J. Little, and D. O'Shea, *Ideals, varieties, and algorithms*, 4th ed., Undergraduate Texts in Mathematics, Springer, Cham, 2015. An introduction to computational algebraic geometry and commutative algebra. MR3330490

[11] Marc Culler, Nathan M. Dunfield, Matthias Goerner, and Jeffrey R. Weeks, *SnapPy, a computer program for studying the geometry and topology of 3-manifolds*, Available at `http://snappy.computop.org`.

[12] M. Culler, C. McA. Gordon, J. Luecke, and P. B. Shalen, *Dehn surgery on knots*, Ann. of Math. (2) **125** (1987), no. 2, 237–300, DOI 10.2307/1971311. MR881270

[13] M. Culler and P. B. Shalen, *Varieties of group representations and splittings of* 3-*manifolds*, Ann. of Math. (2) **117** (1983), no. 1, 109–146, DOI 10.2307/2006973. MR683804

[14] N. M. Dunfield and S. Garoufalidis, *Incompressibility criteria for spun-normal surfaces*, Trans. Amer. Math. Soc. **364** (2012), no. 11, 6109–6137, DOI 10.1090/S0002-9947-2012-05663-7. MR2946944

[15] W. Floyd and A. Hatcher, *The space of incompressible surfaces in a* 2-*bridge link complement*, Trans. Amer. Math. Soc. **305** (1988), no. 2, 575–599, DOI 10.2307/2000879. MR924770

[16] D. Futer and F. Guéritaud, *From angled triangulations to hyperbolic structures*, Interactions between hyperbolic geometry, quantum topology and number theory, Contemp. Math., vol. 541, Amer. Math. Soc., Providence, RI, 2011, pp. 159–182, DOI 10.1090/conm/541/10683. MR2796632

[17] Daniel R. Grayson and Michael E. Stillman, *Macaulay2, a software system for research in algebraic geometry*, Available at `https://faculty.math.illinois.edu/Macaulay2/`.

[18] W. Haken, *Theorie der Normalflächen* (German), Acta Math. **105** (1961), 245–375, DOI 10.1007/BF02559591. MR141106

[19] C. D. Hodgson, G. R. Meyerhoff, and J. R. Weeks, *Surgeries on the Whitehead link yield geometrically similar manifolds*, Topology '90 (Columbus, OH, 1990), Ohio State Univ. Math. Res. Inst. Publ., vol. 1, de Gruyter, Berlin, 1992, pp. 195–206. MR1184411

[20] J. Hoste and P. D. Shanahan, *Computing boundary slopes of 2-bridge links*, Math. Comp. **76** (2007), no. 259, 1521–1545, DOI 10.1090/S0025-5718-07-01936-9. MR2299787

[21] Gabriel Indurskis, Culler-Shalen seminorms of fillings of the Whitehead link exterior, *Characters in Low-Dimensional Topology* (O. Collin, S. Friedl, C. Gordon, and S. Tillmann, eds.), Contemporary Mathematics, vol. 760, Amer. Math. Soc., Providence, RI, 2020, pp. 167–205.

[22] Anders N. Jensen, *Gfan, a software system for Gröbner fans and tropical varieties*, Available at `http://home.imf.au.dk/jensen/software/gfan/gfan.html`.

[23] E. Kang, *Normal surfaces in non-compact 3-manifolds*, J. Aust. Math. Soc. **78** (2005), no. 3, 305–321, DOI 10.1017/S1446788700008557. MR2142159

[24] E. Kang and J. H. Rubinstein, *Note on spun normal surfaces in 1-efficient ideal triangulations*, Proc. Japan Acad. Ser. A Math. Sci. **91** (2015), no. 8, 118–122, DOI 10.3792/pjaa.91.118. MR3403943

[25] A. E. Lash, *Boundary curve space of the Whitehead link complement*, ProQuest LLC, Ann Arbor, MI, 1993. Thesis (Ph.D.)–University of California, Santa Barbara. MR2689967

[26] F. Luo and S. Tillmann, *Angle structures and normal surfaces*, Trans. Amer. Math. Soc. **360** (2008), no. 6, 2849–2866, DOI 10.1090/S0002-9947-08-04301-8. MR2379778

[27] J. W. Morgan, *The Smith conjecture*, The Smith conjecture (New York, 1979), Pure Appl. Math., vol. 112, Academic Press, Orlando, FL, 1984, pp. 3–6, DOI 10.1016/S0079-8169(08)61632-3. MR758460

[28] J. W. Morgan and P. B. Shalen, *Valuations, trees, and degenerations of hyperbolic structures. I*, Ann. of Math. (2) **120** (1984), no. 3, 401–476, DOI 10.2307/1971082. MR769158

[29] W. D. Neumann and A. W. Reid, *Arithmetic of hyperbolic manifolds*, Topology '90 (Columbus, OH, 1990), Ohio State Univ. Math. Res. Inst. Publ., vol. 1, de Gruyter, Berlin, 1992, pp. 273–310. MR1184416

[30] W. D. Neumann and D. Zagier, *Volumes of hyperbolic three-manifolds*, Topology **24** (1985), no. 3, 307–332, DOI 10.1016/0040-9383(85)90004-7. MR815482

[31] H. Segerman and S. Tillmann, *Pseudo-developing maps for ideal triangulations I: essential edges and generalised hyperbolic gluing equations*, Topology and geometry in dimension three, Contemp. Math., vol. 560, Amer. Math. Soc., Providence, RI, 2011, pp. 85–102, DOI 10.1090/conm/560/11093. MR2866925

[32] P. B. Shalen, *Representations of 3-manifold groups*, Handbook of geometric topology, North-Holland, Amsterdam, 2002, pp. 955–1044. MR1886685

[33] J.-P. Otal, *William P. Thurston: "Three-dimensional manifolds, Kleinian groups and hyperbolic geometry"*, Jahresber. Dtsch. Math.-Ver. **116** (2014), no. 1, 3–20, DOI 10.1365/s13291-014-0079-5. MR3179624

[34] W. P. Thurston, *A norm for the homology of* 3-*manifolds*, Mem. Amer. Math. Soc. **59** (1986), no. 339, i–vi and 99–130. MR823443

[35] S. Tillmann, *On the Kinoshita-Terasaka knot and generalised Conway mutation*, J. Knot Theory Ramifications **9** (2000), no. 4, 557–575, DOI 10.1142/S0218216500000311. MR1758873

[36] S. Tillmann, *Character varieties of mutative 3-manifolds*, Algebr. Geom. Topol. **4** (2004), 133–149, DOI 10.2140/agt.2004.4.133. MR2059186

[37] S. Tillmann, *Boundary slopes and the logarithmic limit set*, Topology **44** (2005), no. 1, 203–216, DOI 10.1016/j.top.2004.07.002. MR2104008

[38] S. Tillmann, *Normal surfaces in topologically finite 3-manifolds*, Enseign. Math. (2) **54** (2008), no. 3-4, 329–380. MR2478091

[39] S. Tillmann, *Degenerations of ideal hyperbolic triangulations*, Math. Z. **272** (2012), no. 3-4, 793–823, DOI 10.1007/s00209-011-0958-8. MR2995140

[40] ———, *Geometric splittings at ideal points*, in preparation.

[41] G. S. Walsh, *Incompressible surfaces and spunnormal form*, Geom. Dedicata **151** (2011), 221–231, DOI 10.1007/s10711-010-9529-0. MR2780747

[42] T. Yoshida, *On ideal points of deformation curves of hyperbolic 3-manifolds with one cusp*, Topology **30** (1991), no. 2, 155–170, DOI 10.1016/0040-9383(91)90003-M. MR1098911

School of Mathematics and Statistics F07, The University of Sydney, NSW 2006 Australia

Email address: stephan.tillmann@sydney.edu.au

Contemporary Mathematics
Volume **760**, 2020
https://doi.org/10.1090/conm/760/15297

A note on the coherence of 3-manifold groups

Daniel T. Wise

ABSTRACT. We explain that the coherence of 3-manifold groups is a consequence of the coherence of free-by-cyclic groups.

1. Introduction

DEFINITION 1. A group G is *coherent* if each finitely generated subgroup of G is finitely presented.

There are two especially striking results about coherence:
The first was proven by Scott [**9**] and Shalen (unpublished).

THEOREM 2. *Let M be a 3-manifold. Then $\pi_1 M$ is coherent.*

The second was proven by Feighn-Handel in [**3**]:

THEOREM 3. *Every Free-by-Cyclic group is coherent.*

Coherence holds for "small" groups like finite groups, virtually abelian groups, or more generally virtually polycyclic groups. However, there is currently a limited number of known classes of coherent groups, and we refer to [**12**] for a survey of what is known and plausible ways forward. Moreover, coherence appears to be a 2-dimensional phenomenon in the sense that the known coherent higher dimensional groups tend to be manufactured from polycyclic groups using free constructions (see e.g. Theorem 7). The coherence of 3-manifold groups appears to contraindicate this trend.

The goal of this note is to explain that Theorem 2 is actually a consequence of Theorem 3. The original proof of Theorem 2 uses classical 3-manifold topology ideas e.g. the disk theorem, coupled with some interesting group theoretical ideas that are simple but subtle. The proof of Theorem 3 is elementary and self-contained. However, the transition described here between the two results is built upon our modern comprehensive understanding of 3-manifolds: The JSJ decomposition, Geometrization, and Virtual Fibering. For our objective, these advances serve to explicate the structure of a 3-manifold group, and our interest is to extract coherence from this structure and pinpoint its source as supported by some simple controlling algebra. We are certainly not offering a simpler overall proof of Theorem 2. The purpose of the proof is rather to support the idea that 3-manifold

2010 *Mathematics Subject Classification.* Primary 20F06, 05C38.
Research supported by NSERC.

coherence has a 2-dimensional nature. There is no novelty here, except for this contention and the assemblage of its supporting ingredients.

2. Preliminaries

LEMMA 4. *If G has a coherent finite index subgroup G', then G is itself coherent.*

It follows that coherence is a commensurability invariant.

PROOF. Without loss of generality, we may assume G' is normal. Indeed, the intersection of all conjugates of G' is normal and of finite index in G. And it is coherent since G' is coherent.

Let H be a f.g. subgroup of G. Let $H' = H \cap G'$. Then H' is f.g. since $[H : H'] < \infty$. Hence H' is f.p. since G' is coherent. Thus H is f.p. since $[H : H'] < \infty$. Indeed, there is a short exact sequence $1 \to H' \to H \to H/H' \to 1$. And H/H' is f.p. since it is finite. Thus H is f.p. since finite presentability is preserved under extension. □

LEMMA 5. *Free groups are coherent. More generally, $\pi_1 S$ is coherent for any surface S.*

PROOF. The first holds since a subgroup of a free group is itself free. The second holds since a subgroup of $\pi_1 S$ is either of finite index (and hence f.p.) or is of infinite index, and hence free. □

Note that $\pi_1 S$ is free-by-cyclic unless S is a projective plane, so Lemma 5 is a trivial special case of Theorem 3.

Polycyclic groups are coherent, as can be proven by induction using the following:

LEMMA 6. *Consider a short exact sequence $1 \to N \to G \to Q \to 1$. If N is Noetherian and coherent, and Q is coherent then G is coherent.*

PROOF. Let $H \subset G$ be a f.g. subgroup. Observe that there is a short exact sequence $1 \to (N \cap H) \to H \to P \to 1$ where P is the image of H in Q. Then Q is f.g. and hence f.p. by hypothesis. And $N \cap H$ is f.g. by hypothesis, and hence hence f.p. by hypothesis. Finally, H is f.p. since finite presentability is preserved by extensions. □

The following theorem is essentially due to Karrass-Solitar [**6**].

THEOREM 7. *A graph of groups with coherent vertex groups, and Noetherian edge groups is coherent.*

PROOF. Let X be a graph of spaces with underlying graph Γ corresponding to the graph of groups decomposition of G. Let $\widehat{X}$ be the based covering space of X corresponding to a finitely generated subgroup of G, and let $\widehat{\Gamma}$ be the underlying graph of the graph of spaces structure of $\widehat{X}$.

Since $\pi_1 \widehat{X}$ is finitely generated, for some finite subgraph Λ of $\widehat{\Gamma}$, the graph of spaces Y corresponding to Λ has the property that its inclusion in $\widehat{X}$ induces a π_1-isomorphism $\pi_1 Y \cong \pi_1 \widehat{X}$.

By hypothesis, each edge group of Λ is finitely generated. Since $\pi_1 Y$ is f.g. it then follows that each vertex group of Λ is f.g. and hence f.p. by hypothesis.

Combining the vertex groups along these f.g. edge groups yields a finite presentation for $\pi_1\widehat{X}$. □

Theorem 7 implies that coherence is determined by the rigid subgroups in a JSJ decomposition for G. A quick description of the classical JSJ decomposition for 3-manifolds is given in [**5**, Sec 2]. The JSJ decomposition for f.p. groups over cyclic and then polycyclic groups is given in [**8**] and [**2**, **4**].

PROPOSITION 8. *Let M be a finite volume hyperbolic 3-manifold. Then $\pi_1 M$ is coherent.*

PROOF. Note that M has a finite cover $\widehat{M}$ that is homeomorphic to a surface bundle over a circle, and hence $\pi_1\widehat{M} \cong \pi_1 S \rtimes \mathbb{Z}$. This is proven in [**11**] when M has cusps, and in [**1**] when M is closed.

By Lemma 4, $\pi_1 M$ is coherent if $\pi_1\widehat{M}$ is coherent.

Finally, we verify that $\pi_1\widehat{M}$ is coherent. Let $H \subset \pi_1\widehat{M}$ be f.g. If $[\pi_1\widehat{M} : H] < \infty$ then H is f.p. since H equals π_1 of a finite cover of $\widehat{M}$. If $H \subset \pi_1 S$ then H is f.p. by Lemma 5. Finally, we are left with the case where $[\pi_1 S : H \cap \pi_1 S] = \infty$ and H has infinite image $\bar{H}$ in the quotient $\pi_1 M \to \mathbb{Z}$. Note that $H \cap \pi_1 S$ is free. The short exact sequence $1 \to (H \cap \pi_1 S) \to H \to \bar{H} \to 1$ shows that H is free-by-cyclic. Thus H is f.p. by Theorem 3. □

LEMMA 9. *Let N be a Seifert-fibered 3-manifold. Then $\pi_1 N$ is coherent.*

PROOF. There is a short exact sequence $1 \to \mathbb{Z} \to \pi_1 N \to K \to 1$ where K is virtually a free group or a surface group. Thus K is coherent by Lemma 4 and Lemma 5. Finally, $\pi_1 N$ is coherent by Lemma 6. □

We refer to [**10**] for an inviting discussion of Seifert-fibered spaces and the prime and JSJ decompositions which are used below, as well as an account of Thurston's geometrization conjecture, which was proven by Perelman [**7**].

3. Free-by-$\mathbb{Z}$ $\Rightarrow$ 3-manifold

We now explain that Theorem 3 implies Theorem 2.

Let M be a compact 3-manifold. By doubling M along its boundary, we may assume that $\partial M = \emptyset$. Note that $\pi_1 M \hookrightarrow \pi_1(M \cup_{\partial M} M)$, and it suffices to prove coherence of a larger group containing $\pi_1 M$. It suffices to understand the prime case (since the coherence of a free product is implied by coherence of its free factors - a trivial case of Theorem 7). The JSJ decomposition of M provides a splitting of $\pi_1 M$ as a graph of groups where each edge group is $\mathbb{Z}^2$ and each vertex group is geometric.

Thus by Theorem 7 it suffices to show that each vertex group is coherent. Each non-hyperbolic piece has coherent π_1 by Lemma 9. Each hyperbolic piece is either a finite volume hyperbolic manifold with one or more cusps, i.e. it is compact atoroidal and boundary consisting of one or more tori, hence its π_1 is virtually free-by-cyclic by [**11**].

Suppose the JSJ decomposition of M is trivial. Then either $\pi_1 M$ is polycyclic and hence coherent by Lemma 6, or $\pi_1 M$ is finite, or M is Seifert-fibered so $\pi_1 M$ is coherent by Lemma 9, or M is hyperbolic so $\pi_1 M$ is coherent by Proposition 8.

References

[1] I. Agol, *The virtual Haken conjecture*, Doc. Math. **18** (2013), 1045–1087, DOI 10.1016/j.procs.2013.05.269. With an appendix by Agol, Daniel Groves, and Jason Manning. MR3104553

[2] M. J. Dunwoody and M. E. Sageev, *JSJ-splittings for finitely presented groups over slender groups*, Invent. Math. **135** (1999), no. 1, 25–44, DOI 10.1007/s002220050278. MR1664694

[3] M. Feighn and M. Handel, *Mapping tori of free group automorphisms are coherent*, Ann. of Math. (2) **149** (1999), no. 3, 1061–1077, DOI 10.2307/121081. MR1709311

[4] K. Fujiwara and P. Papasoglu, *JSJ-decompositions of finitely presented groups and complexes of groups*, Geom. Funct. Anal. **16** (2006), no. 1, 70–125, DOI 10.1007/s00039-006-0550-2. MR2221253

[5] A. Hatcher. Notes on basic 3-manifold topology. `https://www.math.cornell.edu/~hatcher/3M/3M.pdf`, 2007.

[6] A. Karrass and D. Solitar, *The subgroups of a free product of two groups with an amalgamated subgroup*, Trans. Amer. Math. Soc. **150** (1970), 227–255, DOI 10.2307/1995492. MR260879

[7] B. Kleiner and J. Lott, *Notes on Perelman's papers*, Geom. Topol. **12** (2008), no. 5, 2587–2855, DOI 10.2140/gt.2008.12.2587. MR2460872

[8] E. Rips and Z. Sela, *Cyclic splittings of finitely presented groups and the canonical JSJ decomposition*, Ann. of Math. (2) **146** (1997), no. 1, 53–109, DOI 10.2307/2951832. MR1469317

[9] G. P. Scott, *Finitely generated 3-manifold groups are finitely presented*, J. London Math. Soc. (2) **6** (1973), 437–440, DOI 10.1112/jlms/s2-6.3.437. MR380763

[10] P. Scott, *The geometries of 3-manifolds*, Bull. London Math. Soc. **15** (1983), no. 5, 401–487, DOI 10.1112/blms/15.5.401. MR705527

[11] D. T. Wise. *The structure of groups with a quasiconvex hierarchy.* Annals of Mathematics Studies, to appear.

[12] D. T. Wise. An Invitation to Coherent Groups. In D. Thurston, editor, *What's Next?: The Mathematical Legacy of William P. Thurston*, pages 1–89. Princeton University Press, 2020. To appear.

Dept. of Math. & Stats, McGill University, Montreal, Quebec, Canada H3A 2K6
Email address: `wise@math.mcgill.ca`

Contemporary Mathematics
Volume **760**, 2020
https://doi.org/10.1090/conm/760/15298

Remarks on $SU(2)$-simple knots and $SU(2)$-cyclic 3-manifolds

Xingru Zhang

Dedicated to Steve Boyer on the occasion of his 65th birthday

ABSTRACT. We give some remarks on two closely related issues as stated in the title. In particular we show that a Montesinos knot is $SU(2)$-simple if and only if it is a 2-bridge knot, extending a result of Zentner for 3-tangle summand pretzel knots. We conjecture with some evidence that an $SU(2)$-cyclic rational homology 3-sphere is an L-space.

For a knot K in S^3, $M_{\rm K}$ will be its exterior and μ a meridian slope of K. Up to a choice of an orientation for μ and a choice of the base point for $\pi_1(M_{\rm K})$, we may also consider μ as an element of $\pi_1(M_{\rm K})$. A representation $\rho : \pi_1(M_{\rm K}) \to SU(2)$ is called trace free if the trace of $\rho(\mu)$ is zero (which is obviously well defined). An $SU(2)$-representation of $\pi_1(M_{\rm K})$ is called binary dihedral if its image is isomorphic to a binary dihedral group. Note that every binary dihedral representation of $\pi_1(M_{\rm K})$ is trace free [**K**, Proof of Theorem 10]. A knot K is called $SU(2)$-*simple* if every irreducible trace free $SU(2)$-representation of $\pi_1(M_{\rm K})$ is binary dihedral.

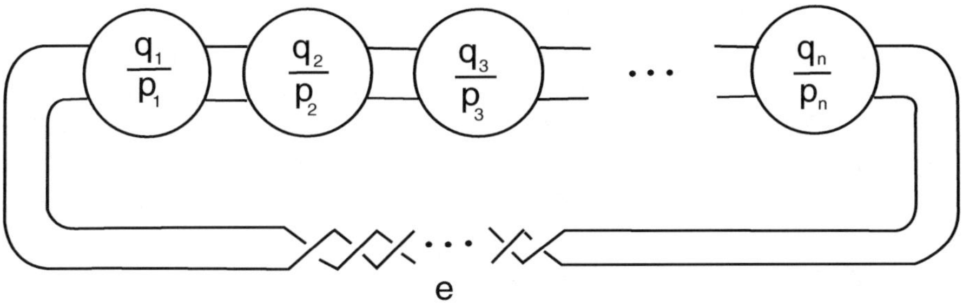

FIGURE 1. A Montesinos link $K(e; q_1/p_1, q_2/p_2, ..., q_n/p_n)$.

A Montesinos link is usually denoted by $K(e; q_1/p_1, q_2/p_2, ..., q_n/p_n)$ where q_i/p_i represents a rational tangle, $|p_i| > 1$ and $(q_i, p_i) = 1$ for all i (see Figure 1). By combining the e twists in the figure with one of the tangles, we may assume that $e = 0$, and we will simply write a Montesinos link as $K(q_1/p_1, \cdots, q_n/p_n)$ and sometimes we refer it as a cyclic tangle sum of n rational tangles. When $q_i = 1, i = 1, ..., n$, we get a pretzel link. In [**Z1**] it was shown that every pretzel

2010 *Mathematics Subject Classification.* Primary 57M25.

The author thanks Steven Sivek and the referee for helpful comments.

knot $K(1/p, 1/q, 1/r)$, p, q, r pairwise coprime, is not $SU(2)$-simple. In this paper we extend this result to all Montesinos knots of at least three rational tangle summands.

THEOREM 1. *Every Montesinos knot $K(q_1/p_1, \cdots, q_n/p_n)$, where $n \geq 3$ and $|p_i| > 1$ for all $i = 1, ..., n$, is not $SU(2)$-simple.*

For any $SU(2)$-representation $\rho : \pi_1(M_K) \to SU(2)$, let $\bar{\rho} : \pi_1(M_K) \to PSU(2)$ be the induced $PSU(2)$-representation. If $\rho : \pi_1(M_K) \to SU(2)$ is a trace free representation, then $\rho(\mu)$ is an order 4 matrix and $\rho(\mu^2) = -I$, where I is the identity matrix of $SU(2)$. So $\bar{\rho} : \pi_1(M_K) \to PSU(2)$ factors through the quotient group $\pi_1(M_K)/\langle \mu^2 \rangle$, where $\langle \mu^2 \rangle$ denotes the normal subgroup of $\pi_1(M_K)$ generated by μ^2. Let $\Sigma_2(K)$ denote the double branched cover of (S^3, K), then $\pi_1(\Sigma_2(K))$ is an index two subgroup of $\pi_1(M_K)/\langle \mu^2 \rangle$. It is known that a trace free irreducible representation $\rho : \pi_1(M_K) \to SU(2)$ is binary dihedral if and only if the restriction of $\bar{\rho}$ on $\pi_1(\Sigma_2(K))$ has nontrivial cyclic image [**K**, Section I.E]. For any 2-bridge knot K, $\Sigma_2(K)$ is a lens space and so every irreducible trace free SU(2)-representation of $\pi_1(M_K)$ is binary dihedral, that is, every 2-bridge knot is $SU(2)$-simple. As a Montesinos knot is a 2-bridge knot if and only if it has lass than three rational tangle summands, we have

COROLLARY 2. *A Montesinos knot is $SU(2)$-simple if and only if it is a 2-bridge knot.*

By [**KM**, Corollary 7.17], every nontrivial knot in S^3 has an irreducible trace free $SU(2)$-representation. It follows that if the double branched cover of a nontrivial knot K is a homology 3-sphere, i.e. if the knot determinant $|\Delta_K(-1)| = 1$ where $\Delta_K(t)$ is the Alexander polynomial of K, then K is not $SU(2)$-simple.

A 3-manifold Y is called $SU(2)$-cyclic (resp. $PSU(2)$-cyclic) if every $SU(2)$-representation (resp. $PSU(2)$-representation) of $\pi_1(Y)$ has cyclic image. In general $PSU(2)$-cyclic is a stronger condition than $SU(2)$-cyclic, that is, $PSU(2)$-cyclic implies $SU(2)$-cyclic but not the other way around. Since $\Sigma_2(K)$ is an $\mathbb{Z}_2$-homology 3-sphere [**R**, Cor 3 of 8D], every $PSU(2)$-representation of $\pi_1(\Sigma_2(K))$ lifts to an $SU(2)$-representation [**BZ**, Page 752] and thus $\Sigma_2(K)$ is $SU(2)$-cyclic if and only if it is $PSU(2)$-cyclic. So if $\Sigma_2(K)$ is $SU(2)$-cyclic, then K is an $SU(2)$-simple knot. The following question concerns the converse.

QUESTION 3. Is there an $SU(2)$-simple knot K in S^3 whose double branched cover $\Sigma_2(K)$ is not $SU(2)$-cyclic (that is, the double branched cover $\Sigma_2(K)$ has irreducible $PSU(2)$-representations but none of them extend to M_K)?

One may consider an $SU(2)$-cyclic 3-manifold as an $SU(2)$-representation L-space. The following conjecture suggests that for a rational homology 3-sphere being $SU(2)$-cyclic is more restrictive than being a usual L-space in the Heegaard Floer homology sense.

CONJECTURE 4. *If a rational homology 3-sphere is $SU(2)$-cyclic, then it is an L-space.*

Certainly the converse of Conjecture 4 does not hold; there are many L-spaces which are not $SU(2)$-cyclic. For instance, the double branched covers of all alternating Montesinos knots of at least three tangle summands are not $SU(2)$-cyclic but are L-spaces.

Here are some evidences for the conjecture. Let $K_1 = T(p_1, q_1)$ and $K_2 = T(p_2, q_2)$ be two torus knots in S^3, and let M_1 and M_2 be their exteriors. Let $Y(T(p_1, q_1), T(p_2, q_2))$ be the graph manifold obtained by gluing M_1 and M_2 along their boundary tori by an orientation reversing homeomorphism $h : \partial M_1 \to \partial M_2$ which identifies the meridian slope in ∂M_1 to the Seifert fiber slope in ∂M_2 and identifies the Seifert fiber slope in ∂M_1 with the meridian slope in ∂M_2. By [**Mot**, Proposition 5] $Y(T(p_1, q_2), T(P_2, q_2))$ has only cyclic $PSL_2(\mathbb{C})$-representations. (Although it was assumed in [**Mot**] that all p_1, q_1, p_2, q_2 are positive, the same argument with obvious modification works without this assumption). Therefore $Y(T(p_1, q_1), T(p_2, q_2))$ is $SU(2)$-cyclic.

PROPOSITION 5. *$Y(T(p_1, q_1), T(p_2, q_2))$ is an L-space.*

PROOF. We prove this assertion by applying [**HW**, Theorem 1.6]. By that theorem, we just need to verify that $h(\mathcal{L}^{\circ}_{M_1}) \cup \mathcal{L}^{\circ}_{M_2} \cong \mathbb{Q} \cup \{1/0\}$, where $\mathcal{L}^0_{M_i}$ is the interior of the set of L-space filling slopes of M_i, $i = 1, 2$. Note that a general torus knot can be expressed as $T(p, q)$ with $(p, q) = 1$ and $|p|, q \geq 2$. By [**OS**, Corollary 1.4]

$$\mathcal{L}^0(M_i) = \begin{cases} \text{slopes in the open interval } (p_iq_i - p_i - q_i, \infty), & \text{if } p_i > 0, \\ \text{slopes in the open interval } (-\infty, p_iq_i - p_i + q_i), & \text{if } p_i < 0. \end{cases} \tag{1}$$

Let μ_i, λ_i be the meridian and longitude of K_i. Note that p_iq_i is the Seifert fiber slope in ∂M_i. We have $h(\mu_1) = \mu_2^{p_2q_2}\lambda_2$ and $h(\mu_1^{p_1q_1}\lambda_1) = \mu_2$. Hence for a general slope m/n in ∂M_1, where m, n are relative prime,

$$\begin{aligned} h(\mu_1^m\lambda_1^n) = h(\mu_1^{m-p_1q_1n}(\mu_1^{p_1q_2}\lambda_1)^n) &= (\mu_2^{p_2q_2}\lambda_2)^{m-p_1q_1n}\mu_2^n \\ &= \mu_2^{p_2q_2(m-p_1q_1n)+n}\lambda_2^{m-p_1q_1n}. \end{aligned}$$

Now suppose a/b is a slope in ∂M_2, where a, b are relatively prime. Choose $n = a - p_2q_2b$ and $m = p_1q_1(a - p_2q_2b) + b$, then m, n are relatively prime, $h(m/n) = a/b$, and

$$\frac{m}{n} = p_1q_1 + \frac{b}{a - p_2q_2b} = p_1q_1 + \frac{1}{\frac{a}{b} - p_2q_2}. \tag{2}$$

Case 1. $p_1 > 0$ and $p_2 > 0$.

For any $a/b \notin \mathcal{L}^0(M_2)$, i.e. either $a/b = 1/0$ or a/b is finite and $a/b \leq p_2q_2 - p_2 - q_2$ by (1), choose correspondingly in ∂M_1 the slope $m/n = p_1q_1$ or as in (2) which yields $m/n \geq p_1q_1 + \frac{1}{-p_2-q_2} > p_1q_1 - 1$. So in either case $m/n \in \mathcal{L}^0(M_1)$ by (1) and $h(m/n) = a/b$, which means $h(\mathcal{L}^{\circ}_{M_1}) \cup \mathcal{L}^{\circ}_{M_2} \cong \mathbb{Q} \cup \{1/0\}$ in this case.

Case 2. $p_1 > 0$ and $p_2 < 0$.

For any $a/b \notin \mathcal{L}^0(M_2)$, we may assume that a/b is finite and so $a/b \geq p_2q_2 - p_2 + q_2$ by (1). So $\frac{a}{b} - p_2q_2$ is positive. Choose the slope m/n in ∂M_1 as in (2) which yields $m/n > p_1q_1$ in this case. So $m/n \in \mathcal{L}^0(M_1)$ by (1) and $h(m/n) = a/b$. Thus $h(\mathcal{L}^{\circ}_{M_1}) \cup \mathcal{L}^{\circ}_{M_2} \cong \mathbb{Q} \cup \{1/0\}$ holds in this case.

Case 3. $p_1 < 0$ and $p_2 > 0$.

This case is really Case 2 if we switch K_1 and K_2.

Case 4. $p_1 < 0$ and $p_2 < 0$.

For any $a/b \notin \mathcal{L}^0(M_2)$, again we may assume a/b is finite and so $a/b \geq p_2q_2 - p_2 + q_2$ by (1). Choose the slope m/n in ∂M_1 as in (2) which yields $m/n \leq p_1q_1 + \frac{1}{-p_2+q_2} < p_1q_1 + 1$. So $m/n \in \mathcal{L}^0(M_1)$, $h(m/n) = a/b$ and we have $h(\mathcal{L}^{\circ}_{M_1}) \cup \mathcal{L}^{\circ}_{M_2} \cong \mathbb{Q} \cup \{1/0\}$.

The proof of Proposition 5 is now completed. □

It was shown in [**Z2**] that if $p_1q_1p_2q_2 - 1$ is odd, then $Y(T(p_1, q_1), T(p_2, q_2))$ is the double branched cover of an alternating knot in S^3, so $Y(T(p_1, q_1), T(p_2, q_2))$ is an L-space and the knot in S^3 is an $SU(2)$-simple knot (and is an arborescent knot) but is not a 2-bridge knot.

REMARK 6. It is pointed out by Steven Sivek and the referee that Proposition 5 actually follows from [**Z2**]; the proof of the result of [**Z2**] cited above generalizes immediately to give the conclusion that $Y(T(p_1, q_1), T(p_2, q_2))$ is the double branched cover of an alternating link in S^3 and thus is an L-space. Proving Proposition 5 this way gives a little more information for free, because branched double covers of alternating links are known to be L-spaces in pretty much every version of Floer homology including monopole Floer homology and (framed) instanton homology. By contrast, the proof using [**HW**] does not apply in instanton homology.

There are also examples of hyperbolic rational homology 3-spheres which are $SU(2)$-cyclic [**C**]. These examples are also double branched covers of alternating knots in S^3 and thus are L-spaces. These alternating knots are thus $SU(2)$-simple but are not arborescent.

Conjecture 4 can be equivalently stated as: if an rational homology 3-sphere is not an L-space, then it has an irreducible $SU(2)$-representation. There are evidences supporting the conjecture from this point of view. For instances Dehn surgery on any nontrivial knot in S^3 with any slope in the interval $(-1, 1)$ yields a manifold which is not an L-space [**OS**] and is not $SU(2)$-cyclic either [**KM**].

Steven Sivek and the referee provide the following remark with further evidence for Conjecture 4.

REMARK 7. A rational homology 3-sphere Y which is $SU(2)$-cyclic is conjecturally an instanton homology L-space (meaning $I^{\#}(Y)$ has rank $|H_1(Y)|$). On the other hand $I^{\#}(Y)$ is conjecturally isomorphic to $\widehat{HF}(Y)$. It was shown in [**BS**, Theorem 4.6] that if Y is a $SU(2)$-cyclic rational homology 3-sphere whose fundamental group is cyclically finite, then Y is an instanton homology L-space. Here the notion of cyclically finite was first defined in [**BN**] meaning that as ρ ranges over reducible representations of $\pi_1(Y) \to SU(2)$, all of the finite cyclic covers of Y corresponding to subgroups $ker(ad(\rho)) \lhd \pi_1(Y)$ are rational homology 3-spheres.

Proof of Theorem 1. Let $K = K(q_1/p_1, \cdots, q_n/p_n)$ be a Montesinos knot with $n \geq 3$. We need to show that $\pi_1(M_K)$ has an irreducible trace free $SU(2)$-representation which is not binary dihedral. Here is an outline of how the proof goes. We show that the double branched cover $\Sigma_2(K)$ has an irreducible $PSU(2)$-representation $\bar{\rho}_0$ which can be extended to an $PSU(2)$-representation $\bar{\rho}$ of $\pi_1(M_K)$ up to conjugation. This $PSU(2)$-representation $\bar{\rho}$ lifts to an $SU(2)$-representation ρ of $\pi_1(M_K)$ which is automatically trace free. Since $\bar{\rho}_0$ is an irreducible representation, ρ is not binary dihedral. The existence of $\bar{\rho}_0$ is provided by [**B**]. We first apply some ideas from [**Mat**] to show that $\bar{\rho}_0$ extends to a unique $PSL_2(\mathbb{C})$-representation $\bar{\rho}$ of $\pi_1(M_K)$. Then we further show that this $\bar{\rho}$ is conjugate to an $PSU(2)$-representation by applying some results from [**HP**][**CD**].

Now we give the details of the proof. For a finitely generated group Γ, $\bar{R}(\Gamma) = Hom(\Gamma, PSL_2(\mathbb{C}))$ denotes the $PSL_2(\mathbb{C})$ representation variety of Γ and $\bar{X}(\Gamma)$ the

$PSL_2(\mathbb{C})$ character variety of Γ. Let $t:\bar{R}(\Gamma)\to\bar{X}(\Gamma)$ be the map which sends a representation $\bar{\rho}$ to its character $\chi_{\bar{\rho}}$. We shall write an element in $PSL_2(\mathbb{C})$ as $\bar{A}$ which is the image of an element A in $SL_2(\mathbb{C})$ under the quotient map $SL_2(\mathbb{C})\to PSL_2(\mathbb{C})$ and for convenience we sometimes call elements in $PSL_2(\mathbb{C})$ as matrices. For any $\bar{A}\in PSL_2(\mathbb{C})$ define $tr^2(\bar{A})=(trace(A))^2$ which is obviously well defined. Recall that the character $\chi_{\bar{\rho}}$ of an $PSL_2(\mathbb{C})$-representation $\bar{\rho}$ is the function $\chi_{\bar{\rho}}:\Gamma\to\mathbb{C}$ defined by $\chi_{\bar{\rho}}(\gamma)=tr^2(\bar{\rho}(\gamma))$.

A character $\chi_{\bar{\rho}}$ is real if $\chi_{\bar{\rho}}(\gamma)\in\mathbb{R}$ for all $\gamma\in\Gamma$. If we consider $\bar{X}(\Gamma)$ as an algebraic subset in $\mathbb{C}^n$ (for some n), then real characters of $\bar{X}(\Gamma)$ correspond to real points of $\bar{X}(\Gamma)$, i.e. points of $\bar{X}(\Gamma)\cap\mathbb{R}^n$. If $\sigma:\mathbb{C}^n\to\mathbb{C}^n$ (for each $n\geq 1$) denotes the operation of coordinatewise taking complex conjugation, then any complex affine algebraic set Y in $\mathbb{C}^n$ defined over $\mathbb{Q}$ is invariant under σ and the set of real points of Y is precisely the fixed point set of σ in Y. Note that $\bar{R}(\Gamma)$ and $\bar{X}(\Gamma)$ are both algebraic sets defined over $\mathbb{Q}$ and that the map $t:\bar{R}(\Gamma)\to\bar{X}(\Gamma)$ is an algebraic map defined over $\mathbb{Q}$, we thus have the following commutative diagram of maps:

$$\begin{array}{ccc}\bar{R}(\Gamma) & \stackrel{\sigma}{\longrightarrow} & \bar{R}(\Gamma)\\ \downarrow t & & \downarrow t\\ \bar{X}(\Gamma) & \stackrel{\sigma}{\longrightarrow} & \bar{X}(\Gamma).\end{array}$$

It follows that $\sigma(\chi_{\bar{\rho}})=\chi_{\sigma(\bar{\rho})}$.

Recall that a representation $\bar{\rho}\in\bar{R}(\Gamma)$ is called irreducible if the image of $\bar{\rho}$ cannot be conjugated into the set $\{\bar{A};\ A \text{ upper triangular}\}$ ([**BZ**, Definition on page 752]). Two irreducible representations in $\bar{R}(\Gamma)$ are conjugate if and only if they have the same character (This property is proved in [**BZ**, the second paragraph on page 753]).

If W is a compact manifold, $\bar{R}(W)$ and $\bar{X}(W)$ denote $\bar{R}(\pi_1 W)$ and $\bar{X}(\pi_1 W)$ respectively.

Let $K=K(q_1/p_1,\cdots,q_n/p_n)$ and $M=M_K$. We may assume that all p_i are positive (by changing the sign of q_i if necessary). Let $p:\tilde{M}\to M$ be the 2-fold cyclic covering and let $\tilde{\mu}=p^{-1}(\mu)$ which is a connected simple closed essential curve in $\partial\tilde{M}$ which double covers μ. Then $p_*:\pi_1(\tilde{M})\to\pi_1(M)$ is an injection and we may consider $\pi_1(\tilde{M})$ as an index two normal subgroup of $\pi_1(M)$, in which $\tilde{\mu}=\mu^2$. Dehn filling $\tilde{M}(\tilde{\mu})$ of $\tilde{M}$ with the slope $\tilde{\mu}$ is the double branched cover $\Sigma_2(K)$ of (S^3,K). The covering involution τ on $\tilde{M}$ extends to one on $\tilde{M}(\tilde{\mu})$ which we still denote by τ. Montesinos proved in [**Mont1**][**Mont2**] that $\tilde{M}(\tilde{\mu})$ admits a Seifert fibering invariant under the covering involution τ, the base orbifold of the Seifert fibred space is $S^2(p_1,...,p_n)$ which is the 2-sphere with n cone points of orders $p_1,...,p_n$, and τ descends down to an involution $\bar{\tau}$ on $S^2(p_1,...,p_n)$ which is a reflection in a circle passing through all the cone points (see Figure 2).

We denote the orbifold fundamental group of $S^2(p_1,...,p_n)$ by $\Delta(p_1,...,p_n)$ which has the following presentation:

$$\Delta(p_1,...,p_n)=\langle a_1,...,a_n;a_i^{p_i}=1,i=1,...,n,a_1a_2\cdots a_n=1\rangle.$$

Geometrically the element a_i is represented by the loop $b_{i-1}^{-1}b_i$ shown in Figure 2 (b_0 is the trivial loop). Note that there is a quotient homomorphism from $\pi_1(\tilde{M}(\tilde{\mu}))$ onto $\Delta(p_1,...,p_n)$.

It was shown in [**Mat**, Section 3.3] that when $n=3$ any irreducible $PSL_2(\mathbb{C})$-representation of $\pi_1(\tilde{M})$ which factors through $\pi_1(\tilde{M}(\tilde{\mu}))$ has a unique extension to

$\pi_1(M)$. Note that this extended representation can be lifted to a trace free $SL_2(\mathbb{C})$-representation of $\pi_1(M)$. We shall slightly extend this result to the following

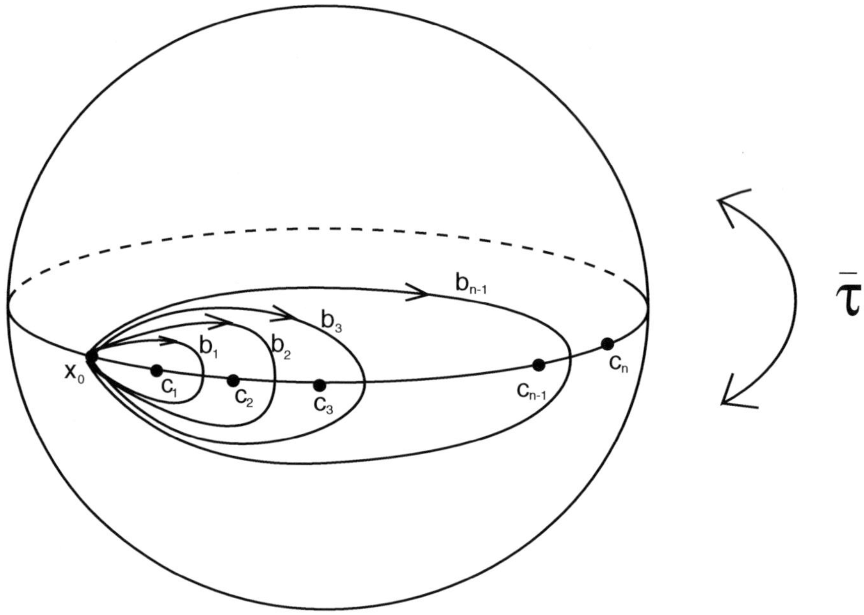

FIGURE 2. The orbifold $S^2(p_1, \cdots, p_n)$, its involution $\bar{\tau}$ and the generating set $b_1, ..., b_{n-1}$ for $\Delta(p_1, ..., p_n)$, where x_0 is the base point and $c_1, ..., c_n$ are cone points of orders $p_1, ..., p_n$ respectively.

PROPOSITION 8. *Let δ be the composition of the three quotient homomorphisms*

$$\pi_1(\widetilde{M}) \to \pi_1(\widetilde{M}(\tilde{\mu})) \to \Delta(p_1, p_2, ..., p_n) \to \Delta(p_1, p_2, p_3).$$

Let $\phi : \Delta(p_1, p_2, p_3) \to PSL_2(\mathbb{C})$ be any irreducible representation. Then $\bar{\rho}_0 = \phi \circ \delta$ has a unique extension to $\pi_1(M_K)$.

PROOF. The proof for uniqueness is verbatim as that given in [**Mat**] on page 38-39. We need to note that as $K(q_1/p_1, \cdots, q_n/p_n)$ is a knot at most one of p_i's is even. So [**Mat**, Lemma 2.4.9] still applies to our current case, i.e. the center of the image group of $\bar{\rho}_0$ is the trivial group.

CLAIM 9. There will be an extension $\bar{\rho}$ if and only if there is $\bar{A} \in PSL_2(\mathbb{C})$ such that $\bar{A}^2 = \bar{I}$ (where I is the identity matrix of $SL_2(\mathbb{C})$) and $\bar{A}\bar{\rho}_0(\beta)\bar{A}^{-1} = \bar{\rho}_0(\mu\beta\mu^{-1})$ for all $\beta \in \pi_1(\tilde{M})$.

Again this claim can be proved verbatim as that of [**Mat**, Claim 3.3.2].

So to finish the proof of Proposition 8, we just need to find an $\bar{A} \in PSL_2(\mathbb{C})$ with the properties stated in Claim 9, which is what we are going to do in the rest of the proof of Proposition 8. Recall that $\tilde{M}(\tilde{\mu})$ is the Dehn filling of $\tilde{M}$ with a solid torus N whose meridian slope is identified with the slope $\tilde{\mu}$. The core circle of N is the fixed point set of τ in $\tilde{M}(\tilde{\mu})$. Let D be a meridian disk of N such that the fixed point of τ in D (the center point of D) is disjoint from the singular fibers of the Seifert fibred space $\tilde{M}(\tilde{\mu})$. Choose a point $\tilde{x}$ in ∂D and let $\tilde{x}_0$ be the center point of D. Then arguing as on [**Mat**, Page 40] we have the following commutative diagram:

$$\begin{array}{ccccccc}
\pi_1(\tilde{M},\tilde{x}) & \longrightarrow & \pi_1(\tilde{M}(\tilde{\mu}),\tilde{x}) & \longrightarrow & \pi_1(\tilde{M}(\tilde{\mu}),\tilde{x}_0) & \longrightarrow & \Delta(p_1,....,p_n) \\
\downarrow(\cdot)^{\mu} & & & & \downarrow\bar{\tau}_* & & \downarrow\bar{\tau}_* \\
\pi_1(\tilde{M},\tilde{x}) & \longrightarrow & \pi_1(\tilde{M}(\tilde{\mu}),\tilde{x}) & \longrightarrow & \pi_1(\tilde{M}(\tilde{\mu}),\tilde{x}_0) & \longrightarrow & \Delta(p_1,....,p_n)
\end{array}$$

where $(\cdot)^{\mu}:\pi_1(\tilde{M},\tilde{x})\to\pi_1(\tilde{M},\tilde{x})$ corresponds to the conjugation action by μ, i.e. $(\beta)^{\mu}=\mu\beta\mu^{-1}$ and $\Delta(p_1,...,p_n)$ is the orbifold fundamental group of $S^2(p_1,...,p_n)$ whose base point is the image x_0 of the point $\tilde{x}_0$ under the quotient map

$$\tilde{M}(\tilde{\mu})\to S^2(p_1,...,p_n).$$

Figure 2 shows the generating set $b_1,...,b_{n-1}$ of the orbifold fundamental group $\Delta(p_1,...,p_n)$ of $S^2(p_1,...,p_n)$. In fact we have

$$a_1=b_1, a_2=b_1^{-1}b_2, a_3=b_2^{-1}b_3,\cdots,a_{n-1}=b_{n-2}^{-1}b_{n-1}, a_n=b_{n-1}^{-1}$$

and conversely

$$b_1=a_1, b_2=a_1a_2, b_3=a_1a_2a_3,\cdots,b_{n-1}=a_1a_2\cdots a_{n-1}, b_{n-1}=a_n^{-1}.$$

Obviously from Figure 2 the induced isomorphism $\bar{\tau}_*:\Delta(p_1,...,p_n)\to\Delta(p_1,...,p_n)$ sends b_i to b_i^{-1}, $i=1,...,n-1$. So we have
$\bar{\tau}_*(a_1)=a_1^{-1}$, $\bar{\tau}_*(a_2)=b_1b_2^{-1}=a_1a_2^{-1}a_1^{-1}$, $\bar{\tau}_*(a_3)=b_2b_3^{-1}=a_1a_2a_3^{-1}a_2^{-1}a_1^{-1}$, $\cdots$, $\bar{\tau}_*(a_{n-1})=b_{n-2}b_{n-1}^{-1}=a_1a_2\cdots a_{n-2}a_{n-1}^{-1}a_{n-2}^{-1}\cdots a_2^{-1}a_1^{-1}$, $\bar{\tau}_*(a_n)=\bar{\tau}_*(b_{n-1}^{-1})=b_{n-1}=a_n^{-1}$.

Since the quotient homomorphism

$$\begin{array}{l}\Delta(p_1,\cdots,p_n)=\langle a_1,...,a_n; a_i^{p_i}=1, i=1,...,n, a_1a_2\cdots a_n=1\rangle \\ \longrightarrow\Delta(p_1,p_2,p_3)=\langle \bar{a}_1,\bar{a}_2,\bar{a}_3; \bar{a}_i^{p_i}=1, i=1,2,3, \bar{a}_1\bar{a}_2\bar{a}_3=1\rangle\end{array}$$

sends a_i to $\bar{a}_i$, $i=1,2,3$, and send a_i to 1, $i=4,...,n$. we see that $\bar{\tau}_*$ descents to an isomorphism $\bar{\tau}_{\#}:\Delta(p_1,p_2,p_3)\to\Delta(p_1,p_2,p_3)$ such that $\bar{\tau}_{\#}(\bar{a}_1)=\bar{a}_1^{-1}$, $\bar{\tau}_{\#}(\bar{a}_2)=\bar{a}_1\bar{a}_2^{-1}\bar{a}_1^{-1}$, $\bar{\tau}_{\#}(\bar{a}_3)=\bar{a}_1\bar{a}_2\bar{a}_3^{-1}\bar{a}_2^{-1}\bar{a}_1^{-1}$ and we have the following commutative diagram:

$$\begin{array}{ccccccccc}
\pi_1(\tilde{M}) & \longrightarrow & \pi_1(\tilde{M}(\tilde{\mu})) & \longrightarrow & \Delta(p_1,....,p_n) & \longrightarrow & \Delta(p_1,p_2,p_3) & \xrightarrow{\phi} & PSL_2(\mathbb{C}) \\
\downarrow(\cdot)^{\mu} & & \downarrow\tau_* & & \downarrow\bar{\tau}_* & & \downarrow\bar{\tau}_{\#} & & \\
\pi_1(\tilde{M}) & \longrightarrow & \pi_1(\tilde{M}(\tilde{\mu})) & \longrightarrow & \Delta(p_1,....,p_n) & \longrightarrow & \Delta(p_1,p_2,p_3) & \xrightarrow{\phi} & PSL_2(\mathbb{C})
\end{array}$$

So $\bar{\tau}_{\#}(\bar{a}_1\bar{a}_2)=\bar{a}_2^{-1}\bar{a}_1^{-1}=(\bar{a}_1\bar{a}_2)^{-1}$. Since $\Delta(p_1,p_2,p_3)$ is generated by $\bar{a}_1,\bar{a}_2$, we see by applying [**BZ**, Lemma 3.1] that ϕ and $\phi\circ\bar{\tau}_{\#}$ have the same $PSL_2(\mathbb{C})$ character. (In fact if $\phi(\bar{a}_1)=\bar{A}_1$ and $\phi(\bar{a}_2)=\bar{A}_2$, then $\phi(\bar{a}_1\bar{a}_2)=\bar{A}_1\bar{A}_2=\overline{A_1A_2}$, $(\phi\circ\bar{\tau}_{\#})(\bar{a}_1)=(\bar{A}_1)^{-1}=\overline{A_1^{-1}}$, $(\phi\circ\bar{\tau}_{\#})(\bar{a}_2)=\bar{A}_1(\bar{A}_2)^{-1}(\bar{A}_1)^{-1}=\overline{A_1A_2^{-1}A_1^{-1}}$ and $(\phi\circ\bar{\tau}_{\#})(\bar{a}_1\bar{a}_2)=(\bar{A}_2)^{-1}(\bar{A}_1)^{-1}=\overline{(A_1A_2)^{-1}}$. Now let F_2 be the free group on two generators ξ_1 and ξ_2. Let ρ_1 and ρ_2 be the $SL_2(\mathbb{C})$ representations of F_2 defined by $\rho_1(\xi_i)=A_i$, $i=1,2$, and $\rho_2(\xi_1)=A_1^{-1}$, $\rho_2(\xi_2)=A_1A_2^{-1}A_1^{-1}$. Then one can easily verify that $tr(\rho_1(\xi_1))=tr(\rho_2(\xi_1))$, $tr(\rho_1(\xi_2))=tr(\rho_2(\xi_2))$ and $tr(\rho_1(\xi_1\xi_2))=tr(\rho_2(\xi_1\xi_2))$. So [**BZ**, Lemma 3.1] applies.) So ϕ and $\phi\circ\bar{\tau}_{\#}$ are conjugate $PSL_2(\mathbb{C})$ representations, that is, there is $\bar{A}\in PSL_2(\mathbb{C})$ with $\bar{A}\phi\bar{A}^{-1}=$

$\phi \circ \bar{\tau}_{\#}$. Combining this with the definition of $\bar{\rho}_0$ and the last commutative diagram, we see that $\bar{A}\bar{\rho}_0(\beta)\bar{A}^{-1} = \bar{\rho}_0(\mu\beta\mu^{-1})$ for each $\beta \in \pi_1(\tilde{M})$. The proof of Proposition 8 is now finished. □

Now by [**B**], every triangle group $\Delta(p_1, p_2, p_3)$ has an irreducible $SO(3) \cong PSU(2)$-representation. Therefore there is an irreducible representation $\bar{\rho}_0$ as given in Proposition 8 with its image contained in $PSU(2)$. So the character $\chi_{\bar{\rho}_0}$ of $\bar{\rho}_0$ is real valued. Let $\bar{\rho}$ be the unique extension of $\bar{\rho}_0$ to $\pi_1(M)$ as guaranteed by Proposition 8. The rest of proof is to show that $\bar{\rho}$ is also a $PSU(2)$-representation.

CLAIM 10. The character $\chi_{\bar{\rho}}$ of $\bar{\rho}$ is real valued.

Suppose otherwise. Recall that $\sigma : \bar{X}(M) \to \bar{X}(M)$ is the operation of taking complex conjugation and a character is real valued if and only if it is a fixed point of σ. So $\chi_{\bar{\rho}} \neq \sigma(\chi_{\bar{\rho}}) = \chi_{\sigma(\bar{\rho})}$ are two different characters of irreducible representations and thus $\bar{\rho}$ and $\sigma(\bar{\rho})$ are non-conjugate representations. But $\chi_{\bar{\rho}_0} = \sigma(\chi_{\bar{\rho}_0}) = \chi_{\sigma(\bar{\rho}_0)}$ and $\bar{\rho}_0$ is irreducible. Hence $\bar{\rho}_0$ and $\sigma(\bar{\rho}_0)$ are conjugate representations, that is, there is $\bar{B} \in PSL_2(\mathbb{C})$ such that $\bar{\rho}_0 = \bar{B}\sigma(\bar{\rho}_0)\bar{B}^{-1}$. Hence $\bar{\rho}$ and $\bar{B}\sigma(\bar{\rho})\bar{B}^{-1}$ are non-conjugate representations which have the same restriction on $\pi_1(\tilde{M})$. We get a contradiction with Proposition 8.

By [**HP**, Lemma 10.1] an $PSL_2(\mathbb{C})$-character $\chi_{\bar{\rho}}$ of a finitely generated group is real valued if and only if the image of $\bar{\rho}$ can be conjugated into $PSU(2)$ or $PGL_2(\mathbb{R})$. So our current representation $\bar{\rho}$ can be conjugated into $PSU(2)$ or $PGL_2(\mathbb{R})$. If it can be conjugated into $PSU(2)$, then we are done because this conjugated representation lifts to a trace free $SU(2)$-representation which is not binary dihedral. So suppose that $\bar{\rho}$ is conjugate to an $PGL_2(\mathbb{R})$-representation $\bar{\rho}'$. As noted in [**HP**] right after Lemma 10.1, $PGL_2(\mathbb{R}) \subset PGL_2(\mathbb{C}) \cong PSL_2(\mathbb{C})$ has two components, the identity component is $PSL_2(\mathbb{R})$ and the other component consists of matrices of determinant -1 (which in $PSL_2(\mathbb{C})$ are represented by matrices with entries in $\mathbb{C}$ with zero real part). Considering the action of $PSL_2(\mathbb{C})$ on hyperbolic space $\mathbb{H}^3$ by orientation preserving isometries, the group $PGL_2(\mathbb{R})$ is the stabilizer of a total geodesic plane in $\mathbb{H}^3$, and an element of $PGL_2(\mathbb{R})$ is contained in $PSL_2(\mathbb{R})$ if and only if it preserves the orientation of the plane (cf [**JS**, Section 2.8]). Since $\pi_1(\tilde{M})$ is the unique index two normal subgroup of $\pi_1(M)$, $\pi_1(\tilde{M})$ is generated by elements γ^2, $\gamma \in \pi_1(M)$. (To see this assertion holds, it is easy to verify that the subgroup G of $\pi_1(M)$ generated by squares is a normal subgroup and the quotient group $\pi_1(M)/G$ is an abelian group in which every element has order two. Since $H_1(M) = \mathbb{Z}$, $\pi_1(M)/G$ has to be the cyclic group of order two.) Therefore the image of the restriction $\bar{\rho}_0'$ of $\bar{\rho}'$ on $\pi_1(\tilde{M})$ consists of elements preserving the orientation of the total geodesic plane mentioned above, i.e. the image of $\bar{\rho}_0'$ is contained in $PSL_2(\mathbb{R})$. So the image of $\bar{\rho}_0$ can be conjugated into $PSL_2(\mathbb{R})$. But the image of $\bar{\rho}_0$ is contained in $PSU(2)$. [**CD**, Lemma 2.10] says that if an $PSU(2)$-presentation can be conjugated into an $PSL_2(\mathbb{R})$-representation, then it is a reducible representation. So our $\bar{\rho}_0$ is a reducible representation. We arrive at a contradiction, which completes the proof of Theorem 1.

References

[BS] J. A. Baldwin and S. Sivek, *Stein fillings and* SU(2) *representations*, Geom. Topol. **22** (2018), no. 7, 4307–4380, DOI 10.2140/gt.2018.22.4307. MR3890778

[B] S. Boyer, *On proper powers in free products and Dehn surgery*, J. Pure Appl. Algebra **51** (1988), no. 3, 217–229, DOI 10.1016/0022-4049(88)90061-8. MR946573

[BN] S. Boyer and A. Nicas, *Varieties of group representations and Casson's invariant for rational homology 3-spheres*, Trans. Amer. Math. Soc. **322** (1990), no. 2, 507–522, DOI 10.2307/2001712. MR972701

[BZ] S. Boyer and X. Zhang, *On Culler-Shalen seminorms and Dehn filling*, Ann. of Math. (2) **148** (1998), no. 3, 737–801, DOI 10.2307/121031. MR1670053

[C] C. Cornwell, *Character varieties of knot complements and branched double-covers via the cord ring*, `arXiv:1509.04962`.

[CD] M. Culler and N. M. Dunfield, *Orderability and Dehn filling*, Geom. Topol. **22** (2018), no. 3, 1405–1457, DOI 10.2140/gt.2018.22.1405. MR3780437

[HW] J. Hanselman and L. Watson, *A calculus for rordered Floer homology*, `arXiv:1508.05445`.

[HK] M. Heusener and E. Klassen, *Deformations of dihedral representations*, Proc. Amer. Math. Soc. **125** (1997), no. 10, 3039–3047, DOI 10.1090/S0002-9939-97-04195-6. MR1443155

[HP] M. Heusener and J. Porti, *Deformations of reducible representations of 3-manifold groups into* $PSL_2(\mathbb{C})$, Algebr. Geom. Topol. **5** (2005), 965–997, DOI 10.2140/agt.2005.5.965. MR2171800

[JS] G. A. Jones and D. Singerman, *Complex functions: An algebraic and geometric viewpoint*, Cambridge University Press, Cambridge, 1987. MR890746

[K] E. P. Klassen, *Representations of knot groups in* SU(2), Trans. Amer. Math. Soc. **326** (1991), no. 2, 795–828, DOI 10.2307/2001784. MR1008696

[KM] P. Kronheimer and T. Mrowka, *Knots, sutures, and excision*, J. Differential Geom. **84** (2010), no. 2, 301–364. MR2652464

[KM2] P. B. Kronheimer and T. S. Mrowka, *Dehn surgery, the fundamental group and* $SU(2)$, Math. Res. Lett. **11** (2004), no. 5-6, 741–754, DOI 10.4310/MRL.2004.v11.n6.a3. MR2106239

[Mat] T. W. Mattman, *The Culler-Shalen seminorms of pretzel knots*, ProQuest LLC, Ann Arbor, MI, 2000. Thesis (Ph.D.)–McGill University (Canada). MR2702771

[Mont1] J. M. Montesinos, *Seifert manifolds that are ramified two-sheeted cyclic coverings* (Spanish), Bol. Soc. Mat. Mexicana (2) **18** (1973), 1–32. MR341467

[Mont2] J. Montesinos, *Revêtements ramifiés de noeuds, espaces fibrês de Seifert et scindements de Heegaard*, Orsay Lecture Notes, 1976.

[Mot] K. Motegi, *Haken manifolds and representations of their fundamental groups in* $SL(2, \mathbf{C})$, Topology Appl. **29** (1988), no. 3, 207–212, DOI 10.1016/0166-8641(88)90019-3. MR953952

[OS] P. S. Ozsváth and Z. Szabó, *Knot Floer homology and rational surgeries*, Algebr. Geom. Topol. **11** (2011), no. 1, 1–68, DOI 10.2140/agt.2011.11.1. MR2764036

[R] D. Rolfsen, *Knots and links*, Publish or Perish, Inc., Berkeley, Calif., 1976. Mathematics Lecture Series, No. 7. MR0515288

[Z1] R. Zentner, *Representation spaces of pretzel knots*, Algebr. Geom. Topol. **11** (2011), no. 5, 2941–2970, DOI 10.2140/agt.2011.11.2941. MR2869448

[Z2] R. Zentner, *A class of knots with simple* $SU(2)$*-representations*, Selecta Math. (N.S.) **23** (2017), no. 3, 2219–2242, DOI 10.1007/s00029-017-0314-x. MR3663606

Mathematics Department, University at Buffalo, Buffalo, New York 14260
Email address: `xinzhang@buffalo.edu`

Published Titles in This Subseries

760 **Olivier Collin, Stefan Friedl, Cameron Gordon, Stephan Tillmann, and Liam Watson, Editors,** Characters in Low-Dimensional Topology, 2020

750 **A. Bourhim, J. Mashreghi, L. Oubbi, and Z. Abdelali, Editors,** Linear and Multilinear Algebra and Function Spaces, 2020

743 **H. Garth Dales, Dmitry Khavinson, and Javad Mashreghi, Editors,** Complex Analysis and Spectral Theory, 2020

680 **Sergei Gukov, Mikhail Khovanov, and Johannes Walcher, Editors,** Physics and Mathematics of Link Homology, 2016

655 **A. C. Cojocaru, C. David, and F. Pappalardi, Editors,** SCHOLAR—a Scientific Celebration Highlighting Open Lines of Arithmetic Research, 2015

654 **Carlo Gasbarri, Steven Lu, Mike Roth, and Yuri Tschinkel, Editors,** Rational Points, Rational Curves, and Entire Holomorphic Curves on Projective Varieties, 2015

638 **Javad Mashreghi, Emmanuel Fricain, and William Ross, Editors,** Invariant Subspaces of the Shift Operator, 2015

630 **Pierre Albin, Dmitry Jakobson, and Frédéric Rochon, Editors,** Geometric and Spectral Analysis, 2014

622 **S. Ejaz Ahmed, Editor,** Perspectives on Big Data Analysis, 2014

606 **Chantal David, Matilde Lalín, and Michelle Manes, Editors,** Women in Numbers 2, 2013

605 **Omid Amini, Matthew Baker, and Xander Faber, Editors,** Tropical and Non-Archimedean Geometry, 2013